W9-BTD-935

Liu
Ying
Rowly 150

ENVIRONMENTAL TRANSPORT PROCESSES

ENVIRONMENTAL TRANSPORT PROCESSES

BRUCE E. LOGAN
Kappe Professor of Environmental Engineering
The Pennsylvania State University
University Park, PA

A Wiley-Interscience Publication
JOHN WILEY & SONS, INC.
New York / Chichester / Weinheim / Brisbane / Singapore / Toronto

Library of Congress Cataloging-in-Publication Data:

Logan, Bruce E.
    Environmental transport processes / Bruce E. Logan.
      p. cm.
    Includes bibliographical references and index.
    ISBN 0–471–18871–9 (alk. paper)
    1. Ecological engineering. 2. Transport theory. 3. Environmental
chemistry. 4. Chemical engineering. I. Title.
GE350.L64   1999
628—dc21   98-8018

To Angela, Alex, and Maggie

CONTENTS

PREFACE   **xi**

1 INTRODUCTION   **1**

    1.1 Background / 1

    1.2 Notation for Chemical Transport / 4

    1.3 Simplifications for Environmental Systems / 7

    1.4 Review of Mass Balances / 13

         References / 23

2 EQUILIBRIUM CALCULATIONS   **24**

    2.1 Introduction / 24

    2.2 Thermodynamic State Functions / 28

    2.3 Chemical Potentials / 29

    2.4 Gibbs Free Energy and Equilibrium Constants / 33

    2.5 Distribution of Chemicals Based on Fugacities / 35

         References / 57

3 DIFFUSIVE TRANSPORT   **58**

    3.1 Introduction / 58

    3.2 Diffusion / 58

3.3   Calculation of Molecular Diffusion Coefficients / 60

3.4   Experimental Determination of Diffusivities / 81
References / 109

4 THE CONSTITUTIVE TRANSPORT EQUATION   111

4.1   Introduction / 111

4.2   Derivation of the General Transport Equation / 112

4.3   Special Forms of the General Transport Equation / 114

4.4   Similarity of Mass, Momentum, and Heat Dispersion Laws / 117

4.5   Transport Relative to a Fixed or Moving Coordinate System / 120

4.6   Special Forms of the Constitutive Transport Equation / 123

4.7   The Constitutive Transport Equation in Spherical and Cylindrical
Coordinates / 125

5 CONCENTRATION PROFILES AND CHEMICAL FLUXES   130

5.1   Introduction / 130

5.2   The Three Theories of Mass Transport / 130

5.3   Mass Transport in Radial and Cylindrical Coordinates Using Shell
Balances / 153
References / 164

6 MASS TRANSPORT COEFFICIENTS: FROM THEORY
TO EMPIRICISM   165

6.1   Definition of a Mass Transport Coefficient / 165

6.2   The Three Theories / 166

6.3   Multiple Resistances During Interphase Mass Transport / 171

6.4   Empirical Correlations for Mass Transport Coefficients / 180

6.5   Mass, Energy, and Momentum Transport Analogies / 186
References / 192

7 CHEMICAL TRANSPORT IN SHEARED REACTORS   193

7.1   Introduction / 193

7.2   Fluid Shear and Turbulence / 194

7.3   Mass Transport Correlations for Sheared Systems (General) / 200

7.4   Mean Shear Rates in Reactors / 204

7.5   Maximum Size of Aggregates in Sheared Fluids / 215

7.6   Chemical Transport from Bubbles / 220
References / 229

8 SUSPENDED UNATTACHED AND AGGREGATED MICROORGANISMS   **231**

   8.1 Introduction / 231
   8.2 Chemical Transport to Cells at Rest / 231
   8.3 Effect of Fluid Motion on Microorganisms / 235
   8.4 Transport to Microbial Aggregates / 241
   8.5 Effectiveness Factors for Mass Transport / 252
   8.6 Relative Uptake Factors for Mass Transport / 255
        References / 262

9 BIOFILMS   **263**

   9.1 Introduction / 263
   9.2 Transport in the Fluid Layer above a Biofilm / 264
   9.3 Biofilm Kinetics / 271
   9.4 Bioreactor Modeling / 287
   9.5 Recent Developments in the Study of Biofilms / 325
        References / 331

10 DISPERSION   **333**

   10.1 Introduction / 333
   10.2 Averaging Properties to Derive Dispersion Coefficients in Turbulent Fluids / 337
   10.3 Dispersion in Nonbounded Turbulent Sheared Fluids / 342
   10.4 Longitudinal Dispersion Coefficients for Defined Systems / 348
   10.5 Dispersion in Porous Media / 358
         References / 375

11 RIVERS, LAKES, AND OCEANS   **377**

   11.1 Introduction / 377
   11.2 Chemical Transport in Rivers / 378
   11.3 Mixing in Lakes / 390
   11.4 Mixing in Estuaries / 395
   11.5 Mixing in the Ocean / 396
   11.6 Transport of Chemicals Present as Pure Phases / 399
         References / 410

12 CHEMICAL TRANSPORT IN POROUS MEDIA   **412**

   12.1 Introduction / 412
   12.2 Porous Media Hydraulics / 412
   12.3 Contaminant Transport of Conservative Tracers / 415
   12.4 Transport with Reaction / 418

12.5   Transport with Chemical Adsorption / 423

12.6   Formation of Ganglia of Non-Aqueous Phase Liquids / 432

12.7   Mass Transport Calculations of Chemical Fluxes from NAPL
Ganglia / 444

References / 464

13 PARTICLES AND FRACTALS   466

13.1   Introduction / 466

13.2   Solid Particles and Fractal Aggregate Geometries / 467

13.3   Particle Size Spectra / 485

13.4   Measuring Particle Size Distributions / 490

13.5   Calculating Fractal Dimensions from Particle Size
Distributions / 493

References / 503

14 COAGULATION IN NATURAL AND ENGINEERED SYSTEMS   505

14.1   Introduction / 505

14.2   The General Coagulation Equations: Integral and Summation
Forms / 506

14.3   Factors Affecting the Stability of Aquasols / 507

14.4   Coagulation Kinetics: Collision Kernels for Spheres / 519

14.5   Fractal Coagulation Models / 536

14.6   Coagulation in the Ocean / 549

References / 562

15 PARTICLE TRANSPORT IN POROUS MEDIA   564

15.1   Introduction / 564

15.2   A Macroscopic Particle Transport Equation / 565

15.3   Clean-Bed Filtration Theory / 568

15.4   Discrete Particle Size Distributions Prepared by Filtration / 584

15.5   The Dimensionless Collision Number / 594

15.6   Pressure Drops in Filters / 597

15.7   Filter Ripening, Blocking, and Clogging / 599

15.8   Particle Transport in Groundwater Aquifers / 604

APPENDIXES

1   Notation / 614

2   Transport Equations / 622

3   Chemical Properties / 625

4   Mathematical Functions and Solution Techniques / 636

INDEX   649

PREFACE

This book has been written for environmental engineering graduate students who have never had a course in mass transfer, and chemical engineering students who have had a mass transport course but have no substantial experience in environmental topics. For others interested in the subject of transport phenomena, this book may serve as a reference. The subject of chemical transport is integral to several engineering fields of study, and there are many excellent textbooks available on this subject. However, most textbooks are written for chemical engineering students and, as a result, focus on describing transport processes from the perspective of chemical production. The chemical engineer must consider the behavior of concentrated solutions of chemicals, often under conditions of high temperature and pressure, where the behavior of the solutions is nonideal. Mathematical models developed for these circumstances are necessarily complex but are restricted to systems that are defined and well controlled.

In this text, the focus is on environmental systems—both environmentally engineered treatment processes and natural systems. These systems are generally at ambient temperature and pressure and are exceedingly complex in terms of the number of phases and chemicals involved. As an example, consider the components of a typical water sample. The main components are humic and fulvic acids, which have no exact chemical structure but rather are classified based on their characteristics and resistance to biodegradation. Such a poor definition of system components would be unacceptable to a chemical engineer designing a chemical reactor, but the situation is unavoidable in an environmental system. Thus transport problems derived for environmental systems must focus on making transport calculations in highly heterogeneous, and often poorly defined, environments.

In order to combine information from many different fields, I have built upon the approach used by L. J. Thibodeaux in his text *Chemodynamics*. In Chapters 1 and

2 in this text the description of chemical partitioning will seem familiar to those who have used Thibodeaux's book. Thereafter, material presented is quite different. While Thibodeaux assumed the reader was well versed in transport phenomena, in this book much of the material is devoted to covering transport fundamentals. In Chapter 3 I therefore present the fundamentals of molecular diffusion and emphasize how diffusivities and molecular size distributions are determined for heterogeneous samples typical of waters and wastewaters. The concept of dispersion is dealt with in subsequent chapters.

In Chapter 4, the constitutive transport equation is derived and applied to the solution of classical transport calculations. Those readers who have taken an undergraduate course in transport phenomena will find, on the surface, little new in Chapter 4, except that here the equations used to describe the motion of concentrated solutions of chemicals are cast in appropriate forms for dilute solutions. Chapter 5 builds upon the constitutive transport equations showing how chemical fluxes can be derived once the transport equations are solved to obtain concentration profiles. Many first-year graduate students in EnvE often have forgotten much of their calculus. The main purpose of the detailed derivations is to give the student some exposure to solving differential equations and applying these solutions to the calculation of chemical fluxes.

Once the mathematics of mass transport calculations have been covered, the transition can be made to the use of mass transport correlations. In Chapter 6 it is shown that for more complex, and therefore more realistic, systems the mass transport coefficients are more difficult to derive analytically. For many systems we must therefore rely upon empirically derived correlations to quantify chemical fluxes. The case of a moving sphere serves as a primary example. For a sphere at rest, we can exactly derive a mass transport coefficient. As we introduce flow past the sphere, and the situation changes from laminar to turbulent flow, the equations describing transport become increasingly complex, and we find it convenient to move from analytical to empirical solutions.

Chapters 7 and 8 deal with mass transport in sheared systems. The mass transport calculations in these two chapters are developed around the concept of a Kolmogorov microscale. That is, most of the particles in a sheared fluid exist in a laminar shear field. In Chapter 7 the correlations are presented in a general context, applicable to a wide range of conditions (natural systems and engineering reactors). In Chapter 8, however, the subject is restricted to considerations of mass transport to suspended and attached microbes. The subject of transport to aggregates occurs quite frequently in environmental systems and requires special attention.

Mass transport to biofilms is a wonderful subject for bridging classical mass transport theories on wetted walls with concepts related to microbial kinetics. In Chapter 9, biofilm kinetic models are presented and then merged with mass transport equations. Several models have been presented in the literature, but these are summarized here for the first time in a textbook along with mass transport equations. As an example of an important application for EnvE students, trickling filters are described in detail to demonstrate how theoretical models may be transferred to practice, much like absorption or stripping unit processes are used to relate to mass transport calculations in chemical engineering textbooks.

Chapter 10 describes chemical dispersion in a manner that provides a nice contrast to simple chemical diffusion. While molecular diffusion processes are quite slow, dispersion processes can be (relatively) quite fast. It is shown how chemical dispersion coefficients are analogous to diffusion coefficients, how dispersion coefficients can be derived from chemical diffusivities based on knowledge of the bulk flow (for example, laminar flow in a pipe), and several example problems are provided based on chemical dispersion in natural systems.

Chemical dispersion is the basis of calculations of chemical movement in lakes, rivers, and subsurface systems in Chapters 11 and 12. Chemical movement in fluid systems is a well-studied area, and much of the information presented here is obtained from specialized textbooks on this subject. In Chapter 12, however, the content of chemical transport specifically deals with a subject of great importance to engineers who will need to remediate subsurface environments: nonaqueous phase liquids (NAPLs). In the past ten years there have been new approaches and calculations published in the literature on the dissolution of NAPLs in groundwater aquifers. This chapter consolidates the findings on this subject and provides many useful transport example calculations.

One of the most common features to engineered systems for water and wastewater treatment is the focus on the removal of particles. Most transport texts approach the subject of transport phenomena from the perspective that all chemicals are dissolved in a solvent. However, a large proportion of unit processes in water and wastewater treatment plants are specifically designed to remove particles. Chapter 13 is a summary of particles, both from the perspective of Euclidean geometry (where particle is treated as a sphere), and from the assumption of fractal geometry. The theory behind the motion, transport, filtration, and coagulation of particles is covered in Chapters 14 and 15. Within the context of coagulation, fractals are incorporated into the discussion (where possible). These last chapters therefore provide a gateway to specialized studies in particle removal and transport, and the foundation of mass transport calculations important in virtually all water and wastewater treatment systems.

An enormous number of people have contributed to this text directly and indirectly. My mentors in transport phenomena at Berkeley were James Hunt and Hugo Fisher; in my research in natural and engineered systems, inspiration was provided through research projects with my colleagues Alice Alldredge, Robert Arnold, David Kirchman, and Charlie O'Melia. I also thank the reviewers of sections and chapters in this book, James Hunt, Bill Ball, James Farrell, Xiaoyan Li, Susan Powers, Lily Sehayek, Kenny Unice, and Terri Camesano. I am indebted to the hundreds of graduate students that have taken my Transport class over the years; they suffered through early drafts and typographical errors, but found the time to provide comments, feedback, and constructive criticism.

BRUCE LOGAN
State College, PA

ENVIRONMENTAL TRANSPORT PROCESSES

CHAPTER 1

INTRODUCTION

1.1 BACKGROUND

During the twentieth century, the creation of new chemicals and chemical production in the United States grew exponentially. Unfortunately, this rapid increase in chemical manufacturing was not accompanied by a corresponding increase in responsible disposal practices, resulting in the accidental and intentional release of many chemicals into the environment. The quantity and toxicity of chemicals released frequently exceeded the ability of the biotic environment to degrade these chemicals, and many ecological processes were disrupted causing vast changes in the normal cycles of terrestrial and aquatic wildlife.

In the United States alone there are over 15,000 registered chemicals, and only a small fraction of these have been researched for their effects on human health and the environment. On a mass production basis, the 50 chemicals shown in Table 1.1 are the most common. However, this mass basis does not directly translate to an impact and toxicity of each chemical in the environment. Many of these chemicals are not regulated by the Safe Drinking Water Act, a law that requires limits on the concentrations of toxic chemicals in drinking water (Table 1.2). Although other laws have been enacted to control the release of chemicals into the environment, many sites are already chemically contaminated, and chemicals are continuing to appear in drinking water aquifers.

Environmental engineers (EnvE) are usually called upon to design systems either to treat water and remove toxic chemicals, or to clean up soils contaminated with chemicals. The engineering of these treatment systems requires a good understanding of the chemicals properties and transport in the system. Predicting the fate and transport of these chemicals in both natural and engineered systems can be difficult, but fortunately, the same engineering disciplines that have been used to manufacture

TABLE 1.1 Chemical Production in 1988

Rank	Chemical	lb $\times 10^{-9}$	Rank	Chemical	lb $\times 10^{-9}$
1	Sulfuric acid	85.56	26	Hydrochloric acid	5.87
2	Nitrogen	52.10	27	p-Xylene	5.60
3	Oxygen	37.09	28	Ethylene oxide	5.37
4	Ethylene	36.56	29	Ethylene glycol	4.90
5	Ammonia	33.89	30	Cumene	4.80
6	Lime	32.34	31	Methyl tert-butyl ether	4.68
7	Sodium hydroxide	23.97	32	Ammonium sulfate	4.67
8	Phosphoric acid	23.43	33	Phenol	3.53
9	Chlorine	22.66	34	Potash	3.35
10	Propylene	19.97	35	Butadiene	3.19
11	Sodium carbonate	19.10	36	Acetic acid	3.16
12	Nitric acid	15.78	37	Propylene oxide	3.11
13	Urea	15.76	38	Carbon black	2.92
14	Ammonium nitrate	14.38	39	Aluminum sulfate	2.59
15	Ethylene dichloride	13.65	40	Acrylonitrile	2.58
16	Benzene	11.84	41	Vinyl acetate	2.56
17	Ethylbenzene	9.94	42	Cyclohexane	2.32
18	Terephthlalic acid	9.60	43	Acetone	2.29
19	Carbon dioxide	9.38	44	Titanium dioxide	2.04
20	Vinyl chloride	9.06	45	Sodium sulfate	1.69
21	Styrene	8.59	46	Sodium silicate	1.64
22	Methanol	7.34	47	Adipic acid	1.59
23	Formaldehyde	6.73	48	Isopropyl alcohol	1.42
24	Toluene	6.47	49	Calcium chloride	1.31
25	Xylene	5.93	50	Caprolactam	1.26

Totals

organic: 213.79
inorganic: 395.76

Source: *C & EN*, 1989.

chemicals can be used to understand chemical transport in engineered systems and the natural environment. For example, in order to design chemical reactor and production facilities, chemical engineers have been trained in the fields of equilibrium thermodynamics, chemical reaction kinetics, and mass transfer. These same fields of study can be applied to solve environmental problems. In every case, the final fate of a chemical in any system is dictated by the laws of thermodynamics. The rate of chemical transformation, reaction, and transport, however, may be less related to equilibrium conditions. Several good textbooks have been developed to focus on chemical equilibrium and reaction rates applicable to environmental processes, but the only textbooks on chemical transport phenomena have been developed from a

TABLE 1.2 Chemicals Covered under Reauthorized Safe Drinking Water Act (excludes radionuclides and microbiological contaminants)

Volatile Organic Chemicals	Organics	Inorganics
Benzene	Acrylamide	Aluminum
Carbon tetrachloride	Adipates	Antimony
Chlorobenzene	Alachlor	Arsenic
Dichlorobenzene	Aldicarb	Asbestos
1,2-Dichlorobenzene	Atrazine	Barium
1,1-Dichloroethylene	Carbofuran	Beryllium
cis-1,2-Dichloroethylene	Chlordane	Cadmium
trans-1,2-Dichloroethylene	Dalapon	Chromium
Methylene chloride	Dibromochloropropane (DBCP)	Copper
Tetrachloroethylene	Dibromomethane	Cyanide
Trichlorobenzene	1,2-Dichloropropane	Fluoride
1,1,1-Trichloroethane	Dinoseb	Lead
Trichloroethylene	Diquat	Mercury
Vinyl chloride	Endothall	Molybdenum
	Endrin	Nickel
	Epichlorohydrin	Nitrate
	Ethylene dibromide (EDB)	Selenium
	Glyphosphate	Silver
	Hexachlorocyclopentadiene	Sodium
	Lindane	Sulfate
	Methoxychlor	Thallium
	Pentachlorophenol	Vanadium
	Phthalates	Zinc
	Pichloram	
	Polychlorinated biphenyls (PCBs)	
	Polynuclear aromatic hydrocarbons (PAHs)	
	Simazine	
	2,3,7,8-Tetrachlorodibenzodioxin (dioxin)	
	Toluene	
	Toxaphene	
	2,4,5-TP (Silvex)	
	1,1,2-Trichloroethane	
	Vydate	
	Xylene	

Source: Pontius, 1990.

chemical engineering perspective of developing concentrated solutions for chemical production.

Although chemical and environmental transport processes share a common mathematical foundation (and perhaps framework), an environmental engineer often has little control over the conditions of the transport calculation or the complexity of the system. The natural variability of ecological processes normally gives rise to situations with too many equations and too few coefficients. Thus, while there may be only one table needed for chemical engineers to describe the properties of a benzene–water system, the environmental engineer may find that the controlling factor is not the solution properties but the adsorptive capacity of different soils in a groundwater aquifer. Therefore, the information contained in this book has been assembled in an effort to teach transport phenomena at a depth and breadth consistent with our understanding of how chemicals are transported in engineered reactors, for water, wastewater, and hazardous waste treatment, and in the natural environment.

1.2 NOTATION FOR CHEMICAL TRANSPORT

In order to develop a mathematical description of chemical distribution in the environment, a consistent notation system is helpful to specify the concentrations, phases, degradation rates, and transport rates of the chemical of interest. Unfortunately, there are too few constants and too many equations needed to describe chemical transport in the systems we shall consider. For example, k has been used as a mass transport coefficient, rate constant, permeability constant, and Boltzmann's constant, just to name a few. In order to distinguish many of these different factors, we can use upper- and lower-case letters, different fonts, Greek letters, and a string of subscripts. In particular, it may seem like there is always a large number of subscripts attached to a variable. However, a system of subscripts will be set up so that their use can be abbreviated or eliminated depending on the extent to which the use of the variable is unclear, as will be shown in the following. This notation scheme, and the general organization of Sections 1.2 and 1.3, were inspired by Thibodeaux (1979).

Mole and Mass Concentrations

Both mass and molar chemical concentrations are defined here using the letter c. The two different units of concentration are related by the molecular weight, M, or

$$c[\text{moles L}^{-3}] = \frac{c[\text{mass L}^{-3}]}{M[\text{mass moles}^{-1}]} \tag{1-1}$$

where $[\text{L}^{-3}]$ indicates inverse length cubed. Sometimes, in order to distinguish between the two concentration units, for example, when variables with different concentration units are being used in the same equation, ρ will be used for mass concentration, and c for mole concentration. The two are related by

$$c = \frac{\rho}{M} \tag{1-2}$$

When there is more than one chemical or phase in the system, subscripts will be added. Each chemical will be indicated using a capital letter, such as B, C, D, etc., with the phase indicated with a small letter such as a = air, g = gas, l = liquid, w = water, and s = soil. Thus we can use a double subscript with an upper- and lower-case letter to indicate the concentration of chemical C in air, water, or soil as

$$c_{Ca} = \text{concentration of species } C \text{ in air}$$

$$c_{Cw} = \text{concentration of species } C \text{ in water} \qquad (1\text{-}3)$$

$$c_{Cs} = \text{concentration of species } C \text{ in soil}$$

If additional phases are present, we can use other lower-case letters. For example, the concentration of chemical X in phase o = oil would be c_{Xo}. If we need to specify mass concentration units for additional clarity, the same notation applies when using ρ. Thus, in mass concentration units, we would write in general, ρ_{ij} for any chemical i in phase j, or ρ_{Co} for chemical C in oil.

The sum of the concentrations of all chemicals in the same phase must equal the total phase concentration. For example, the molar and mass concentrations of several chemicals in water would be

$$c_w = c_{Bw} + c_{Cw} + c_{Dw} + \cdots + c_{Ww} \qquad (1\text{-}4)$$

$$\rho_w = \rho_{Bw} + \rho_{Cw} + \rho_{Dw} + \cdots + \rho_{Ww} \qquad (1\text{-}5)$$

where c_{Ww} is the concentration of water in the phase we have designated as the water phase. For dilute solutions (see the following), we will see that $c_{Ww} \approx c_w$. The total phase concentration is also known as the phase density. For just pure water, for example, the mass density of water is $\rho_w = 1 \text{ kg L}^{-3}$, or since water has a molecular weight of 18 g mole^{-1}, the molar density of water is $c_w = (1000 \text{ g L}^{-1})/(18 \text{ g mol}^{-1}) = 55.6 \text{ mol L}^{-1}$.

When there is only one phase present, the double subscript notation can be simplified to a single subscript, or the subscripts may be entirely omitted. For example, for the concentration of two different chemicals in water, we need only to write c_B or c_C, and for a single chemical in water we can write just c. If the density of water is used in the same calculation, it should be written as ρ_w (even if c is being used in mass concentration units). The elimination of too many subscripts can cause confusion. If c_{Cw} is simplified to c, and c_w simplified to c, then we would have two different variables using the same symbol! Thus the phase molar density (or molar density or concentration) should always carry a subscript.

Mole and Mass Fractions

In many instances it is more convenient to express concentrations as mole or mass fractions. It is customary from the chemistry field to designate mole fractions of chemical species i in air, water, and soil as

$$x_i = \text{mass fraction of species } C \text{ in water}$$

$$y_i = \text{mass fraction of species } C \text{ in air}$$

$$z_i = \text{mass fraction of species } C \text{ in soil} \qquad (1\text{-}6)$$

The second subscript representing the phase is omitted for chemicals in air, water, and soil, since we have adopted different letters for each phase. For a system consisting of only chemical C and water, the mole fraction of chemical C in water, x_C, can be calculated from the moles of different chemicals using

$$x_C = \frac{n_C}{n_C + n_W} \qquad (1\text{-}7)$$

For a multicomponent system of N species, the mole fraction is calculated using

$$x_C = \frac{n_C}{\sum\limits_{i=1}^{N} n_i} \qquad (1\text{-}8)$$

If there are other phases, a second subscript will be necessary. The choice of x, y, or z for the mole fraction for other phases depends on whether the chemical is in the gas $= y$, liquid $= x$, or solid $= z$ phase. For example, the mole fraction of chemical C in phase o $=$ oil would be written using x for the oil (liquid) phase as x_{Co}.

Mass fractions chemicals can be designated in the same fashion as molar fractions, again with different letters for each phase, according to

$$\psi_i = \text{mole fraction of species } i \text{ in air}$$

$$\phi_i = \text{mole fraction of species } i \text{ in water}$$

$$\omega_i = \text{mole fraction of species } i \text{ in soil} \qquad (1\text{-}9)$$

Additional species in other phases, for example, TCE in oil, would be ϕ_{To}, where ϕ is used for an aqueous phase, $T = $ TCE and oil $= $ o.

The advantage of using mass and mole fractions is that the summation of all terms is equal to unity, or

$$\sum_{i=1}^{N} x_i = \sum_{i=1}^{N} y_i = \sum_{i=1}^{N} z_i = 1 \qquad (1\text{-}10)$$

$$\sum_{i=1}^{N} \phi_i = \sum_{i=1}^{N} \psi_i = \sum_{i=1}^{N} \omega_i = 1 \qquad (1\text{-}11)$$

1.3 SIMPLIFICATIONS FOR ENVIRONMENTAL SYSTEMS

The properties of concentrated gases, liquids, and solids can be difficult to describe when temperature and pressures are highly variable. In chemical engineering processes these phase properties can be tightly controlled to improve process performance, chemical yields, and overall costs. In environmental systems, temperature and pressure oftentimes cannot be controlled at all, although they will vary over a considerably smaller range compared to chemical production reactors. For example, while a chemical process may occur in an engineered reactor at elevated pressures and temperatures of 5 atm and 470 K with less than 1% variation in either condition, pressures in natural systems are almost always very close to 1 atm. Temperatures can average 15°C, but on an absolute temperature scale variations of ± 20°C represent <10% change in temperature at 293 K. An important component of chemical transport models for the natural environment is the variability of the system. Natural systems can differ only slightly, for example, slight variations in water quality or soil structure, or they can be substantially different due to large natural changes such as the change in seasons (or sudden storms). Given the difficulties in precisely defining all the characteristics of a natural system (versus a better defined chemical reactor), it often makes more sense to use simpler chemical models to describe natural systems. The main challenge in using even these simpler models will be quantifying the constants and parameters necessary for these models.

One assumption that will simplify models of chemical transport is to assume that all chemicals behave as ideal solutions. When chemical solutions are mixed, it is therefore assumed that they are mutually soluble and that the production or consumption of heat can be ignored. Molecular diameters are also assumed to be similar, as are intermolecular forces of attraction and repulsion in the bulk phase. With these assumptions there are no volume changes when two (or more) solutions are combined. The volume v produced by the mixing of two solutions of v_B and v_C is the sum $v = v_B + v_C$.

The assumption of ideal solutions greatly simplifies calculations between chemical mole fractions and concentrations. For example, consider the case of a binary mixture of chemical C in water. The mole fraction of chemical C in water W is

$$x_C = \frac{n_C}{n_C + n_W} \tag{1-12}$$

Since the solution is ideal, it is also true that $c_C = n_C/v$ and $c_W = n_W/v$. Substituting these into Eq. 1-12 produces

$$x_C = \frac{c_C v}{c_C v + c_W v} \tag{1-13}$$

or simplifying,

$$x_C = \frac{c_C}{c_C + c_W} \tag{1-14}$$

Thus mole fractions for these solutions can be calculated either on the basis of concentrations or moles.

The assumption of ideal solutions extends to gas-phase components. The ideal gas law for all components in the air phase is

$$Pv_a = n_a \mathcal{R} T \tag{1-15}$$

where v_a is the volume of air, P the total pressure, $\mathcal{R}$ the universal gas constant, and T the absolute temperature. For a single chemical C in air, with partial pressure p_{Ca}, the ideal gas law becomes

$$p_{Ca} v_a = n_{Ca} \mathcal{R} T \tag{1-16}$$

If air is the only phase present in our system, we can simplify our notation and rewrite the same expression as

$$p_C v = n_C \mathcal{R} T \tag{1-17}$$

The concentration of C in the air is

$$c_C = \frac{n_C}{v} = \frac{p_C}{\mathcal{R} T} \tag{1-18}$$

In most instances it can further be assumed that all mixtures are dilute solutions. Operationally, a dilute solution can be defined as any system when the concentration of chemicals in the phase is less than 5% of the total concentration, or

$$c_{Cj} < 0.05 c_j, \qquad \text{dilute solution} \tag{1-19}$$

This assumption allows much simpler calculations of mole and mass fractions. For example, for a dilute solution of C in water, $c_{Cw} \ll c_{Ww}$, so that Eq. 1-14 becomes

$$x_C \simeq \frac{c_{Cw}}{c_{Ww}} \tag{1-20}$$

The notation of c_{Ww} now becomes redundant, and we can write Eq. 1-20 more simply as

$$x_C \simeq \frac{c_C}{c_w} \tag{1-21}$$

Simplified relationships for chemicals in dilute solutions are summarized in Tables 1.3 and 1.4.

TABLE 1.3   Notation and Relationships for Multiphase and Multicomponent Systems

Gas	Liquid	Solid
$y_C = c_{Ca}/c_a$	$x_C = c_{Cw}/c_w$	$z_C = c_{Cs}/c_s$
$\rho_{Ca} = c_{Ca}M_C$	$\rho_{Cw} = c_{Cw}M_C$	$\rho_{Cs} = c_{Cs}M_C$
$\psi_C = \rho_{Ca}/\rho_a$	$\phi_C = \rho_{Cw}/\rho_w$	$\omega_C = \rho_{Cs}/\rho_s$
$1 = y_B + y_C + \cdots + y_N$	$1 = x_B + x_C + \cdots + x_N$	$1 = z_B + z_C + \cdots + z_N$
$\rho_a = \rho_{Ba} + \rho_{Ca} + \cdots + \rho_{Na}$	$\rho_w = \rho_{Bw} + \rho_{Cw} + \cdots + \rho_{Nw}$	$\rho_s = \rho_{Bs} + \rho_{Cs} + \cdots + \rho_{Ns}$
$c_a = c_{Ba} + c_{Ca} + \cdots + c_{Na}$	$c_w = c_{Bw} + c_{Cw} + \cdots + c_{Nw}$	$c_s = c_{Bs} + c_{Cs} + \cdots + c_{Ns}$
$\rho_a = c_a M_a$	$\rho_w = c_w M_w$	$\rho_s = c_s M_s$
$1 = \psi_B + \psi_C + \cdots + \psi_N$	$1 = \phi_B + \phi_C + \cdots + \phi_N$	$1 = \omega_B + \omega_C + \cdots + \omega_N$

TABLE 1.4   Additional Relationships between Mole Fractions, Mass Fractions, and Molecular Weight

Mole fraction	$x_C = \dfrac{\phi_C/M_C}{\phi_B/M_B + \phi_C/M_C + \cdots + \phi_N/M_N}$
Mass fraction	$\phi_C = \dfrac{y_C/M_C}{y_B/M_B + y_C/M_C + \cdots + y_N/M_N}$
Molecular weight	$M_w = x_B M_B + x_C M_C + \cdots + x_N M_N$
	$\dfrac{1}{M_w} = \dfrac{\phi_B}{M_B} + \dfrac{\phi_C}{M_C} + \cdots \dfrac{\phi_N}{M_N}$

Example 1.1

The concentrations of most chemicals in water are much less than the <5% criterion for a dilute solution. Show that relatively soluble compound such as benzene in water at 100 mg L^{-1} would be a dilute solution.

The molar concentration of a mass concentration solution of 100 mg L^{-1} of benzene is

$$c_{Bw} = \frac{\rho_{Bw}}{M_B} = \frac{0.1 \ [\text{g L}^{-1}]}{78 \ [\text{g mol}^{-1}]} = 0.0013 \ \frac{\text{mol}}{\text{L}} \qquad (1\text{-}22)$$

Since the molar concentration of water is 55.6 mol L^{-1}, the solution can be dilute since $0.0013 \ll 55.6$. Furthermore, even at the solubility limit of benzene in water, the solution is dilute. At the solubility limit or equilibrium concentration of benzene in water of $\rho_{Bw,eq} = 1.78$ g L^{-1}, the molar concentration is

$$c_{Bw} = \frac{\rho_{Bw}}{M_B} = \frac{1.78 \ [\text{g L}^{-1}]}{78 \ [\text{g mol}^{-1}]} = 0.228 \ \frac{\text{mol}}{\text{L}} \qquad (1\text{-}23)$$

TABLE 1.5 Molecular Weights and Pure Phase Concentrations of Air, Water, and Soil

Phase[a]	Molecular Weight M_i[g mol^{-1}]	Pure-Phase Concentrations	
		ρ_i[g L^{-1}]	c_i[mol L^{-1}]
Air	29	1.203	0.0415
Water—fresh	18	998	55.4
—sea	18	1021	56.7
Soil (SiO$_2$)	60	2650	44.1

[a]Assumes 20°C and 1 atm.

According to the 5% criterion established for a dilute solution, any aqueous solution that is less than 2.8 mol L^{-1} can be considered a dilute solution. Similar criteria can be developed for chemical concentrations in air. The molar concentration in air at 20°C can be calculated using

$$c_a = \frac{P}{\mathcal{R}T} = \frac{1 \text{ atm}}{0.0821 \text{ [L atm/mol K] } 293 \text{ K}} = \frac{\text{mol}}{24.1 \text{ L}} \tag{1-24}$$

The mass concentration or air can be obtained from the relationship $\rho_a = c_a M_a$. The molecular weight of air can be approximated as 29 g mol^{-1} by assuming that air is 79% nitrogen and 21% oxygen. This results in $\rho_a = 1.203$ g L^{-1}.

For soils, the assumption of an ideal solution is not strictly valid since soil is not homogeneous and chemical components in soil are not likely to be homogeneously distributed. Although we will discuss the implications of this situation at greater length in our presentation of groundwater transport, in some situations (as a first approximation) we can consider the soil to be well mixed and, for the chemicals in the soil, to be evenly distributed.

Useful characteristics of air, water, and soils are summarized in Table 1.5 for use in calculations of chemical concentrations. The calculated densities are based on 20°C and 1 atm, and for soil assuming pure silica (SiO$_2$).

Example 1.2

A large volume of chloroform = C is spilled in a closed room. For air at 26°C and 1 atm, calculate: (a) mole fraction, (b) mole concentration, (c) mass concentration, and (d) mass fraction of chloroform in air.

  a. At 26°C, the vapor pressure of chloroform, defined as C, is $p_C^\circ = 200$ mm Hg. The mole fraction of chloroform in air is:

$$y_C = \frac{p_C^\circ}{P} = \frac{200 \text{ mm Hg}}{760 \text{ mm Hg}} = 0.263 \tag{1-25}$$

Note that this is *not* a dilute solution since 0.263 > 0.05.

b. Before we calculate the mole concentration, we first need to calculate the total molar concentration of air, c_a. One approach is to calculate $c_a = P_a/\mathcal{R}T$. However, we can see in Table 1.5 that at 20°C and 1 atm $c_a = 0.0415$ mol l^{-1}. To correct this concentration to our temperature of 26°C, we can multiply by the ratio of absolute temperatures, or

$$c_{a,T} = c_a \frac{T}{T_T} = (0.0415 \text{ mol L}^{-1}) \frac{293 \text{ K}}{299 \text{ K}} = 0.0407 \text{ mol L}^{-1} \qquad (1\text{-}26)$$

The molar concentration of chloroform in air can be calculated from the mole fraction in part (a), or

$$c_{Ca} = y_C c_a = (0.263)(0.0407 \text{ mol L}^{-1}) = 0.0107 \text{ mol L}^{-1} \qquad (1\text{-}27)$$

c. Based on the molecular weight of chloroform of 119.4 g mole^{-1}, the mass concentration of chloroform is

$$\rho_{Ca} = c_{Ca} M_C = (0.0107 \text{ mol L}^{-1})(119.4 \text{ g mol}^{-1}) \times 10^{-3} \text{ L m}^{-3} \qquad (1\text{-}28)$$

$$= 1280 \text{ g m}^{-3} \qquad (1\text{-}29)$$

d. The mass fraction of C in the air–chloroform mixture cannot be calculated by assuming that the solution is dilute. Although the mass fraction $\psi_C = m_C/m_a$ is still valid, it must be recognized that the molecular weight of air is no longer equal to 29 g mol^{-1}, as previously calculated, since the air now contains a large fraction of chloroform. The molecular weight of air, defined as M_a, is calculated from the mole fractions of air = A and chloroform = C as

$$M_a = y_C M_C + y_A M_A = y_C M_C + (1 - y_C)M_A \qquad (1\text{-}30)$$

$$= (0.263)(119.4) + (1 - 0.263)(29) = 52.8 \text{ g mol}^{-1} \qquad (1\text{-}31)$$

It is left to the reader (Problem 1.2) to show that ψ_C can be calculated from the identity

$$\psi_C = \frac{y_C M_C}{M_a} \qquad (1\text{-}32)$$

Using this definition of ψ_C, we have

$$\psi_C = \frac{y_C M_C}{M_a} = \frac{(0.263)(119.4 \text{ g mol}^{-1})}{52.8 \text{ g mol}^{-1}} = 0.595 \qquad (1\text{-}33)$$

Conventions on Concentrations

Depending on the phase, different concentration units are preferred in the literature. The conventions for different units have emerged partially as a function of convenience, and partly as a function of properties of the system. The potentially most confusing set of units arises for liquid concentrations. For dilute liquid solutions, concentrations are usually reported in parts per million (ppm). While this choice of units contains good imagery, that is, a concentration based on "one-in-a-million," it is confusing in terms of analytical concentrations and solution preparations. The definition of ppm is always

$$\text{ppm} = \frac{1 \text{ part } C}{10^6 \text{ parts } B} \tag{1-34}$$

Assuming "parts" refers to "mass" in grams, this can be converted to mass concentration from the calculation

$$\text{ppm} = \frac{1 \text{ g } C}{10^6 \text{ g } B} \times \frac{10^3 \text{ mg } C}{\text{g } C} \times \frac{10^3 \text{ g } B}{\text{mg } B} \times \frac{1 \text{ kg } B}{\text{L}} = 1 \frac{\text{mg } C}{\text{L } B} \tag{1-35}$$

Example 1.3

If 1 ppm = 1 mg L^{-1}, or 1 mg/10^6 mg, in English units, does 1 ppm = 1 lb/10^6 gal, or 1 lb/10^6 lb?

In order to establish this relationship in English units, let us start with an equivalence that we are sure of, namely, that 1 ppm = 1 mg l^{-1}. Converting to lb/gal

$$1 \text{ ppm} = 1 \frac{\text{mg}}{\text{L}} \times \frac{1 \text{ kg}}{10^6 \text{ mg}} \times \frac{3.7854 \text{ L}}{\text{gal}} \times \frac{2.2 \text{ lb}}{1 \text{ kg}} = \frac{8.34 \text{ lb}}{10^6 \text{ gal}} \tag{1-36}$$

Since 1 ppm = 8.34 lb/10^6 gal, and water has a density of 8.34 lb gal^{-1}, it follows that 1 ppm = 1 lb/10^6 lb. As this example shows, the ppm calculation must always be made in terms of identical units in both the numerator and denominator. The only reason that a one-to-one conversion from ppm to mg L^{-1} works in metric units is that the density of water in metric units is approximately unity (1 g mL^{-1}). If we had a concentrated solution where this density was not correct, then it would follow that 1 ppm would not equal 1 mg L^{-1}.

Solid–water solutions, such as slurries and sludges, are often given in units of percent. The conventions with percent are the same as ppm: The ratio assumes the same units in the numerator and denominator. Since a percent is a ratio of 1 to 100,

TABLE 1.6 Preferred Basis of Units for Concentrations in Different Phases

Phase	Symbol	Units
Gas	y_C	$\dfrac{\text{volume}}{\text{volume}}$
Liquid	ρ_{Cw}	$\dfrac{\text{mass}}{\text{volume}}$
Solid	ω_{Cs}	$\dfrac{\text{mass}}{\text{mass}}$

a 5% solids suspension is

$$5\% = \frac{5\ \text{g}}{100\ \text{g}} \times \frac{10^3\ \text{g}}{1\ \text{L}} \times \frac{10^3\ \text{mg}}{\text{g}} = 50{,}000\ \frac{\text{mg}}{\text{L}} \qquad (1\text{-}37)$$

The preferred units for concentrations in the gas, liquid, and solid phases are shown in Table 1.6. When ppm is used for gas solutions, it is calculated on a volume basis, and therefore ppm in this context is often written as ppmv.

Dilutions   Two conventions are used to make dilution calculations. The two definitions differ in whether the volume of a solution is considered relative to the total volume or to the volume of the solvent. The actual dilution, **D**, is calculated based on the total volume produced as

$$\mathbf{D} = \frac{\text{volume of total}}{\text{volume of } C} \qquad (1\text{-}38)$$

but the relative dilution, $\mathbf{D}_R$, is calculated from the ratio of the two volumes, according to

$$\mathbf{D}_R = \frac{\text{volume of } B}{\text{volume of } C} \qquad (1\text{-}39)$$

A comparison of these two dilution definitions reveals that $\mathbf{D} = 1 + \mathbf{D}_R$. For concentrated solutions, the two dilutions are significantly different, but as the solution becomes more dilute, $\mathbf{D}_R \gg 1$, and so the values of **D** and $\mathbf{D}_R$ become essentially equal.

1.4   REVIEW OF MASS BALANCES

The mass balance approach is the foundation of mass transport calculations. In every system mass must be conserved. Defining the boundaries of a system, and writing the equations of conservation, we begin the process of translating abstract ideas into mathematical form. The general transport equation (see Chapter 4) is the most rigorous form of a mass balance. Many transport problems, however, can be solved more simply. The approach used in a mass balance is:

1. Convert a word problem to a sketch.
2. Define a control volume. More complicated problems may require several control volumes and subsketches.
3. Identify where mass enters or leaves the control volume, or where mass within the control volume may be altered or destroyed.
4. Write down the mass balance equations.
5. When the number of equations equals the number of unknowns, solve the equations. If unknowns exceed the knowns, return to step 2.

Simply expressed, the mass balance equations start with

$$\text{accumulation} = \text{in} - \text{out} \pm \text{reaction} \tag{1-40}$$

For the mass balances examples considered in this section, we focus on mass in fluids (water). All terms in the mass balance are usually written in terms of rates, which can be expressed in several forms. Accumulation is usually written as

$$\frac{dm}{dt} = \frac{d(cV)}{dt} \tag{1-41}$$

where c can be considered to have units of mass or moles. Flow entering into the control volume containing mass contributes mass to the system at the rate w, or based on the flow rate Q

$$w = Qc \tag{1-42}$$

The intrinsic rate r of chemical creation or destruction due to a reaction within the control volume must be multiplied by the total volume in which it occurs, or

$$rV = \left(\frac{dc}{dt}\right)V \tag{1-43}$$

In order to increase the number of known quantities, the mass balance equations are applied both to the chemical and the liquid phase, producing two equations. For example, the first equation would be written about chemical C in water, with the rate the chemical enters into the control volume defined as $Q_{w,in}c_{Cw,in}$, or more simply as $Q_{in}c_{in}$. The second equation would balance water in the control volume, with the rate water entering as $Q_{w,in}$ or just Q_{in}.

The use of the mass balance approach to solving equations is best illustrated by example. Below, we consider three cases: constant flow with no reaction, constant flow with reaction, and unsteady flow with no reaction. Problems at the end of the chapter provide further examples.

Example 1.4   Constant Flow With No Reaction

A waste stream ($Q_{in,1}$ = 0.62 m³ s⁻¹) contains 35 mg L⁻¹ of chemical C. How much of a second waste stream containing only 1.7 mg C L⁻¹ must be combined with it to achieve a final concentration of 20 mg L⁻¹?

In order to solve this problem using a mass balance, we can sketch the problem showing knowns and unknowns as shown in Figure 1.1.

Setting up an equation on just the flow streams, we have going into the control volume $Q_{in,1}$ + $Q_{in,2}$, and leaving the control volume Q_{out}, with no flow accumulated or destroyed. Using these terms in Eq. 1-40, we have for the balance on the phase flow

$$0 = (Q_{in,1} + Q_{in,2}) - Q_{out} \pm 0 \tag{1-44}$$

Each of the flow streams contains mass, entering or leaving at a rate Qc. For the control volume shown above, the mass balance is

$$0 = (Q_{in,1}c_{in,1} + Q_{in,2}c_{in,2}) - (Q_{out}c_{out}) \pm 0 \tag{1-45}$$

Rearranging Eq. 1-44, and substituting in the given information, we have

$$Q_{in,2} = Q_{out} - 0.62 \text{ m}^3 \text{ s}^{-1} \tag{1-46}$$

and from Eq. 1-45 we have

$$(0.62 \text{ m}^3 \text{ s}^{-1})(35 \text{ g m}^{-3}) + Q_{in,2} (1.7 \text{ g m}^{-3}) = Q_{out} (20 \text{ g m}^{-3}) \tag{1-47}$$

Substituting Eq. 1-46 into Eq. 1-47 and solving, we obtain the solution

$$Q_{in,2} = 0.51 \text{ m}^3 \text{ s}^{-1}$$
$$Q_{out} = 1.13 \text{ m}^3 \text{ s}^{-1} \tag{1-48}$$

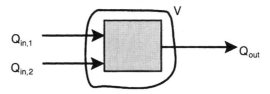

Figure 1.1   Control volume for Example 1.4.

Example 1.5   Constant Flow With Reaction

It is found that the chemical in the previous example can be degraded biologically according to first-order kinetics, with a rate constant $17.5 \ \text{h}^{-1}$. (a) Derive an equation for the concentration of chemical leaving the constant flow, stirred tank reactor (CSTR). (b) What volume V would you choose for a CSTR to degrade the chemical from 35 to 20 mg L^{-1}?

a. By definition, the flow in a CSTR is constant, so that $Q_{in} = Q_{out}$; so we can just designate the flow as Q to and from the system. From the mass balance equation, we have mass entering at rate Qc_{in}, and leaving at a rate Qc_{out}, or (Fig. 1.2)

$$0 = (Qc_{in}) - (Qc_{out}) + rV \tag{1-49}$$

Since the rate of reaction in the tank is first order, we can substitute $r = -kc$, producing

$$(Qc_{in}) = (Qc_{out}) + kcV \tag{1-50}$$

This result contains two unknowns, c_{out} and c. However, since the reactor is completely mixed, the concentration in the reactor is exactly the same in the fluid leaving the reactor, or $c_{out} = c$. Using this equality and rearranging, the concentration leaving the reactor is

$$c_{out} = \frac{c_{in}}{1 + kV/Q} \tag{1-51}$$

b. The volume of the reactor is obtained by rearranging Eq. 1-51 to produce

$$V = \left(\frac{c_{in}}{c_{out}} - 1 \right) \frac{Q}{k} \tag{1-52}$$

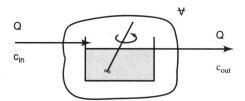

Figure 1.2   Control volume for Example 1.5.

Substituting in the given information

$$V = \left(\frac{35 \text{ mg L}^{-1}}{20 \text{ mg L}^{-1}} - 1 \right) \frac{0.62 \text{ m}^3 \text{ h}^{-1}}{17.5 \text{ h}^{-1}} \times \frac{1000 \text{ L}}{\text{m}^3} \qquad (1\text{-}53)$$

$$= 27 \text{ L} \qquad (1\text{-}54)$$

Example 1.6   Unsteady Flow Without Reaction

We now consider a case where a vessel containing a selective membrane is initially filled to a volume V_o with solution at a concentration of chemical able to pass through the membrane, c_{r0}. At $t = 0$ fluid begins leaving the vessel at a constant rate Q. The concentration in the effluent (or membrane permeate), c_p, is always less than the concentration in the reactor, c_r, by the constant ratio $p_c = c_p/c_r$. Because water is leaving faster than the mass of chemical, the concentrations in the permeate and retentate will increase with time. Derive an expression for c_p as a function of the fraction of volume filtered F, defined as $F = 1 - (V_f/V_o)$, where V_f is the total volume that has left the vessel.

The system shown in Fig. 1.3 is not particularly informative, but helps to define the variables used in the equation derivation. From the mass balance equation, we can write

$$\frac{d(V_r c_r)}{dt} = 0 - Q c_p + 0 \qquad (1\text{-}55)$$

Using the given information that $c_p = p_c c_r$, the above equation can be written only in terms of c_r as

$$\frac{d(V_r c_r)}{dt} = -p_c Q c_r \qquad (1\text{-}56)$$

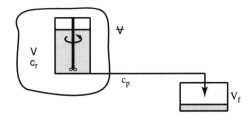

Figure 1.3   Control volume for Example 1.6.

Using the chain rule, the derivative $d(V_r c_r)/dt$ is

$$\frac{d(V_r c_r)}{dt} = V_r \frac{d(c_r)}{dt} + c_r \frac{dV_r}{dt} \qquad (1\text{-}57)$$

To simplify, we need to recognize that $Q = -dV/dt$. Combining this with Eqs. 1-56 and 1-57 yields

$$\frac{d(c_r)}{dt} = \frac{Q}{V_r}(1 - p_c)c_r \qquad (1\text{-}58)$$

Before Eq. 1-58 can be integrated and solved, V_r must be explicitly expressed in terms of time, t. This can be done by substituting $V_r = V_o - Qt$ into Eq. 1-58 to obtain

$$\frac{d(c_r)}{dt} = \frac{Q}{V_0 - Qt}(1 - p_c)c_r \qquad (1\text{-}59)$$

Separating terms,

$$\int_{c_{ro}}^{c_r} \frac{d(c_r)}{c_r} = Q(1 - p_c) \int_0^t \frac{1}{V_o - Qt}\, dt \qquad (1\text{-}60)$$

This is an integration of the form $\int (du)/(a + bu) = (1/b)\ln(a + bu)$. Solving Eq. 1-60 and simplifying

$$\ln \frac{c_r}{c_{r0}} = (p_c - 1)\ln \frac{V_0 - Qt}{V_0} \qquad (1\text{-}61)$$

Since $Qt = V_f$

$$\ln \frac{c_r}{c_{r0}} = (p_c - 1)\ln \frac{V_0 - V_f}{V_0} \qquad (1\text{-}62)$$

Substituting in $F = (V_o - V_f)/V_f$ and taking the exp of both sides of the equation produces

$$c_r = c_{r0}F^{p_c-1} \qquad (1\text{-}63)$$

Since $c_p = p_c c_r$, we have the solution

$$c_p = p_c c_{r0}F^{p_c-1} \qquad (1\text{-}64)$$

PROBLEMS

1.1 A sludge digester gas is analyzed as: CO_2, 35.4%; CH_4, 64.0%; H_2, 0.5%; and N_2, 0.1%. Determine the following properties of the gas mixture: (a) Mole fraction of methane; (b) Weight fraction of methane; (c) Average molecular weight of the natural gas mixture; (d) Partial pressure of each component if the total pressure is 1500 mm Hg

1.2 It was stated that the mass fraction of a chemical in air, ψ_C, can be calculated for a nondilute solution from

$$\psi_C = \frac{M_C y_C}{M_a} \tag{1-65}$$

where M_a is the molecular weight of the air containing chemical C (Eq. 1-32). Show that this is correct by deriving Eq. 1-65.

1.3 Figure P-1.3 shows a typical arrangement used to recycle cooling water in the power manufacturing industry. Three percent of the water entering the tower is lost by evaporation (consumptive use). Blowdown (at location b) is required to prevent a buildup of unreasonable salt concentrations in the recycled cooling water (adapted from a problem of R. Selleck). (a) What is the ratio of blowdown to makeup for this system? (b) Chlorine is added before the condenser to control the growth of biological slimes in the condenser. Assuming that all of the chlorine is reduced to chloride, compute the amount of chloride added per day if: $Cl^- = 550$ ppm in the condensor effluent (location cd), $Cl^- = 80$ ppm in makeup water (location m), and $Q_b = 10$ m³/day. Ignore the increase in TDS caused by adding the chlorine. Also, carry a few extra (non) significant figures.

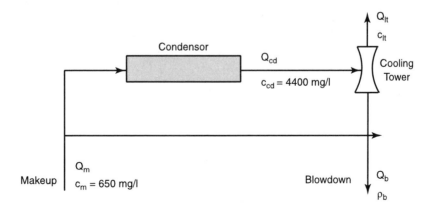

Figure P-1.3

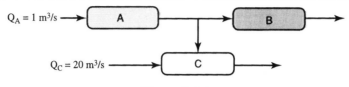

Figure P-1.4

1.4 You are a hazardous waste engineer at a chemical plant that uses chemical X in some of their aqueous processes (Fig. P-1.4). Chemical X is highly volatile, and you know that 10% of the mass entering into plant A and 12% entering plant B is lost to the air with a negligible loss of water. The process stream entering plant A (1 m^3 s^{-1}) contains 200 g m^{-3} of X, the stream entering plant C (20 m^3 s^{-1}) contains only 20 g m^{-3} of X, and 15% of the stream leaving plant A is diverted to plant C. (a) What is the flow rate and concentration of X leaving plant B? (b) What is the concentration of X leaving plant C?

1.5 A new treatment system (Fig. P-1.5) has been designed to reduce the concentration of solids in mine drainage wastes. The advantage of the system is that, independent of the influent solids concentration, the total solids are reduced to 100 g m^{-3} for 75% of the wastewater that is treated. In order to minimize the amount of wastewater generated that must be further treated, it is proposed to combine some fraction of the untreated wastewater (containing 10,000 mg L^{-1} of solids) with the treated wastewater (adapted from a problem by R. Selleck). (a) For an influent flow of 10 m^3 h^{-1}, calculate the flow rate of drainage that does not have to be treated assuming that the mixture must meet the discharge limit of 500 g m^{-3} of solids. (b) What are the concentrations and flow rates of the wastewater that must be further treated?

1.6 As shown in Figure P-1.6, 10^3 gd^{-1} of chemical polymer C is produced using 10^3 Ld^{-1}(Q_{in}) of ultrapure water in a reactor using a patented process, and separated in a continuous centrifuge, so that C is lost in the effluent at an effluent concentration of 0.022%. Assuming polymer is recovered in the product stream (Q_p) flowing at 10^2 Ld^{-1}, and the concentration of C leaving the reactor is 3.3 gL^{-1}, what is the recycle flow rate and concentration of polymer in the product stream (Q_p)?

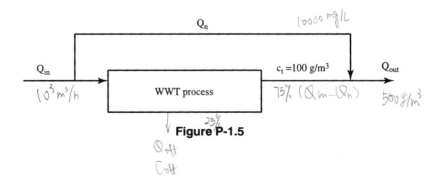

Figure P-1.5

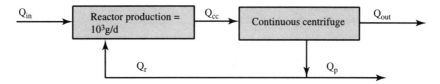

Figure P-1.6

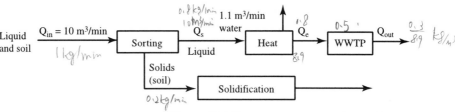

Figure P-1.7

1.7 Shown in Figure P-1.7 is a waste treatment scheme for chemical Z. There is a total addition of chemical Z to system of 1 kg min^{-1}. Assume 20% of the chemical is adsorbed on the soil. The waste stream goes through a sedimentation process that removes all the soil and a negligible amount of water. During heating of the wastewater, 1.1 m^3 min^{-1} of distilled water is lost. (a) Calculate the rate (kg min^{-1}) of chemical lost in the soil. (b) Calculate the concentration of chemical leaving the WWTP if 0.5 kg min^{-1} of the chemical is destroyed there.

1.8 The process shown in Figure P-1.8 has been suggested to destroy chemical X to a safe level of 1 ppm. The process treats 89 m^3 month^{-1} of water containing 53 mg L^{-1} of X, although 13 m^3 month^{-1} of water are lost (to evaporation) in the process of destroying X. Also, a flow equal to 10% of the flow leaving the plant is recycled and combined with influent flow, as shown in Figure P-1.8 (i.e., $Q_r = 0.1 Q_o$). What is the concentration of material entering the process at point e in the figure?

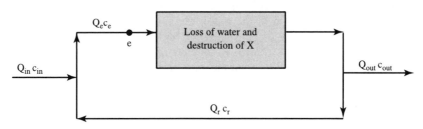

Figure P-1.8

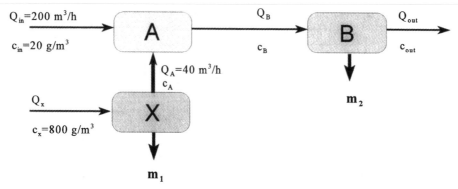

Figure P-1.9

1.9 Your technician made measurements of metal concentrations in a metals re-covery facility as shown in the flow sheet of Figure P-1.9. Unfortunately, several of the data points were missing. You know from past experience that 90% of the mass of metals entering a treatment system are recovered as a metals sludge containing a *negligible* amount of water. Assuming this percent recovery for the streams treated in systems *B* and *X*, calculate (a) the metal concentration in the water leaving the plant, c_{out}, and (b) the rate that sludges $(m_1 + m_2)$ are generated (g hr^{-1}) in the facility from the contaminated water streams. (*Hint*: the units of *m* and the product Qc are the same.)

1.10 In order to predict DDT profiles in ocean sediments off the Palos Verdes coast, researchers (Logan et al., 1989) first used a mass balance to reproduce profiles of volatile solids in the sediments based on historic wastewater discharges. This question explores how the yearly thickness (*L*) of the sediment core can be calculated from the effluent discharge and estimated sedimentation rates in the area. Effluent discharge to a site (0.4 km^2) was 0.5% of the total yearly discharge of 1.7×10^5 Mton yr^{-1}, where 1 Mton = 10^3 kg. The original wastewater has 70% volatile solids, and 35% of these volatile solids will be degraded before the wastewater solids reach the sediment site. The nonvolatile solids are not altered. This degraded wastewater solids, combined with a sed-iment flux of 400 mg cm^{-2} yr^{-1} (of which only 5% is volatile solids), forms the sediment.

Calculate the thickness (*L*) of the sediment layer for this year at this site. Recall that the definition of the mass fraction of volatile solids (chemical *V*) in the sediment (phase s) consisting of both sewage solids (phase g) and natural solids (phase n) is $\omega_{Vs} = m_{Vs}/m_s$. (*Hint*: within this context, $\omega_{Vs} = J_{Vs}/J_s$, where *J* is the mass flux having units mg cm^{-2} yr^{-1}). You may wish to use the following equations.

$$L[\text{cm yr}^{-1}] = \frac{J_s[\text{mg cm}^{-2} \text{ yr}^{-1}] \times 10^{-3}}{(1 - \omega_{Ds})\rho_s[\text{g cm}^{-3}]}$$

$$\omega_{Ds} = 1.15\omega_{Vs} + 0.2624$$

$$\rho_s = -1.67\omega_{Vs} + 1.794$$

REFERENCES

Chemical & Engineering News. 1989. April 10, p. 12.

Logan, B. E., A. J. Steele, and R. G. Arnold. 1989. *J. Environ. Eng.* **115**(1):221–38.

Pointius. 1990. *J. Amer. Water Works Assoc.* **82**(2):32–52.

Thibodeaux, L. J. 1979. *Chemodynamics*. Wiley.

CHAPTER 2

EQUILIBRIUM CALCULATIONS

2.1 INTRODUCTION

The distribution of components in a system is a function of two characteristics of the system: equilibrium and rate. Once the system is defined, there is only one equilibrium condition. However, the amount of time necessary for the system to reach equilibrium is a function of rates that can consist of both intrinsic rates (kinetics), which are a function of time, and rates that are a function of spatial distribution (mass transport rates). Equilibrium conditions establish the driving forces, since the system will always tend to move towards equilibrium. In a closed system, the equilibrium conditions are relatively easy to calculate. In open systems, however, equilibrium conditions may never be met everywhere in the system. The situation becomes extremely complicated in natural systems. We may seek to isolate a part of a system for study, but natural perturbations can prevent a static equilibrium condition from occurring. More likely, a system subject to natural cycles of change forced by temperatures and solar periods will reach a dynamic equilibrium.

An example of equilibrium in a natural system arises for the classical case of a single predator and prey. There is some static equilibrium point where the number of prey is exactly balanced with the number of predators (Fig. 2.1). However, that exact point is rarely achieved in the system since other factors such as disease, environmental stresses, and other predators and prey disturb the system from the static equilibrium condition. As a result, the number of predators and prey often oscillate around the static equilibrium point, producing a *dynamic equilibrium*. Many natural and engineered systems display a tendency towards a dynamic equilibrium. In recent years, the static equilibrium point has become known as a "strange attractor," since the system is constantly drawn back to that point. A large disruption of

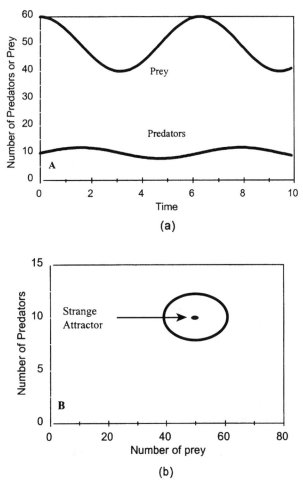

Figure 2.1 Behavior of predator–prey relationships. (A) Predator and prey numbers versus time; (B) predator–prey phase plane. Strange attractor is the point of dynamic equilibrium.

the system can cause the system to oscillate out of control and to fail—which in the case of our predator–prey model can mean loss of one or both of the species, and the establishment of a new strange attractor. A more detailed analysis of the motions of systems has indicated that what appears to be a completely random response to a perturbation actually reaches a predictable outcome. A description of such a system, although not exact at every point, is referred to as a *chaotic system*. Our ability to describe chaotic systems mathematically is improving, but is not yet well developed (Gleick 1987).

Chemical Equilibrium

The distribution of a chemical among different phases in a system can be expressed as a function of two characteristics of the system: chemical equilibrium and mass transport rate. Equilibrium conditions can be determined from thermodynamic calculations. The amount of time necessary for the system to reach equilibrium will be determined by describing the chemical concentrations in the system as a function of phase, time, and spatial distribution. Chemical reactions will be incorporated using two approaches. Either the kinetics will be fast enough to be incorporated into a local equilibrium condition, or a reaction rate expression will be identified in the governing transport equation.

Before the techniques to calculate chemical equilibrium in a system are presented, it is important to distinguish chemical equilibrium from steady state. Equilibrium is a condition that indicates that a system, or a point in the system, is at its thermodynamically calculated lowest state of energy. Steady state in a system means that a condition of the system is time invariant. Equilibrium and steady-state conditions may or may not coincide at different points within a system.

In order qualitatively to distinguish equilibrium and steady state in a system, consider the distribution of chemical C as a function of distance from the air–liquid interface (i.e., in one dimension) in the liquid and gas phases in the system shown in Figure 2.2 at three times: $t = 0$, δ, and ∞. Air is blown above the vent to remove chemical C from the tube containing chemical C such that when the vent is opened there is an insignificant concentration of chemical in the horizontally moving air. The reservoir of liquid chemical C is infinite, and the chemical does not react with any other components in the system. At time $t \leq 0$ the vent at point c is closed, and the concentration everywhere in the air in the sealed portion is at equilibrium concentration, $c_{C,eq}$. The concentration of C in the air is at the equilibrium condition since sufficient time has passed for chemical C to volatilize into the gas phase and

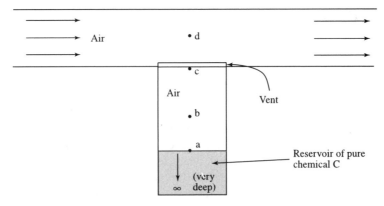

Figure 2.2   Chemical distribution in a tube initially covered with closed vent. At time $t = 0$ the vent is opened.

to have reached its equilibrium partial pressure. The system is at steady state since there are no changes in concentration with time at any point in the system.

At time $t = \delta$ the vent has just been opened. At this point, only point a in the system can be at equilibrium. Point a is the concentration in the air at the air–liquid interface, or the point in the air immediately above the liquid. The concentration at other points in the air are dependent on the amount of time that has passed and the mass transfer conditions in the vessel. Since the time $t = \delta$ is small, it is likely that the concentration of chemical in the tube is changing with time, and therefore the system is not at steady state.

After a long period of time (as $t \rightarrow \infty$), a concentration profile will develop in the tube above the chemical reservoir. If there are no disturbances in the system, the system will reach a steady state and the concentration of C will not change with time at any specific location in the vessel. Equilibrium conditions will only arise at a single point in the system (point a). All other points in the system will have $c_C \leq c_{C,\text{eq}}$. The exact shape of the concentration profile in the air at steady state is a function of the rate of mass transfer of chemical C out of the system and the equilibrium conditions at the air–liquid interface. In this chapter, we will review methods to characterize equilibrium conditions in the tube. In following chapters we will learn how to develop mass transport equations to describe the spatial and temporal concentrations of C in the system.

Calculating Equilibrium

There are three methods typically used to make equilibrium calculations:

- Chemical potentials
- Gibbs free energies
- Fugacities

These three approaches are all based on the same thermodynamic principles. The selection of a method is based on the nature of the calculation and a preference for units. Gibbs free energy calculations are usually employed to determine if a specific reaction is spontaneous. Chemical potential is a partial molar property, and chemical potentials are well suited for examining *phase distributions*, or, for example, whether concentrations of chemicals in different phases are in equilibrium, and if they are not, to identify the direction of spontaneous change. When solutions are nonideal, chemical fugacities are used to define the concentrations of the chemicals. Although it is assumed that chemical solutions in most environmental conditions act as an ideal solution, the use of fugacities conveys a simplicity to chemical partitioning calculations that makes fugacity calculations preferable, as will be further discussed.

Most readers will be familiar with equilibrium calculations using Gibbs free energy and chemical potentials. However, the use of fugacities is less common. The purpose of this chapter is therefore to present fugacities as a preferred calculation method for partitioning calculations. In order to relate fugacities to more familiar calculations using chemical potentials and Gibbs free energy, these latter approaches are reviewed prior to presentation of chemical fugacities.

2.2 THERMODYNAMIC STATE FUNCTIONS

The fundamental thermodynamic variables necessary to describe a system are temperature, T, entropy, S, pressure, P, and volume, V. Using these variables, thermodynamic functions can be developed based on the four extensive state properties of the system: internal energy, E, enthalpy, H, Helmholtz free energy, F, and Gibbs free energy, G. In a closed system of fixed concentration, these four thermodynamic state functions can be related to the system properties according to

$$dE = T\,dS - P\,dV \qquad (2\text{-}1)$$

$$dH = T\,dS + V\,dP \qquad (2\text{-}2)$$

$$dF = -S\,dT + P\,dV \qquad (2\text{-}3)$$

$$dG = -S\,dT + V\,dP \qquad (2\text{-}4)$$

Free energy calculations are more complicated when the system is open and may have variable composition. An additional term must be added to Eqs. 2-1–2-4 to account for changes in the energy of the system resulting from changes in chemical concentrations and abundances. Gibbs introduced the concept of chemical potential, μ_{ij}. The change in free energy can be related to the chemical potentials of i species using the term $\Sigma_i\,\mu_i\,dn_i$ since the infinitesimal change in the number of moles of species i, or dn_i, is multiplied by the chemical potentials of that species. Summation of chemical potentials accounts for the change in potential of the system, and therefore the change in the energy of the system.

Inclusion of chemical potentials into the four thermodynamic state functions in Eqs. 2-1–2-4 for species distributed among j phases produces

$$dE = T\,dS - P\,dV + \sum_j \sum_i \mu_{ij}\,dn_{ij} \qquad (2\text{-}5)$$

$$dH = T\,dS + V\,dP + \sum_j \sum_i \mu_{ij}\,dn_{ij} \qquad (2\text{-}6)$$

$$dF = -S\,dT - P\,dV + \sum_j \sum_i \mu_{ij}\,dn_{ij} \qquad (2\text{-}7)$$

$$dG = -S\,dT + V\,dP + \sum_j \sum_i \mu_{ij}\,dn_{ij} \qquad (2\text{-}8)$$

Calculations based on free energies indicate whether a process is possible. When a system is at equilibrium, the Helmholz and Gibbs free energies are zero. Spontaneous processes have negative changes in free energies, while processes requiring energy will have positive free energies. Although both Helmholz and Gibbs free energies can be used, Gibbs free energy calculations are more common and are the type discussed further in this text.

Environmental systems are almost always open and are generally quite complex in composition. For simplicity, only a few components are usually considered in an equilibrium calculation. The reduction of components and our inability to control inputs into the open system means that these assumptions will introduce some error into the resulting calculations. Since we must accept some error due to these unavoidable constraints, we can usually relax other factors in our calculations to make the computations easier, and hopefully no less descriptive. The most useful assumption in equilibrium calculations is that we have constant temperature and pressures in the system. As a result, the Gibbs free energy calculation for an open system from Eq. 2-8 is simplified to

$$dG = \sum_j \sum_i \mu_{ij} \, dn_{ij} \qquad (2\text{-}9)$$

Integration of Eq. 2-9 produces

$$G = \sum_j \sum_i \mu_{ij} n_{ij} \qquad (2\text{-}10)$$

This expression shows that Gibbs free energy is an extensive property of the system, while chemical potential is a partial molar property. The chemical potential and Gibbs free energies are equal in a system with only one phase and one component.

2.3 CHEMICAL POTENTIALS

From the definition in Eq. 2-10 it is clear that equilibrium conditions of a system can be calculated using either chemical potentials or Gibbs free energies. In order to calculate whether a system is at equilibrium, the chemical potential must be calculated relative to a reference state. We define μ_i° as the chemical potential of a 1 M solution of species i at a specified set of conditions, usually 1 atm and 0°C. The chemical potential of the chemical at any other condition is calculated from

$$\mu_i = \mu_i^{\circ} + \mathscr{R}T \ln\{i\} \qquad (2\text{-}11)$$

where $\{i\}$ is the activity of species i. The magnitude of $\{i\}$ depends on the chosen standard conditions of the system. The activity of a species is equal to the product

$$\{i\} = x_i \gamma_i \qquad (2\text{-}12)$$

where x_i is the mole fraction of i in the phase (here chosen as the liquid phase) and γ_i is the activity coefficient. If chemical i does not behave as an ideal fluid, the activity coefficient varies as a function of the chemical concentration and system properties such as temperature and pressure.

We can combine Eqs. 2-11 and 2-12 to obtain a general equation for chemical potential of species i for a chemical at any concentration, temperature, and pressure as

$$\mu_i = \mu_i^° + \mathcal{R}T \ln x_i \gamma_i \qquad (2\text{-}13)$$

We can further simplify this equation by carefully choosing the system *reference* and *standard* states. The choice of these two states is completely arbitrary, but once chosen they must remain consistently defined for all calculations. For example, Stumm and Morgan (1970) defined the reference state as

$$\gamma_i = 1 \qquad \text{and} \qquad \{i\} = x_i \qquad (2\text{-}14)$$

and the standard state as

$$\{i\} = 1 \qquad \text{and} \qquad \mu_i = \mu_i^° \qquad (2\text{-}15)$$

It must be emphasized that different choices of the reference and standard states are equally valid for thermodynamic calculations as long as the values used for the concentration units are consistent with the thermodynamic constants taken from the literature.

In chemical calculations based on chemical potential and Gibbs free energies, we will assume the following conditions consistent with those used by Snoeyink and Jenkins (1980):

1. The activity of water is unity, or $\{i\} = x_w \gamma_w = 1$.
2. Activity coefficients for chemicals in dilute liquid solutions are unity.
3. The activity coefficient for any gas component is unity since all gases are ideal under typical environmental conditions. This is true even if the gas is not a dilute solution.
4. The activity of pure solids or liquids is unity.
5. Chemical concentrations are expressed in terms of molar concentrations.

Using these simplifications, the activities of chemicals in solutions become easily calculated values of chemical molar concentrations (for liquids) or partial pressures (for gases). The chemical equations are further simplified by defining the concentrations of water and solid in the system as unity since their concentrations (in ideal solutions) will not affect the chemical activities. For example, the activity of a gas is $\{i\} = \gamma_i y_i$. Using the chemical partial pressure and assuming the activity coefficient is unity produces $\{C\} = 1 \times y_C$. The general equations used for chemical potentials as a function of the different phases are summarized in Table 2.1.

The choice of concentration units has no effect on the system since any constants included in the activity term can be included in the definition of the reference state. First, let us compare a system where the activity coefficient for a dilute solution is

TABLE 2.1   Chemical Potentials of Different Phases

Phase	Nonideal Solutions (in general)	Ideal Solution (as used here)
Gas	$\mu_i = \mu_i^\circ + \mathscr{R}T \ln \gamma_i y_i$	$\mu_i = \mu_i^\circ + \mathscr{R}T \ln p_i$
Liquid	$\mu_i = \mu_i^\circ + \mathscr{R}T \ln \gamma_i x_i$	$\mu_i = \mu_i^\circ + \mathscr{R}T \ln c_i$
(water)		$\mu_i = \mu_i^\circ$
Solid^a	$\mu_i = \mu_i^\circ + \mathscr{R}T \ln \gamma_i z_i$	$\mu_i = \mu_i^\circ + \mathscr{R}T \ln 1$
(soil)		$\mu_i = \mu_i^\circ$

^{a}Assumes a pure solid.

assumed to be constant with the choice of an activity coefficient of unity. For the former case, we can separate the activity coefficient and concentration terms to produce

$$\mu_i = u_i^\circ + \mathscr{R}T \ln \gamma_i + \mathscr{R}T \ln x_i \qquad (2\text{-}16)$$

For the latter case, we can write the same equation, noting $\gamma_i = 1$, or

$$\mu_i = \mu_i^{\circ\circ} + \mathscr{R}T \ln 1 + \mathscr{R}T \ln x_i \qquad (2\text{-}17)$$

or since $\ln 1 = 0$,

$$\mu_i = \mu_i^{\circ\circ} + \mathscr{R}T \ln x_i \qquad (2\text{-}18)$$

Comparing Eqs. 2-16 and 2-18 shows that the two definitions are related by the reference state constants. If γ_i is a constant, then the term $\mathscr{R}T \ln \gamma_i$ is easily included into the reference chemical potential. From the comparison shown above, we can see that the two reference potentials are related by

$$\mu_i^{\circ\circ} = \mu_i^\circ + \mathscr{R}T \ln \gamma_i \qquad (2\text{-}19)$$

Similarly, the choice of concentration units does not affect the outcome of the calculation as long as the tabulated reference chemical potentials are consistent with the definitions of activity and chemical concentration used for calculating the chemical potentials at other states. Since $x_i = c_{iw}/c_w$ for dilute solutions, we can rewrite Eq. 2-19 as

$$\mu_i = \mu_i^{\circ\circ} - \mathscr{R}T \ln c_w + \mathscr{R}T \ln c_{iw} \qquad (2\text{-}20)$$

or defining a reference potential as $\mu_i' = \mu_i^{\circ\circ} - \mathscr{R}T \ln c_w$, we have

$$\mu_i = \mu_i' + \mathscr{R}T \ln c_{iw} \qquad (2\text{-}21)$$

Comparison of Eqs. 2-16, 2-19, and 2-21 demonstrates that the chemical potential expressions are all equivalent, but depend on the choice of a reference chemical and activity coefficient.

Example 2.1   Phases

Given

$$c_{Ca} \leftrightarrow c_{Cw} \qquad (2\text{-}22)$$

if the partial pressure of CO_2 (chemical C) is 0.001 atm, and its concentration in the liquid phase is 1×10^{-5} moles L^{-1}, is the system at equilibrium? If not, what is the direction of change? The following data are obtained from tables of thermodynamic data for chemical potentials at the reference state (25°C, 1 atm) (example from Stumm and Morgan 1970).

$$\mu_{Ca}^{\circ} = -94.26 \text{ kcal/mol} \qquad (2\text{-}23)$$

$$\mu_{Cw}^{\circ} = -99.13 \text{ kcal/mol} \qquad (2\text{-}24)$$

Chemical potentials can be used to determine if a system is at equilibrium. Assuming for CO_2 a concentration basis for the liquid phase and partial pressures for the gas phase, and from the given information, $p_C = 0.001$ atm and $c_{Cw} = 1 \times 10^{-5}$ mol L^{-1}. Substituting this into the equation for chemical potential in the gas phase, or

$$\mu_{Ca} = \mu_{Ca}^{\circ} + \mathcal{R}T \ln(p_C) \qquad (2\text{-}25)$$

yields

$$\mu_{Ca} = -94.23 + \left(1.987 \, \frac{\text{cal}}{\text{mol K}}\right)(298 \text{ K}) \, \frac{1 \text{ kcal}}{1000 \text{ cal}} \ln(0.001) \qquad (2\text{-}26)$$

$$= -98.32 \, \frac{\text{kcal}}{\text{mole}} \qquad (2\text{-}27)$$

for the concentration in the gas phase. For the concentration in the liquid phase

$$\mu_{Cw} = \mu_{Cw}^{\circ} + \mathcal{R}T \ln c_{Cw} \qquad (2\text{-}28)$$

$$= -99.13 \, \frac{\text{kcal}}{\text{mole}} + \mathcal{R}T \ln 10^{-5} \qquad (2\text{-}29)$$

or

$$\mu_{Cw} = -99.13 \, \frac{\text{kcal}}{\text{mole}} \tag{2-30}$$

Therefore, the system as defined is *not* at equilibrium since $\mu_{Ca} > \mu_{Cw}$. Since the liquid phase is at the lower energy (more negative), the direction of change is gas into liquid.

2.4 GIBBS FREE ENERGY AND EQUILIBRIUM CONSTANTS

Gibbs free energies are usually used in order to examine if *reactions* are spontaneous in the indicated direction (from left to right). Since Gibbs free energies and chemical potentials are related by Eq. 2-10, either can be used. Chemical reactions can be incorporated into equilibrium calculations by comparing the energies of the system with a system of the same components at equilibrium. Consider the case of the reversible reaction

$$\nu_A\{A\} + \nu_B\{B\} = \nu_C\{C\} + \nu_D\{D\} \tag{2-31}$$

where ν_i are the stoichiometric coefficients of chemical i in phase j. The chemical potential of each species in Eq. 2-31 can be calculated using Eq. 2-21 as

$$\mu_{ij} = \mu'_{ij} + \Re T \ln c_{ij} \tag{2-32}$$

From Eq. 2-9, the Gibbs free energy is

$$\Delta G = \sum_i \sum_j \mu_{ij} \, dn_{ij} \tag{2-33}$$

where a difference in free energy is assumed for the derivative. Combining Eqs. 2-32 and 2-33, we have

$$\Delta G = \sum_i \sum_j \nu_i \mu^\circ_{ij} + \Re T \sum_i \sum_j \nu_i \ln\{i\} \tag{2-34}$$

$$\Delta G^\circ = \sum_i \sum_j \nu_i G^\circ_{ij} - \sum_i \sum_j \nu_i G^\circ_{ij} \tag{2-35}$$

The stoichiometric coefficient for all chemicals on the right-hand side of Eq. 2-31 are positive, while those on the left-hand side are negative. If we explicitly show the signs in the equations, we can rewrite Eq. 2-35 in terms of the products and reactants as

$$\Delta G^\circ = \left(\sum_i \sum_j \nu_i G^\circ_{ij} \right)_{\text{products}} - \left(\sum_i \sum_j \nu_i G^\circ_{ij} \right)_{\text{reactants}} \tag{2-36}$$

The term $\Re T \sum v_i \ln\{i\}$ on the right-hand side of Eq. 2-34 can be simplified by recognizing that

$$\sum v_i \ln\{i\} = \ln \prod \{i\}^{v_i} \tag{2-37}$$

Substituting Eq. 2-37 into Eq. 2-34, assuming the activity coefficients are unity, and from the definition of $\Delta G°$, we have

$$\Delta G = \Delta G° + \ln \prod c_i^{v_i} \tag{2-38}$$

or with $i = A, B, C,$ and D, and assuming all chemicals are in the liquid phase,

$$\Delta G = \Delta G° + \Re T \ln \frac{c_C^{v_C} c_D^{v_D}}{c_A^{v_A} c_B^{v_B}} \tag{2-39}$$

An equilibrium constant for the given equation is defined as

$$K_{eq} = \frac{c_C^{v_C} c_D^{v_D}}{c_A^{v_A} c_B^{v_B}} \tag{2-40}$$

As a special case of Eq. 2-39, we recognize that at equilibrium, $\Delta G = 0$. Therefore, we see that $\Delta G°$ and K_{eq} are related by

$$\Delta G° = -\Re T \ln K_{eq} \tag{2-41}$$

The actual ratios of chemical concentrations are often defined as Q, or

$$Q = \frac{c_C^{v_C} c_D^{v_D}}{c_A^{v_A} c_B^{v_B}} \tag{2-42}$$

so that Eq. 2-39 can also be written in the form

$$\Delta G = \Re T \ln \frac{Q}{K} \tag{2-43}$$

Example 2.2   Chemical Reactions

Bacteria produce methane (M) according to the chemical equation below. From a thermodynamic view, is methane production likely for at an acetate (A) concentration of 10^{-3} M, when the bicarbonate (B) concentration is 0.05 M and 0.5 atm of methane are present?

$$C_2H_3O_2^- + H_2O \rightleftharpoons CH_4 + HCO_3^- \tag{2-44}$$

In order to answer this question, we need to determine whether the reaction is spontaneous by calculating whether $\Delta G°$ is negative. The given chemical equation can be used with Eq. 2-38 to determine the Gibbs free energy as

$$\Delta G = \Delta G° + \mathcal{R}T \ln \frac{\{CH_4\}\{HCO_3^-\}}{\{C_2H_3O_2^-\}\{H_2O\}} = \Delta G° + \mathcal{R}T \ln Q \qquad (2\text{-}45)$$

This equation can be solved by substitution of the appropriate concentration terms and reference state Gibbs free energies. Assuming methane is an ideal gas, $p_M = 0.5$ atm. For the chemicals in water, the activities of $\{HCO_3^-\}$ and $\{C_2H_3O_2^-\}$ can be approximated by the concentrations of the chemicals in solution, or $c_B = 0.05$ M, and $c_A = 10^{-3}$ M. The activity of water, by convention, is defined as unity or $\{H_2O\} = 1$. Substitution of these values into the reaction quotient Q

$$Q = \frac{p_M c_B}{c_A \times 1} = \frac{(0.5)(0.05)}{(10^{-3})} = 25 \qquad (2\text{-}46)$$

The value of $\Delta G°$ can be computed from data in standard thermodynamic reference tables. Using Eq. 2-36, the Gibbs free energies of the two products at their reference states are

$$\Delta G°_{products} = \Delta G°(CH_4) + \Delta G°(HCO_3^{-1}) \qquad (2\text{-}47)$$

$$= -12.14 - 140.3 = -152.44 \text{ kcal mol}^{-1} \qquad (2\text{-}48)$$

and the Gibbs free energies of the reactants are

$$\Delta G°_{reactants} = \Delta G°(C_2H_3O_2^-) + \Delta G°(H_2O) \qquad (2\text{-}49)$$

$$= -88.19 - 56.69 = -144.88 \text{ kcal mol}^{-1} \qquad (2\text{-}50)$$

Summing up, $\Delta G° = (-152.44) - (-144.88)$, or $\Delta G° = -7.56$ kcal mol^{-1}. Substituting in $T = 298$ K, $\mathcal{R} = 1.987$ cal/mol K into the original equation

$$\Delta G = -7.56 \frac{\text{kcal}}{\text{mol K}} + 1.987 \frac{\text{cal}}{\text{mol K}} \ 298 \text{ K} \ \frac{1 \text{ kcal}}{10^3 \text{ cal}} \ln 25 \qquad (2\text{-}51)$$

$$= -7.56 + 0.592 \ln 25 \qquad (2\text{-}52)$$

$$= -5.65 \text{ kcal mol}^{-1} \qquad (2\text{-}53)$$

Since $\Delta G < 0$, the reaction is spontaneous.

2.5   DISTRIBUTION OF CHEMICALS BASED ON FUGACITIES

The distribution of chemicals in multiple *phases* at equilibrium can easily be measured and related to distribution coefficients using fugacities. We will consider several examples of distribution coefficients in this section. Chemical fugacities were originally developed to deal with nonideal solutions, for example, gases at high pressures. In the context of environmental calculations, the more rigorous approach of chemical fugacities might appear unwarranted. However, chemical equilibrium calculations based on fugacities have two advantages over the use of chemical potentials or Gibbs free energies. First, the reference state is incorporated directly into the fugacity, making the reference states easier to use in calculations. Second, since the fugacity is expressed in units of pressure, the direction of change is always easy to identify since the system will move towards an equilibrium point of equal pressures.

The approach used here for fugacity calculations is based on reference states and notation conventions developed by Thibodeaux (1979) in his text *Chemodynamics*. According to this approach, the fugacity of a chemical is written as

$$f_{Cj} = x_C \gamma_{Cj} f^\circ_{Cj} \tag{2-54}$$

According to this definition of fugacity, the reference state of a chemical is directly incorporated in the expression as a reference fugacity, f°_{Cj}. As discussed, the definition of a reference state for thermodynamic calculations is arbitrary. The definition of the reference fugacity used here is in agreement with preceding chemical potential calculations.

The use of a reference fugacity does not affect the outcome of an equilibrium calculation for the same reasons that the concentration units do not affect a thermodynamic calculation based on chemical potentials. This can be seen by comparing the definitions of a fugacity and chemical potential. From the definition of a chemical potential, we can write for species C in the j phase

$$\mu_{Cj} = \mu^\circ_{Cj} + \mathcal{R}T \ln\{C\} \tag{2-55}$$

The reference state can be moved within the ln term by rewriting Eq. 2-55 as

$$\mu_{Cj} = \mathcal{R}T \ln e^{\mu^\circ_{Cj}/\mathcal{R}T} + \mathcal{R}T \ln\{C\} \tag{2-56}$$

and by combining terms

$$\mu_{Cj} = \mathcal{R}T \ln(e^{\mu^\circ_{Cj}/\mathcal{R}T}\{C\}) \tag{2-57}$$

Rearranging terms, and substituting in $\{i\} = x_A \gamma_{Aj}$ produces

$$e^{\mu_{Cj}/\mathcal{R}T} = e^{\mu^\circ_{Cj}/\mathcal{R}T} = x_C \gamma_{Cj} \tag{2-58}$$

Comparison of Eq. 2-58 with the fugacity definition will show that at constant temperatures the fugacity and chemical potential expressions are equivalent. At different temperatures, the reference fugacities must be changed. The expression for changing these reference pressures differs for the different phases.

The approach used for chemical potential calculations was that at equilibrium the chemical potentials of C in two phases were equal. Similarly, the fugacities of chemical C in two phases must also be equal at equilibrium. For example, the fugacities of chemical C in the air and water states would be written as

$$f_{Ca} = f_{Cw} \tag{2-59}$$

Using the definition of the fugacity, we can expand this to write

$$y_C \gamma_{Ca} f^\circ_{Ca} = x_C \gamma_{Cw} f^\circ_{Cw} \tag{2-60}$$

From this simple equation equilibrium conditions at interfaces or between phases can easily be derived. The direction of change in a system not at equilibrium is always from the higher to lower pressure. If $f_{Ca} > f_{Cw}$, then the direction of spontaneous motion is from the gas to the liquid. As will be shown in the sections below, fugacity expressions can be used to easily derive common equilibrium partitioning coefficients, such as distribution coefficients and Henry's Law constants, from thermodynamic data.

Simplifications of Fugacity Expressions

In order to develop fugacity expressions for chemicals in air, water, and soils, we will define activity coefficients and reference fugacities that simplify fugacity calculations. For fugacity calculations, the following are assumed for activity coefficients:

1. For a chemical in air, $\gamma_{ia} = 1$.
2. For any pure phase, $\gamma_j = 1$.
3. For a chemical in water, γ_{iw} is constant when the solution is dilute.

Activity coefficients are used to adjust the properties of a phase for nonideal behavior. Since all calculations made for air assume that air is an ideal gas, no adjustment to the activity coefficient is necessary. Similarly, no activity corrections are necessary for a pure phase, since all intramolecular forces are similar in the phase. The activity coefficient of a chemical in water at any concentration is

$$\gamma_{iw} \approx \exp[b(1 - x_i)^2] \tag{2-61}$$

where b is a constant. When x_i is unity, $\gamma_{iw} = 1$. For a dilute solution, $x_i \ll 1$, and Eq. 2-61 can be simplified to show that the activity coefficient is a constant, or

$$\gamma_{iw} \approx e^b, \qquad \text{dilute solution} \tag{2-62}$$

The following are assumed here for reference fugacities for chemicals in different phases.

1. For chemicals in air, $f_{ia}^{\circ} = P$, where P is the total air pressure.
2. For chemicals in water, $f_{iw}^{\circ} = p_i^{\circ}$, where p_i° is the vapor pressure of the pure liquid i.
3. For any solid, $f_{is}^{\circ} = p_i^{\circ}$, where p_i° is the vapor pressure of the pure solid i.

A difficulty with the reference fugacities is that the chemical may not exist in the specified phase at the temperature in the system. In such cases, the vapor pressures must be extrapolated to the system temperature before the calculation can be made. Vapor pressures at typical environmental temperatures of 20 to 25°C are contained in the appendix.

With these assumptions, we can now derive more simple expressions for chemical fugacities in the air, water and soil, phases.

Air   The fugacity of a chemical C in air (phase a) is

$$f_{Ca} = y_C \gamma_{Ca} f_{Ca}^{\circ} \tag{2-63}$$

where y_C is the mole fraction of C in air, γ_{Ca} is the activity coefficient, and f_{Ca}° is a reference fugacity. Using the assumptions above, $\gamma_{Ca} = 1$ and $f_{Ca}^{\circ} = P$, or

$$f_{Ca} = y_C P \tag{2-64}$$

For a pressure of 1 atm this relationship is simply $f_{Ca} = y_C$. From the definition of partial pressures, Eq. 2-64 can also be written as $f_{Ca} = p_C$. The reference fugacity for air at different temperatures and reference pressures (or altitudes) can easily be obtained from the ideal gas law.

Water   The fugacity of a chemical C in water (phase w) is

$$f_{Cw} = x_C \gamma_{Cw} f_{Cw}^{\circ} \tag{2-65}$$

where x_C is the mole fraction of C in water, γ_{Cw} is the activity coefficient, and f_{Cw}° a reference fugacity. Using the assumptions above, the reference fugacity is replaced by the vapor pressure of pure liquid chemical C, or

$$f_{Cw} = x_C \gamma_{Cw} p_C^{\circ} \tag{2-66}$$

The vapor pressure of a liquid can be adjusted for temperature changes using the Clausius–Clapeyron equation,

$$\ln\left(\frac{p^{\circ}_{Cw}}{p^{\circ}_{Cw,T}}\right) = \frac{\lambda_C}{\mathcal{R}}\left(\frac{1}{T_T} - \frac{1}{T}\right) \tag{2-67}$$

where λ_C is the molal heat of vaporization, $\mathcal{R}$ the universal gas constant, and $p^{\circ}_{Cw,T}$ and p°_{Cw} the vapor pressures at T_T and T. Therefore, once the vapor pressure is defined at one temperature, it can be extrapolated to other temperatures based on the heat of vaporization. Values of λ_C are available in the appendix as well as standard thermodynamics reference texts.

Soils   The fugacity of a chemical C in soil (phase s) is

$$f_{Cs} = z_C \gamma_{Cs} f^{\circ}_{Cs} \tag{2-68}$$

where z_C is the mole fraction of C in the soil, γ_{Cs} the activity coefficient, and f°_{Cs} a reference fugacity. The reference fugacity for this system is the vapor pressure of pure solid C at the indicated temperature. The main difficulty with this reference fugacity is that the chemical may not be a solid at this temperature. Since a chemical adsorbed to a soil particle can be considered to be in the solid state, this definition of the reference fugacity is unavoidable. For chemicals that are not solids, the reference vapor pressures must be extrapolated to the reference temperature and state using thermodynamic relationships beyond our level of interest here. Reference fugacities for some solid chemicals have been included in the appendix, although fugacity calculations are usually not used for systems containing solids.

Distribution Coefficients

Under equilibrium conditions, the extent that a chemical will partition among two different phases can be quantified in terms of a partition, or distribution coefficient, defined as

$$K^{*}_{Cjk} = \frac{\text{concentration of } C \text{ in phase } j}{\text{concentration of } C \text{ in phase } k} \tag{2-69}$$

where j and k are two separate phases and the asterisk is included when the dimensions of K_{Cjk} are such that K_{Cjk} is dimensionless. For a chemical distributed between air and water, the distribution coefficient, K^{*}_{Caw} is

$$K^{*}_{Caw} = \frac{y_C}{x_C} \tag{2-70}$$

There is no convention for which phase is defined first, so that definition of a distribution coefficient as the reciprocal of the one in Eq. 2-70, or $K^{*}_{Cwa} = (x_C/y_C) =$

$(K_{Caw}^{*})^{-1}$, is equally valid. Distribution coefficients may also be expressed in other units. For example

$$K_{Cws}^{*} = \frac{\phi_C}{\omega_C} \qquad (2\text{-}71)$$

is the equilibrium distribution of chemical C between water and soil, where the concentrations in the two phases are in mass fractions. The units of concentration in the ratio of the concentrations in the different phases are often mismatched, producing distribution coefficients such as

$$K_{Csw} = \frac{\omega_C}{\rho_{Cw}} \qquad (2\text{-}72)$$

since partitioning of C between the water and soil often relies on liquid-phase mass concentrations and soil mass fractions. It is important to check the units of K_{Cij} in order to know which set of units has been used.

The definition of the distribution coefficient given in Eq. 2-69 incorrectly implies that the distribution coefficient is constant over the whole concentration range for all chemicals and all phases. While this may be true under some conditions, it is not correct for all cases. The presence of other chemicals, phases or particles may influence partitioning. For example, the concentration of organic matter in soils may influence soil–water partitioning, and the presence of highly reactive colloids may influence air–water partitioning. Conditions where the distribution coefficient can be assumed constant need to be carefully evaluated over the concentration range of interest before a constant distribution coefficient can be applied to partitioning calculations.

Equilibrium between Gases and Liquids

The distribution of chemicals among different phases often is expressed using a variety of constants such as Henry's Law and relative volatility. The relationship between these constants and the fugacity relationships derived above, some simplified forms of fugacity equations, and practical forms of the distribution coefficients are presented in the remainder of this section.

Henry's Law Constants   The most common approach to calculating equilibrium concentrations between gases and liquids is to use a Henry's Law constant, defined as

$$H_{ijk} = \frac{\text{concentration in gas phase}}{\text{concentration in liquid phase}} \qquad (2\text{-}73)$$

The units of both concentration terms frequently vary, producing Henry's Law constants with different units. One common definition of the Henry's Law constant is

$$H_{Cpx} = \frac{p_C}{x_C} \qquad (2\text{-}74)$$

resulting in a Henry's Law constant in units of pressure. Using fugacities, we can derive a Henry's law constant in terms of thermodynamic relationships. Starting with the equilibrium condition $f_{Ca} = f_{Cw}$, we have

$$y_C \gamma_{Ca} f_{Ca}^\circ = x_C \gamma_{Cw} f_{Cw}^\circ \qquad (2\text{-}75)$$

Substituting in the assumption of $\gamma_{Ca} = 1$, $f_{Ca}^\circ = P$, and $f_{Cw}^\circ = p_C^\circ$ produces

$$y_C P = x_C \gamma_{Cw} p_C^\circ \qquad (2\text{-}76)$$

Recognizing that $p_C = y_C P$, and rearranging,

$$\frac{p_C}{x_C} = \gamma_{Cw} p_C^\circ \qquad (2\text{-}77)$$

Comparing Eqs. 2-74 and 2-77, we see that

$$H_{Cpx} = \gamma_{Cw} p_C^\circ \qquad (2\text{-}78)$$

This relationship points out that the Henry's Law constant is just a combination of two thermodynamic constants, the activity coefficient of C in water and vapor pressure of pure chemical C. Henry's law constants are given for several chemicals in Table 2.2 and in Appendix 3.

Example 2.3

A vinyl chloride concentration of 2.5 mg L^{-1} is found in the aeration tank of an activated sludge wastewater treatment system. Assuming the wastewater and air are in equilibrium, what is the vinyl chloride concentration in the air?

    Although fugacities could be used to solve this problem, the most direct solution is to use the Henry's Law constant $H_{Ccc} = 50$ mg L^{-1} air/mg L^{-1} water. Solving for c_{Ca},

$$c_{Ca} = H_{Ccc} c_{Cw} = \left(50 \; \frac{\text{mg L}^{-1}}{\text{mg L}^{-1}} \right) (2.5 \text{ mg L}^{-1}) \qquad (2\text{-}79)$$

$$= 125 \text{ mg L}^{-1} \qquad (2\text{-}80)$$

TABLE 2.2   Vapor–Liquid Equilibria of Selected Gases and Liquids in Water at 25°C

Components	Normal Boiling Point (°C)	Henry's Law Constant, H_{Cpx} (atm/mol fraction)	Relative Volatility, α^*_{Cw}
Nitrogen	−195.8	86,500	2,768,000
Hydrogen sulfide (H$_2$S)	−59.6	64,500	1,744,000
Oxygen (O$_2$)	−183	43,800	1,402,000
Ethane (C$_2$H$_6$)	−88.6	30,200	966,400
Propylene (C$_3$H$_6$)	−48	5,690	182,100
Carbon dioxide (CO$_2$)	−78.5	1,640	52,480
Acetylene (C$_2$H$_2$)	−84	1,330	42,560
Bromine (Br$_2$)	−58.8	73.7	2,358
Ammonia (NH$_3$)	−33.4	0.843	27.0
Acetylaldehyde	20.2	5.88	188
Acetone	56.4	1.99	63.7
Isopropanol	82.5	1.19	38.1
n-Propanol	97.8	0.471	15.1
Ethanol	78.4	0.363	11.6
Methanol	67.7	0.300	9.60
n-Butanol	117	0.182	5.82
Acetic acid	118.1	0.0627	2.01
Formic acid	100.8	0.0247	0.790
Propionic acid	141.1	0.0130	0.416
Phenol	181.4	0.0102	0.326

Source: Thibodeaux (1979).

Relative Volatility   It is often useful to compare the volatility of two chemicals to evaluate whether one chemical will be preferentially volatilized from a well-aerated solution. Water treatment by air stripping, for example, is applied to the removal of volatile organic compounds (VOCs) such as tricholoroethylene (TCE) that are frequent contaminants of groundwater. The relative volatility, α^*_{BC} of chemicals B and C is defined as

$$\alpha^*_{BC} = \frac{K^*_{Baw}}{K^*_{Caw}} \tag{2-81}$$

Chemical removal from a wastewater stream by air stripping requires that the chemical be substantially more volatile than water. The relative volatility of a chemical in water is

$$\alpha^*_{CW} = \frac{K^*_{Caw}}{K^*_{Waw}} \tag{2-82}$$

where W is water. From the definition of the distribution coefficient, we can simplify this expression as

$$\alpha^*_{CW} = \frac{y_C/x_C}{y_W/x_W} \tag{2-83}$$

For a dilute solution, $x_W = 1$, $y_W = p^\circ_W/P_T$, and using Eq. 2-76 for y_C/x_C, we have

$$\alpha^*_{CW} = \frac{\gamma_C p^\circ_C}{p^\circ_W} \tag{2-84}$$

If $\alpha^*_{CW} > 1$, then C will preferentially associate in the air. Conversely, if $\alpha^*_{CW} < 1$, air will tend to evaporate water more than C in an air stripping process, although the rates of chemical movement will be a function of the mass transfer coefficients of the chemicals.

Pure Phases   If a pure phase is present, the fugacity expression is simplified since the mole fraction of the pure compound is unity. For a pure gas in contact with water, the fugacity balance becomes

$$P = x_C \gamma_{Cw} f^\circ_{Cw} \tag{2-85}$$

where P is the total air pressure. Solving for the mole fraction in water and substituting in the liquid vapor pressure for the reference fugacity produces

$$x_C = \frac{P}{\gamma_{Cw} p^\circ_C} \tag{2-86}$$

For a pure liquid in contact with air, both activity coefficients and the mole fraction of the liquid are unity, resulting in

$$y_C P = p^\circ_C \tag{2-87}$$

Substituting $p_C = y_C P$ produces the equality of $p_C = p^\circ_C$, which is of course Raolt's Law. This result should be obvious given our definition that the reference fugacity is the vapor pressure of a pure liquid.

Example 2.4

Methanotrophs are microorganisms that grow aerobically on methane gas. These microbes have been shown to be able to degrade chlorinated compounds such as chloroform and trichloroethylene, two common contaminants of groundwater. Unfortunately, methane is sparingly soluble in water, and gas concentrations of methane larger than 4% are explosive. Using fugacities, calculate the solubility (g L^{-1}) of (a) pure methane gas in water at 25°C at a total pressure of 1 atm, and (b) methane at a partial pressure of 0.04 atm.

   (a)   Using Eq. 2-86 with $P = 1$ atm and for M, methane, we have

$$x_M = \frac{1}{\gamma_{Mw}p_M^\circ} \tag{2-88}$$

In order to solve for the solubility in mass concentration units of g L^{-1}, we need to derive a solution in terms of ρ_{Mw}. Recognizing that $\rho_{Mw} = c_{Mw}M_M$, and that $c_{Mw} = x_M c_w$, where c_w is the molar concentration of water (55.6 mol L^{-1}), Eq. 2-88 becomes

$$\rho_{Mw} = \frac{c_w M_C}{\gamma_{Cw}p_C^\circ} \tag{2-89}$$

From Appendix 3, we have $\gamma_{Cw} = 137.5$ and $p_C^\circ = 269$ atm. Substituting into Eq. 2-89 produces

$$\rho_{Cw} = \frac{(55.6 \text{ mol L}^{-1})(16 \text{ g mol}^{-1})(1 \text{ atm}) \times 10^3 \text{ mg m}^{-1}}{(137.5)(269 \text{ atm})} = 24.1 \text{ mg L}^{-1} \tag{2-90}$$

   (b)   If methane is present at a partial pressure of 0.04 atm, we can use the partial pressure, p_C, instead of the total pressure P in Eq. 2-85. This produces the same result as multiplying ρ_{Cw} by p_C, yielding 0.96 mg L^{-1}.

Example 2.5

What volume (μL) of benzene (chemical B) should be injected into a 55-ml test tube containing 25 ml of water to obtain a 10^{-4} M concentration of benzene in the liquid phase? Benzene is very volatile, has a Henry's Law constant of 5.49 $\times$ 10^{-3} atm m^3 mol^{-1}, a density of 0.8787 g cm^{-3}, and a molecular weight of 78.

   To solve this, we will calculate the moles of benzene in the liquid phase, and then using Henry's Law, calculate the total moles that must be present in the gas phase at equilibrium. The total volume of benzene is then the sum of the total moles converted to volume. The moles of benzene in water are

$$n_{Bw} = c_{Bw}V_w = 10^{-4}\left(\frac{\text{mol}}{\text{L}}\right) \times 0.025 \text{ L} = 2.5 \times 10^{-6} \text{ mol} \tag{2-91}$$

The total moles in the gas phase is similarly

$$n_{Ba} = c_{Ba}V_a = y_B c_a V_a = \frac{p_B c_a V_a}{P} \tag{2-92}$$

Substituting in Henry's Law in the form $p_B = H_{Bpc}c_{Bw}$,

$$n_{Ba} = \frac{H_{Bpc}c_{Bw}c_aV_a}{P} \tag{2-93}$$

and using $c_a = 0.0415$ mol L^{-1} (Table 1.5)

$$n_{Ba} = \frac{(5.49 \times 10^{-3} \text{ atm m}^3 \text{ mol}^{-1})(10^{-4} \text{ mol L}^{-1})(0.0415 \text{ mol L}^{-1})(0.03 \text{ L})}{1 \text{ atm}}$$

$$\times \frac{10^3 \text{ L}}{\text{m}^3} \tag{2-94}$$

$$= 0.7 \times 10^{-6} \text{ mol}$$

The total moles in both phases is

$$n_B = n_{Ba} + n_{Bw} = 3.2 \times 10^{-6} \text{ mol} \tag{2-95}$$

From the density of pure benzene, we calculate the volume as

$$V_B = \frac{n_B}{c_B} = \frac{n_B M_B}{\rho_B} = \frac{(3.2 \times 10^{-6} \text{ mol})(78.11 \text{ g mol}^{-1})}{(0.8787 \text{ g cm}^{-3})} \times \frac{10^3 \text{ } \mu\text{L}}{\text{cm}^3} \tag{2-96}$$

$$= 0.28 \text{ } \mu\text{L} \tag{2-97}$$

Multiple Liquid Phases and Water

It is possible that a pure phase of a chemical can exist in equilibrium with water, and that contained in that new phase is a chemical that is distributed between the two phases. An example might be an oil, which can be immiscible in water, containing aromatic compounds such as benzene and xylene. Alternately, the organic matter in soil can be considered to be a separate phase since the partitioning of hydrophobic chemicals in soils is more a function of organic matter content than mineral type.

In the presence of multiple liquid phases, the fugacity balance $f_{Cw} = f_{Co}$, where w is water and o oil (or some other separate phase), can be simplified to

$$x_C \gamma_{Cw} = x_{Co} \gamma_{Co} \tag{2-98}$$

since the reference phase for both sides of the equation is the vapor pressure of pure liquid C, or $f^{\circ}_{Cw} = f^{\circ}_{Co} = p^{\circ}_C$. A pure phase of oil containing large amounts of different, and perhaps volatile, hydrocarbons may not act as an ideal fluid. This can result from interactions of different hydrocarbons with each other. The result of such interaction can be a solubility enhancement of compounds in water, particularly compounds that are hydrophobic.

Example 2.6

Assuming that a chemical dissolved in oil is soluble in water. Starting with Eq. 2-98, show that the concentration of the chemical in water is given by

$$c_{Cw} = c_{Cw,eq} x_{Co} \gamma_{Co} \qquad (2\text{-}99)$$

where $c_{Cw,eq}$ is the solubility of the pure chemical in water, x_{Co} is the mole fraction, and γ_{Co} is the activity coefficient of the chemical in the oil slick (oil, phase o) (adapted from Thibodeaux 1979).

To solve this problem, we apply Eq. 2-98 twice. First, we apply Eq. 2-98 to a two-phase system consisting of a pure chemical phase in water. Defining pure chemical as phase c, we have from Eq. 2-98

$$x_c \gamma_{Cw} = x_{Cc} \gamma_{Cc} \qquad (2\text{-}100)$$

For a pure chemical phase $x_{Cc} = 1$ and $\gamma_{Cc} = 1$. Since water is in equilibrium with pure chemical, we designate the mole fraction x_{Cw} as $x_{Cw,eq}$, producing

$$x_{Cw,eq} = \frac{1}{\gamma_{Cw}} \qquad (2\text{-}101)$$

Recognizing that $x_{Cw,eq} = c_{Cw,eq}/c_w$, and rearranging

$$\gamma_{Cw} = \frac{c_w}{c_{Cw,eq}} \qquad (2\text{-}102)$$

Combining this result with the system for chemical C in both water and oil (Eq. 2-98), produces

$$x_C \left(\frac{c_w}{c_{Cw,eq}} \right) = x_{Co} \gamma_{Co} \qquad (2\text{-}103)$$

Using the identity that $x_{Cw} = c_{Cw}/c_w$, results in

$$\left(\frac{c_{Cw}}{c_w} \right) \left(\frac{c_w}{c_{Cw,eq}} \right) = x_{Co} \gamma_{Co} \qquad (2\text{-}104)$$

and produces the desired result

$$c_{Cw} = c_{Cw,eq} x_{Co} \gamma_{Co} \qquad (2\text{-}105)$$

Equilibrium Expressions with Solids

Pure Solids Fugacity expressions can be developed for solids in contact with liquids and gases, but it can be argued that such expressions have limited applications to environmental systems. One difficulty with fugacity expressions for chemicals in the solid phase is that the reference fugacity is the pure compound vapor pressure. Since many compounds may not exist at ambient temperatures as pure solids, the reference fugacities are difficult to calculate for solid phases. For pure solids, the fugacity equilibrium expression $f_{Cs} = f_{Cw}$ yields

$$x_C = \frac{p^\circ_{Cs}}{\gamma_{Cw} p^\circ_C} \tag{2-106}$$

where p°_{Cs} is the vapor pressure of pure solid C. For a pure solid and air, $f_{Cs} = f_{Ca}$ can be simplified to

$$y_C = \frac{p^\circ_{Cs}}{P} \tag{2-107}$$

Soils   A more useful environmental situation is chemical partitioning onto soils. Once a chemical is adsorbed onto a soil particle, the chemical is considered to be in the solid phase. Within this soil–chemical context, developing equations for chemical "solids" at ambient temperatures becomes more relevant. Even in this context, however, a real limitation of the fugacity approach is that many aspects of chemical partitioning onto soils makes it difficult to assume ideal "solutions" for the mixture. Chemical aspects that can affect the extent of solid partitioning include:

- van der Waals attractive forces
- hydrophobic bonding
- hydrogen bonding
- ligand exchange
- ion exchange
- chemisorption

In the fugacity equation approach used here, all these factors must be incorporated into the activity coefficient of the chemical in solid form, γ_C.

In the chemical industry, thermodynamic expressions are useful since the system is completely described once the thermodynamic properties have been calculated. Unfortunately for environmental situations, soils are extremely heterogeneous, making it impossible to calculate a constant for an activity coefficient that describes the behavior of more than one specific soil. Even the simplest conversion of a mole fraction into a mass concentration is frustrated by an inaccuracy of any assumed molecular weight for a soil. As explained in the first chapter, we have assumed a molecular weight of 60 g mol^{-1}, but that choice was based on a "soil" being pure silica (sand). Soils have substantially different compositions, and therefore, molecular weights, but a convention is useful for mole–mass conversions. Since the extent of chemical partitioning onto soil can be considered dilute ($<5\%$), the selection of a molecular weight for soils is relatively unimportant. What is more important is how the composition of the soil affects chemical partitioning.

Distribution Coefficients   A distribution coefficient for chemical c partitioning between soil and water can be derived from fugacities using

$$K_{Csw}^* = \frac{z_C}{x_C} = \frac{\gamma_{Cw} f_{Cw}^o}{\gamma_{Cs} f_{Cs}^o} \qquad (2\text{-}108)$$

As pointed out above, activity coefficients are difficult to determine for soils since they vary for each soil and chemical. As a result, fugacities are not generally employed in partitioning calculations for soil–water systems. Instead, a partition coefficient is measured on a case-by-case basis.

The partitioning of a chemical between soil and water is most commonly described in terms of the concentration in water and a mass fraction in the soil phase based on dry solids. These concentration units are used since liquid concentrations can be measured by centrifugation and filtration of muds and sediments, and the weight of soils can be gravimetrically determined after soil drying. Using these units, a distribution coefficient is defined as

$$K_{Csw} = \frac{\omega_C}{\rho_{Cw}} \qquad (2\text{-}109)$$

This coefficient is not always constant. Constant distribution coefficients are typically found only for highly hydrophobic compounds in groundwater and marine sediments (DiToro et al., 1991).

Example 2.7

The partitioning of a chemical is strongly affected by the nature of the soil. An experiment was conducted with the chemical 4-amino-3,5,6-trichloropicolinic acid (TCPA = T) to determine if the partitioning was affected more by the organic content of the soil than by the type of soil. For five different soils, 4 ml of a 1.0 ppm solution of TPCA was added to 1 g of soil, and the pH adjusted to 2 (using HNO_3). After 1 h, the data shown in Table 2.3 was measured on the distribution of TPCA in the water and soils. Using these data, calculate the liquid concentration of TPCA (μg T cm^{-3}) and the soil concentration (μg T g total soil^{-1}). Then, calculate equilibrium distribution coefficients for TPCA based on the soil (s) phase and organic matter (o) phases, K_{Tws} and K_{Two}, for the different soils. Present your results in a summary table. What factor (soil type or organic content) most affects the partitioning of TPCA to soil? (Adapted from Thibodeaux, 1979, Problem 2.1I).

TABLE 2.3 Distribution of TPCA Sorbed onto Soils as a Function of Organic Matter Content

Soil	Organic Matter (% by weight)	Distribution of TPCA	
		TCPA in Water (%)	TPCA in Soil (%)
C1	1.0	51	49
C2	2.7	23	77
C3	4.1	11	89
C4	10.7	5.8	94
C5	32.2	2	98

Source: Thibodeaux (1979).

The total mass of T added to each soil, m_T, is calculated from the total mass concentration added initially to the soil, $\rho_{Tw,i}$, and the total liquid volume, V_w, as

$$m_T = \rho_{Tw,i} V_w \tag{2-110}$$

$$= (1 \text{ mg L}^{-1})(4 \text{ mL}) \times \frac{1}{10^3 \text{ mL}} \times \frac{10^3 \text{ } \mu g}{\text{mg}} = 4 \text{ } \mu g \tag{2-111}$$

The weight percents given in the table are converted to mass fractions by dividing the weight percents by 100, where the mass fractions are m_{Tw}/m_T for the water, and m_{Ts}/m_T for the soil. The water concentration at equilibrium, $\rho_{Tw,eq}$ is obtained from the mass ratio m_{Tw}/m_T and the volume V_w

$$\rho_{Tw} = \left(\frac{m_{Tw}}{m_T}\right) \frac{m_T}{V_w} \tag{2-112}$$

Substituting in values for soil C1

$$\rho_{Tw} = (0.51) \frac{(4 \text{ } \mu g)}{(4 \text{ cm}^3)} = 0.51 \text{ } \mu g \text{ cm}^{-3} \tag{2-113}$$

Similarly, the mass fraction ω_{Ts} is calculated from the mass fraction m_{Ts}/m_T and the mass of soil using

$$\omega_{Ts} = \left(\frac{m_{Ts}}{m_T}\right) \frac{m_T}{m_s} \tag{2-114}$$

which for soil C1 becomes

$$\omega_{Ts} = (0.49) \frac{(4 \text{ } \mu g)}{(1 \text{ g})} = 1.96 \text{ } \mu g \text{ g}^{-1} \tag{2-115}$$

The mass fraction of T based on only organic matter, ω_{To}, can be obtained from ω_{Ts} and the mass fraction of organic matter, m_{os}/m_s, using

$$\omega_{To} = \omega_{Ts} \frac{m_s}{m_{os}} \qquad (2\text{-}116)$$

For soil C1, this becomes

$$\omega_{To} = (1.96 \ \mu\text{g g}^{-1}) \frac{1}{0.01} = 196 \ \mu\text{g g}^{-1} \qquad (2\text{-}117)$$

To solve for the two equilibrium partition coefficients, we use the relationships

$$K_{Tws} = \frac{\rho_{Tw}}{\omega_{Ts}} \frac{[\mu\text{g } T \text{ in water/(cm}^3 \text{ water)}]}{[\mu\text{g } T \text{ in soil/(g soil)}]} \qquad (2\text{-}118)$$

$$K_{Two} = \frac{\rho_{Tw}}{\omega_{To}} \frac{[\mu\text{g } T \text{ in water/(cm}^3 \text{ water)}]}{[\mu\text{g } T \text{ in organic matter/(g organic matter)}]} \qquad (2\text{-}119)$$

Combining these results into a single table for all soils produces the results shown in Table 2.4. Comparison of the two distribution coefficients shows that the distribution of T based only on the soil type varies over two orders of magnitude, but that distribution coefficients based only on the organic matter content are nearly similar. Thus we conclude that the concentration of organic matter is a reasonable predictor of the partitioning of T into the soil.

TABLE 2.4 Equilibrium Concentrations and Partition Coefficients for TPCA as a Function of Total Soil and Soil Organic Matter Content

Soil	ρ_{Tw} ($\mu\text{g cm}^{-3}$)	ω_{Ts} ($\mu\text{g g}^{-1}$)	ω_{To} ($\mu\text{g g}^{-1}$)	K_{Tws} (g cm^{-3})	K_{Two} (g cm^{-3})
C1	0.51	1.96	196	0.26	0.0026
C2	0.23	3.08	114	0.075	0.0022
C3	0.11	3.56	86.8	0.031	0.0013
C4	0.058	3.77	35.2	0.015	0.0016
C5	0.02	3.92	12.2	0.005	0.0016

Adsorption Isotherms Nonlinearity of chemical adsorption onto soils with chemical concentration is usually modeled using an empirical Freundlich adsorption isotherm

$$\omega_{Ts} = Bc_{Tw}^{b} \qquad (2\text{-}120)$$

where ω_{Ts} is the mass fraction of chemical on the soil in equilibrium with the concentration of chemical in the liquid phase, and c_{Tw} is usually in units of mass concentration. The two constants b and B are empirical and must be determined experimentally. The Freundlich isotherm can be linearized by a taking the natural log of both sides of the equation, producing

$$\ln \omega_{Ts} = \ln B + b \ln c_{Tw} \qquad (2\text{-}121)$$

A plot of $\ln \omega_{Ts}$ versus $\ln c_{Tw}$ produces a straight line with slope b and y-intercept $\ln B$.

Example 2.8

The data shown in Table 2.5 is for TCE adsorption on Calgon Filtrasorb 300 activated carbon (phase g) produced in an experiment where the initial concentration in each flask was 100 mg L^{-1}. Determine the Freundlich adsorption isotherm constants.

To calculate the mass of adsorbed TCE, we subtract the final mass of TCE in water from the initial mass assuming a liter of sample. For the first experiment, the mass of TCE adsorbed, m_{Tg}, is $(c_{Tg,0} - c_{Tw,eq})V_w = [(100 - 45)$ mg L$^{-1}] \times$ 1 L = 55 mg. Taking the natural log of both ω_{Ts} and c_{Tw}, produces the values in Table 2.6. Plotting $\ln \omega_{Tg}$ as a function of $\ln c_{Tw}$ produces Figure 2.3. The slope is 0.60, and the y intercept -3.6, producing the Freundlich isotherm equation $\omega_{Tg} = 0.028 \, \rho_{Tw}^{0.60}$.

TABLE 2.5   TCE Absorption on Activated Carbon

Mass of GAG (g L^{-1})	$\rho_{Tw,eq}$ (mg L^{-1})
0.00	100
0.20	45
1.00	7.3
4.82	0.6
14.2	0.1

TABLE 2.6   Data for Adsorption Isotherm Plot

ω_{Tg} (mg mg^{-1})	$\ln \omega_{Tg}$	$\ln c_{Tw}$
55/200 = 0.275	-1.29	3.81
92.7/1000 = 0.0927	-2.38	1.99
99.4/4820 = 0.0206	-3.88	-0.51
99.9/14200 = 0.007	-4.96	-2.30

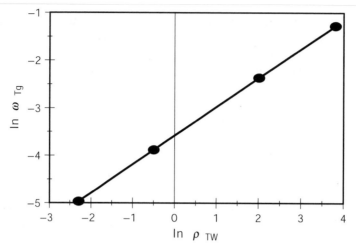

Figure 2.3   TCE adsorption isotherm.

PROBLEMS

√ **2.1**   Henry's Law for expressing the equilibrium distribution of a chemical between dilute gas and liquid phases enjoys several formulations. The mole fraction form is dimensionless and can be expressed as

$$y_C = H_{Cyx} x_C \qquad\qquad \text{(a)}$$

A common form is also

$$p_C = H_{Cpx} x_C \qquad\qquad \text{(b)}$$

where H_{Cpx} is in atm. Other forms are

$$p_C = H_{Cpp} \rho_{Cw} \qquad\qquad \text{(c)}$$

where H_{Cpp} is in atm cm^3 g^{-1}, and

$$\rho_{Ca} = H_{Cpp} \rho_{Cw} \qquad\qquad \text{(d)}$$

where H_{Cpp} is in cm^3 water/cm^3 air STP, y_C and x_C are mole fractions, and the other variables have units of: p_C [atm], ρ_{Ca} [g cm^{-3}], and ρ_{Cw} [g/cm^3]. Find the multipliers b_C, b, and b_p that satisfy the equality

$$H_{Cyx} = b_C H_{Cpx} = b H_{Cpp} = b_p H_{Cpp} \qquad\qquad \text{(e)}$$

Watch units! (Adapted from Problem 2.1 K in Thibodeaux, 1979).

$$\alpha_{Cw}^{*} = \frac{\gamma_{w} p_{w}^{0}}{\gamma_{w} p_{w}^{0}}$$

2.2 Certain chemicals can only be analyzed in the gas phase (using a gas chromatograph), and therefore their concentrations in another phase must be calculated using a partitioning coefficient. Let us assume you have a 25-mL test tube containing chemical C, which has had sufficient time to partition and achieve equilibrium concentrations in both the gas and liquid phases. You have available the following thermodynamic data: air at this temperature (25°C) can hold 0.0189 lb H_2O/lb air; the relative volatility of the chemical is 1.2×10^6; the activity coefficient in the liquid phase is 137.5, the molecular weight 16 g mol^{-1}, and $\mathcal{R} = 0.08206$ L atm mol^{-1} K^{-1}. (a) How much water would you add to a second 25-mL test tube (sealed and initially empty), if you wanted to add just enough water to vaporize the water fully? (b) What is the vapor pressure of chemical C? (c) If you add 1 mL of water containing 20 mg/L of chemical C to the second 25-mL test tube (sealed and empty), what is the mole fraction of C in the gas phase in the second test tube at equilibrium? (d) What is the mass concentration of C in solution in the second test tube at equilibrium?

2.3 Gas measurements of the composition of air above contaminated groundwaters are used to determine the levels of contaminant in the soil. For this problem, we will consider solid DDT ($C_{14}H_9Cl_5$) in the soil at 20°C, with $(\gamma_{Ds}f_{Ds}^{\circ}) = 2.6 \times 10^{-6}$ atm units missing. (a) Derive an expression for the mass fraction of DDT in soil ($\omega_D = \mu g$ DDT/g soil) in terms of the following variables: ρ_{Da}, c_a, M_a, γ_{Ds}, f_{Ds}°, and any other variables or constants you think you may need. (b) Solve this expression for the case where we measure 11.3 μg DDT/m^3 air. (c) Sometimes, other material in groundwater can increase the solubility of a chemical. Let us say that the solubility of DDT in pure water is 1.2 μg L^{-1}. Calculate the concentration of DDT in water containing 5 mg humics/L, using the expression: $\omega_H = 0.42 \, \rho_{Dw}^{0.93}$, where ρ_{Dw} is the concentration of DDT in water [μg L^{-1}], and ω_H is the mass fraction of humics, in units of μg humics/g soil.

2.4 In the not-so-distant future, you are a consultant evaluating two treatment processes for the removal of a new chemical C with formula $C_5H_8O_{14}$. Two companies have bid on treatment schemes. Company X wants to treat the wastewater containing C biologically in an activated sludge treatment system, since they claim C is readily biodegradable and can be completely mineralized to CO_2 and H_2O in a bioreactor of this type (lots of air, lots of cell recycle, lots of money). Company Y wants to remove chemical C physically using an air stripper, and says it can save you 50% of the cost that company X would charge. Based on the thermodynamic data below, which treatment scheme would you choose and why? You must show all calculations necessary to support your position. What information might be useful in addition to the information below?

ΔG°[kcal mol^{-1}]	$p_C^{\circ} = 20$ atm	$\{C_5H_8O_{14}\} = 10^{-2}$
$C_5H_8O_{14}$, -601	$p_w^{\circ} = 0.04$ atm (water)	$\{CO_2\} = 0.05$
CO_2, -92.3	$\gamma_{Cw} = 4 \times 10^{-4}$	
H_2O, -56.6		

2.5 We are interested in whether the reaction:

$$C_5H_6O_2 + O_2 \rightarrow CO_2 + H_2O$$

is thermodynamically feasible. We initially have a 500 mL bottle *open* to the air, filled with 420 ml of fluid containing 20 mg/L of the first compound and 0.02 atm of carbon dioxide. Oxygen is at its equilibrium concentration. You may wish to use these data: $\Delta G°$[kcal mol^{-1}]: $C_5H_6O_2$, -612; O_2, 0; CO_2, -92.3; H_2O, -56.6; $\mathcal{R} = 1.987$ cal mol^{-1} K^{-1}, and $\mathcal{R} = 0.08206$ L atm mol^{-1} K^{-1}; $\gamma_{Ow}p_O° = 4.01 \times 10^4$ atm at 20°C. (a) Balance this equation and determine the equilibrium constant K_{eq}. (b) Assume the reaction goes to completion (is this likely?) Calculate the mass of oxygen (mg) required. (c) Calculate the volume of air (mL) required (i.e., containing) the mass of oxygen in (b).

2.6 In this problem, we now place a cap on the bottle in Problem 2.5, such that the volumes of air and water are fixed (*closed* bottle). We will assume that a material has been added to the bottle that instantly strips out CO_2 from the gas phase that is initially present or generated thereafter. The *initial* pressure in the bottle is 1 atm; the vapor pressure of water is negligible. Assume air consists only of oxygen and nitrogen in the mole ratio of 0.21 : 0.79. (a) Starting with the fugacity identity, $f_{Oa} = f_{Ow}$, derive an expression for the moles of O oxygen in the gas phase (n_{Oa}) in terms of (at most) the following parameters: n_{Ow}, c_w, V_a, V_w, $\mathcal{R}$, T, γ_{Ow}, $p_O°$. (b) Let us now assume that the equation in Problem 2.5 went to completion in the *closed* bottle. Use the results from part (a) of this question to determine the concentration of oxygen in solution (c_{Ow}) after the reaction is complete. (c) What is the total pressure (atm) in the gas phase when the reaction is complete? (d) How much would the height (cm) of a mercury manometer change if it was hooked up to measure pressure changes in the bottle? (1 atm = 760 mm Hg).

2.7 Compute the equilibrium concentration of chemical A at 25°C, given the following reaction, concentrations (molar), and thermodynamic data for $\Delta G°$[kcal mol^{-1}]: $A = -40.2$, $B = -98.2$, $C = -150$, $D = -10.2$.

$$\begin{array}{ccccc} 2A & + B & \rightarrow & C & + & 3D \\ (3\text{ M}) & & & (10^{-1}\text{ M}) & & (1.59\text{ M}) \end{array}$$

2.8 Toluene (molecular weight of 92 g mol^{-1}) is to be placed in a microcosm (a fancy name for a test tube containing air, water, and dirt) for a biological degradation experiment. The 50-mL sealed test tube contains 39 mL of air, 10 mL of water, and 1 g of soil. You find that toluene has a relative volatility of 15,600 at 20°C. The total pressure in the microcosm is 1 atm, which includes a vapor pressure of water of 0.024 atm. You have previously determined the empirical constants $b = 1$ and $B = 50$ for the Freundlich adsorption isotherm ($\omega_T = B\phi^b$) for toluene (chemical $= T$) and this soil. (a) Derive an expression for the total mass of toluene (T) in the headspace (test tube air space) as a function only of the mass concentration in the liquid and other known quantities. You may wish to use the following: $n_{Ta} = n_{Tw}V_a\gamma_{Tw}p_T^o/(c_wV_w\mathcal{R}T)$, where n refers to moles. Assume the system is at equilibrium. (b) Starting with the expression $\omega_T = B\phi^b$, derive an expression for the mass of T on the soil as a function of the total mass of soil, the mass concentration in the liquid phase, and other known constants. (c) How much toluene should you add to the microcosm if you want a final aqueous concentration of 200 µg L^{-1}?

2.9 The chemical U is spilled (1.2 m^3) in a completely enclosed and sealed room that expands to maintain constant pressure of 1 atm. Also in the room is a large tank of water (30 m^3) kept isolated from direct contact with the liquid spill. The room volume, not including the tank, is 100 m^3. The chemical has a molecular weight of 199, liquid density of 0.92 g cm^{-3}, vapor pressure of 0.026 atm, and a liquid activity coefficient of 7.4×10^5. (a) Starting with the fugacity relationship $f_{Ua} = f_{Uw}$, derive an expression for the Henry's Law constant in units of atm cm^3 g^{-1}. Then calculate the quantity of the constant. (b) How much of the chemical will be in each phase (grams) when equilibrium is reached?

2.10 This question considers the differences and similarities of distribution coefficients and adsorption isotherms. (a) Starting with the expression for the distribution coefficient $K_{Csw}^* = \omega_C/\phi_C$, derive an expression for ρ_{Cw} in terms of m_{Cs}. (b) What is the value of K_{Csw}^* in the system with 10 kg of soil, 90 L of water, 15 mg L^{-1} of the chemical in solution, 50% of the chemical is adsorbed, and a chemical density and molecular weight of 0.92 g cm^{-3} and 150 g mol^{-1}? (c) In many texts, you often find an Freundlich adsorption isotherm in the form of

$$\frac{x}{m} = Kc^{1/n}$$

where x is the mass of chemical on the solid, m is the mass of solid, and c is the mass concentration of chemical in the liquid. Write the above adsorption isotherm in the notation used here. Under what conditions are the isotherm and distribution coefficient equivalent? (d) Using data from part (b) and assuming that $n = 2$ for the isotherm, how much mass would be adsorbed in the system above if only 10 mg L^{-1} of C is in solution.

2.11 Sometimes the pressure of water in the gas phase is not included in calculations since in many places air is usually nearly completely saturated with water. However, in Tucson, Arizona, the relative humidity (percent saturation of air with water vapor) can be as low as 5%. Assume that dry air is 80% nitrogen and 20% oxygen, and that in Tucson the air has a 5% relative humidity and total pressure of 700 mm Hg, and that water has a vapor pressure of 15 mm Hg. (a) What is the final pressure in a test tube after the air in the tube comes into equilibrium with the water in the tube? The tube volume is

50 mL, and the water volume is 30 mL. (b) Will the pressure change in part (a) alter the solubility of oxygen in the water, which is initially 8.5 mg L^{-1}? Be quantitatively persuasive.

2.12 We have a suspension of pure liquid o-xylene (M_E = 106 g mole^{-1}), with a liquid vapor pressure of 8.7 $\times$ 10^{-3} atm, and an activity coefficient in water of 3.37 $\times$ 10^4. Starting with the fugacity equality $f_{Ew} = f_{Ee}$, where e is the pure xylene phase, derive and solve for an expression for the solubility (mg L^{-1}) of o-xylene in water.

2.13 You are designing a contaminated soil biodegradation experiment to determine the degradation rate of a chemical at different bulk fluid concentrations of a chemical. You start with a chemical concentration of 15 mg L^{-1} in solution, but after you add soil that concentration will change. There is no headspace in the system, a Freundlich adsorption isotherm constant is 0.035 g/g (L/g)$^{0.5}$, the volume is 0.4 liters, and volume changes occurring from adding the soil are considered negligible. What mass of soil (in grams) must be added to the water to get 10 mg L^{-1} of chemical C in solution?

√ **2.14** For this question, assume that all solutions are ideal (even though under the conditions of the question that may not be true) and that the temperature is 25°C. Data on o-xylene (E) and water (W) are given below the problem. (a) A 50-mL test tube is sealed with 20 mL of water. At time 0, the air in the test tube is completely removed, leaving a vacuum in the tube. When the system comes to equilibrium and the temperature returns to room temperature, what will be the total gas pressure in the tube? Neglect volume changes in the liquid. (b) If pure o-xylene liquid is now added to the system so that it just saturates the gas and water phases, what will be the total gas pressure in the tube? (c) What volume, V_E, (μL) of o-xylene must be added to just saturate the gas and liquid phases with o-xylene?

γ_{Ew} = 3.37 $\times$ 10^4 γ_{Ww} = 1.0

p_E° = 8.71 $\times$ 10^{-3} atm p_W° = 23.76 mm Hg

ρ_{Ew}^* = 175 mg L^{-1} c_w = 55.6 mol L^{-1}

ρ_E = 0.88 g cm^{-3} ρ_w = 1.0 g cm^{-3}

M_E = 106 g mol^{-1} M_w = 18 g mol^{-1}

2.15 Toluene (100 μL) is added at room temperature (20°C) to a test tube (55 mL) containing 20 mL of water and 20 mL of octanol, and the tube sealed. In order to bring the distribution of toluene to equilibrium in all phases rapidly, the tube is agitated for 24 h. Following this period, sufficient time is allowed for the phases to separate (water phase w, octanol phase o). The following data are available for toluene: ρ_T = 0.87 g mL^{-1}, M_T = 92.14 g mol^{-1}, p_T° = 3.7 $\times$ 10^{-2} atm, ρ_{Tw}^* = 515 mg L^{-1}, H_{Tpx} = 6.6 $\times$ 10^{-3} atm m^3 mol^{-1}, and log K_{Tow} = 2.69 (mg L^{-1})/(mg L^{-1}). (a) Derive an equation for the concentration (mg L^{-1}) in the liquid phase, ρ_{Tw}, in terms of only the given information above. Use only variables given above (i.e., V_a, M_T, etc.) or commonly known constants (for example the gas constant $\mathcal{R}$). Your final equation should contain only conversion factors (based on the given units) and variables. (b) Solve this equation for the given conditions.

2.16 For this question, assume that all solutions are ideal (even though under the conditions of the question that may not be true) and that the temperature is 25°C. (a) A 50-mL test tube is sealed with 20 mL of water and 30 mL of air. At time 0, the air in the test tube is completely removed, leaving a vacuum in the tube. When the system comes to equilibrium and the temperature returns to room temperature, what will be the total gas pressure in the tube? Neglect volume changes in the liquid. (b) A small volume of pure *o*-xylene liquid is now added to the system such that all the chemical goes into the gas and liquid phases (i.e., that there is no pure xylene phase). Starting with a fugacity equilibrium condition, derive an expression for the total number of moles in the liquid phase, n_{Ew}, in terms of *at least* the following: n_E (total moles of E), $\mathcal{R}$, T, V_w, γ_{Ew}, and p_E^o. (c) Assuming that 5 μL of pure *o*-xylene was added to the system, what would be the total gas pressure in the tube?

2.17 Phenol has an octanol–water partition coefficient of log K_{Pow} = 1.46, when the partition coefficient is calculated on a mass concentration basis, where 1-octanol ($C_8H_{18}O$) has a density of 0.83 kg L^{-1}, and a molecular weight of 130 g moL^{-1}. (a) Derive a constant to convert K_{Pow} on a mass concentration basis to a mole fraction basis. (b) Given a Henry's Law constant of $H_{P,yx}$ and K_{Pow} for phenol (mole fraction basis), derive an expression for a partition coefficient between the octanol and air, or K_{Poa} (mole fraction basis). (c) If phenol is measured at a concentration of 2 mg L^{-1} in water, calculate the number of moles of phenol in the octanol in a sealed 55-ml tube filled with 20 mL of water and 1 mL of octanol at equilibrium. (d) Calculate the number of moles of phenol in air and the total moles of phenol in the tube if $H_{P,yx}$ = 0.0102.

2.18 Solubilities are often given in a variety of ways. For example, the Merck Index indicates that 6.5 volumes of propane gas are soluble in 100 volumes of water at 17.8°C, while 3.5 mL of methane are soluble in 100 mL of water at 17°C. Using this information, calculate the solubilities of these two gases in water in units of mg L^{-1}.

REFERENCES

DiToro, D. M., et al. 1991. *Environ. Toxicol. Chem.* **10**(12), 1541.

Gleick, J. 1987. *Chaos*. Penguin. New York, NY.

Snoeyink, and Jenkins. 1980. *Water Chemistry*. Wiley. New York, NY.

Stumm, W., and J. J. Morgan. 1970. *Aquatic Chemistry*. Wiley. New York, NY.

Thibodeaux, L. J. 1979. *Chemodynamics*. Wiley. New York, NY.

Thibodeaux, L. J. 1996. *Chemodynamics*, 2nd edition. Wiley. New York, NY.

CHAPTER 3

DIFFUSIVE TRANSPORT

3.1 INTRODUCTION

Transport of material by molecular diffusion is a relatively slow process compared to other bulk transport processes such as mixing and advection. If we were to place a drop of red dye in an undisturbed bowl of water, the dye would diffuse slowly away from the point of injection, all the time decreasing in concentration. Given sufficient time, dye in an undisturbed bowl will reach a final and equal concentration throughout. However, if we were to mix in the dye with a spoon, the dissolution of the dye within the cup would become complete within seconds. This example demonstrates what we intuitively know: Bulk transport processes, such as mixing, speed up transport processes. For the case of dye transport within a cup, transport within the cup is increased by moving packets of liquid containing dye around the cup and shearing these packets of liquid.

The goal of our transport calculations here is to describe the motion of one or more chemicals in both natural and engineered systems. Mathematically describing the transport of a chemical within a system is easier if the different transport mechanisms (diffusion and advection) are handled as separate terms in a governing equation. In this chapter we shall consider only the mechanism of diffusive transport. Advection, mixing, and transport facilitated by bulk motion will be considered in later chapters when we can combine these processes with diffusion into a more comprehensive transport equation.

3.2 DIFFUSION

In order to understand how an equation is derived to describe diffusive transport in a system, we should consider exactly how chemical transport by molecular diffusion arises. The motion of a single molecule, driven by thermal energy, is called *Brownian motion.* The path that any specific molecule follows is unique and completely random. Although we cannot describe the exact path of any one molecule, we can quantify the net effect of Brownian motion on the transport of all the molecules.

Let us consider the average motion of all molecules of chemical C introduced at a single spot into a uniform fluid containing only one other type of molecule (A) (Fig. 3.1). Since the motion of all molecules is random, there is no net change in concentration of A anywhere in the control volume. Thus we say there is no chemical gradient of A in the system. The concentration of C is not uniform: It is high near the point of injection, and zero a short distance away. A short time after injection the concentration of C will decrease at the injection point as the random motion of some molecules carries them away from their initial location. Some molecules will move a long distance, while the path of other molecules will produce only a minor net change in distance. The random motion of C molecules has produced a change in concentration in the fluid with distance from the point of injection, dc/dx.

Fick's First Law

We can use this basic view of diffusion to quantify the rate of chemical transport in this system. If we consider any plane in the system in Figure 3.1, we can see that the net flux of chemical C across that plane must be proportional to the concentration that we put into the system. For example, if we have 15 molecules in the system, at a time t later 3 molecules might diffuse across the line e. However, if we place 30 molecules in the system, we would have 6 molecules crossing line e. The net transport must be proportional to the concentration of the chemical in the system, but the concentration at any point changes. At point f in Figure 3.1 the concentration of chemical C is changing more slowly, and thus the flux of C across point f is lower.

These observations can be used to recognize that the flux of the chemical at any point is proportional to the concentration gradient at that point, or

$$j_{Cw,z} = -D_{Cw} \frac{dc_{Cw}}{dz} \tag{3-1}$$

Figure 3.1 Diffusion of chemical C (●) in chemical A (■) as a function of time: (a) chemical injected at one location moves by random motion; (b) chemical gradient forms for C in A.

where $j_{Cw,z}$ is the molar or mass flux of C through phase w (water) in the z direction, dc_{Cw}/dz is the gradient of C at any point z, and D_{Cw} is a constant of proportionality known as the diffusion coefficient or diffusivity of chemical C in phase w. Equation 3-1 is known as Fick's first law. The diffusion constant is different for every chemical and every phase, and must be a function of temperature in the system since thermal energy drives the rate of Brownian motion. The negative sign in Eq. 3-1 reflects that the net flux is in the opposite direction of the chemical gradient. In our example, the flux is in the positive z direction, but the concentration increases in the negative z direction.

The diffusion equation can be expanded to cover other dimensions and physical conditions. Equation 3-1 is only valid under isothermal (and for air, isobaric) conditions, or where the total mass in a volume is unchanged. For nonisothermal and nonisobaric conditions, we can write the more general expression of Fick's Law as

$$j_{Cw,z} = -c_w D_{Cw} \frac{dx_C}{dz} \tag{3-2}$$

where x_C is the mole fraction of C in the system of total concentration c_w. Assuming the diffusivity is constant in every direction, the flux of C can be calculated in three dimensions as

$$\mathbf{j}_{Cw} = -c_w D_{Cw} \, \nabla x_C \tag{3-3}$$

where the flux is now indicated in boldface to indicate that it is a vector quantity.

Chemical diffusion will be used here to describe the transport of a chemical due to its thermal energy and should not be confused with chemical *dispersion*. When a chemical is transported over distances larger than those accomplished solely by molecular diffusion (due to either bulk motion or turbulent motion), the chemical transport mechanism is defined as dispersion. In this chapter we will deal almost exclusively with defining the diffusivity of a molecule in different phases. In one case, however, we will consider the case of Taylor dispersion in order to describe how the diffusivity of a chemical can be calculated from its dispersion in a capillary tube.

3.3   CALCULATION OF MOLECULAR DIFFUSION COEFFICIENTS

The molecular diffusion coefficient is a basic property of a chemical. As a result, chemical diffusivities have been measured for many compounds using a variety of techniques. These values are available from many sources such as *The Chemical Engineering Handbook* (Perry and Chilton) and the *CRC Handbook of Chemistry and Physics*. If the diffusivity of a chemical cannot be found in a standard reference, however, it can be estimated using one of several equations each suited for different conditions. In general, the diffusivity of a chemical is inversely proportional to its size, and therefore its molecular weight. Diffusion rates increase with temperature but decrease with the viscosity of the medium. Since there are various chemical

溶质 溶剂

interactions between the solute and solvent, and since chemicals of different molecular weights differ in size, correlations are often used to calculate molecular diffusivities.

Chemicals in Gases

The diffusion of a chemical in a gas can be affected by intermolecular forces. In low-density gases, and when intermolecular forces are absent, the diffusion coefficient can be approximated using the kinetic theory of gases (Welty et al., 1976). It must be assumed that the molecules are of equal size and mass, they have equal average and linear velocities, and the gas is ideal. The diffusion of an isotope (C^*) in a mixture of C meets these criteria. The diffusivity of C^* in C is

$$D_{C^*C} = \frac{2}{3\pi^{3/2}d^2P} \left(\frac{k_B^3 T^3}{m}\right)^{1/2} \tag{3-4}$$

where d is the molecule diameter, P the total pressure, k_B the Boltzmann constant (1.38×10^{-16} erg K^{-1}, where 10^7 erg = 1 kg m^2 s^{-2}), T the absolute temperature, and m the mass of the molecule. The diffusivity of each molecule of C depends entirely on the properties of the gas and the size of the molecules.

   More useful expressions of diffusion coefficients consider the forces of attraction and repulsion between the molecules. Using the Lennard–Jones equation, Hirshfelder et al. (1949) derived an expression for the diffusion of C in a gas (phase g), D_{Cg}, assuming nonpolar, nonreacting molecules, as

$$D_{Cg} = \frac{0.001858T^{3/2}}{P\sigma_{Cg}^2\Omega_D} \left(\frac{1}{M_C} + \frac{1}{M_g}\right)^{1/2} \tag{3-5}$$

where T is the absolute temperature [K], M_C and M_g are the molecular weights of the chemical and gas, and P is the total pressure [atm]. The Lennard–Jones parameter, σ_{Cg} [Å], is the collision diameter that can either be obtained from tabulated values (see Appendix 3) or can be calculated from the properties of the chemical and gas using one of the equations:

$$\sigma_C = 1.18V_{C,b}^{1/3} \tag{3-6}$$

$$\sigma_C = 0.841V_{C,c} \tag{3-7}$$

where $V_{C,b}$ is the molecular volume at the normal boiling point of chemical C, and $V_{C,c}$ is the critical molecular volume. The collision diameter σ_{Cg} can be calculated as the average of the individual diameters, or $\sigma_{Cg} = (\sigma_C + \sigma_g)/2$. The collision integral for molecular diffusion, Ω_D, is a dimensionless number that is a function of the temperature and the intermolecular potential field of one molecule of C for one molecule of g. Values of Ω_D are listed in Appendix 3 as a function of k_BT/ε_{Cg}, where ε_{Cg} (ergs) is another Lennard–Jones parameter describing the energy of molecular interaction for the binary system of C and phase g. Values of ε_{Cg} can be obtained from the properties of the gas using

$$\frac{\varepsilon_C}{k_B} = 0.77 T_{C,c} \tag{3-8}$$

$$\frac{\varepsilon_C}{k_B} = 1.15 T_{C,b} \tag{3-9}$$

where $T_{C,b}$ is the temperature at the normal boiling point, and $T_{C,c}$ is the critical temperature. The bulk property ε_{Cg} is calculated from $\varepsilon_{Cg} = (\varepsilon_C \varepsilon_g)^{1/2}$. Equation 3-5 must be further modified for polar–polar and polar–nonpolar gases as described by Hirshfelder et al. (1954); comparison of Eq. 3-5 with data on 80 gas pairs determined that there was an average error of 6%.

For chemicals in air, tabulated values of the properties of air can be used in Eqs. 3-2–3-6 to calculate D_{Ca} instead of D_{Cg}. However, if the composition of air is substantially altered, for example, by low concentrations of oxygen in a biological reactor, the diffusion coefficients may also need to be adjusted based on the composition of the new gas. With a gas is designated as phase g, the diffusion coefficient for C in phase g is

$$D_{Cg} = \left(\frac{y_{Bg,C}}{D_{CB}} + \frac{y_{Dg,C}}{D_{CD}} + \cdots + \frac{y_{ig,C}}{D_{Ci}} \right)^{-1} \tag{3-10}$$

where D_{CB} is the diffusivity of C in pure B, and $y_{Bg,C}$ is the mole fraction of the gas on a component C free basis, or

$$y_{Bg,C} = \frac{y_{Bg}}{y_{Bg} + y_{Dg} + \cdots + y_{ig}} \tag{3-11}$$

While the Hirshfelder correlation (Eq. 3-5) is sufficiently accurate for most calculations here, the amount of data necessary for its use makes it highly improbable that such data would be available for chemicals for which D_{Ca} was not already well known.

When Lennard–Jones parameters σ_{ij} and ε_{ij} are either not available or not well known, the Fuller correlation (Fuller et al., 1966) can be used

$$D_{Cg} = \frac{10^{-3} T^{1.75}}{P(V_{C,d}^{1/3} + V_{g,d}^{1/3})^2} \left(\frac{1}{M_C} + \frac{1}{M_g} \right)^{1/2} \tag{3-12}$$

where D_{Cg} is in $cm^2 \ s^{-1}$, T in K, and P in atm. The atomic diffusion volumes can be determined by summation of the volumes for each element in C and g using Table 3.1.

TABLE 3.1 Atomic Diffusion Volumes for Use in Estimating D_{Cg} by Method of Fuller et al.

Atomic and Structure Diffusion–Volume Increments, v					
C	16.5	Cl		19.5	
H	1.98	S		17.0	
O	5.48	Aromatic ring		−20.2	
N	5.69	Heterocyclic ring		−20.2	

Diffusion Volumes for Simple Molecules, v					
H_2	7.07	Ar	16.1	H_2O	12.7
D_2	6.70	Kr	22.8	$CClF_2$	114.8
He	2.88	CO	18.9	SF_6	69.7
N_2	17.9	CO_2	26.9	Cl_2	37.7
O_2	16.6	N_2O	35.9	Br_2	41.1
Air	20.1	NH_3	14.9	SO_2	41.1

Source: Welty et al., 1984.

Since most diffusion coefficients in environmental systems are for compounds in air, the Fuller correlation can be simplified by substitution of known values for M_a and $V_{C,d}$ for air, resulting in

$$D_{Ca} = \frac{10^{-3}T^{1.75}}{P(V_{C,d}^{1/3} + 2.72)^2}\left(\frac{1}{M_C} + 0.0345\right)^{1/2} \tag{3-13}$$

Example 3.1

Calculate the diffusion constant for methane in air at 0°C and 1 atm using both the Hirshfelder and Fuller correlations. Calculate the percent error compared to the reported value of 0.16 cm^2 s^{-1}.

For the Hirshfelder correlation, we will need the values for σ_{Ma} and ε_{Ma}/k_B. From Appendix 3 we find that for air (phase a) $\sigma_a = 3.167$ Å and $\varepsilon_a/k_B = 97$ K, and for methane (methane M) $\sigma_M = 3.822$ Å and $\varepsilon_M/k_B = 136.5$ K. For methane in air, $\sigma_{Ma} = (\sigma_a + \sigma_M)/2 = 3.495$ Å, and $\varepsilon_{Ma}/k_B = (\varepsilon_M/k_B + \varepsilon_a/k_B)^{1/2} = 115.1$ K. To obtain Ω_D, we calculate the product $Tk_B/\varepsilon_{Ma} = 2.37$ and from tabulated values calculate that $\Omega_D = 1.016$. The Hirshfelder equation is

$$D_{Ma} = \frac{0.001858T^{3/2}}{P\sigma_{Ma}^2\Omega_D}\left(\frac{1}{M_M} + \frac{1}{M_a}\right)^{1/2} \tag{3-14}$$

For air and methane, $M_a = 29$ and $M_M = 16$. Using the above values, we find

$$D_{Ma} = \frac{0.001858(273)^{3/2}}{(1)(3.495)^2(1.016)} \left(\frac{1}{16} + \frac{1}{29} \right)^{1/2} \tag{3-15}$$

$$D_{Ag} = 0.21 \text{ cm}^2 \text{ s}^{-1} \tag{3-16}$$

Comparison with the reported value of $0.16 \text{ cm}^2 \text{ s}^{-1}$ shows that the error is

$$\text{Error} = \frac{0.21 - 0.16}{0.16} \times 100\% = 31\% \tag{3-17}$$

The Fuller correlation for methane in air is

$$D_{Ma} = \frac{10^{-3}T^{1.75}}{P(V_{M,d}^{1/3} + 2.72)^2} \left(\frac{1}{M_M} + 0.0345 \right)^{1/2} \tag{3-18}$$

All values are already calculated for the Hirshfelder equation, except for $V_{M,d}$, which from Table 3.1 is $V_{M,d} = (16.5 \times 1 + 1.98 \times 4) = 24.4$. Substitution into Eq. 3-18 produces

$$D_{Ma} = \frac{10^{-3}273^{1.75}}{P(24.4^{1/3} + 2.72)^2} \left(\frac{1}{16} + 0.0345 \right)^{1/2} \tag{3-19}$$

$$= 0.18 \text{ cm}^2 \text{ s}^{-1} \tag{3-20}$$

with an error of 13%. Thus, in this case, the Fuller correlation is more accurate than the Hirshfelder equation.

Temperature Effects   Quite often diffusion coefficients for chemicals of interest in air will be known and tabulated. The main use of the Hirshfelder and Fuller equations is to adjust the diffusion coefficients to temperatures other than those reported in the tables. Equation 3-5 can be rearranged so that all variables (D_{Ca}, P, and T) are on one side of the equation, and all constants on the other, or

$$\frac{D_{Ca}P\Omega_D}{T^{3/2}} = \text{const} \tag{3-21}$$

Since this grouping of variables equals a constant, we can write that at another temperature T_T and pressure P_P, the diffusivity $D_{Ca}(P_P, T_T)$ is

$$D_{Ca}(P_P, T_T) = D_{Ca} \frac{P}{P_P} \left(\frac{T_T}{T} \right)^{3/2} \frac{\Omega_D}{\Omega_D(T_T)} \tag{3-22}$$

The temperature dependence of Ω_D is very small, and therefore the ratio of collision integrals is usually assumed to be unity for temperature changes in environmental systems, producing

$$D_{Ca}(T_T, P_P) \cong D_{Ca} \frac{P}{P_P} \left(\frac{T_T}{T} \right)^{3/2} \tag{3-23}$$

If the Fuller correlation is used to adjust for temperature changes, a slightly different equation is produced. Following the same procedure used for the Hirshfelder correlation of isolating all constants onto the right-hand side of the equation, D_{Ca} is calculated at other temperatures and pressures using the Fuller equation as

$$D_{Ca}(P_P, T_T) \cong D_{Ca} \frac{P}{P_P} \left(\frac{T_T}{T} \right)^{1.75} \tag{3-24}$$

This equation is the same as Eq. 3-23 except that the exponent on the temperature ratio is 1.75 versus 1.5. Given the accuracy of these calculations, this difference produces a negligible difference between the two correlations over the range of typical environmental temperatures.

Chemicals in Water

There is a substantial range of molecular weights of material in water, ranging from simple compounds such as ammonia (NH_3, $M_A = 17$ dalton) to a single virus ($M_V = 10^6$ dalton). Given the range of 10^1 to 10^6 in molecular weight, the variation in diffusivities is much smaller, decreasing from 10^{-4} to 10^{-6} cm^2 s^{-1}, as shown in Table 3.2.

The relationship between the diffusion of a chemical and the size of the molecule resulted in a hydrodynamic relationship relating the diffusion velocity to the resis-

流体力学

TABLE 3.2 Diffusion Coefficients as a Function of Molecular Weight

Substance Molecular Weight	Diameter (μm)	Diffusivity (cm^2 s^{-1} $\times$ 10^8)
10^1	0.00029	2200
10^2	0.00062	700
10^3	0.00132	250
10^4	0.00285	110
10^5	0.0062	50
10^6	0.0132	25

Source: Perry and Chilton, 1973.

tance of the liquid (phase l), known as the Stokes–Einstein equation

$$D_{Cl} = \frac{k_B T}{6\pi r_C \mu_l} \qquad (3\text{-}25)$$

The liquid viscosity, μ_l, is usually given in centipoise, where 1 cp = 0.01 g s^{-1} cm^{-1}. This equation works for colloidal and large, round molecules in any dilute solution. It is unchanged when applied to diffusion constants for chemicals in water, except the notation is changed to produce

$$D_{Cw} = \frac{k_B T}{6\pi r_C \mu_w} \qquad (3\text{-}26)$$

For small compounds in aqueous solution, the Wilke–Chang correlation is preferred

$$D_{Cl} = \frac{7.4 \times 10^{-8} T (\Phi_l M_l)^{1/2}}{\mu_l V_{C,b}^{0.6}} \qquad (3\text{-}27)$$

where D_{Cw} is the diffusivity of the chemical in cm^2 s^{-1}, μ_w the viscosity of the solvent in centipoise, T the absolute temperature in K, M_l the molecular weight of the liquid, $V_{C,b}$ the molal volume of chemical C at the normal boiling point in cm^3 g-mol^{-1}, and Φ_l the "association" parameter for the liquid phase l. This equation is only valid for $V_{C,b} < 0.27 \, (\Phi_l M_l)^{1.87}$.

The two constants in the Wilke–Chang correlation are easily estimated. Since there are generally few phases of interest in environmental applications, the association parameter will usually be known (Table 3.3). Molal volumes at normal boiling points, $V_{C,b}$, are estimated from tabulated value for each element shown in Table 3.4. The atomic volumes for each element in the compound are summed, and then corrections are made for ring structures as recommended in Table 3.4.

The Wilke–Chang correlation can be greatly simplified for chemicals in water. The association constant for water is $\Phi_w = 2.6$, and $M_w = 18$. If we further assume

TABLE 3.3   Association Parameter, Φ_l, for a Few Common Solvents

Solvent	Φ_4
Water	2.6
Methanol	1.9
Ethanol	1.5
Benzene, ether, heptane, and other unassociated solvents	1.0

Source: Welty et al., 1984, p. 495.

TABLE 3.4   Atomic Volumes for Complex Molecular Volumes for Simple Substances

Element	$V_{C,b}$ (cm^3/g-mol^{-1})	Element	$V_{C,b}$ (cm^3/g-mol^{-1})
Bromine	27.0	Oxygen, except as noted	
Carbon	14.8	below	7.4
Chlorine	21.6	Oxygen, in methyl esters	9.1
Hydrogen	3.7	Oxygen, in methyl ethers	9.9
Iodine	37.0	Oxygen, in higher ethers	
Nitrogen, double bond	15.6	and other esters	11.0
Nitrogen, in primary amines	10.5	Oxygen, in acids	12.0
Nitrogen, in secondary amines	12.0	Sulfur	25.6
for three-membered ring, as ethylene oxide			deduct 6
for four-membered ring, as cyclobutane			deduct 8.5
for five membered ring, as furan			deduct 11.5
for six-membered ring, as pyridine			deduct 15
for six-membered ring, as benzene ring			deduct 15
for naphthalene ring			deduct 30
for anthracene ring			deduct 47.5

Source: Welty et al., 1984, p. 495.

a temperature of 20°C, then $\mu_w = 1.00$ cp and $T = 293$ K, producing

$$D_{Cw} = 1.48 \times 10^{-4} V_{C,b}^{-0.6}, \qquad 20°C \tag{3-28}$$

This equation is valid for chemicals in water where $V_{C,b} < 359$ cm^3 g mol^{-1}.

Several other diffusion correlations have been developed for different conditions. The Reddy–Doraiswamy correlation covers a wider range of chemicals than the Wilke–Chang correlation, but instead of an association parameter this model uses molal volumes at the boiling point of both the chemical and solvent, when $V_{C,b} > 22$ cm^3 g mol^{-1}. The general expression for any chemical is

$$D_{Cl} = \frac{10 \times 10^{-8} T M_i^{1/2}}{\mu_l V_{C,b}^{1/3} V_{l,b}^{1/3}} \tag{3-29}$$

This expression can also be simplified for water. Using $M_w = 18$, $V_{w,b} = 14.8$, and assuming 20°C, produces

$$D_{Cw} = 5.07 \times 10^{-5} V_{C,b}^{-1/3}, \qquad 20°C \tag{3-30}$$

For nonelectrolytes in water, Hayduk and Laudie (1974) proposed

$$D_{Cw} = 13.26 \times 10^{-5} \mu_w^{-1.14} V_{C,b}^{-0.589} \tag{3-31}$$

where D_{Cw} is the diffusivity of the chemical in water in cm^2 s^{-1}, μ_w the viscosity of water in centipoise, and $V_{C,b}$ the molal volume of chemical C at the normal boiling point in cm^3 g mol^{-1}. This equation yields values similar to those obtained using the Wilke–Chang correlation. For a chemical at 20°C, it reduces to

$$D_{Cw} = 15.14 \times 10^{-5} V_{C,b}^{-0.589}, \qquad 20°C \qquad (3\text{-}32)$$

While Eqs. 3-27–3-32 are useful for small compounds, and the Stokes–Einstein equation (Eqs. 3-25 and 3-26) for large compounds, there exists in water many compounds with sizes that do not fall into either category. Examples of diffusivities of different proteins and colloidal particles are shown in Table 3.5. The most frequently used correlation to describe diffusivities of proteins was developed by Polsen (1950). By correlating molecular weights of a large number of proteins to their diffusion constants, Polsen proposed the relationship

$$D_{Pw} = 2.74 \times 10^{-5} M_P^{-1/3}, \qquad 20°C; \text{ proteins} \qquad (3\text{-}33)$$

where the molecular diffusivity of the protein, D_{pw}, is in cm^2 s^{-1}.

Using a series of dextrans with a wide range of molecular weights, Frigon et al. (1983) derived the relationship

$$D_{Dw} = 7.04 \times 10^{-5} M_D^{-0.47}, \qquad 20°C; \text{ dextrans} \qquad (3\text{-}34)$$

where D_{Dw} is in cm^2 s^{-1}. The differences in these two equations reflects the extent to which the different molecules are folded. Proteins are somewhat hydrophobic and are more tightly coiled in water than hydrophilic dextrans. Thus the hydrodynamic

TABLE 3.5   **Diffusion Constants of Proteins**

Protein	$M_P \times 10^{-3}$ g mol^{-1}	$D_{Pw} \times 10^8$ cm^2 s^{-1}
Myoglobin (beef heart)	16.9	113
Lysozyme (egg yellow)	16.4	112
Pepsin (pig)	35.5	90
Insulin (beef)	46	75.3
Hemoglobin (man)	63	69
Hemoglobin (horse)	68	63
Serum albumin (horse)	70	61
Serum albumin (man)	72	59
Serum globin (man)	153	40
Urease (jack bean)	480	34.6
Edestin	381	31.8
Botulinus toxin A	810	21
Hemocyanin (octopus)	2800	16.5
Tobacco mosaic virus	31400	5.3

Source: Moore, 1972.

resistance to diffusion for a protein is much less than for a dextran of similar weight, as evidenced by Eq. 3-25, the Stokes–Einstein relationship.

Temperature changes can be incorporated into all these equations using the same approach. The diffusivity of a chemical in solution can be expressed as

$$\frac{D_{Cw}\mu_w}{T} = \text{const} \tag{3-35}$$

where the constant varies depending on the correlation above. Once a diffusion coefficient is known, the diffusion constant $D_{Cw,T}$ at the new temperature T_T and solution viscosity $\mu_{w,T}$

$$D_{Cw}(T) = D_{Cw}\frac{\mu_w}{\mu_{w,T}}\frac{T_T}{T} \tag{3-36}$$

Chemicals in water span a wide range of interest, including defined chemicals such as benzene or ammonia, to other less defined classes of naturally occurring chemicals such as humic and fulvic acids. The whole ensemble of organic chemicals in water is called dissolved organic matter (DOM), with only a small fraction (<4%) of chemicals known in exact composition. As a result, chemical molecular weights and diffusivities are usually measured experimentally (see Section 3.3).

Example 3.2

Calculate the diffusion constant for glucose G in water at 25°C using the Wilke –Chang and Reddy–Doraiswamy correlations. Calculate the percent error compared to the tabulated value of 690 $\mu m^2 \, s^{-1}$.

In order to use these two correlations, we need to calculate the atomic volume, $V_{G,b}$. Glucose ($C_6H_{12}O_6$) has an oxygen in a five-carbon furan ring. Using values in Table 3.4, the volumes for each element are:

C (6 × 14.8) +

H (6 × 3.7) +

O (1 × 11.) +   {O in higher ether}

   (5 × 7.4) +   {other oxygens}

   (−15) =   {deduction for six-membered ring}

166.2

The Wilke–Chang correlation at 20°C is

$$D_{Gw}(20°C) = 1.48 \times 10^{-4}V_{G,b}^{-0.6} \tag{3-37}$$

Substituting in $V_{G,b}$ for glucose, we have

$$D_{Gw}(20°C) = 1.48 \times 10^{-4}(166.2)^{-0.6}$$

$$= 690 \times 10^{-8} \text{ cm}^2 \text{ s}^{-1} \tag{3-38}$$

The Reddy–Doraiswamy correlation at 20°C is

$$D_{Gw}(20°C) = 5.07 \times 10^{-5}V_{G,b}^{-1/3} \tag{3-39}$$

Substituting in $V_{G,b}$ for glucose into this expression, we have

$$D_{Gw}(20°C) = 5.07 \times 10^{-5}(166.2)^{-1/3}$$

$$= 920 \text{ cm}^2 \text{ s}^{-1} \tag{3-40}$$

Both of these results can be corrected to 25°C using Eq. 3-36

$$D_{Gw}(T_T) = D_{Gw} \frac{\mu_w}{\mu_{w,T}} \frac{T_T}{T} \tag{3-41}$$

Substituting in the water viscosities at 20 and 25°C of 1.0 and 0.89 cp produces the correction factor of

$$D_{Gw}(T_T) = D_{Gw} \frac{(1.0)}{(0.89)} \frac{(298)}{(293)} = D_{Gw}(1.14) \tag{3-42}$$

Multiplication of the diffusion constants at 20°C by a factor of 1.14 produces

$$D_{Gw}(25°C) = 790 \times 10^{-8} \text{ cm}^2 \text{ s}^{-1}, \quad \text{Wilke–Chang} \tag{3-43}$$

$$= 1050 \times 10^{-8} \text{ cm}^2 \text{ s}^{-1}, \quad \text{Reddy–Doraiswamy} \tag{3-44}$$

Both results overpredict the diffusivity of glucose, with an error of 14% for the Wilke–Chang correlation, and 52% for the Reddy–Doraiswamy correlation.

Example 3.3

In order to see how the diffusion constants of proteins compare in magnitude with those for polysaccharides, graph the data in the Table 3.5 for proteins, and Table 3.5A for dextrans (from Frigon et al., 1983) versus the Polsen and Frigon correlations.

TABLE 3.5A   Diffusivities of Dextrans

$M_D \times 10^{-3}$ g mol^{-1}	$D_{Dw} \times 10^{8}$ cm^2 s^{-1}	$M_D \times 10^{-3}$ g mol^{-1}	$D_{Dw} \times 10^{8}$ cm^2 s^{-1}
437	14.0	52.	42.4
200	21.1	35.2	49.1
181	22.0	27.4	55.8
159	22.2	19.8	62.0
129	25.2	11.2	88.0
133	25.7	125	23.0
113	26.2	84	29.5
110	28.4	60.2	35.2
92.5	28.9	49.6	39.5
67.8	36.5	28.8	54.0

Source: Frigon et al., 1983.

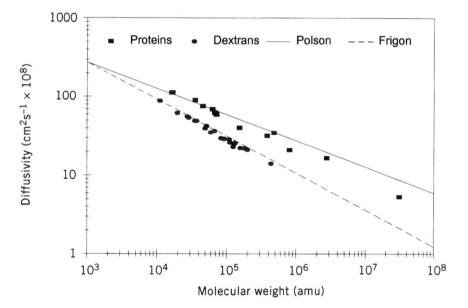

Figure 3.2   Comparison of protein and dextran diffusivities.

The results graphed in Figure 3.2 show that the diffusion constants of proteins and dextrans are similar at ~1000 daltons, but that at higher molecular weights proteins have larger diffusivities than dextrans. This result is consistent with the Stokes–Einstein equation since dextrans are relatively hydrophilic and less coiled than proteins of similar molecular weight, and as a result, have a larger radius.

Chemicals in Solids

Chemical diffusivities in solids cover a wide range, but in general the rate of chemical diffusion in a solid is quite small. Chemical diffusion coefficients for gases, such as hydrogen and helium in metals, are about a factor of 10^3 smaller than those in liquids (Table 3.6). Metal diffusion coefficients in metals are even lower; for aluminum in copper the chemical diffusivity is 1.3×10^{-30} cm^2 s^{-1}. The methods to predict diffusivity in solids are not very accurate; so diffusivities must usually be experimentally determined. One approach to estimating metal diffusivities is based on assuming that the solid has a face-centered-cubic arrangement of molecules in a metal lattice. The diffusivity is calculated as

$$D_{Cs} = b_l f_v e_{\text{jump}} \tag{3-45}$$

where b_l is the length between atoms in the lattice, f_v is the fraction of sites that are vacant, and e_j is the jump frequency. Data for b_l are obtained using crystallographic techniques, and f_v is calculated from Gibbs free energy of mixing. One approach for calculating e_{jump} is to use the reaction rate relationship

$$e_{\text{jump}} = \left(\frac{\Delta H}{2 M_c b_l^2} \right)^{1/2} \tag{3-46}$$

where $\Delta H = (36 \text{ cal g mole}^{-1} \text{ K}^{-1}) T_{\text{melt}}$ and T_{melt} is the melting temperature of the metal.

The wide range of these diffusivities and the generally low values usually result in diffusivities of chemicals in solids being neglected in most environmental transport calculations. Of greater interest than the diffusion of a chemical within the solid matrix of another chemical is the chemical diffisivity on a surface of a solid or within a porous matrix.

TABLE 3.6   Diffusivities of Chemicals in Solids

System	Temperature (°C)	Diffusivity (cm^2 s^{-1})
Helium in SiO$_2$	20	4.0×10^{-10}
Pyrex	20	4.5×10^{-11}
Hydrogen in iron	10	1.66×10^{-9}
iron	50	1.14×10^{-8}
nickel	85	1.16×10^{-9}
Aluminum in copper	20	1.3×10^{-30}
Antimony in silver	20	3.5×10^{-21}
Bismuth in lead	20	1.1×10^{-16}
Cadmium in copper	20	2.7×10^{-15}
Silver in aluminum	50	1.2×10^{-9}

Source: Cussler, 1984.

Multiphase Diffusion

In many environmental situations molecules must diffuse within one phase, such as air or water, constrained by a second phase such as a soil. For example, benzene evaporating from a pool of gasoline in the subsurface will diffuse from the surface of the gasoline into the air contained in the porous medium. Benzene may adsorb onto mineral surfaces, diffuse along this surface, and perhaps partition into some organic layer, or it may diffuse into the ground water. In each case the diffusivity of benzene will be different from that for benzene in the pure phase unobstructed by the porous media.

In the sections to follow, it is shown how to calculate chemical diffusion coefficients in porous media, on surfaces, and within macroporous structures. In all cases diffusivities are decreased relative to their pure phase values. The subject of chemical dispersion, where the spread of the chemical is increased due to turbulence of the phase or a mutiplicity of paths (as in a solid) are considered as separate subjects in other chapters.

Diffusion in Pores   The IUPAC definition of pore diameters is meant to differentiate the types of forces that control adsorption processes in pores. This leads to the definitions of

$$d_h < 2 \text{ nm,} \qquad \text{micropores}$$

$$2 < d_h < 50 \text{ nm,} \qquad \text{mesopores}$$

$$d_h > 50 \text{ nm,} \qquad \text{macropores} \qquad (3\text{-}47)$$

where d_h is the pore or hole diameter. Molecules adsorbed in micropores are trapped by surface forces and are unable to escape the pore. In mesopores, capillary forces become increasingly important, and macropores contribute less significantly to the overall chemical adsorption capacity of the material (Kärger and Ruthven, 1992).

As a first approximation of the behavior of a chemical in a pore, we can examine the case of a long straight pore having a diameter larger than the diameter of the molecule. In pores filled with water the diffusivity of the chemical is the same as that in the pure liquid, or $D_{Cw,h} = D_{Cw}$. In gas-filled pores, however, the overall diffusivity can be strongly influenced by the size of the pore. For a situation in very small pores where the mean free path of the molecule may become equal to or greater than the pore diameter, there are more frequent collisions between the molecule and the pore surface than between just the molecules. In this situation, known as Knudsen diffusion, the rebound of a molecule from a surface produces a new random direction and velocity based on the exchange of energy between the molecule and the surface. The resulting chemical flux is described by a pore diffusivity that now depends on the pore size and mean molecular velocity.

The Knudsen diffusion coefficient can be derived from the kinetic theory of gases by taking into account the rate at which molecules collide with the pore wall, the molecular concentration, and the distribution of velocities of the molecules, to yield

$$D_{Ca,K} = d_h \left(\frac{8k_B T}{9\pi m_C} \right)^{1/2} = 4850 d_h \left(\frac{T}{M_C} \right)^{1/2} \tag{3-48}$$

where $D_{Ca,K}$ [cm^2 s^{-1}] is the Knudsen diffusion coefficient, d_h [cm] the pore diameter, m_C the mass of a molecule of C, k_B the Boltzmann constant, T [K] the temperature, and M_C the molecular weight of C. The Knudsen diffusion constant is independent of pressure and varies only slightly with temperature.

At intermediate pore diameters, where the mean free path of the molecule is comparable with the pore diameter, both molecular and Knudsen diffusion will contribute to the overall diffusion constant of the molecule. The combined effective diffusion coefficient can be calculated from

$$\frac{1}{D_{Ca,p}} = \frac{1}{D_{Ca,K}} + \frac{1}{D_{Ca}} \tag{3-49}$$

which is always applicable to dilute solutions. An example of the relative importance of molecular and Knudsen diffusion to the total diffusion constant is shown in Table 3.7.

Diffusion in Macroporous Solids   The most common approach for calculating diffusion in a porous matrix is to view the porous matrix as a series of pores. The diffusivity is then calculated from the pore diffusivity as

$$D_{Cj,pm} = D_{Cj,h} \frac{\theta}{\tau_f} \tag{3-50}$$

where $D_{Cj,pm}$ is the diffusion constant for C in the porous matrix filled with phase j, θ is the porosity of the porous medium, defined as a ratio of the volume of voids to the total volume, and τ_f is an empirical factor called the tortuosity factor. Since both of the constants, θ and τ_f, are a function of the structure of the media, they should not be a function of the solvent or solute if there are no chemical interactions between them. Typical values of $\theta = 0.3$ and $\tau_f = 3$ result in the reduction of the molecular diffusivity by about an order of magnitude.

TABLE 3.7   Relative Contributions of Molecular and Knudsen Diffusivities to the Overall Effective Diffusion Coefficient for a Chemical in a Pore ($D_{Ca,h}$) of Radius R_h

R_h (cm)	D_{Ca} (cm^2 s^{-1})	$D_{Ca,K}$ (cm^2 s^{-1})	$D_{Ca,h}$ (cm^2 s^{-1})	$D_{Ca,h}/D_{Ca}$
10^{-6}	0.2	0.03	0.027	0.14
10^{-5}	0.2	0.3	0.121	0.61
10^{-4}	0.2	3	0.19	0.95

Source: Kärger and Ruthven, 1992.

Another approach used to calculate diffusion in a macroporous matrix is to consider the porous matrix composed of large and small sizes of pores, or macropores and micropores. The overall structure of the porous media consists of stacked layers of microparticles with macroporosity between the particles. According to this random pore model, the overall diffusivity is

$$D_{Cj,pm} = D_{Cj,h}\theta^2 + D_{Cj,\mu h}\theta_\mu^2 \frac{(1 + 3\theta)}{(1 - \theta)} \tag{3-51}$$

where $D_{Cj,\mu h}$ is the diffusivity in the micropores and θ_μ is the porosity of the microporous structure. The main disadvantage of this model is essentially its advantage: more fitting factors to explain the overall diffusion coefficient. As the micropore porosity goes to zero, Eq. 3-51 reduces to

$$D_{Cj,h} = D_{Cj}\theta^2 \tag{3-52}$$

Since θ is usually ~0.3, this result for $D_{Cj,pm}$ in a porous matrix calculated by the random pore model for a single type of pore becomes nearly the same as that calculated using a tortuosity factor model.

Macroporous Solids Filled with Air   In some cases it is possible to derive a tortuosity factor based on the properties of a specific medium, but the large amount of effort involved in characterizing the media makes such an approach impractical on a routine basis. Instead, the tortuosity is treated as an empirical "constant" that can fall in the range of $2 < \tau_f < 5$, but for chemicals in air τ_f is typically in the range of 3 to 4. Tortuosity is somewhat correlated to porosity, where the highest tortuosity factors measured for highly compressed porous pellets with the slope of a plot of porosity versus θ/τ_f is 1.5 (Fig. 3.3). The random pore diffusivity model better accounts for the increase in diffusivity as porosity increases.

Macroporous Solids Filled with Water   Since the pore diffusivity is the same as the molecular diffusivity for liquid-filled macroporous solids, the analysis of diffusion is much easier than for gases. The tortuosity model (Eq. 3-50) can be used, where τ_f is usually in the range of 2 to 6, although in some studies values of τ_f less than unity have been implied.

One special situation that arises is when the diameter of the chemical, d_C, approaches that of the pore diameter, d_h, so that diffusion of the molecule is hindered by the pore walls (steric hindrance). A semiempirical steric hindrance model (Kärger and Ruthven, 1992), based on the ratio of these pore diameters of $d_h^* = d_C/d_h$, is

$$D_{Cw,pm} = D_{Cw} \frac{(1 - d_h^*)^2}{1 + b_\lambda d_h^*} \tag{3-53}$$

where b_λ is a constant depending on the pore size and fluid viscosity.

Surface Diffusion   The diffusion of a chemical adsorbed onto a surface may occur in the same manner as chemical diffusion in a solid. Molecules held on the surface

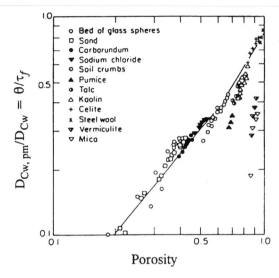

Figure 3.3 The effect of porosity on tortuosity. (From Currie, 1960.)

may have to "jump" from one favorable site to another. The flux of molecules from a surface can be the sum of the chemical flux out of a pore and the flux from the surface, or

$$D_{Dw,h+\text{sur}} = D_{Dw,h} + \frac{(1-\theta)K^*D_{Cw,\text{sur}}}{\theta} \tag{3-54}$$

where $D_{Cw,h+\text{sur}}$ is the overall diffusion coefficient for chemical diffusion from the pores and the surface, and K^* is a dimensionless adsorption partitioning coefficient. In practice it is difficult to separate the different diffusion coefficients, but in general we expect $D_{Cw,\text{sur}} < D_{Cw,h}$.

Bacterial Membranes One interesting example of very rapid surface diffusion is encountered on bacterial membranes for proteins imbedded within the membrane that protrude out through the membrane surface. The membranes of microorganisms (and by analogy perhaps any micellular structure) are composed of lipid bilayers that are in constant lateral motion. Estimates of diffusion constants for lipids in a variety of membranes are $^-10^{-8}$ cm^2 s^{-1}. Assuming the simple diffusion relationship that the distance of diffusion, L_D, is related to the diffusivity according to

$$L_D = (4D_{Cj}t)^{1/2} \tag{3-55}$$

a lipid could therefore diffuse over a distance of 2 μm in approximately one second. Thus lipids could move from one side of a bacterium to the other once every second (Stryer, 1981).

The diffusion of proteins in lipid membranes is generally slower than that of lipids, and protein diffusivities can vary by orders of magnitude. For example, the photoreceptor protein rhodopsin has a lateral diffusion coefficient of 4×10^{-9} cm^2 s^{-1} and is quite mobile in the membrane, while fibrinonectin, a peripheral glycoprotein involved in cell–substratum interactions, has a diffusion coefficient $<10^{-12}$ cm^2 s^{-1} and is therefore relatively immobile in the membrane.

The transverse diffusion of molecules from the outer side to inner side of the membrane can be substantially slower than their lateral diffusion. A phospholipid molecule, for example, takes orders of magnitude longer to flip within a 5-nm membrane than to traverse the same distance along the surface of the membrane (Stryer, 1981).

Diffusion of Chemicals in Different Types of Porous Media

There are many other cases of chemical diffusion in environmental systems where the chemical diffuses through a porous medium partially filled with water or some type of semisolid. Soil systems can be variably saturated with water and therefore fall into a region where all three phases (air, water, and soil) can affect the overall diffusion coefficient.

Chemical diffusion in microbial biofilms is of considerable interest to environmental engineers due to their use in fixed film systems in wastewater treatment systems. Chemical diffusivity is affected by the porosity and structure of the cells in the biofilm in much the same manner as observed for inorganic porous media, except that in general the biofilms have higher porosities than these other systems and density of the biofilm may not be constant. Sediments in lakes and oceans provide another unique environment because this porous medium is somewhat fluid, because it consists of both inorganic and biotic components, and because the activity of organisms living in the sediment can affect the overall dispersion of the chemical in the media. In a few limited instances, diffusivities in semisolids become important. These cases are discussed in the sections below.

Gas Diffusion in Variably Saturated Soils

Several different correlations have been proposed to model gas diffusion in air-filled media. One of the first correlations was the Buckinham equation

$$D_{Ca,s} = D_{Ca} \frac{\theta_a^2}{\tau_B} \tag{3-56}$$

where $\tau_B = 1.18$. The porosity of the gas phase, θ_a, is the void space filled with air and is obtained from $\theta = \theta_a + \theta_w$, where θ_w is the porosity of the water phase in the soil. Penman suggested the equation

$$D_{Ca,s} = D_{Ca} \frac{\theta_a}{\tau_P} \tag{3-57}$$

where τ_P varies from 1.25 to 1.64 and is typically 1.49.

The most common correlation for describing gas diffusivities in porous media is the Millington–Quirk equation,

$$D_{Ca,s} = D_{Ca}\,\frac{\theta_a^{10/3}}{\theta^2} \tag{3-58}$$

As the soils become dry, $\theta_a \to \theta$ and the Millington–Quirk equation reduces to

$$D_{Ca,s} = D_{Ca}\theta^{4/3}, \qquad \text{dry soils} \tag{3-59}$$

A comparison of the Penman and Millington–Quirk equations suggests that the Millington–Quirk equation works well over the 0–40% saturation range, and that the Penman model compares better to data at higher moisture contents. At water contents >40%, calculated diffusivities are much higher than those observed (Fig. 3.4).

Liquid Diffusion in Biofilms   It has often been assumed that the diffusivity of a chemical in a biofilm, D_{Cb}, is smaller than the molecular diffusivity by a constant ratio such as $D_{Cb}/D_{Cw} = 0.8$ (Williamson and McCarty, 1976) or 0.5 (Siegrist and Gujer, 1985). Further investigations have yielded a wide range in this diffusivity ratio of 0.04 to 1.4 (Fan et al., 1990) suggesting that diffusion coefficients in biofilms are highly variable. The range in chemical diffusivities in biofilms can be attributed to a variable biofilm porosity. Zhang and Bishop (1994) measured a decrease in biofilm porosities from 0.68 to 0.81 at the biofilm surface to 0.38 to 0.45 at the biofilm bottom (the biofilm–support interface). They described the tortuosity using two models, the random porous cluster model (RM) and the cylindrical pore model

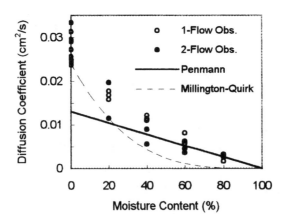

Figure 3.4 Comparison of experimentally measured diffusivities of TCE in air with the Penman and Millington–Quirk equations (adapted from Batterman et al., 1996).

(CM) based on the relationship between molecular and biofilm diffusivities of

$$D_{Cb} = D_{Cw} \frac{\theta}{\tau_f} \tag{3-60}$$

Comparing Eq. 3-60 to the porous matrix model (Eq. 3-50), we see that if the biofilm is defined as the porous matrix ($D_{Cb} = D_{Cw,pm}$), Eqs. 3-50 and 3-60 are the same.

According to Zhang and Bishop (1994), diffusion in a two-dimensional lattice leads to the observation that for a completely random arrangement of filled and empty pores that for the RPCM

$$\tau_f = \frac{1}{\theta^2} \tag{3-61}$$

Combining this result with Eq. 3-60, we have

$$D_{Cb} = D_{Cw}\theta^3 \tag{3-62}$$

This result is clearly different from that obtained for the porous matrix analysis. Zhang and Bishop note that there is some confusion in the chemical engineering literature concerning the tortuosity, τ, and the tortuosity factor, τ_f, where $\tau^2 = \tau_f$ and τ is defined as a ratio of the length of the pore to the length of the porous medium. However, it does not appear that such confusion could account for biofilm diffusivities proportional to θ^3 versus a factor of θ^2 implied by the porous matrix model.

The CM model for biofilm consists of a geometrical analysis of the pores as cylinders of length L_h and radius R_h, leading to the solution of the tortuosity of

$$\tau_f = \tau^2 = \frac{\pi^2}{4} \left(1 - 1.32 \frac{R_h}{L_h} \right) \tag{3-63}$$

The radius of pores in biofilms ranged from 0.3 to 2.75 μm, with pore lengths of 1.08 to 6.79 μm, with the largest and longest pores near the biofilm surface.

Tortuosities and the ratios of diffusion coefficients calculated using the RM and CM models for a single biofilm are shown in Table 3.8 for a biofilm grown in an annular reactor (Zhang and Bishop, 1994). Tortuosity factors ranged from 1.22 to 1.58, resulting in diffusion coefficients that were smaller than the molecular diffusivities by factors of 0.37 to 0.71.

Liquid Diffusivities in Sediments   Diffusion in sediments is different from the above situations since the porous medium is not fixed. Benthic organisms living in the sediments move around in the medium and redistribute components, and some organisms ingest and excrete part of the material. This increases the dispersion of chemicals and particles and produces higher effective diffusivities compared to those

TABLE 3.8 Tortuosities and Relative Diffusion Coefficients for Chemicals in Biofilms Grown in Annular Biofilm Reactors

Biofilm depth (μm)	Porosity, θ	Pore radius, R_h (μm)	Pore length, L_h (μm)	Tortuosity factor, τ_f		D_{Cb}/D_{Cw}	
				CM	RM	CM	RM
0–550	0.879	1.90	4.98	1.22	1.29	0.71	0.68
550–800	0.725	0.92	2.86	1.42	1.90	0.51	0.38
800–1050	0.609	0.52	1.84	1.55	2.70	0.39	0.23
1050–1420	0.582	0.33	1.21	1.58	2.95	0.37	0.20

Source: Zhang and Bishop, 1994.

for molecular diffusion in water. This means that some bulk motion of the sediment is involved, and, therefore, that chemicals are moving by chemical dispersion. For example, the diffusivity of a 10-μm particle is 0.1×10^{-8} cm^2 s^{-1} according to the Stokes–Einstein equation (Eq. 3-25). However, effective diffusion coefficients in marine sediments (the Santa Catalina Basin in the eastern Pacific) for glass beads 8–16 μm in diameter were measured as 3.3×10^{-8} cm^2 s^{-1}, a value 33 times larger than the molecular diffusivity. Even though the fluid nature of the sediment results in higher diffusion rates, there still appears to be some effect of particle size on dispersion. Larger particles (126–420 μm) had a lower diffusivity of 0.15×10^{-8} cm^2 s^{-1} (Wheatcroft, 1992). In general, horizontal mixing rates are larger than vertical mixing rates with mean horizontal dispersion coefficients of $(16 \pm 5) \times 10^{-8}$ cm^2 s^{-1} and vertical dispersion coefficients of $(1.6 \pm 0.08) \times 10^{-8}$ cm^2 s^{-1} (Wheatcroft, 1991).

Unless there is high activity of invertebrate populations in freshwater sediments, the sediment diffusivity is similar to molecular diffusivity. Sweerts et al. (1991) defined the diffusion of a chemical in sediment as

$$D_{Cw,s} = \frac{D_{Cw}}{\tau^2} \qquad (3\text{-}64)$$

The tortuosity, τ, was expressed in terms of the porosity and a constant F using measurements of tritiated pore waters from several lakes as $\tau^2 = \theta F$, where θ is the sediment porosity, so that Eq. 3-64 becomes

$$D_{Cw,s} = \frac{D_{Cw}}{\theta F} \qquad (3\text{-}65)$$

Sweerts et al. (1991) correlated porosity with F, as shown in Figure 3.5 over a porosity range of 0.4 to 0.9.

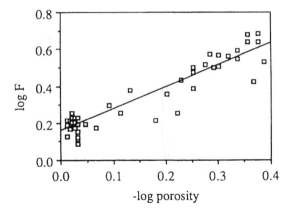

Figure 3.5 Relationship between porosity and F derived from tritiated water studies on freshwater lake sediments. (Reprinted from Sweerts et al. (1991), with permission by the American Society of Limnology and Oceanography.)

Example 3.4

Calculate the ratio of sediment to molecular diffusivities for a sediment porosity of 0.63 using the data of Sweerts et al. (1991) shown in Figure 3.5.

At a sediment porosity of $0.63 = -\log_{10} 0.2$, we obtain from Figure 3.5 $F = 2.51$. The product of θF is $0.63 \times 2.51 = 1.58$. Thus the ratio of sediment to molecular diffusivity in a freshwater sediment of porosity is

$$\frac{D_{Cw,s}}{D_{Cw}} = \frac{1}{1.58} = 0.63 \tag{3-66}$$

This result indicates that sediment diffusivity is only reduced by 63% compared molecular diffusivities in the water column.

3.4 EXPERIMENTAL DETERMINATION OF DIFFUSIVITIES

There are a variety of methods used to measure molecular diffusion constants. While a full review of different techniques cannot be presented here, it is instructive to review some of the more common methods for measuring diffusivities for conditions with environmental transport, and not chemical reactor, relevance.

Gases

Diffusion constants for chemicals in gases can easily be determined by measuring the evaporation rate of a chemical from a small tube. This device, known as the Arnold diffusion cell, is operated by flowing pure gas over a small-diameter tube so that stagnant gas conditions are maintained in the tube while chemical diffusing from the tube is pulled away rapidly enough to maintain the maximum chemical concentration gradient in the tube. A reservoir of pure chemical at the bottom of the tube maintains a supply of chemical over the duration of the experiment. Under constant temperature and pressure conditions, and assuming the gas is not soluble in the chemical, the evaporation rate is constant. The diffusion constant for the binary mixture of chemical C in a gas A can be calculated from the change in height of the fluid using the equation

$$D_{CA} = \frac{\rho_C y_{A,lm}(z^2 - z_o^2)}{2M_C c_g t(y_{C,i} - y_{C,t})} \tag{3-67}$$

where z_o is the initial distance of pure chemical C from the top of the tube, z is the height at time t, c_g the total molar gas concentration of the binary mixture of air and chemical, $y_{A,i}$ and $y_{A,t}$ the mole fractions of A in the gas at the chemical–gas interface and at the top of the tube, and $y_{A,lm}$ is the log-mean concentration of the gas over the height of the tube, calculated from

$$y_{A,lm} = \frac{y_{A,t} - y_{A,i}}{\ln(y_{A,t}/y_{A,i})} \tag{3-68}$$

For many volatile chemicals of interest, the mole fraction in the gas is not a dilute solution. If the gas is air, and the solution is dilute, we can replace A everywhere with the phase a, and $y_{A,lm} = 1$.

Example 3.5

The evaporation rate of chloroform (chemical C) in air (chemical A) has been experimentally measured in an Arnold diffusion cell at 25°C and 1 atm. Over a period of 10 h, the chloroform level dropped 0.44 cm from the initial surface 7.40 cm from the top of the tube. If the density and vapor pressure of pure chloroform are 1.485 g cm^{-3} and 200 mm Hg, calculate the diffusion constant of chloroform in air (adapted from Welty et al., 1984).

Since the vapor pressure of chloroform is so high, we cannot assume a dilute solution here. At the top of the tube, we assume that chemical is rapidly removed so that $y_{C,t} = 0$. Since this is a binary mixture, $y_{A,t} = 1$. At the chemical–air interface at the bottom of the tube, we assume there is an equilibrium concentration of C in air, so that we can calculate the mole fraction of C in air from the vapor pressure of the chemical, or

$$y_{C,i} = \frac{p_C^o}{P_T} = \frac{200 \text{ mm Hg}}{760 \text{ mm Hg}} = 0.263 \tag{3-69}$$

For the binary mixture, $y_{A,i} = 1 - y_{C,i} = 0.737$. From Eq. 3-68, we calculate $y_{A,lm}$, as

$$y_{A,lm} = \frac{y_{A,t} - y_{A,i}}{\ln(y_{A,t}/y_{A,i})} = \frac{1.0 - 0.737}{\ln(1.0/0.737)} = 0.862 \qquad (3\text{-}70)$$

The total molar concentration, c_g, is

$$c_g = \frac{1 \text{ atm}}{\left(82.06 \, \dfrac{\text{atm cm}^3}{\text{mol K}}\right)(298K)} = 4.09 \times 10^{-5} \, \frac{\text{mol}}{\text{cm}^3} \qquad (3\text{-}71)$$

Substituting these values into Eq. 3-67, we have

$$D_{CA} = \frac{(1.485 \text{ g cm}^{-3})(0.862)(7.84^2 - 7.40^2) \text{ cm}^2}{2(119.37 \text{ g mol}^{-1})(4.09 \times 10^{-5} \text{ mol cm}^{-3})(10 \text{ h})(3600 \text{ s h}^{-1})(0.263 - 0)} \qquad (3\text{-}72)$$

$$= 0.093 \text{ cm}^2 \text{ s}^{-1} \qquad (3\text{-}73)$$

Liquids

The diffusivities of common environmental chemicals of concern, such as chlorinated solvents, aromatic compounds, or even common molecules in bioreactors such as ammonia and oxygen, are well known and tabulated. For any chemical whose structure is known, D_{Cw} is easily calculated from a correlation or obtained from a table of values. For large, poorly classified chemicals, whose size and shape may strongly be related to the specific conditions of the water or wastewater, experimental measurement of the size distribution of chemicals is often necessary. For example, the size distribution of dissolved organic matter (DOM) can affect the potential for the formation of trihalomethanes (THMs) in drinking water (Collins et al., 1986). The range of different experimental techniques that can be used to classify the sizes and molecular weights of compounds in water samples is summarized in Figure 3.6.

Centrifugal Techniques   The diffusion constants of individual large molecules has most often been determined by the Svedberg method in ultracentrifuges. The settling velocity of large molecules due to gravity, u_C, is determined by the balance between gravitational forces and particle drag, or frictional resistances by the fluid. A particle of mass m_C will experience a gravitational force

$$(1 - V_C\rho_w)m_C g \qquad (3\text{-}74)$$

where V_C is the partial specific volume of a molecule of C, or its volume at infinite dilution in water. Both V_C and m_C must be determined independently by other tests. The term $V_C\rho_w m_C g$ is the weight of the fluid displaced by a single molecule or particle

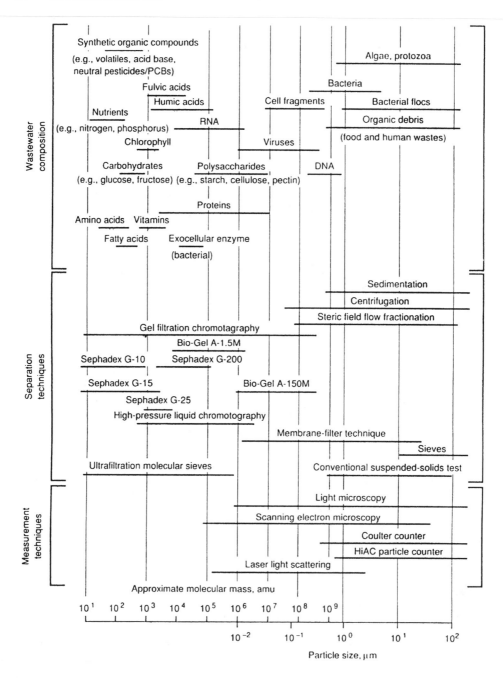

Figure 3.6 Different analytical techniques that can be used to characterize particles (<100 μm) in water samples. (Reproduced with permission of the McGraw-Hill companies from Tchobanoglous and Burton, 1991.)

of C. From a force balance, the terminal settling velocity can be calculated as

$$b_f u_C = (1 - V_C \rho_w) m_C g \qquad (3\text{-}75)$$

where b_f is the frictional coefficient, and $b_f u_C$ is the frictional force.

In an ultracentrifuge, the centrifugal force is much larger than the gravitational force, and we can replace g in Eq. 3-75 by $\omega^2 x$, producing

$$b_f u_C = (1 - V_C \rho_w) m_C \omega^2 x \qquad (3\text{-}76)$$

where ω is the angular velocity and x is the distance from the center of rotation. The sedimentation properties of many chemicals are tabulated in terms of a sedimentation constant s, defined as

$$s = \frac{u_C}{\omega^2 x} \qquad (3\text{-}77)$$

This constant is the sedimentation rate for a unit centrifugal acceleration, and is a constant for a given temperature, molecule, and solvent. In honor of the pioneer of this method, s is often expressed in terms of Svedburg units, equal to $10^{-13} s$.

In order to relate the sedimentation properties of the molecule to its diffusivity, the relationship derived by Einstein in 1905 is used

$$D_{Cw} = \frac{\mathscr{R}T}{b_f} \qquad (3\text{-}78)$$

where $\mathscr{R}$ is the gas constant and T absolute temperature. By eliminating b_f, Eqs. 3-54–3-56 can be combined to produce the final equation for determining diffusion constants in centrifuges

$$D_{Cw} = \frac{\mathscr{R}Ts}{m_C(1 - V_C \rho_w)} \qquad (3\text{-}79)$$

Typical values of diffusion constants for common proteins were listed in Table 3.5.

Light Scattering Techniques   Tyn and Gusek (1990) have proposed a more direct approach for calculating diffusion constants based on direct observation of particle sizes. Their method is a modification of the Stokes–Einstein approach. The main limitation of the Stokes–Einstein equation (Eq. 3-25) is that its derivation is based on the rigid-sphere model. Tyn and Gusek contended that the Stokes–Einstein equation could be used if the radius of gyration of the molecule was used instead of a hard-sphere radius. The radius of gyration, $r_{C,g}$, is related to the hydrodynamic sphere radius in the Stokes–Einstein equation by the simple relationship $r_C = b_g r_{C,g}$, where b_g is a constant that depends on the configuration of the molecule, producing the modified Stokes–Einstein equation

$$D_{Cw} = \frac{k_B T}{6\pi\mu_w b_g r_{C,g}}$$

(3-80)

The radius of gyration for molecules and particles can easily be determined using light scattering techniques, such as small-angle X-ray scattering. By examining the diffusivities of a variety of proteins and particles, such as viruses, Tyn and Gusek produced the correlation

$$D_{Cw} = \frac{1.69 \times 10^{-5}}{r_{C,g}}$$

(3-81)

Their correlation provided good agreement ($\pm 20\%$) with 87.4% of the proteins examined.

This same approach was used by Thurman et al. (1982) to measure the sizes of humic and fulvic acids in natural waters. Humic and fulvic acids account for 30 to 50% of dissolved organic carbon (DOC) in water and can act as buffers, can solubilize other organic compounds and chemical pollutants, and are known to generate carcinogenic compounds (trihalomethanes, or THMs) during chlorination of water samples. Their sizes range from 4.7 to 33 $\times$ 10^{-6} μm, based on their radius of gyration using small-angle X-ray scattering, or 500 to 10,000 daltons (Table 3.9). Fulvic acids have molecular weights of 500 to 2,000, and are therefore smaller than humic acids, which range in size from 1,000 to 10,000 daltons.

Chromatographic Techniques   Components in water samples can be separated according to their diffusivities using size exclusion chromatography (SEC). In this technique a sample (solute) is injected into a solvent that is flowing through a small-diameter chromatography column packed with a porous stationary phase. Small molecules can diffuse into the porous particles and are retarded in their elution compared to larger particles that can be excluded from some or all of the pores. Very large molecules cannot enter any particles and define the upper size limit of the column. The largest particles therefore exit first at one pore volume (the column void volume), while smaller particles elute at longer times. The smallest pores define the lowest size of molecules that can be separated in the column.

Two forms of SEC are common. Gel permeation chromatography (GPC) is based on the use of Sephadex gels in low-pressure columns. Due to the long detention times and dilution of samples, many samples need to be concentrated prior to injection into the column. Typical concentration factors are 100:1 by processes such as roto-evaporation, freeze-drying, adsorption–desorption using resins, and ultrafiltration. The main drawback of this method is that these concentration procedures can alter the sample size distribution either by changing the sizes of compounds or through species loss. It must also be assumed that compounds do not react with each other, through coagulation or bonding, leading to an alteration in component sizes. The eluent can also strongly affect the size and shapes of molecules. GPC chromat-

TABLE 3.9 Water Sources, Radii of Gyration ($r_{C,g}$), Concentration of Aquatic Humic Substances (c_{Cw}), and Percent of DOC for Which Molecular Weight Is Determined (%DOC)

Source	$r_{C,g}$ (Å ± 10%)	c_{Cw} (mg-DOC L^{-1})	%DOC
Ground waters			
Trona water near Rock Springs, Wyoming (humic acid)	12	44,000	95
Biscayne aquifer near Miami, Florida (humic acid)	8.8	0.8	6
Biscayne aquifer near Miami, Florida (fulvic acid)	5.3	8.5	60
Madison aquifer, near Belle Fourche, South Dakota (fulvic acid)	9.8	0.20	30
St. Peter aquifer in southeastern Minnesota (fulvic acid)	4.7	0.04	10
Coal field in western North Dakota (humic acid)	8.1, 12.3, 17, 26	10	90
Surface waters			
Tap water Arvada, Colorado (humic acid)	11, 17, 33	0.05	3
Tap water Arvada, Colorado (fulvic acid)	6.8	0.8	40
Deer Creek, Montezuma, Colorado (fulvic acid)	7.7	0.35	50
Colorado River, Yuma, Arizona (fulvic acid)	6.1	1.2	35
Suwanee River, Gargo, Georgia (humic acid)	12.8	1.5	5
Alpine Creek, Ward, Colorado (fulvic and humic acid)	8.6, 11.3	1.0	50
Thoreau's Bog, Concord, Massachusetts (fulvic and humic acid)	6.3, 11.7	21.0	69
Hawaii wetland, Hawaii, Hawaii (fulvic acid)	11.5	7.0	60
Como Creek, Ward, Colorado (fulvic acid)	10.5	3.5	55
Yampa River, Yampa, Colorado (fulvic acid)	7.9	1.0	50

Source: Thurman et al., 1982.

ograms for water samples have been prepared in alkaline buffers, such as 0.02 M borate at a pH = 9.1 (Amy et al., 1987) or distilled water buffered with 10^{-3} NaHCO$_3$ (El-Rehaili and Weber, 1987). The alternative to GPC is high-performance liquid chromatography SEC (HPLC-SEC). This newer method is identical in concept to GPC, but uses high-pressure pumps to move the material through substantially smaller-bore columns, achieving reduced retention times and therefore faster separation times. Unfortunately, HPLC-SEC has not been successful at separating many water samples due to the interaction of water components with column packing material. As a result of this interaction, compounds may be delayed in their elution times, indicating lower molecular weights than their true size. Due to the high pressures, packing materials used in GPC cannot be used in HPLC-SEC.

Typical GPC chromatograms, such as those shown in Figure 3.7, show the heterogeneous nature of dissolved organic matter (DOM). Humic and fulvic acids are distinguished from other compounds in samples by their absorption of light at 254

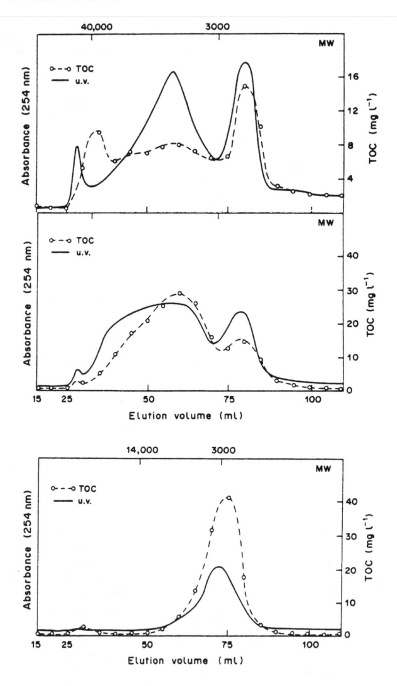

Figure 3.7 Chromatograms using gel permeation chromatography (GPC) to characterize dissolved organic matter in water samples. (a) Aldrich humic acid; (b) Contech fulvic acid; (c) Huron River water. (Reprinted from El-Rehaili and Weber, 1987, with kind permission from Elsevier Science Ltd., The Boulevard, Landford Lane, Kidlington, OX5 1GB.)

nm. Molecular sizes measured in this manner are referred to as "apparent molecular weights" (AMWs) since compounds in water samples may interact with column packings and each other, and may not be the true molecular weights. In addition, the molecular weight is not being measured, but is calculated from calibration with proteins and other compounds that may have a different relationship between molecular weight and diffusivity than the DOM in the sample.

The main use of these size chromatograms is to study the effect of different treatment processes on the size distribution of molecules in order to understand the effectiveness of a process for treating different water samples. When the model Aldrich humic acid was subjected to treatment by granular activated carbon (GAC), alum coagulation, and lime softening, different AMWs evolved (Fig. 3.8). The organic matter with a high molecular weight (>40,000) was the least adsorbable onto GAC, likely as a result of pore exclusion. Alum coagulation was most efficient at removing UV-absorbing compounds, but removal by both alum and lime softening was a function of dose. Thus the most effective treatment process is a function of the water sample and the economics of the treatment scheme, and if applicable, chemical dose.

Capillary Tube Technique (Taylor–Aris Dispersion)   When a small pulse of a chemical solution is injected into capillary column flowing under laminar conditions, the chemical concentration becomes Gaussian after a relatively short period of time. This spreading out of the chemical in the axial direction is known as *longitudinal dispersion*. Taylor (1953) recognized that by measuring the spread of the chemical at the tube exit and the mean velocity of the solvent in the tube, it was possible to calculate the diffusion coefficient of the chemical in the solvent. This method has been demonstrated to be inexpensive and easy to conduct in a laboratory, requiring only a pump, capillary tube, and detector (Bello et al., 1994). It is applicable to a wide range of chemical sizes, including small molecules, macromolecules, colloids, and even bacteria-sized particles.

The velocity of a chemical in a tube under laminar flow conditions is a parabolic function of the tube radius with a maximum at the tube center, and a minimum (zero velocity for a no-slip condition) at the tube wall. When a chemical (or particle) solution (solute) is introduced into the fluid (solvent) individual molecules move at different velocities depending on their location in the tube. Molecular diffusion moves the molecules from high- to low-velocity regions, causing the chemical to be dispersed in the longitudinal (axial) direction. The chemical diffusivity therefore influences both the time of elution and the width of the eluting plume. Molecules with low diffusion coefficients have shorter elution times but broad peaks. Chemicals with high diffusivities take longer to elute, since they will rapidly diffuse into low-velocity regions, but they will have narrower profiles when they elute since they will tend to sample slow- and fast-moving streamlines equally (Fig. 3.9).

The flux of a chemical, $J_{Cw,z}$ in a capillary tube containing a liquid in the z direction relative to a fixed location (see Eq. 4-59) is

$$ J_{C,z} = -E_C \frac{dc_C}{dz} + c_C u \tag{3-82} $$

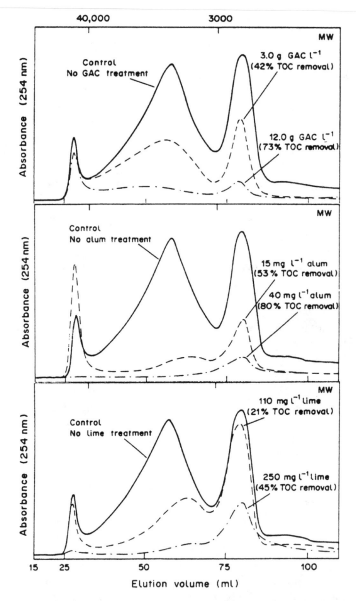

Figure 3.8   Chromatograms using gel permeation chromatography (GPC) to examine the effect of different water treatment processes on the molecular size distribution of Aldrich humic acids. (a) Granular activated carbon (GAC); (b) alum coagulation; (c) lime softening. (Reprinted from El-Rehaili and Weber, 1987, with kind permission from Elsevier Science Ltd., The Boulevard, Landford Lane, Kidlington, OX5 1GB.)

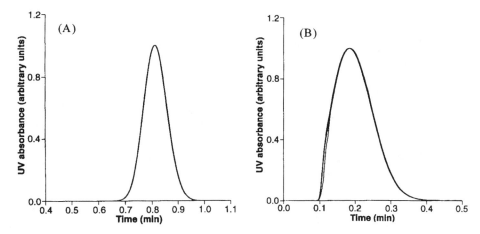

Figure 3.9 Elution profiles from five consecutive runs (images superimposed) of (A) 0.25% DL-phenylalanine in a 98.8-μm-pore-diameter capillary tube and (B) 1% oval-bumin (hen's egg) in 100 mM Tris−borate buffer at pH = 8 in a 75-μm-pore-diameter capillary tube. (Reprinted with permission from Bello et al., 1994. Copyright 1987, American Chemical Society.)

where c_C is the mean chemical concentration across the tube cross section, u the mean velocity in the z direction, and E_c the dispersion coefficient of C in water. The dispersion of the chemical is a function of the chemical diffusivity (Taylor, 1953) according to

$$E_c = D_C + \frac{R_T^2 u^2}{48 D_C} \tag{3-83}$$

where D_C is the molecular diffusivity of C and R_T the tube radius. The time dependence of E_c is negligible for times greater than $0.5 R_T^2 / D_C$.

The zero (m_0), first (m_1), and second (m_2) moments of the chemical plume exiting the tube (Bello et al., 1994) are

$$m_0 = \int_0^\infty c_C \, dt \tag{3-84}$$

$$m_1 = \frac{1}{m_0} \int_0^\infty c_C t \, dt \tag{3-85}$$

$$m_2 = \frac{1}{m_0} \int_0^\infty c_C (t - m_1)^2 \, dt \tag{3-86}$$

For a Gaussian distribution, the time moments are related to chemical dispersion according to

$$m_1 = \frac{L}{u} + \frac{2E_C}{u^2} \tag{3-87}$$

$$m_2 = \frac{2E_C L}{u^3} + \frac{8E_C^2}{u^4} \tag{3-88}$$

The chemical diffusivity in Eq. 3-83 can be calculated by rearranging the above equations to solve for L/D_C and E_C using

$$\frac{L}{u} = \frac{m_1}{2}\left[3 - \left(1 + \frac{4m_2}{m_1^2}\right)^{1/2}\right] \tag{3-89}$$

$$E_C = \frac{u^2}{2}\left(m_1 - \frac{L}{u}\right) \tag{3-90}$$

As an example of the method, the peak shown in Figure 3.9 was measured in a silica capillary tube of diameter 100 μm. The molecular diffusivity of DL-phenylalanine in water was $D_{Pw} = (0.708 \pm 0.012) \times 10^{-5}$ cm^2 s^{-1}, versus a literature value of $D_{Pw} = 0.705 \times 10^{-5}$ cm^2 s^{-1}. For ovalbumin, the molecular diffusivity in water was $D_{Ow} = (0.759 \pm 0.014) \times 10^{-6}$ cm^2 s^{-1}, versus a literature value of $D_{Ow} = 0.776 \times 10^{-6}$ cm^2 s^{-1} (Bello et al., 1994).

Field-Flow Fractionation Field-flow fractionation (FFF) is a relatively new, chromatographic-type technique that can be used to assess the molecular sizes of compounds over a wide range. In this method a crossflow of carrier solution is maintained at a right angle to the flow of sample down through a rectangular channel (Fig. 3.10). This flow orientation is made possible by constructing the sides of the channel from permeable membranes. A cloud of particles is formed against the accumulation wall by the crossflow. The mean thickness of the cloud is l, dependent on the bulk crossflow linear velocity u_x and the diffusivity of the particle, D_C. In this cell, the retention is obtained (Beckett et al., 1987) from

$$\frac{V_o}{V_r} = 6l^*\left(\coth\frac{1}{2l^*} - 2l^*\right) \tag{3-91}$$

where V_r is the sample retention volume. The dimensionless retention parameter l^* is calculated from

$$l^* = \frac{l}{w} = \frac{D_C}{u_x w} = \frac{D_C V_o}{Q_c w^2} \tag{3-92}$$

where V_o is the channel void volume, Q_c the volumetric flow rate, and w the channel width.

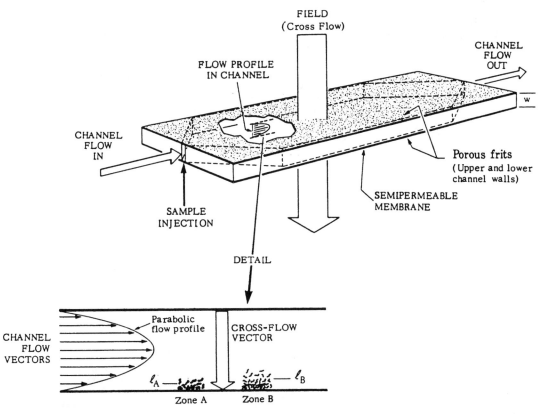

Figure 3.10 Schematic of a field-flow fractionation (FFF) cell, with a detail of the cross section of the channel indicating migrating sample clouds and parabolic velocity profile. (Reprinted with permission from Beckett et al. (1987). Copyright 1987, American Chemical Society.)

During flow through the channel, the parabolic velocity profile across the thin dimension of the channel moves thicker clouds of large compounds down the channel more quickly than more compact clouds of smaller species in the sample. Separation of the cloud occurs since smaller species diffuse more rapidly than larger species with lower molecular weights. Using the above equations and calibration with species of known diffusivities allows determination of the size distributions of molecules or particles in water samples.

The sizes of humic and fulvic acids, determined by FFF, are similar to sizes of molecules indicated by other methods, indicating results from other techniques (which may require changes in solution chemistry or pH or concentration through evaporation) are accurate indicators of size (Table 3.10). Fulvic acids from different sources were found to increase in size in the order aquatic < soil < peat bog < lignite coal. A correlation between molecular weight and diffusivity for humic and fulvic

TABLE 3.10 Comparison of Molecular Weights of Reference Humic Substances Using Field-Flow Fractionation (FFF) Versus Other Methods. Vapor Pressure Osmometry and Low-Angle X-Ray Scattering: D_{Cw} Is the Diffusion Coefficient at the Maximum of the FFF Peak

Sample	Molecular weight M_C		Diffusivity $D_{Cw} \times 10^6 \ cm^2 \ s^{-1}$
	FFF	Other methods	
Suwannee stream fulvate	860	750	4.10
Suwannee stream humate	1490	1500	3.23
Mattole soil fulvate	1010	1200	3.76
Mattole soil humate	1750	~2000	3.01

Source: Beckett et al., 1987.

acids in natural waters yielded

$$D_{Cw} = 1.42 \times 10^{-4} M_C^{-0.422} \tag{3-93}$$

where D_{Cw} is in $cm^2 \ s^{-1}$. This correlation was found to be in agreement with those from other studies for macromolecular diffusion constants.

Ultrafiltration Techniques Ultrafiltration (UF) membranes contain pores that permit the passage of particles based only on their size. Particles larger than the pore are rejected, while smaller particles can pass through the membrane. Ideally, there is no interaction between the solute and membrane. In practice, however, particles adsorb to the membrane, reducing the sizes of the pores or preventing the passage of other molecules with chemical charges that are repulsed by either the membrane or adsorbed particles. As the concentration of material on the membrane increases, the membrane can become polarized, leading to rejection independent of solute concentration. Compound diffusivities must be estimated by calibration since the ultrafiltration technique measures molecule size. Membrane sizes are typically reported in terms of molecular weights of compounds used during calibration, but diffusion coefficients can be calculated from molecular sizes if the type of compound (i.e., protein, polysaccharide, etc.) is known.

Molecular weight size distributions can be determined using ultrafiltration membranes under a variety of hydraulic conditions, including: pressurized batch and continuous-flow stirred cells, centrifuge tubes with UF membranes on the bottom of the tube, and continuous-flow membrane reactors typically consisting of plates or wound tubes. The most common approach in the laboratory is to use batch stirred cells. The advantage of this method is a relatively small volume of sample can be used (100 mL), producing a sample only slightly reduced in volume (~50 to 80%) that can be subjected to further analysis (DOC, UV absorption, etc.). The disadvantage is that membranes may not equally reject all compounds in the sample based only on size, sample loss can occur through adsorption and flocculation on the membrane, and membranes have pore distributions that may vary by batch and manufacturer. Concentrations of DOM in the sample must be minimized to avoid membrane polarization. Changes in the chemical conditions in the retentate, such as ionic strength and pH, can also alter the ability of molecules to pass the membrane as a function of the extent of sample concentration. Dilution or buffering of the sample may alter molecular diameters compared to the initial sample.

Samples analyzed using UF cells provide easily obtained apparent molecular weight distributions that reflect the different nature of water and wastewater samples, and can be used to show changes in molecular sizes prior to and subsequent to treatment. UF has been found to provide slightly low apparent molecular weights than GPC based on the analysis of six different water samples (Amy et al., 1987), as shown in Figure 3.11. It is recognized that such a comparison is highly a function of the buffers used in GPC, the specific UF membranes, etc.

Treatment of water samples with chlorine for disinfection is a standard practice. However, during chlorination halogen-containing chemicals can be formed. Trihalomethanes (THMs) are carcinogenic compounds regulated in drinking water. The THM formation potential (THMFP) of a water is determined by measuring the concentration of THMs as a function of chlorine dose. Research has shown that the THMFP of a water sample is first a function of the concentration of organic matter and second a function of the DOM size distribution. In general, the THM formation increases inversely proportional to apparent molecular weight, with the compounds <10,000 AMW the most consistently reactive compounds in drinking water sources (Table 3.11). Conventional, direct filtration, and lime softening treatment processes removes higher-molecular-weight material to a greater extent than the smallest (<500 AMW) size fractions. Since the THMFP is lower for larger compounds, treatment does not favorably remove THM precursors (Collins et al., 1986).

The size distribution of organic matter is also important for wastewaters treated in biological wastewater treatment processes such as trickling filters and activated sludge reactors. The significance of the size distribution is that while compounds less than ˜1,000 daltons can pass through the membranes of microorganisms, larger compounds must be cleaved or hydrolyzed before assimilation. Since smaller compounds also diffuse in liquid more rapidly than larger compounds, they are likely to be removed faster under mass-transfer-limited conditions. A survey of several treatment plants during winter and summer months indicated that trickling filters efficiently removed small compounds (Table 3.12). Small compounds (<500 amu) in secondary effluents from three different types of treatment plants ranged from ˜10 to 40% of the total DOC, versus 60% of DOC < 500 amu in wastewaters prior to treatment (primary effluent). Chemical treatment of effluents can further alter size distributions. Breakpoint chlorination of activated sludge effluents reduces average molecular sizes and biodegradability, and increases the amount of nonadsorbable material (Trogovcich et al., 1983).

Samples can be processed through UF cells either in series or in parallel (Fig. 3.12). When samples are processed in series, a sample is first passed through the membrane with the largest pores (highest-molecular-weight cutoff). The concentration of material in the filtrate is measured, and then the remaining sample passed through the membrane of the next smaller pore. This procedure is repeated, but since

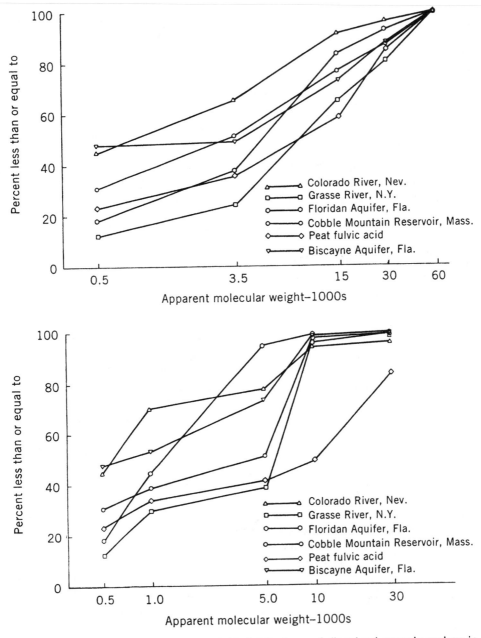

Figure 3.11 Apparent molecular weight distributions of dissolved organic carbon in water samples analyzed using either (a) gel permeation chromatography, or (b) ultra-filtration (Amy et al., 1987). (Reprinted from Journal AWWA, Vol. 79, No. 1, January 1987, by permission. Copyright 1987, American Water Works Association.)

TABLE 3.11 Ratio of Trihalomethane Reactivities to Initial Mass of DOC in Various Untreated and Treated Waters as a Function of Molecular Weight Cutoff of UF Membranes. Samples Were Processed in Parallel through UF Membranes

Treatment Plant	Water	THM Yield (μg/mg C) of AMW Fractions					
		<0.5K	<1K	<5K	<10K	<30K	<0.45 μm
Springfield	untreated	55.4	61.3	72.0	81.0	78.6	77.3
	treated	53.4	60.7	66.1	65.1	61.7	66.0
Canton	untreated	56.5	68.2	75.5	93.0	92.8	93.3
	treated	56.8	55.7	56.6	60.2	61.0	62.2
Daytona	untreated	54.1	55.2	59.4	62.1	62.4	63.0
	treated	49.8	50.0	51.0	52.1	52.1	52.8
Las Vegas	untreated	55.7	58.3	56.7	56.6	55.6	55.3
	treated	48.7	47.8	48.0	47.5	46.6	47.9

Source: Collins et al., 1985.

cells cannot be run to dryness each time, some portion of the sample is lost. The number of membranes that can be used is limited by sample size. When samples are processed in parallel, a larger initial sample volume is used, but filtrate does not need to be collected and reused. The concentrations in the cell permeates are subtracted from each other to produce the final molecular size distribution.

In many UF molecular weight distribution studies, investigators have neglected to adjust size distributions for membrane rejection. Some molecules, even though they are smaller than the pore size of the membrane, are partially rejected by the membrane. If a 100 mg L^{-1} solution of sucrose (M_S = 342 daltons), for example, is

TABLE 3.12 Molecular Weight Distributions of DOC in Primary and Secondary Effluents at Different Times of Year[a]

Plant	Season	DOC in apparent molecular weight range (1%)						Total DOC (mg l^{-1})
		<0.5K	0.5–5K	5–10K	10–30K	30–100K	>100K	
Primary	Summer	60	16	6	7	4	7	53.9
TF	Winter	18	39	5	2	20	16	12.50
	Summer	13	37	6	2	11	31	13.90
AS- pure O_2	Winter	15	46	11	9	1	18	11.00
	Summer	27	19	16	5	15	18	10.2
AS	Winter	38	3	26	7	4	22	8.86
	Summer	34	28	17	0	11	10	7.78
	Summer	33	31	18	0	9	9	7.25

[a]Primary = effluent from primary clarifier, TF = trickling factor, AS- pure O_2 = Pure oxygen activated sludge system, AS = conventional activated sludge. Samples from wastewater treatment plants in Tucson, AZ.
Source: Amy et al., 1987.

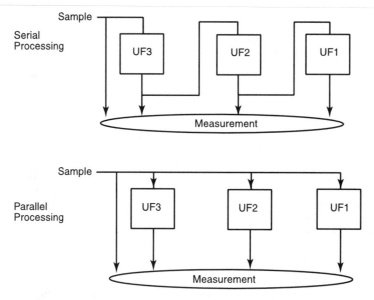

Figure 3.12 Molecular size distributions can be determined either by serial or parallel processing through membranes in batch ultrafiltration (UF) cells. Membrane pore sizes decrease in order UF3 > UF2 > UF1.

passed through a 500 amu membrane, only 50 mg L^{-1} of sucrose may initially be present in the cell permeate. The high rejection of compounds with molecular weights of sufficient size to pass the membrane can produce erroneous size distributions.

One method proposed to correct for membrane rejection was to reprocess a sample several times to completely rinse out the mass of material able to pass the membrane (Bryant et al., 1990). For example, a 100 mL sample would be ultrafiltered to 20 mL, resuspended in distilled water, ultrafiltered again, and this process repeated as many as 5 times. Each time the total mass recovered in the filtrate would decrease. By addition of the total mass recovered, the concentration of material in the original sample of size able to pass the membrane could be back calculated. The disadvantages of this technique are the time necessary for each filtration step and the potential for changes in the solution chemistry during the redissolution of the remaining sample in distilled water since the conformation, and therefore size, of the molecules likely changes as the pH and ionic strength of the sample is changed (see Problem 3.9).

A more direct approach for calculating the size distribution of compounds in waters and wastewaters is based on a permeation coefficient model (Logan and Jiang, 1990). When a sample is passed through a UF membrane, the concentration in the permeate (c_p) will always be less than the concentration in the retentate (c_r) due to membrane rejection. As the sample is processed through the membrane, this results in the concentration of molecules in the retentate of sufficient size to pass through the membrane increasing with time. Since the retentate concentration increases with time, so does the permeate concentration. A typical permeate concentration profile through a 1000 molecular weight cutoff membrane is shown in Figure 3.13 for a sample containing 105 mg DOC L^{-1} of material, with 50 mg DOC L^{-1} of sucrose (M_S = 342), 50 mg DOC L^{-1} Vitamin B-12 (M_V = 1,192). At the beginning of the ultrafiltration, the total mass leaving the cell was ~45 mg L^{-1}, but after 90 mL had passed, the permeate concentration was >150 mg L^{-1}.

In order to calculate the initial concentration of material of sufficient size to pass through the UF membrane, it is assumed that the membrane rejection is constant during the filtration procedure, or that

$$p_c = \frac{c_p}{c_r} \tag{3-94}$$

where p_c is the permeation coefficient. The notation used to describe the concentrations in the permeate, retentate, and filtrate are summarized in Figure 3.14. A mass balance conducted around the UF cell is for unsteady flow with no reaction (see Example 1.6). The governing equation for a constant flow out of Q_w is

$$\frac{d(V_r c_r)}{dt} = -p_c c_r Q_w \tag{3-95}$$

where V is the volume in the cell at any time t. Recognizing that $Q_w = -dV_r/dt$, and

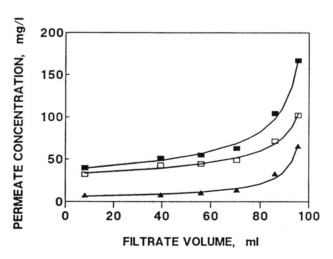

Figure 3.13 DOC concentrations measured in the permeate from a batch UF cell containing a 1,000 apparent molecular weight cutoff (YM2, Amicon) membrane: sucrose (□), Vitamin B-12 (▲), and total DOC (■). Initial concentrations were 50 mg L^{-1} of sucrose, and 55 mg L^{-1} of Vitamin B-12. Line calculated from permeation coefficient model (Logan and Jiang, 1990).

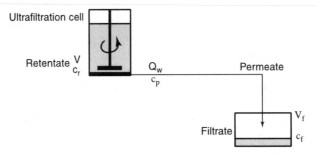

Figure 3.14 Notation used to describe concentrations in the UF cell retentate, permeate, and filtrate. (Adapted from Logan and Jiang, 1990.)

integrating from $V_{r,0}$ to V_r, the concentration of material of sufficient size able to pass the membrane present in the original sample, $c_{r,0}$, is

$$c_r = c_{r,0} \left(\frac{V_r}{V_{r,0}} \right)^{p_c - 1} \tag{3-96}$$

Defining F to be the fractional reduction of retentate, $F = 1 - (V_f/V_{r,0})$, where $V_f = (V_{r,0} - V_r)$ is the total volume of filtrate to pass through the membrane, and substituting in Eq. 3-96, we have the permeation coefficient model

$$c_p = p_c c_{r,0} F^{p_c - 1} \tag{3-97}$$

The permeate coefficient model can be linearized by taking the logs of both sides of Eq. 3-97 to obtain

$$\ln c_p = \ln(p_c c_{r,0}) + (p_c - 1)\ln F \tag{3-98}$$

The value of $c_{r,0}$ and p_c can be calculated by measuring the concentration of material in the permeate as a function of time, and plotting $\ln c_p$ versus $\ln F$, where $(p_c - 1)$ is the slope and $\ln(p_c c_{r,0})$ the y intercept. The molecular size distribution of a sample is calculated from the initial total concentration of material in the sample and the $c_{r,0}$s from separations in different UF cells. It is not recommended that samples be processed in series since loss of chemicals through membrane rejection can alter the final calculated size distribution.

Once the values of p_c and $c_{r,0}$ are known, the permeation coefficient model can be used to estimate errors in the size distribution based on a typically employed approach of assuming the concentration in a volume of filtrate collected over time,

c_f, is equal to $c_{r,0}$. Integrating permeate concentrations over the time of collection yields

$$c_f = c_{r,0} \frac{(1 - F^{p_c})}{(1 - F)} \qquad (3\text{-}99)$$

The difference between $c_{r,0}$ and c_f is calculated based on the volume of sample that would be collected (defined in terms of F). Since c_f is the cumulative concentration in all filtrate, $c_f \le c_{r,0}$.

If the size distribution is not adjusted for membrane rejection, the concentration of material in larger size fractions will be overestimated. The molecular size distributions of two groundwaters calculated using the permeate coefficient model were compared with the apparent size distributions based only on the collected filtrate (Fig. 3.15). Compound concentrations were calculated by UV absorption, since most of the material in the groundwater sample was assumed to be humic and fulvic acids that strongly absorb UV light (254 nm). The apparent size distribution indicated that 66% of material was >5,000 daltons, whereas the corrected size distribution indicated only 28% of the material was larger than this size. Similarly, 84% of material in Biscayne aquifer water was <5,000 daltons, while the unadjusted sample contained 59%. Previous measurements on Biscayne groundwater found that humic and fulvic acids, with molecular weights of 800 to 2,000 daltons, accounted for 66% of the DOC of this water (Thurman et al., 1982). Correcting the size distribution for membrane rejection decreased the average molecular weights of the sample and provided a size distribution more consistent with other measurements.

Adjusting the size distribution of DOC in wastewater samples using the permeation coefficient model similarly decreased the average size distribution. The permeation coefficient for the 1K molecular weight cutoff membrane was 0.75 ± 0.07,

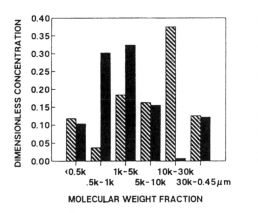 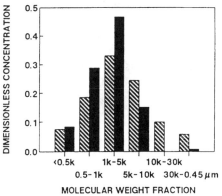

Figure 3.15 Molecular size distributions of UV (254 nm) absorbing compounds in (a) Orange County Water District and (b) Biscayne groundwaters before (hatched boxes) and after (solid boxes) adjustment using the permeation coefficient model. Concentrations normalized to original sample. (Reproduced by permission of ASCE from Logan and Jiang, 1990.)

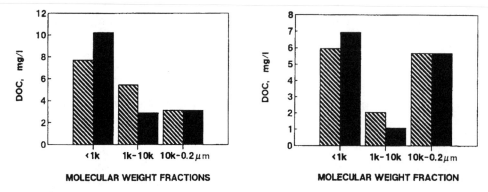

Figure 3.16 Size distributions of DOC in wastewater effluent from a trickling filter before (hatched boxes) and after (solid boxes) adjustment using the permeation coefficient model. (Reproduced by permission of ASCE from Logan and Jiang, 1990.)

but was unity for the 10K amu membrane (Logan and Jiang, 1990). After correction for membrane rejection, DOC in the <1K size fraction increased from <8 mg L^{-1} to >10 mg L^{-1} (Fig. 3.16). It is particularly important to measure permeation coefficients when examining changes in size distributions as a function of treatment processes, since changes in chemical composition can alter membrane rejection properties.

Example 3.6

Data on permeate concentration have been obtained from an ultrafiltration separation (1,000 amu cutoff) as shown in Table 3.13. The values shown are for the instantaneous permeate (c_p) concentrations of dissolved organic carbon (DOC) in the water sample as a function of the volume that passed the membrane (V_f). The initial volume of retentate is 100 mL, and the total DOC of the sample is 12 mg L^{-1}. (a) What is the permeation coefficient, p_c, and the concentration of DOC that is actually smaller than the molecular weight cutoff of the membrane, $c_{r,0}$? (b) What would the concentration of the sample be if you collected all 60 mL of the

TABLE 3.13   Ultrafiltration Data

c_p	V_f
5.9	5
6.1	10
6.4	20
7.5	50
9.0	70
12.9	90

TABLE 3.14   Calculated Ultrafiltration Parameters

F	$\ln c_p$	$\ln F$
0.95	1.77	−0.051
0.90	1.81	−0.105
0.8	1.86	−0.22
0.5	2.01	−0.69
0.3	2.19	−1.20
0.1	2.56	−2.30

sample? What would be your error (in percent) if you assumed that the total filtrate sample DOC was actually the DOC of the sample less than the membrane cutoff?

(a)   Using the given data, we calculate $\ln c_p$ and $\ln F$ as shown in Table 3.14. From a linear regression of $\ln c_p$ as a function of $\ln F$, we obtain a slope of −0.345 and a y intercept of 1.77. Therefore, $p_c = 1 - 0.34 = 0.66$, and $c_{r,0} = e^{1.77}/0.66 = 9.0$ mg L^{-1}.

(b)   We can calculate the concentration that we would have collected, c_f, using Eq. 3-99 and the values obtained in this problem, or

$$c_f = 9.0 \frac{(1 - F^{0.66})}{(1 - F)} \tag{3-100}$$

While the concentration of DOC able to pass the membrane is really 9 mg L^{-1}, if we had not corrected for the membrane rejection of the DOC we would have measured only 6.4 mg L^{-1} in the filtrate. This is an error of 29%.

PROBLEMS

3.1   The diffusivity of chloroform in air at 25°C and 1 atm as 0.095 cm^2/s was reported by Welty et al. (1984) to have been measured in an Arnold diffusion cell. Calculate the diffusion coefficient using the Fuller correlation, and compare it to that obtained in Example 3.5. What is the percent difference between these results?

3.2   Calculate the diffusivities of thymidine, leucine (amino acids), and rhodamine WT in water at 25°C using (a) Wilke−Chang and (b) Reddy−Doraiswamy correlations.

3.3   Calculate the diffusivities of arginine in water at 25°C using the following methods: (a) Wilke−Chang; (b) Reddy−Doraiswamy; (c) Polson equation; (d) Wilke−Chang, but with $V_{C,b}$ defined as the molecular weight. Present your results in a table. What do you observe from this comparison?

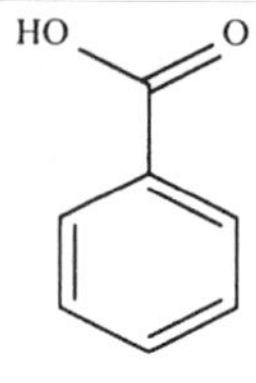

Figure P-3.4

3.4 Calculate the liquid diffusion coefficient $[\text{cm}^2\ \text{s}^{-1}]$ of the chemical in Figure P-3.4 in water at 5°C, using the Wilke–Chang correlation. (You may wish to know that $\rho_w(20°C) = 0.998\ \text{g cm}^{-3}$, $\mu_w(20°C) = 0.01002\ \text{g cm}^{-1}\ \text{s}^{-1}$, and $\rho_w(5°C) = 0.999\ \text{g cm}^{-3}$, $\mu_w(5°C) = 0.01519\ \text{g cm}^{-1}\ \text{s}^{-1}$.)

3.5 Data on generalized diffusion coefficients were given in Table 3.2. Compare data in this table with the two correlations developed by Polson and Frigon correlation in a graph (use log–log coordinates).

3.6 (a) Fill in Table P-3.6 for particle diffusivities at 20°C. (b) Graph the diffusivities calculated above ($\log_{10} D_{Cw}$) as a function of $\log_{10} M_C$. (c) Add the following observed diffusivities to your plot from part (b). Comment on the results (i.e., which equation seems more reasonable for various observed molecular weights?).

3.7 Wastewater from a domestic wastewater treatment plant has long been considered (at least in the laboratory) to be composed of small molecular weight compounds, such as glucose and acetate, which have diffusion coefficients around 700 $\mu\text{m}^2\ \text{s}^{-1}$. However, much of the organic matter in wastewater has molecular weights substantially higher than compounds like glucose. Levine et al. (1985) indicate the distribution of soluble organics in wastewaters can

TABLE P-3.6a.   Diffusivities as a Function of Molecular Weight

Substance Molecular Weight	Diameter (μm)	Diffusivity ($\text{cm}^2\ \text{s}^{-1} \times 10^8$)		
		Perry and Chilton	Stokes– Einstein	Polson
10^1	0.00029	2200		
10^2	0.00062	700		
10^3	0.00132	250		
10^4	0.00285	110		
10^5	0.0062	50		
10^6	0.0132	25		

TABLE P-3.6c   Diffusion Coefficients of Selected Compounds

Substance	Molecular Weight	Observed Diffusivity ($D_{Cw} \times 10^8$ cm^2 s^{-1})
Oxygen	32	1800
Glucose	180	600
Raffinose	504	360
Ribonuclease	12,700	136
Myoglobin	17,500	113
Hemoglobin	68,000	63
Edestin	309,000	39
Virus	6,800,000	11

be divided as shown below, where the compositions in brackets are assumed here as typical.

$$\text{Grease } (C_{51}H_{99}O_6) - 60\%$$

$$\text{Protein } (C_4H_{6.1}O_{1.2}N)_{600} - 20\%$$

$$\text{Complex carbohydrate } (C_6H_{12}O_6)_{200} - 20\%$$

(a) What are the diffusivities of the individual components, assuming that the Polson correlation is valid for all molecules? (b) What are average molecular weights and average diffusivities that could be used to calculate the flux of components in a typical wastewater?

3.8 You and your co-worker in the Environmental Engineering Laboratory are determining the concentration of material in a sample that is less than 10,000 daltons. The total concentration of all material in the sample is 100 mg L^{-1}. You use the permeation coefficient method, and find out that only 50% of the material is smaller than the cutoff of the membrane, and that the membrane has a permeation coefficient of 0.65 for the material. (a) What would concentration of material (<10K) would your co-worker determine if a 120-mL sample was ultrafiltered, and the first 80 mL of the sample collected and analyzed? (b) Your co-worker tells you that they *discarded the first 20 mL* (presumably to avoid contamination in sampling lines), and only collected and analyzed the next 60 mL. What concentration of material would your co-worker determine to be <10K from measurement of the sample by this second method?

3.9 The size distribution of adsorbable organic halides (AOX) in kraft mill wastewaters was determined by Bryant et al. (1990) for several samples using a dilution procedure. The true AOX concentration, AOX$_t$, was calculated from the concentration in the first 20 mL of a 40 mL sample, AOX$_{20}$, using the equation AOX$_t$ = p_{100}AOX$_{20}$, where p_{100} is a correction factor based on dilution

TABLE P-3.9 AOX in Permeates Collected from 40 mL of a Kraft Mill Wastewater Processed in Parallel through Three Membranes with Different Molecular Weight Cutoffs (YCO5 = 500 Dalton, YM2 = 1,000 Dalton, and YM5 = 5,000 Dalton)

Permeate fraction (mL)	AOX (μg L^{-1})		
	YCO5	YM2	YM5
5–10	5,650	7,250	27,750
10–15	6,350	7,550	30,160
15–20	6,940	—	34,040
20–25	7,310	8,900	35,640
25–30	8,340	1,870	35,130
30–35	11,130	13,400	44,120

Source: Bryant et al., 1990.

and refiltration of the sample 5 times. Values calculated for membranes included $p_{100}(0.5K) = 1.65$, $p_{100}(1K) = 1.28$, and $p_{100}(5K) = 1.08$. Compare their size distribution with a size distribution using the permeation coefficient model, using the data in Table P-3.9, and by calculating AOX$_{20}$ from c_f for the first 20 mL of sample.

3.10 A wastewater sample sent to your laboratory is analyzed for the molecular weight distribution of DOC. Data from a 1K and 10K ultrafiltration cell are shown in Fig. P-3.10. Samples (120 mL) used in these batch ultrafiltration cells were prepared in parallel. If the original sample had 30 mg L^{-1} of DOC < 0.2 μm, what is the molecular size distribution of the sample based on your analysis of the data below using the permeation coefficient model?

3.11 Sweerts et al. (1991) determined that the diffusivity of a chemical in soil is related to the chemical in water by the relationship $D_{Cs} = D_{Cw}\tau^{-2}$, where τ is a function of the soil porosity, θ, according to $\tau^2 = \theta F$, where F is an empirical function of porosity as shown in Figure 3.5. According to their data, what do you expect the diffusivity of a chemical to be in clay ($\theta = 0.40$) if the diffusivity in water of the chemical was 2×10^{-5} cm^2 s^{-1}.

3.12 A student preparing a molecular size distribution of dissolved organic carbon (DOC) in a wastewater sample (prefiltered through a 0.2 μm filter) takes a sample with a DOC = 150 mg L^{-1}, and passes it in parallel through two ultrafiltration cells containing membranes with 1K (1,000 amu) apparent molecular weight (AMW) and 100K AMW membranes. The DOC of the 90 mL samples he collected from the 120 mL he started with are 75 mg L^{-1} (<1K) and 125 mg L^{-1} (<100K). He shows the results to his professor, who complains that what he has obtained is only an apparent size distribution. The student returns to the laboratory to determine the true size distribution of the

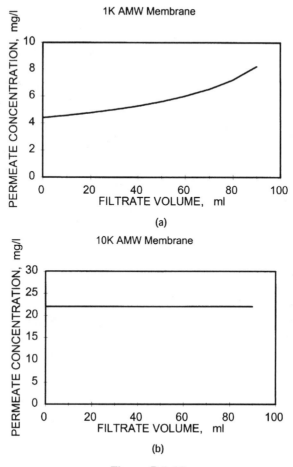

Figure P-3.10

sample. The student still has the 90-mL samples (little was lost for the DOC measurement) and again passes these two samples through the same membrane, collecting 60-ml samples and measuring DOCs of 50 mg L^{-1} (<1K) and 125 mg L^{-1} (<100K) in these filtrates. Based on these experimental results, answer the following questions and present your results in a simple table. (a) What was the "apparent" size distribution of DOC (<1K, 1K–100K, and >100K to <0.2 μm) without correction of the membrane rejection properties based on the first filtration step? (b) Using data from the two filtration steps, what is the *actual* size distribution based on the permeation model? (c) What are the percent errors for each size fraction?

√ **3.13**   Two chemicals, a complex carbohydrate $(C_6H_{12}O_6)_{75}$ and phenol, each present at 100 mg/L in water, are diffusing through a porous membrane so that their concentrations are reduced by 50% over the thickness of the membrane (0.01

cm). (a) Estimate the diffusivities of the two compounds (room temperature of 20°C). (b) Calculate the total flux (mg cm^{-2} s^{-1}) through the membrane. (c) Humic and fulvic acids are now added to the water. These chemicals could also diffuse through the membrane; so we are interested in evaluating whether they could be excluded from the membrane based just on their size. Would you expect these chemicals to be larger or smaller in hydrodynamic radius than complex carbohydrates of similar molecular weight? Based on typical molecular weights of humic and fulvic acids, is there a chance that they will be rejected by this membrane given that the above two chemicals are diffusing through the membrane? You may wish to recall that the diffusivity of humic and fulvic acids is a function of their molecular weight according to D_{Cw} [cm^2 s^{-1}] = 1.42 × 10^{-4}$M_C^{-0.422}$.

3.14   The concentration of DOC in seawater is quite low, averaging 80 to 150 μM (0.96 to 1.8 mg/L as C) in surface waters (Hansell et al., 1993; Guo et al., 1994). Recirculating laminar-flow ultrafiltration membrane systems are commonly used to concentrate organic matter in such dilute suspensions into the retentate fraction and then the retentate is sampled. It is assumed that, by using a high concentration factor, the concentration of DOC in the retentate bigger than the apparent molecular weight (AMW) membrane cutoff, $c_{r,l}$(>AMW) is much larger than the concentration of the material smaller membrane cutoff, c_r(<AMW). This is the same as assuming that all material with an AMW bigger than the membrane is fully rejected, and that all DOC with an AMW smaller than membrane is not rejected at all (p_c = 1). Since a 1,000 amu (1K) membrane is usually used to perform this separation, it is likely that the assumption of complete permeation of material <1K is false. Researchers have concluded that about 50% of the DOC pool is <1,000 amu (<1K) using a recirculating ultrafiltration system. The purpose of this problem is to estimate how far off this estimate could be. (a) Assume that the a sample containing 1 mg/L of DOC contains 0.5 mg/L of material <1K. What concentrations of material would researchers conclude is <1K if they concentrated the sample by a factor of 50 (50 L reduced to 1 L of retentate) if the permeation coefficient p_c = 0.1? Repeat for p_c = 0.9. Is either one of these errors significant? (b) Let us assume that after a fifty-fold concentration of a sample, and that researchers measure 25 mg/L of DOC in the retentate in the sample originally containing 1 mg/l total DOC. Based on this result, what is the actual distribution of the DOC between the <1K and >1K fractions if p_c = 0.5?

3.15   The molecular diffusivities of tetrachloroethylene, and perchloroethylene (PCE) and 1,2,4,5-tetrachlorobenzene (TCB) in water at 20°C are reported to be 800 × 10^{-8} and 640 × 10^{-8} cm^2 s^{-1}. (a) Compare these diffusivities with those predicted by the Wilke–Chang correlation. (b) Calculate the molecular diameters of these chemicals based on the Stokes–Einstein equation and the reported diffusion coefficients. How well do these values compare with those reported by Ball and Roberts (1991) of 0.64 and 0.79 nm for PCE and TCB?

3.16 Another approach to measure molecular size distributions is repeatedly to ultrafilter a sample, rehydrating the retentate each time, in order to recover all the material less than the molecular weight cutoff of the membrane. Assume for this example that the permeation coefficient of the membrane is p_c = 0.65, and that no material adsorbs to the ultrafiltration membrane. (a) If you filter a 150-mL sample you know contains 10 mg L^{-1} of sample of material smaller than the membrane 1K cutoff, and 15 mg L^{-1} larger than the cutoff, what would be the concentration of material <1K in a 135-mL filtrate sample? (b) What is the total mass concentration remaining in the 15 mL of retentate in part (a)? (c) You take this remaining 15 mL of retentate, rehydrate it to 100 mL using ultrapure water (water containing a negligible amount of contaminants), and ultrafilter the sample down to a final retentate volume of 15 mL. What percent of the total material <1K will you recover from the two ultrafiltration separations?

REFERENCES

Amy, G. L., C. W. Bryant, and M. Belani. 1987. *Wat. Sci. Technol.* **19**:529–38.

Amy, G. L., M. R. Collins, C. J. Kuo, and P. H. King. 1987. *J. Amer. Water Works Assoc.* **79**:43–49.

Ball, B., and P. Roberts. 1991. *Environ. Sci. Technol.* **25**(7):1237–49.

Batterman, S., I. Padmanabham, and P. Milne. 1996. *Environ. Sci. Technol.* **30**(3):770–78.

Beckett, R., Z. Jue, and J. C. Giddings. 1987. *Environ. Sci. Technol.* **21**(3):289–95.

Bello, M. S., R. Rezzonico, and P. G. Righetti. 1994. *Science* **266**:773–76.

Bryant, C. W., G. L. Amy, and B. C. Alleman. 1990. *Environ. Technol.* **11**:249–62.

Collins, M. R., G. L. Amy, and P. H. King. 1985. *J. Environ. Eng.* **111**, 850–64.

Collins, M. R., G. L. Amy, and C. Steelink. 1986. *Environ. Sci. Technol.* **20**, 1028–32.

Currie, J. A. 1960. *Br. J. Appl. Phys.* **11**:318–22.

Cussler, E. L. 1984. *Diffusion: Mass Transfer in Fluid Systems*. Cambridge University Press, Cambridge, England.

El-Rehaili, A. M., and W. J. Weber, Jr. 1987. *Wat. Res.* **21**(5):573–82.

Fan, L.-S., R. Leyva-Ramos, K. D. Wisecarver, and B. J. Zehner. 1990. *Biotechnol. Bioeng.* **35**:279–86.

Frigon, R. P., J. K. Leypoldt, S. Uyeji, and L. W. Henderson. 1983. *Anal. Chem.* **55**:1349.

Fuller, E. N., P. D. Schettler, and J. C. Giddings. 1966. *Ind. Eng. Chem.* **58**(5):18.

Guo, L., C. H. Coleman, Jr., and P. H. Santschi. 1984. *Mar. Chem.* **45**:105–19.

Hansell, D. A., P. M. Williams, and B. B. Ward. 1993. *Deep Sea Res. I.* **40**(2):219–34.

Hayduk, W., and H. Laudie. 1974. *A.I.Ch.E.J.* **20**:611.

Hirshfelder, J. O., R. B. Bird, and E. L. Spotz. 1949. *Chem. Revs.* **44**:205–31.

Hirshfelder, J. O., C. F. Curtiss, and R. B. Bird. 1954. *Molecular Theory of Gases and Liquids*. John Wiley, New York.

Karger, J., and D. M. Ruthven. 1992. *Diffusion in Zeolites*. Wiley, NY.

Levine, A. D., G. Tchobanoglous, and T. Asano. 1985. *J. Water Pollut. Control Fed.* **57**: 805–16.

Logan, B. E., and Q. Jiang. 1990. *J. Environ. Engin.* **116**:1046–62.

Moore, W. J. 1972. *Physical Chemistry*, 4th ed. Prentice Hall, NJ. pp. 938–45.

Perry, R. H., and C. H. Chilton. 1973. *Chemical Engineers' Handbook*, 5th ed. McGraw-Hill, New York.

Polsen, A. 1950. *J. Phys. Colloid Chem.* **54**:649.

Siegrist, H., and W. Gujer. 1985. *Wat. Res.* **19**(11):1369–78.

Stryer, L. 1981. *Biochemistry*, 2nd ed. W. H. Freeman and Co., San Francisco, CA.

Sweerts, J-P. R. A., C. A. Kelly, J. W. M. Rudd, R. Hesslein, and T. E. Cappenberg. 1991. *Limnol. Oceanogr.* **36**(2):335–42.

Taylor, G. I. 1953. *Proc. R. Soc. London Ser A*. **219**:186–203.

Tchobanoglous, G., and F. L. Burton. 1991. *Wastewater Engineering: Treatment, Disposal, Reuse*, 3rd ed. McGraw-Hill, NY.

Thurman, E. M., R. L. Wershaw, and D. J. Pinckney. 1982. *Org. Geochem.* **4**:26–35.

Trogovcich, B., E. J. Kirsch, and C. P. L. Grady, Jr. 1983. *J. Water Pollut. Control Fed.* **55**: 966–76.

Tyn, M. T., and T. W. Gusek. 1990. *Biotechnol. Bioeng.* **35**:327–38.

Welty, J. R., C. E. Wicks, and R. E. Wilson. 1984. Fundamentals of Momentum, Heat, and Mass Transfer, 3rd ed. John Wiley, New York, NY.

Wheatcroft, R. A. 1991. *J. Marine Res.* **49**:565–88.

Wheatcroft, R. A. 1992. *Limnol. Oceanogr.* **37**(1):90–104.

Wilke, C. R., and P. Chang. 1955. *A.I.Ch.E.J.* **1**:264.

Williamson, K., and P. L. McCarty. 1976. *J. Water Pollut. Control Fed.* **48**:9–24.

Zhang, T. C., and P. L. Bishop. 1994. *Wat. Res.* **28**(11):2279–87.

CHAPTER 4

THE CONSTITUTIVE TRANSPORT EQUATION

4.1 INTRODUCTION

Development of a mathematical description of chemical transport in a system often begins by writing down the generalized transport equations for all potential transport mechanisms in each phase. If there are a number of different phases (air, water, etc.) or a number of different chemicals, several equations may be necessary to describe the system completely. These equations may then be reduced to simpler (but often times less descriptive) forms. Sometimes these equations are simplified just to make their solution easier (or possible!), but in other cases it is necessary to simplify the equations to reduce the number of coefficients needed to satisfy the equations. With practice, the procedure of selecting important terms in the transport equation can become automatic, leaving only the process of selecting appropriate initial and boundary conditions and solving the equations either analytically or numerically. Even this last step of solving equations is now easier since solutions exist for a large number of different types of equations routinely encountered in mass transport problems. While the choice of the governing equation may become automatic, however, it is still essential that this choice be valid for the system being studied.

There is a similarity of the governing equations describing chemical, momentum, and heat transport that makes solutions of equations transferrable from one medium to the other. Once proficiency is developed in one of these areas, such as mass transport, the mathematical techniques used to solve mass transport equations can be easily applied to study problems in the other two disciplines. The translation of differential equations and solutions from heat to chemical transport is common since heat transport experiments can more easily be conducted than mass transport experiments due to the simplicity of measuring temperature versus chemical concentrations. The outcome of an experimental investigation is usually a mass transport coefficient correlation specific to the specific experiment. In order to translate the results from one medium to the other, similar physical conditions must be obtained. As we shall see, dimensionless scaling parameters can be used to evaluate the similarity of conditions.

The purpose of this chapter is to demonstrate the similarity of the three transport equations describing mass, momentum, and heat transport. Since the main purpose of this text is to describe chemical mass transport, there will be limited development of the other types of transport equations. The main areas of emphasis will be on how to eliminate terms in the transport equation in order to describe a system, and the different forms of the constitutive transport equation when fixed or moving coordinate systems are used.

4.2   DERIVATION OF THE GENERAL TRANSPORT EQUATION

Let ξ be an arbitrary, but continuous, function describing some property of a fluid. This function can be a scalar quantity, such as mass or temperature, a vector such as momentum, or a tensor such as shear stress. In order to describe the properties of this function in a system, we will set up an equation describing the net rate of change of ξ in the control volume shown in Figure 4.1. All surface forces acting on the control volume will be described by the function ω, while all internal forces will be categorized by a function σ. Surface forces will describe those properties that can be translated in a direction other than the direction of motion, such as fluid shear or

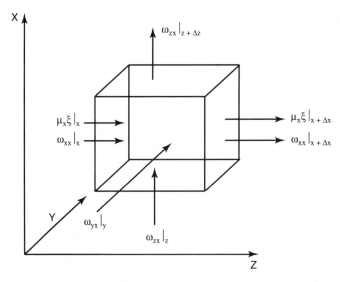

Figure 4.1   Differential control volume used to set up transport equations for describing the transport of ξ.

diffusion. Internal forces include body forces such as gravity, electromagnetic forces, reactions, and other sinks and sources affecting ξ within the control volume.

To derive the general transport equation, we start with the same equation used for the mass balance, or

$$\text{accumulation} = \text{in} - \text{out} + \text{(sinks and/or sources)} \qquad (4\text{-}1)$$

The accumulation of ξ within the control volume is the rate of change of ξ times the differential volume, or

$$\text{accumulation} = \frac{d\xi}{dt} (\Delta x \, \Delta y \, \Delta z) \qquad (4\text{-}2)$$

The rate that mass enters due to advective flow u in the x, y, and z directions is

$$\text{rate in} = u_x \xi|_x (\Delta y \, \Delta z) + u_y \xi|_y (\Delta x \, \Delta z) + u_z \xi|_z (\Delta x \, \Delta y) \qquad (4\text{-}3)$$

and the rate out is

$$\text{rate out} = u_x \xi|_{x+\Delta x} (\Delta y \, \Delta z) + u_y \xi|_{y+\Delta y} (\Delta x \, \Delta z) + u_z \xi|_{z+\Delta z} (\Delta x \, \Delta y) \qquad (4\text{-}4)$$

Surface forces can be composed of nine components, each acting across the face of one of surfaces of the control volume shown in Figure 4.1. These nine components entering the cube are:

$$\text{surface forces in} = (\omega_{xx}|_x + \omega_{yx}|_x + \omega_{zx}|_x)(\Delta y \, \Delta z)$$
$$+ (\omega_{xy}|_y + \omega_{yy}|_y + \omega_{zy}|_y)(\Delta x \, \Delta z)$$
$$+ (\omega_{xz}|_z + \omega_{yz}|_z + \omega_{zz}|_z)(\Delta x \, \Delta y) \qquad (4\text{-}5)$$

with a similar set of forces leaving at $x + \Delta x$, $y + \Delta y$, and $z + \Delta z$.

Internal forces act over the whole volume, and can be summarized as

$$\text{internal forces} = (\sigma_x + \sigma_y + \sigma_z)(\Delta x \, \Delta y \, \Delta z) \qquad (4\text{-}6)$$

Summing up all forces, dividing by $\Delta x \, \Delta y \, \Delta z$, and taking the limit as $\Delta x \, \Delta y \, \Delta z$ goes to zero produces the final general transport equation (GTE)

$$\frac{\partial \xi}{\partial t} = (-\nabla \cdot \xi \mathbf{u}) - \nabla \cdot \vec{\Omega} + \Sigma \qquad (4\text{-}7)$$

where the four terms, from left to right, describe accumulation, convection, dispersion, and net production (or loss) within the system. Fluid velocity is shown in bold, indicating that it is a vector quantity, and Ω has an arrow over it indicating that it is a second-order tensor.

4.3   SPECIAL FORMS OF THE GENERAL TRANSPORT EQUATION

The GTE can be reduced to more useful transport equations by defining the function ξ. Once the choice of ξ is made, the appropriate choices for the surface and body forces will transform the GTE into the specific transport equations such as the continuity, Navier–Stokes, and constitutive transport equations. A similar equation can be developed for heat transport, but since heat transport is considered in any detail in this text, it will not be developed here.

The Equation of Continuity

In the first example, we define $\xi = \rho$, the mass concentration (density) of an incompressible fluid. The advection term $\mathbf{u}\rho$ is the mass flux into the control volume. If we additionally consider the case for which $\omega = \Sigma = 0$, or when we have no reaction and constant surface forces, we obtain from Eq. 4-7

$$\frac{\partial \rho}{\partial t} + \nabla \cdot \rho \mathbf{u} = 0 \tag{4-8}$$

which applies to unsteady, three-dimensional flow. Expanding the derivitive term produces

$$\frac{\partial \rho}{\partial t} + \left(\frac{\partial \rho u_x}{\partial x} + \frac{\partial \rho u_y}{\partial y} + \frac{\partial \rho u_z}{\partial z} \right) = 0 \tag{4-9}$$

If the fluid is incompressible, $\partial \rho / \partial t = \partial \rho / \partial x = \partial \rho / \partial y = \partial \rho / \partial z = 0$, resulting in

$$\left(\frac{\partial u_x}{\partial x} + \frac{\partial u_y}{\partial y} + \frac{\partial u_z}{\partial z} \right) = 0 \tag{4-10}$$

or more simply

$$\nabla \cdot \mathbf{u} = 0 \tag{4-11}$$

Equations 4-8 through 4-11 are different forms of the continuity equation. When the flow is incompressible, Eq. 4-11 applies whether the flow is unsteady or not.

Navier–Stokes Equation

If we define $\xi = \rho \mathbf{u}$, the GTE will describe momentum transport. The rate of momentum transport into the control volume is

$$u_x(\rho u_x)|_x(\Delta y\ \Delta z) + u_y(\rho u_y)|_y(\Delta x\ \Delta z) + u_z(\rho u_z)|_z(\Delta x\ \Delta y) \tag{4-12}$$

For momentum transport, energy is transferred by fluid viscosity. Therefore, the shear stress, τ, becomes the dispersive force, and we can write Eq. 4-5 describing the shear forces entering the three faces of the control volume as

$$(\tau_{xx}|_x + \tau_{xy}|_x + \tau_{xz}|_x)(\Delta y\ \Delta z)$$

$$+ (\tau_{yx}|_y + \tau_{yy}|_y + \tau_{yz}|_y)(\Delta x\ \Delta z)$$

$$+ (\tau_{zx}|_z + \tau_{zy}|_z + \tau_{zz}|_z)(\Delta x\ \Delta y) \qquad (4\text{-}13)$$

Similar equations can be written for momentum and shear forces leaving the control volume. The forces acting on the control volume surface for momentum transport are pressure forces, or

$$(p|_x - p|_{x+\Delta x})(\Delta y \Delta z) + (p|_y - p|_{y+\Delta y})(\Delta x \Delta z) + (p|_z - p|_{z+\Delta z})(\Delta x \Delta y) \qquad (4\text{-}14)$$

and the internal forces are gravity forces, or

$$(\rho g_x + \rho g_y + \rho g_z)(\Delta x\ \Delta y\ \Delta z) \qquad (4\text{-}15)$$

Summing all terms for forces entering and leaving the control volume, taking the limit as $\Delta x\ \Delta y\ \Delta z$ goes to zero, the rate of accumulation of momentum is

$$\frac{\partial(\rho\mathbf{u})}{\partial t} = [-\nabla\cdot(\rho\mathbf{u})\mathbf{u}] - \nabla\cdot\vec{\tau} - \Delta\mathbf{p} + \rho\mathbf{g} \qquad (4\text{-}16)$$

where $\vec{\tau}$ is a second-order tensor. Expanding the first term, and assuming constant ρ and μ (incompressible flow), Eq. 4-16 becomes

$$\rho\,\frac{D\mathbf{u}}{Dt} = \mu\ \nabla^2\mathbf{u} - \nabla\mathbf{p} + \rho\mathbf{g} \qquad (4\text{-}17)$$

where $D\mathbf{u}/Dt$ is the substantial derivative, defined as

$$\frac{D}{Dt} = \frac{\partial}{\partial t} + u_x\frac{\partial}{\partial x} + u_y\frac{\partial}{\partial y} + u_z\frac{\partial}{\partial z} \qquad (4\text{-}18)$$

Equations 4-16 and 4-17 are known as the Navier–Stokes equations.
    If the flow is inviscid ($\mu = 0$), Eq. 4-17 can be simplified to

$$\rho\,\frac{D\mathbf{u}}{Dt} = -\nabla p + \rho\mathbf{g} \qquad (4\text{-}19)$$

which is called Euler's equation.

Constitutive Transport Equation

A general equation describing the mass transport of a chemical C present at a concentration c_C can be obtained by substituting $\xi = c_C$ into the GTE. Following the same steps as above, the accumulation of c_C within the control volume is

$$\frac{\partial c_C}{\partial t}(\Delta x \, \Delta y \, \Delta z) \tag{4-20}$$

and the net rate of advective transport of c_C into the control volume is

$$(u_x c_C|_{x+\Delta x} - u_x c_C|_x)(\Delta y \, \Delta z) + (u_y c_C|_{y+\Delta y} - u_y c_C|_y)(\Delta x \, \Delta z)$$
$$+ (u_z c_C|_{z+\Delta z} - u_z c_C|_z)(\Delta x \, \Delta y) \tag{4-21}$$

As was discussed in Chapter 3, chemical dispersion is calculated from Fick's Law as

$$\mathbf{j}_C = -D_C \left(\frac{\partial c_C}{\partial x} + \frac{\partial c_C}{\partial y} + \frac{\partial c_C}{\partial z} \right) \tag{4-22}$$

or more simply

$$\mathbf{j}_C = -D_C \, \nabla c_C \tag{4-23}$$

The surface force ω in the GTE (Eq. 4-5) becomes the suface flux, $\mathbf{j}_C$. Summing up surface forces entering and leaving the control volume produces

$$(j_C|_{x+\Delta x} - j_C|_x)(\Delta y \, \Delta z) + (j_C|_{y+\Delta y} - j_C|_y)(\Delta x \, \Delta z) + (j_C|_{z+\Delta z} - j_C|_z)(\Delta x \, \Delta y) \tag{4-24}$$

Internal forces will be consolidated into a reaction rate constant, R_C, acting within the control volume, or

$$R_C(\Delta x \, \Delta y \, \Delta z) \tag{4-25}$$

Summing up all terms, and taking the limit as $\Delta x \, \Delta y \, \Delta z$ goes to zero, the transport equation for chemical C is

$$\frac{\partial c_C}{\partial t} = (-\nabla \cdot c_C \mathbf{u}) - \nabla \cdot \mathbf{j}_C + R_C \tag{4-26}$$

Substituting Eq. 4-23 into Eq. 4-26, and simplifying the advection term

$$\frac{\partial c_C}{\partial t} = -(\mathbf{u} \cdot \nabla c_C) - \nabla \cdot (-D_C \, \nabla c_C) + R_C \tag{4-27}$$

For a constant diffusion coefficient, this can be further simplified to

$$\frac{\partial c_C}{\partial t} + \mathbf{u} \cdot \nabla c_C - D_C \nabla^2 c_C - R_C = 0 \tag{4-28}$$

There the four terms, from left to right, represent accumulation, advective transport, diffusive transport, and reaction. Equations 4-26 to 4-28 are known collectively as the constitutive transport equations.

4.4 SIMILARITY OF MASS, MOMENTUM, AND HEAT DISPERSION LAWS

Before continuing with the development of the mass transport equations, it is useful to focus again on the similarity of the different transport equations. Each of the dispersion equations used to describe mass, momentum, and heat transport has been designated as a separate law named after the pioneers in the separate transport fields. While such separation honors the work of these scientists, separation of the laws detracts from the main point that all three of these laws are really the same equation with different proportionality constants specific to the transport functions describing mass, momentum, or heat. This similarity can be appreciated by examining the one-dimensional forms of each of these laws for the magnitude of the proportionality constant for transport in air.

The Three Laws

Fick's First Law describes diffusive mass transport. In one dimension, with constant diffusivity, chemical mass flux is calculated as

$$j_{Ca,x} = -D_{Ca} \frac{dc_{Ca}}{dx} \tag{4-29}$$

where the flux $j_{Ca,x}$ has units of $[\text{M L}^{-2} \text{ T}^{-1}]$. For a typical chemical in air, the molecular diffusivity is

$$D_{Ca} \approx 0.2 \text{ cm}^2 \text{ s}^{-1} \tag{4-30}$$

where the diffusion coefficient has units of $[\text{L}^2 \text{ T}^{-1}]$.

Newton's Law describes the dissipation of momentum through fluid viscosity. The transport of momentum in the y direction due to fluid motion in the x direction is calculated from the fluid shear

$$\tau_{yx} = -\mu \frac{du_x}{dy} \tag{4-31}$$

where τ_{yx} has units of $[M\ L^{-1}\ T^{-2}]$. In order to express the proportionality constant in units similar to diffusion, we multiply both sides of Eq. 4-31 by ρ_a/ρ_a, to obtain

$$\tau_{yx} = -\frac{\mu_a}{\rho_a}\frac{d\rho_a u_x}{dy} \tag{4-32}$$

Since the kinematic viscosity, $\nu_a = \mu_a/\rho_a$, we can equivalently write

$$\tau_{yx} = -\nu_a\frac{d\rho_a u_x}{dy} \tag{4-33}$$

For air, the fluid kinematic viscosity at 20°C is

$$\nu_a \approx 0.2\ \text{cm}^2\ \text{s}^{-1} \tag{4-34}$$

Fourier's Law describes heat conduction. In one dimension, x, it is

$$q_{a,x} = -k_{H,a}\frac{dT}{dx} \tag{4-35}$$

where q is the heat flux, and the thermal heat conductivity, $k_{H,a}$, has units of $[\text{cal}\ L^{-2}\ T^{-1}]$. In order to express this proportionality constant in the same units used for the other proportionality constants, we multiply both sides of Eq. 4-35 by $c_{a,p}\rho_a/c_{a,p}\rho_a$, where $c_{a,p}$ is the heat capacity of air $[\text{cal}\ M^{-1}\ T^{-1}]$ to obtain

$$q_{a,x} = \frac{-k_{H,a}}{c_{a,p}\rho_a}\frac{d(Tc_{a,p}\rho_a)}{dx} \tag{4-36}$$

Since the thermal diffusivity, $\alpha_a = k_{H,a}/c_{a,p}\rho_a$, we can write

$$q_{a,x} = -\alpha_a\frac{d(Tc_{a,p}\rho_a)}{dx} \tag{4-37}$$

For air, the thermal diffusivity at 20°C is:

$$\alpha_a \approx 0.2\ \text{cm}^2\ \text{s}^{-1} \tag{4-38}$$

From this analysis, we can see that for transport in air the equations describing dispersion are all of similar magnitude, or $D_{Ca} \approx \nu_a \approx \alpha_a$ (Table 4.1).

Dimensionless Numbers

The similarity of transport equations makes it possible to translate transport studies from one medium to another. Heat transport correlations are frequently used for mass

TABLE 4.1 Dispersion Equations and Constants for Momentum, Heat, and Constitutive Transport Equations

System	Transport Equation	Diffusion Coefficient	Magnitude of Diffusion Coefficient in Indicated Phase ($cm^2\ s^{-1}$)	
			Air	Water
Momentum	$\tau_{yx} = -v_j \dfrac{d\rho u_x}{dy}$	v_j	0.20	0.01
Heat	$q_{j,x} = -\alpha_j \dfrac{d(Tc_{j,p}\rho_j)}{dx}$	α_j	0.20	1.5×10^{-3}
Chemical	$j_{ij,x} = -D_{ij} \dfrac{dc_{ij}}{dx}$	D_{ij}	0.20	10^{-5}

transport studies due to the relative ease and accuracy of measuring temperature and heat content, compared to chemical concentration. In some systems, there is more than one type of transport occurring, for example, a chemical dissolving into moving water in a pipe or heat conduction into a gas stream.

The rate of momentum diffusivity to thermal diffusivity is characterized by the Prandtl number, Pr, defined as

$$\text{Pr} = \frac{\text{momentum diffusivity}}{\text{thermal diffusivity}} = \frac{v}{\alpha} \tag{4-39}$$

A similar scaling ratio for chemical mass transfer is the Schmidt number, Sc, defined as

$$\text{Sc} = \frac{\text{momentum diffusity}}{\text{chemical diffusivity}} = \frac{v}{D} \tag{4-40}$$

For transport equations involving air, $\text{Pr} \approx \text{Sc} \approx 1$. For water, the situation is more complicated since these numbers are not unity. Typical momentum, mass, and thermal diffusivities are

$$v_w = 0.01\ cm^2\ s^{-1}$$

$$D_{Cw} = 10^{-5}\ cm^2\ s^{-1}$$

$$\alpha_w = 1.5 \times 10^{-3}\ cm^2\ s^{-1} \tag{4-41}$$

For aqueous sytems, $\text{Pr} \approx 10^2$ and $\text{Sc} \approx 10^3$. The fact that these dimensionless numbers are not unity indicates that mass transport correlations developed for different phases will need to be scaled, often in a nonlinear manner, to account for differences due to fluid properties.

4.5 TRANSPORT RELATIVE TO A FIXED OR MOVING COORDINATE SYSTEM

The chemical flux is the mass (or moles) of a chemical that passes through a unit area per unit time. Flux is a vector since the chemical flux cannot be calculated without knowing the direction of transport. Chemical flux can be calculated relative to a moving coordinate system or relative to a fixed point and can include transport due to both diffusion and advection. As we shall see, dilute solutions offer some simplifications to the general transport equations. Describing the dispersion of several different chemicals at dilute concentrations in water reduces to the simple case of a binary solution for each chemical and water.

General Transport Equation Forms

The flux due to only diffusion is given by Ficks first law, or

$$j_{Cj,x} = -D_{Cj} \frac{dc_{Cj}}{dx} \tag{4-42}$$

This flux describes chemical transport in a system without advective motion or a system in which the observer is moving with the fluid. As an example of an observer moving with the fluid, consider a person sitting in a boat moving down the river at the mean river flow rate. The chemical would move away from the observer due to fluid dispersion while the observer moved at the mean fluid velocity. In order to examine the flux of chemical c in a multicomponent mixture that may not be at rest from a fixed reference point, for example, on the bank of the river in our example, the bulk motion of the system must also be considered and incorporated into the governing transport equation.

For a multicomponent system, the molar-average velocity of the phase (or the phase velocity), $\mathbf{u}_j$, is calculated in terms of the velocity of the chemical relative to all other components of the system as

$$\mathbf{u}_j = \frac{\sum\limits_{i=A}^{N} c_{ij}\mathbf{u}_{ij}}{\sum\limits_{i=A}^{N} c_{ij}} = \frac{\sum\limits_{i=A}^{N} c_{ij}\mathbf{u}_{ij}}{c_j} \tag{4-43}$$

where $\mathbf{u}_{ij}$ is the absolute velocity of species i in phase j relative to stationary coordinates. The difference between $\mathbf{u}_{ij}$ and $\mathbf{u}_j$ is the diffusion velocity of species i. In order to simplify notation, let us define the phase $j = w$ (water) in the following discussion, recognizing that the equations apply for air as well. For chemicals in water, Eq. 4-43 becomes

$$\mathbf{u}_w = \frac{\sum_{i=A}^{N} c_{iw}\mathbf{u}_{iw}}{\sum_{i=A}^{N} c_{iw}} = \frac{\sum_{i=A}^{N} c_{iw}\mathbf{u}_{iw}}{c_w} \tag{4-44}$$

For a binary solution consisting of only chemical C in water, the molar flux of C in the x direction can be written as the difference between the velocities of the chemical and the whole phase ($w = C + W$) as

$$j_{Cw,x} = c_{Cw}(u_{Cw,x} - u_{w,x}) \tag{4-45}$$

Since Eqs. 4-42 and 4-45 both describe the same molar flux, we can write

$$j_{Cw,x} = -D_{Cw}\frac{dc_{Cw}}{dx} = c_{Cw}(u_{Cw,x} - u_{w,x}) \tag{4-46}$$

which can be rearranged to produce

$$c_{Cw}u_{Cw,x} = -D_{cw}\frac{dc_{cw}}{dx} + c_{cw}u_{w,x} \tag{4-47}$$

Since this system is a binary solution, $u_{w,x}$ can be simplified using Eq. 4-44 as

$$u_{w,x} = \frac{1}{c_w}(c_{Cw}u_{Cw,x} + c_{Ww}u_{Ww,x}) \tag{4-48}$$

Combining Eqs. 4-46 and 4-48 produces

$$c_{Cw}u_{Cw,x} = -D_{Cw}\frac{dc_{Cw}}{dx} + x_C(c_{Cw}u_{Cw,x} + c_{Ww}u_{Ww,x}) \tag{4-49}$$

Since the velocities $u_{Cw,x}$ and $u_{Ww,x}$ are velocities relative to stationary coordinates, then the products of velocity and concentration are fluxes relative to fixed coordinates. These molar fluxes are designated by the symbol $\mathbf{J}$, which defines the equality

$$\mathbf{J}_{Cj} = c_{Cj}\mathbf{u}_{Cj} \tag{4-50}$$

Using this relationship for both chemical C and water, we can write Eq. 4-49 more simply as

$$J_{Cw,x} = -D_{Cw}\frac{dc_{Cw}}{dx} + x_C(J_{Cw,x} + J_{Ww,x}) \tag{4-51}$$

Simplifications for Dilute Solutions

The equations derived above for nondilute solutions are somewhat complex since the diffusion of a chemical C may affect the bulk properties of the flow. Dilute solutions are easier to describe mathematically because the motion of dilute chemicals will not affect the bulk flow. For a dilute solution of chemical C in water, Eq. 4-43 becomes

$$\mathbf{u}_w = \frac{\displaystyle\sum_{i=A}^{N} c_{iw} u_{iw}}{\displaystyle\sum_{i=A}^{N} c_{iw}} = \frac{c_{Cw} u_{Cw} + c_{Ww} u_{Ww}}{c_w} \tag{4-52}$$

For a dilute solution $c_w \gg c_{Cw}$ and $u_{Ww} \gg u_{Cw}$, so that $c_{Ww} u_{Ww} \gg c_{Cw} u_{Cw}$, and in addition, $c_{Ww} = c_w$ and $u_{Ww} = u_w$. Therefore, Eq. 4-52 can be simplified to

$$\mathbf{u}_w = \frac{c_{Ww} u_{Ww}}{c_w} = \frac{c_w u_w}{c_w} \tag{4-53}$$

It follows from the definition of $\mathbf{J}$ for a dilute solution that $\mathbf{J}_{Ww} \gg \mathbf{J}_{Cw}$, and that $\mathbf{J}_{Ww} = \mathbf{J}_w$. Thus we can write Eq. 4-51 for a dilute solution of C in phase w as

$$J_{Cw,x} = -D_{Cw} \frac{dc_{Cw}}{dx} + x_C J_{w,x} \tag{4-54}$$

The fluid velocity can be explicitly included in Eq. 4-54 from the definition of $\mathbf{J}_w = c_w \mathbf{u}_w$, becoming

$$J_{Cw,x} = -D_{Cw} \frac{dc_{Cw}}{dx} + c_{Cw} u_{w,x} \tag{4-55}$$

In three dimensions for any species i in phase j, Eq. 4-55 can equivalently be written in the form

$$\mathbf{J}_{ij} = -D_{ij} \nabla c_{ij} + c_{ij} \mathbf{u}_j \tag{4-56}$$

For single-phase transport, the double subscript notation is not necessary, and the chemical transport of C in three dimensions could be written as

$$\mathbf{J}_C = -D_C \nabla c_C + c_C \mathbf{u} \tag{4-57}$$

These expressions for the flux relative to moving and fixed coordinates for mass concentrations can be written as a function of either c or ρ, as summarized in Table 4.2.

TABLE 4.2   Flux Equations Written in Terms of a Moving and Fixed Coordinate System, as a Function of Whether the System Is Concentrated (General Case with Variable c_j and ρ_j) or Dilute (Constant c_j and ρ_j)

Conditions	Basis	Coordinate System Basis	
		Moving	Fixed
General conditions variable ρ_j, c_j	Moles	$j_{ij} = -c_j D_{ij} \nabla x_i$	$J_{ij} = -c_j D_{ij} \nabla x_i + x_i(J_{ij} + J_j)$
	Mass	$j_{ij} = -\rho_j D_{ij} \nabla \phi_i$	$J_{ij} = -\rho_j D_{ij} \nabla \phi_i + \phi_i(J_{ij} + J_j)$
Dilute solutions constant ρ_j, c_j	Moles	$j_{ij} = -D_{ij} \nabla c_{ij}$	$J_{ij} = -D_{ij} \nabla c_{ij} + x_i J_j$
			$J_{ij} = -D_{ij} \nabla c_{ij} + c_{ij} \mathbf{u}_j$
	Mass	$j_{ij} = -D_{ij} \nabla \rho_{ij}$	$J_{ij} = -D_{ij} \nabla \rho_{ij} + \phi_i J_j$
			$J_{ij} = -D_{ij} \nabla \rho_{ij} + \rho_{ij} \mathbf{u}_j$

4.6   SPECIAL FORMS OF THE CONSTITUTIVE TRANSPORT EQUATION

The mathematical description of a transport process can begin at many different starting points, but eventually an equation is derived that is specific to the system being modeled. However, even the same transport equation can take on different forms. For example, we have already shown that the constitutive transport equation can be written as

$$\frac{\partial c_{Cj}}{\partial t} + \mathbf{u}_j \cdot \nabla c_{Cj} - D_{Cj} \nabla^2 c_{Cj} - R_{Cj} = 0 \tag{4-58}$$

This equation can also be written in terms of the molar flux, $\mathbf{J}_C$ as

$$\frac{\partial c_{Cj}}{\partial t} + \nabla \mathbf{J}_{Cj} - R_{Cj} = 0 \tag{4-59}$$

In order to see that Eqs. 4-58 and 4-59 describe the same process, we first write out the definition of $\mathbf{J}_C$ using Eq. 4-54 as

$$\mathbf{J}_{Cj} = -D_{Cj} \nabla c_{Cj} + x_C \mathbf{J}_j \tag{4-60}$$

and comparing this with Eq. 4-58, we can see for a chemical C in water that $x_C \mathbf{J}_w = c_{Cw} \mathbf{u}_w$, or

$$\mathbf{J}_{Cj} = -D_{Cj} \nabla c_{Cj} + c_{Cj} \mathbf{u}_j \tag{4-61}$$

Substituting Eq. 4-61 into Eq. 4-59,

$$\frac{\partial c_{Cj}}{\partial t} - \nabla \cdot (D_{Cj} \, \nabla c_{Cj}) + \nabla \cdot (c_{Cj} \mathbf{u}_j) - R_{Cj} = 0 \qquad (4\text{-}62)$$

which can be further simplified for a constant diffusion coefficient to

$$\frac{\partial c_{Cj}}{\partial t} - D_{Cj} \, \nabla^2 c_{Cj} + \nabla \cdot (c_{Cj} \mathbf{u}_j) - R_{Cj} = 0 \qquad (4\text{-}63)$$

Expanding the third term containing the bulk velocity, $\mathbf{u}_j$,

$$\nabla \cdot (c_{Aj} \mathbf{u}_j) = c_{Aj} \, \nabla \cdot \mathbf{u}_j + \mathbf{u}_j \cdot \nabla c_{Aj} \qquad (4\text{-}64)$$

but based on fluid continuity, $\nabla \cdot \mathbf{u}_j = 0$. Combining Eqs. 4-63 and 4-64, we have the result that

$$\frac{\partial c_{Cj}}{\partial t} = \mathbf{u}_j \cdot \nabla c_{Cj} - D_{Cj} \, \nabla^2 c_{Cj} + R_{Cj} \qquad (4\text{-}65)$$

For dilute systems, the constitutive equation in the form shown in Eq. 4-65 is more helpful than the form that specifies the flux in terms of $\mathbf{J}_{ij}$ (Eq. 4-59) since the diffusion and advection components are explicitly identified. Mathematically, however, these equations are equivalent, and the use of one equation or the other is a matter of personal preference.

When the constitutive transport equation is used to describe less complex systems, terms in the equation can be simplified or completely omitted. In the following, this equation is increasingly simplified by a sequence of assumptions.

Transport in Only One Phase   Defining transport only in water (phase w), Eq. 4-65 becomes

$$\frac{\partial c_{Cw}}{\partial t} + \mathbf{u}_w \cdot \nabla c_{Cw} - D_{Cw} \, \nabla^2 c_{Cw} - R_{Cw} = 0 \qquad (4\text{-}66)$$

For just one phase, we can drop the notation for water, and write more simply

$$\frac{\partial c_C}{\partial t} + \mathbf{u} \cdot \nabla c_C - D_C \, \nabla^2 c_C - R_C = 0 \qquad (4\text{-}67)$$

Because there is only one chemical, the subscript C could also be dropped, but we continue to use it here.

Chemical Transport with No Reaction   If chemical transport is conservative, that is, if the chemical does not degrade during transport by abiotic or biotic processes, the reaction term can be omitted to produce

$$\frac{\partial c_C}{\partial t} + \mathbf{u} \cdot \nabla c_C - D_C \, \nabla^2 c_C = 0 \tag{4-68}$$

This can be written more simply in terms of the substantial derivative as

$$\frac{D c_C}{D t} = D_C \, \nabla^2 c_C \tag{4-69}$$

No Advection   If there is no bulk transport of fluid, then the advection term can also be eliminated to produce

$$\frac{\partial c_C}{\partial t} = D_C \, \nabla^2 c_C \tag{4-70}$$

Steady State   Finally, if the system is considered to be at steady state, only diffusion is left in the general equation, and the diffusion constant drops out, leaving

$$\nabla^2 c_C = 0 \tag{4-71}$$

or from the definition of ∇^2

$$\frac{\partial^2 c_C}{\partial x^2} + \frac{\partial^2 c_C}{\partial y^2} + \frac{\partial^2 c_C}{\partial z^2} = 0 \tag{4-72}$$

One Dimension   When the constitutive transport equation is written only for one chemical in a single phase (water) with no advection, no reaction, and only in one dimension, it is reduced to

$$\frac{\partial^2 c_C}{\partial x^2} = 0 \tag{4-73}$$

The first integration of Eq. 4-73 produces the well-known result that under these assumptions the flux is constant (i.e., Fick's First Law).

4.7   THE CONSTITUTIVE TRANSPORT EQUATION IN CYLINDRICAL AND SPHERICAL COORDINATES

In the Section 4.2, we conducted a mass balance around a control volume that was a cube. The same process can be repeated about other shapes. Defining the control volume to be a sphere, the flux into the control volume can now vary as a function of radius, r, and two angles, θ and ϕ (analogous to degrees of latitude and longitude on a globe). When all terms are included, the full constitutive transport equation in radial coordinates is

$$\frac{\partial c_C}{\partial t} + \left(u_r \frac{\partial c_C}{\partial r} + \frac{u_\theta}{r} \frac{\partial c_C}{\partial \theta} + \frac{u_\phi}{r \sin \theta} \frac{\partial c_C}{\partial \phi} \right)$$

$$= D_C \left(\frac{1}{r^2} \frac{\partial}{\partial r} \left(r^2 \frac{\partial c_C}{\partial r} \right) + \frac{1}{r^2 \sin \theta} \frac{\partial}{\partial \theta} \left(\sin \theta \frac{\partial c_C}{\partial \theta} \right) \right.$$

$$\left. + \frac{1}{r^2 \sin^2 \theta} \frac{\partial^2 c_C}{\partial \phi^2} \right) + R_C \tag{4-74}$$

For one-dimensional nonsteady transport, or when c_C is a function only of concentration in the radial direction, this can be simplified to

$$\frac{\partial c_C}{\partial t} + u_r \frac{\partial c_C}{\partial r} - D_C \frac{1}{r^2} \frac{\partial}{\partial r} \left(r^2 \frac{\partial c_C}{\partial r} \right) - R_C = 0 \tag{4-75}$$

and for the case where we can neglect advective transport,

$$\frac{\partial c_C}{\partial t} - D_C \frac{1}{r^2} \frac{\partial}{\partial r} \left(r^2 \frac{\partial c_C}{\partial r} \right) - R_C = 0 \tag{4-76}$$

When the differential control volume is a cylinder (see Problem 4.9), transport is a function of radial distance r, axial distance z, and angle θ, and the constitutive transport equation becomes

$$\frac{\partial c_C}{\partial t} + u_r \frac{\partial c_C}{\partial r} + \frac{u_\theta}{r} \frac{\partial c_C}{\partial \theta} + u_z \frac{\partial c_C}{\partial z}$$

$$= D_C \left(\frac{1}{r} \frac{\partial}{\partial r} \left(r \frac{\partial c_C}{\partial r} \right) + \frac{1}{r^2} \frac{\partial^2 c_c}{\partial \theta^2} + \frac{\partial^2 c_C}{\partial z^2} \right) + R_C \tag{4-77}$$

For the more common case of radial symmetry, and flow only in the axial direction (as in a pipe), this can be simplified to

$$\frac{\partial c_C}{\partial t} + u_z \frac{\partial c_C}{\partial z} - \frac{D_C}{r} \frac{\partial}{\partial r} \left(r \frac{\partial c_C}{\partial r} \right) - R_C = 0 \tag{4-78}$$

Other terms can be eliminated, in the same manner as done above for the constitutive transport equation in Cartesian coordinates, as necessary.

PROBLEMS

4.1 Macromolecules diffuse from a mixed fluid onto a flat surface evenly coated by micoorganisms (Figure P-4.1). The molecules are cleaved by hydrolysis to form two byproducts according to the reaction $M \rightarrow 2B$. (a) Starting with the constitutive transport equation 4-59 in terms of $J_{M,z}$, convert the equation to a form that explicitly shows a velocity term and then reduce terms until you obtain the simplest equation describing the steady-state flux of M to the surface. (b) If the flux is not a dilute system, write out the equation for flux in a fixed coordinate system. (c) Reduce this equation to describe transport if M is a dilute system.

4.2 A nonreactive chemical is spilled into a river. Starting with Eq. 4-59 written in terms of mass concentration in the river, c_{Cw}, derive an expression for one-dimensional nonsteady transport of the chemical in the river in terms of the chemical diffusivity and bulk river velocity.

4.3 A microbe on the submerged surface of a large ship in the water takes up dissolved amino acids at a rate proportional to their concentration in the liquid (a first-order reaction). Assuming only diffusion to the cell surface, derive an equation describing amino acid transport in the liquid. In this calculation, assume the cell forms a hemispherical shape.

4.4 Starting with Eq. 4-59, write the general transport equation in terms of mass concentration units, ρ_C. Then show how this equation would be written with additive simplifications of dilute concentration system, transport in one phase, no reaction, no advection, steady conditions, and one-dimensional transport.

4.5 The wastewater treatment lagoon shown in Figure P-4.5 is slowly accumulating compound C entering in the influent. (a) Write (but do not solve) a differential equation for the accumulation of C in the pond fluid, assuming the system is completely mixed, evaporation can be described in terms of a mass transfer coefficient, and the pond is leaking water and chemical at a rate

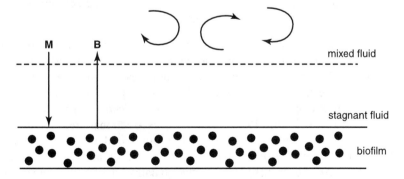

Figure P-4.1

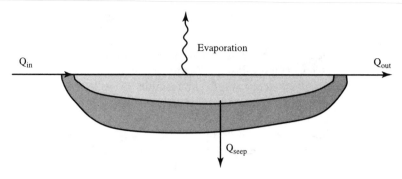

Figure P-4.5

proportional to bc_C, where b is a dimensionless coefficient that varies between 0 and 1, and c_C is the bulk concentration of c in the fluid. Allow for the reactor volume to significantly vary [i.e., accumulation = $\partial(Vc_C)/\partial t$]. (b) Assume that the evaporation rate is small, and that inflow is interrupted for a period of time during which the water continues to leak out the bottom of the pond. Derive an equation that describes the concentration of c in the pond.

4.6 Microbial aggregates consist primarily of water. If an aggregate is placed into a solution of dyed water, initially at concentration c_0, write a differential equation that will describe the rate of chemical diffusion into the aggregate. Consider the aggregate to be spherical in shape with radius R.

4.7 A chemical diffuses from the surface of a pipe and reacts in the soil–water environment around the pipe according to first-order kinetics. Derive a differential equation for the concentration of the chemical outside the pipe at steady state.

4.8 Microbes placed on agar in a petri dish can move by diffusion along the surface of the agar, but also grow while they diffuse. Starting with the general transport equation in the form given in Eq. 4-59, derive a differential equation that describes the diffusion of a microorganism assuming cell division follows the first-order reaction $M \rightarrow 2M$ (adapted from Welty et al., 1984).

4.9 Derivation of the constitutive transport equation in cylindrical coordinates is somewhat more difficult in time and three dimensions (z, r and θ) than in two dimensions, but both derivations can be done by the same shell balance approach. As we have seen, the radial flux due to diffusion is $N_r = -D(dc/dr)$, but the θ flux due to diffusion is $N_\theta = -(D/r)(dc/d\theta)$. (a) Using the control volume shown in Figure P-4.9, draw the cross-sectional areas for transport in the radial, axial, and θ directions. (Hint: for radial diffusion, the area for diffusion into the control volume is $r\,\Delta\theta\,\Delta z$, but the area for diffusion going out is $(r + \Delta r)\,\Delta\theta\,\Delta z$.) (b) Using a shell balance, derive the non-steady-state constitutive transport equation in three dimensions for the case of only ad-

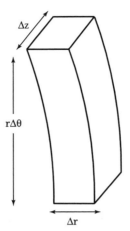

Δz

$r\Delta\theta$

Δr

Figure P-4.9

vective flow in the axial (z) direction, neglecting advection (but not diffusion) in the radial and θ directions and for no reaction.

4.10 A spherical aggregate is produced through coagulation in a solution containing a dilute concentration of dye at a concentration c_0. The aggregate is then removed and placed into clean water. (a) Develop a differential equation that describes the concentration of chemical in the spherical aggregate assuming dye is lost by diffusion. (b) How many boundary/initial conditions are necessary to solve this equation? Suggest ones that could be used.

CHAPTER 5

CONCENTRATION PROFILES AND CHEMICAL FLUXES

5.1 INTRODUCTION

There are several steps that must be followed in order to derive an equation describing the rate of mass transport in a system. First, the system must be defined by a control volume. Often, a simple sketch is made to assist in deciding what parts of the system can be included. This process establishes the number of phases in the system, the components, the system boundaries, and so forth. Second, the governing transport equations are written down for all components and phases using techniques discussed in the previous chapter. Third, boundary conditions are stated for the system. Finally, these equations are solved. As we shall see here, different systems may be described by the same governing equation, but the use of boundary conditions unique to that system can produce different analytical solutions. In this text the main focus is on methods for obtaining analytical solutions for simplified systems. The use of computational techniques for numerically solving problems, beyond those that can be done on a simple spreadsheet, will need to be considered separately elsewhere.

5.2 THE THREE THEORIES OF MASS TRANSPORT

As the description of mass transport in a system becomes more detailed, the equations necessary to describe the system become more complex. However, a great many systems if reduced in complexity simplify to three basic cases: mass transport through a layer assumed to be stagnant (stagnant film theory); the penetration of a chemical from a gas into a falling liquid film (penetration theory); and transport from or to an isolated flat surface in a uniform flow field (boundary-layer theory). In each of these cases, describing the system using a mass transport coefficient k produces a unique relationship between the molecular diffusion coefficient and the mass transport coefficient.

Diffusion Through a Stagnant Film

A stagnant film is usually hypothesized to exist in order to allow very complex systems to be reduced to a simpler system of equations that can be solved analytically. Several examples of systems where stagnant films are assumed are shown in Figure 5.1. These systems have in common a surface, at which the flow is zero, existing in a complex fluid flow environment. When the fluid above a surface is, for example, completely mixed, describing the concentrations of the chemical everywhere in the mixed bulk fluid environment is essentially impossible. However, the rate of mass transport from the surface can still be described in terms of a mass transport coefficient by defining a stagnant film thickness δ over which the concentration of a chemical changes from the surface concentration to the bulk concentration. Setting up this conceptual model therefore makes it possible to perform simple calculations that would otherwise be too difficult to set up and solve.

Although these calculations can be made, it must be realized that stagnant films (with a few notable exceptions) do not exist! Thus any system that is described by such a stagnant film model will contain errors, the severity of which can only be determined by comparing the model predictions to real data. Many stagnant film models should be viewed with caution, since often they are calibrated with actual data, but not verified over the wider conditions where the model may be applied.

Diffusion Into a Stagnant Film With No Reaction One of the situations where a stagnant film truly exists is for chemical evaporation in an Arnold diffusion cell (Fig. 5.2). In this system gas flows over the top of a tube containing pure liquid chemical C. We will consider here that the gas flowing over the tube, and inside the tube, is air. The components of air are assumed to be only sparingly soluble in chemical C, and therefore there is no net flux of air into the liquid. Chemical C evaporates into air, but the distance between the liquid surface and the top of the tube is held constant.

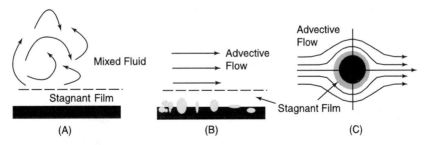

Figure 5.1 Examples of systems where a stagnant film is assumed: (a) transport to a surface in a mixed fluid; (b) transport to an irregular surface; (c) transport to a sphere.

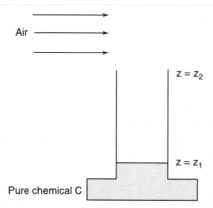

Air

$z = z_2$

$z = z_1$

Pure chemical C

Figure 5.2   Control volume and boundaries defined for the Arnold diffusion cell.

The chemical flux out of the tube can be easily derived for this system, although calculation of the flux requires first that we derive an expression for the concentration profile in the tube above the fluid. The control volume for this calculation is defined as the volume of air in the tube with two boundaries, the air–liquid interface and the plane at the top of the tube. We start with the constitutive transport equation in the form

$$\frac{\partial c_C}{\partial t} = -(\mathbf{u} \cdot \nabla c_C) - \nabla \cdot (-D_C \, \nabla c_C) + R_C \tag{5-1}$$

For this system at steady state, $\partial c_C/\partial t = 0$, and with no reaction, $R_C = 0$, resulting in

$$(\mathbf{u} \cdot \nabla c_C) = \nabla \cdot (D_C \, \nabla c_C) \tag{5-2}$$

In order to write Eq. 5-2 in this form, we assumed constant c_C and D_C. To simplify our derivation, however, we will substitute $\nabla \cdot \mathbf{u} c_C$ for $\mathbf{u} \cdot \nabla c_C$, which is always true when $\nabla \cdot \mathbf{u} = 0$. This produces

$$(\nabla \cdot \mathbf{u} c_C) = \nabla \cdot (D_C \, \nabla c_C) \tag{5-3}$$

For one-dimensional transport in the z-direction, and designating the phase as a air, we can further simplify this to

$$\frac{\partial}{\partial z} (u_{a,z} c_{Ca}) = \frac{\partial}{\partial z} \left(D_{Ca} \frac{dc_{Ca}}{dz} \right) \tag{5-4}$$

Integration of Eq. 5-4 with respect to z and some rearrangement results in

$$b_1 = -D_{Ca} \nabla c_{Ca} + u_{a,z} c_{Ca} \qquad (5\text{-}5)$$

where b_1 is a constant of integration. The flux, $J_{Ca,z}$, is introduced into this equation in two ways. First, from the definition of $u_{a,z}$ using Eq. 4-48, we can write

$$u_{a,z} = \frac{1}{c_a} (c_{Ca} u_{Ca,z} + c_{Aa} u_{Aa,z}) \qquad (5\text{-}6)$$

From the definitions of $J_{Ca,z}$ and $J_{Aa,z}$ (Eq. 4-50), we can also write Eq. 5-6 as

$$u_{a,z} = \frac{1}{c_a} (J_{Ca,z} + J_{Aa,z}) \qquad (5\text{-}7)$$

Since air does not transport into the liquid, $J_{Aa,z} = 0$. Putting Eq. 5-5 into Eq. 5.7, we have

$$b_1 = -D_{Ca} \frac{dc_{Ca}}{dz} + J_{Ca,z} \frac{c_{Ca}}{c_a} \qquad (5\text{-}8)$$

From the definition of y_C, this can be written as

$$b_1 = -D_{Ca} \frac{dC_{Ca}}{dz} + J_{Ca,z} y_C \qquad (5\text{-}9)$$

Equation 5-9 indicates that the sum of the diffusive and advective fluxes is constant. Although the constant of integration appears unknown, b_1 must also be a flux. By comparison of Eq. 5-9 with 4-54, it is evident that the flux $b_1 = J_{Ca,z}$. Including this result into Eq. 5-9 and rearranging produces our final result for the flux

$$J_{Ca,z} = -\frac{D_{Ca}}{(1 - y_C)} \frac{dc_{Ca}}{dz} \qquad (5\text{-}10)$$

This equation for the flux is a general solution valid for any concentration range (it does not require a dilute solution).

Example 5.1

A jar of phenol has been spilled in a room with only one vent 10 cm in diameter and 20 cm deep. Neglecting advective flow through the vent, estimate the rate of phenol transport through the vent if the room concentration is 0.05% ($T = 20°C$).

To solve this problem we will use Eq. 5-10. Since this is just a rough estimate of the flux, we can make some simplifying assumptions for our calculation. When the mole fraction in the air is much less than unity, or $y_P \ll 1$, and the only phase is air, Eq. 5-10 for phenol (chemical P) becomes

$$J_{P,z} = -D_P \frac{dc_P}{dz} \tag{5-11}$$

which is just a statement of Fick's First Law. Let us also assume a linear concentration profile across the vent (this assumption will be discussed further below), so that the flux can be approximated by

$$J_{P,z} = -D_P \frac{\Delta c_P}{\Delta z} \tag{5-12}$$

and that the diffusion constant is that of typical chemical in air, or $D_P \approx 0.1$ cm^2 s^{-1}. The concentration at the vent entrance is

$$c_P = y_P c_a = 0.0005 \times \frac{\text{mol}}{24.1 \text{ L}} = 2.1 \times 10^{-5} \text{ mol L}^{-1} \tag{5-13}$$

Assuming that there is no phenol in the air at the vent exit, we can calculate the flux as

$$J_{P,z} = -0.01 \text{ cm}^2 \text{ s}^{-1} \frac{(2.1 \times 10^{-5} - 0) \text{ mol L}^{-1} \times \dfrac{\text{L}}{10^3 \text{ cm}^3}}{(0 - 20 \text{ cm})} \tag{5-14a}$$

$$= 1 \times 10^{-10} \text{ mol cm}^{-2} \tag{5-14b}$$

The rate chemical leaves the tube is just the flux times the area, or

$$W_{P,z} = 1 \times 10^{-10} \frac{\text{mol}}{\text{cm}^2 \text{ s}} \times \frac{\pi}{4} (10 \text{ cm})^2 = 8 \times 10^{-9} \frac{\text{mol}}{\text{s}} \tag{5-15}$$

Example 5.2

Starting with Eq. 4-59, derive Eq. 5-10.

Equation 4-59 is the generalized constitutive transport equation. For a chemical in air,

$$\frac{\partial c_{Ca}}{\partial t} + \nabla J_{Ca} - R_{Ca} = 0 \tag{5-16}$$

Since we are deriving a steady-state solution, and there is no reaction, this equation can be reduced to

$$\nabla \mathbf{J}_{Ca} = 0 \tag{5-17}$$

From the definition of the del operator

$$\nabla \mathbf{J}_{Ca} = \frac{\partial J_{Ca,x}}{\partial x} + \frac{\partial J_{Ca,y}}{\partial y} + \frac{\partial J_{Ca,z}}{\partial z} \tag{5-18}$$

For one-dimensional transport in the z direction, this simplifies to

$$\frac{dJ_{Ca,z}}{dz} = 0 \tag{5-19}$$

Integrating once, we have

$$J_{Ca,z} = b_1 \tag{5-20}$$

which just indicates that the flux is constant. Using the definition of $J_{Ca,z}$, we can write by inspection (see Eq. 4-51) for the flux in one dimension,

$$J_{Ca,z} = -D_{Ca} \frac{dc_{Ca}}{dz} + y_C(N_{Ca} + N_{Aa}) \tag{5-21}$$

Since there is no net flux of air, $J_{Aa} = 0$. Using this information, and rearranging Eq. 5-21,

$$J_{Ca,z} = -\frac{D_{Ca}}{(1 - y_C)} \frac{dc_{Ca}}{dz} \tag{5-22}$$

which is the same equation as derived above.

In order to solve Eq. 5-10 we must integrate it twice. Expressing the concentration c_{Ca} in terms of a mole fraction and integrating once produces

$$\frac{dy_C}{(1 - y_C) \, dz} = b_1 \tag{5-23}$$

Separating and integrating again produces the general solution

$$-\ln(1 - y_C) = b_1 z + b_2 \tag{5-24}$$

The two constants of integration require two boundary conditions. For the gas in the tube we choose

$$BC\text{-}1: z = z_1, \qquad y_C = y_{C,eq} \tag{5-25}$$

$$BC\text{-}2: z = z_2, \qquad y_C = y_{C,2} \tag{5-26}$$

Substituting these BCs into Eq. 5-24 produces the specific solution for the concentration profile in this system

$$\left(\frac{1 - y_C}{1 - y_{C,eq}} \right) = \left(\frac{1 - y_{C,2}}{1 - y_{C,eq}} \right)^{(z-z_1)/(z_2-z_1)} \tag{5-27}$$

Since there is no reaction and the system has been assumed to reach steady state, the chemical flux is constant throughout the tube. The easiest point to evaluate the flux is at the air–liquid interface, since we know that at this boundary (from BC-1) the mole fraction in the gas phase is in equilibrium with the pure liquid, or $y_C = y_{C,eq}$. Substituting this into our expression for the flux (Eq. 5-10), we have

$$J_{Ca,z} = - \frac{D_{Ca} c_a}{(1 - y_{C,eq})} \frac{dy_C}{dz} \bigg|_{z=z_1} \tag{5-28}$$

where the right vertical line indicates that we are evaluating the flux at z_1. To calculate the gradient at this point we must take the derivative of Eq. 5-27. Recognizing that $d(1 - y_C)/dz = -dy_C/dz$, and that Eq. 5-27 is of the form a^u, from calculus tables we find that $d(a^u) = a^u \ln a \, du$. Applying this to Eq. 5-27 produces

$$\frac{dy_C}{dz} = -\frac{d(1 - y_C)}{dz} = -\frac{(1 - y_{C,eq})}{z_2 - z_1} \ln \left(\frac{1 - y_{C,2}}{1 - y_{C,eq}} \right) \tag{5-29}$$

Combining Eqs. 5-28 and 5-29 produces the final result that the flux across the interface is

$$J_{Ca,z} = \frac{D_{Ca} c_a}{z_2 - z_1} \ln \left(\frac{1 - y_{C,2}}{1 - y_{C,eq}} \right) \tag{5-30}$$

The main application of Eq. 5-29 is in the determination of diffusion constants of chemicals in gases. Rearranging Eq. 5-29 with the diffusion coefficient on the left-hand side of the equation results in

$$D_{Ca} = \frac{J_{Ca,z}(z_1 - z_2)}{c_1 \ln \left(\dfrac{1 - y_{A,2}}{1 - y_{C,eq}} \right)} \tag{5-31}$$

By measuring the flux in a tube with a known height, the diffusion constant is easily calculated.

Many applications of the Arnold diffusion cell are for a binary systems consisting, for example, of a single chemical (c) in air. Since the concentration profile is logarithmic, the log mean concentration of air in the tube is often substituted into Eq. 5-31. The log mean mole fraction of air in the tube is defined as

$$y_{A,\text{lm}} = \frac{\displaystyle\int_{z_1}^{z_2} y_A \, dz}{\displaystyle\int_{z_1}^{z_2} dz} \tag{5-32}$$

From Eq. 5-32 and the assumption of a binary system, so that $y_A + y_C = 1$, we can derive a concentration profile for air as

$$\frac{y_A}{y_{A,1}} = \left(\frac{y_{A,2}}{y_{A,1}}\right)^{(z-z_1)/(z_2-z_1)} \tag{5-33}$$

Substituting Eq. 5-33 into Eq. 5-32 produces

$$y_{A,\text{lm}} = \frac{y_{A,2} - y_{A,1}}{\ln(y_{A,2}/y_{A,1})} \tag{5-34}$$

or equivalently in terms of y_C, since $y_A = 1 - y_C$,

$$y_{A,\text{lm}} = \frac{y_{C,\text{eq}} - y_{C,2}}{\ln \dfrac{(1 - y_{C,2})}{(1 - y_{C,\text{eq}})}} \tag{5-35}$$

where it is assumed that at the air–liquid interface y_C is at its equilibrium concentration, or $y_{C,1} = y_{C,\text{eq}}$. Including $y_{A,\text{lm}}$ into Eq. 5-30

$$J_{Ca,z} = -\frac{D_{Ca}c_a}{z_2 - z_1}\frac{(y_{C,\text{eq}} - y_{C,2})}{y_{A,\text{lm}}} \tag{5-36}$$

and rearranging to obtain a solution for the diffusion coefficient of

$$D_{Ca} = \frac{J_{Ca,z}(z_2 - z_1)y_{A,\text{lm}}}{c_a(y_{C,\text{eq}} - y_{C,2})} \tag{5-37}$$

where $y_{A,\text{lm}}$ is essentially unity for a dilute solution.

Equation 5-37 can be used to measure the diffusion coefficient in an Arnold diffusion cell if the distance of the chemical from the height of the tube ($z_2 - z_1$) is constant during the experiment. However, as long as the change in the diffusion path is small, the diffusivity of the chemical can still be measured in a diffusion cell. Let us assume that the distance $z = z_1$ is defined at the initial time $t = t_0$, and that at some time later the distance of the fluid from the top of the tube has increased to $z = z_1$ at $t = t_1$. That is, we have now defined Δz as the change in the fluid level, not as the distance of the fluid level from the top of the container. As long as the change in fluid height is quite small relative to the total diffusion distance in the tube, it is possible to calculate a diffusion constant. At any instant during the experiment the molar flux in the gas phase is

$$J_{Ca,z} = -\frac{D_{Ca}c_a}{z'}\frac{(y_{C,eq} - y_{C,2})}{y_{A,lm}} \tag{5-38}$$

where we have designated the level change as z'. We can also write that the molar flux is a function of the change in the height of the surface of the fluid, or

$$J_{Ca,z} = c_C \frac{dz'}{dt} \tag{5-39}$$

where c_C is the molar density of pure chemical C. Combining the two preceding equations and separating the variables for integration, we have

$$\int_{t=0}^{t} dt = \frac{c_C y_{A,lm}}{c_a D_{Ca}(y_{C,eq} - y_{C,2})} \int_{z_0'}^{z'} z'\, dz \tag{5-40}$$

Upon integration and rearrangement, this yields

$$D_{Ca} = \frac{c_C y_{A,lm}(z'^2 - z'^2_0)}{2c_a t(y_{C,eq} - y_{C,2})} \tag{5-41}$$

This is the same solution originally presented in Section 3.4 (without derivation) for the diffusion coefficient of a chemical in air.

Effect of Reaction on Mass Transport Rates   The reaction of a chemical in a phase accelerates the rate of transport of that chemical into the phase. When a chemical in one phase is initially brought into contact with another phase, the chemical is assumed to partition into the new phase at its equilibrium concentration at the surface of the new phase. Since the concentration gradient is at its maximum (in fact, at this point it is infinite), the largest rate of chemical transport occurs at this instant. As the concentration of the chemical increases in the new phase, the chemical flux will decrease, since the flux is proportional to the concentration gradient. The removal of the chemical in the new phase due to the reaction will increase the gradient and therefore increase the rate of chemical transport into the phase.

In order to demonstrate the effect of chemical reaction on transport, we will derive and solve the constitutive transport equation describing the flux of chemical into a stagnant, but reactive, fluid (Fig. 5.3). For this example we will assume that there is a chemical C in the gas phase that diffuses into water W containing a chemical B. The two chemicals C and B react according to irreversible first-order kinetics to produce D, or

$$B + C \rightarrow D \tag{5-42}$$

where the rate of reaction, $R_C = -kc_{Cw}$, has units of [moles L^{-3} t^{-1}] and k is the reaction rate constant with units of $[t^{-1}]$. To derive a governing transport equation we start with Eq. 4-59 in the form

$$\frac{\partial c_{Cw}}{\partial t} + \nabla \mathbf{J}_{Cw} - R_{Cw} = 0 \tag{5-43}$$

If we assume steady-state conditions, the first term is zero. For transport only in the z direction, Eq. 5-43 becomes

$$\frac{dJ_{Cw,z}}{dz} = R_{Cw} \tag{5-44}$$

From the definition of $J_{Cw,z}$,

$$J_{Cw,z} = -D_{Cw}\frac{dc_{Cw}}{dz} + x_{Cw}(J_{Cw} + J_{Ww}) \tag{5-45}$$

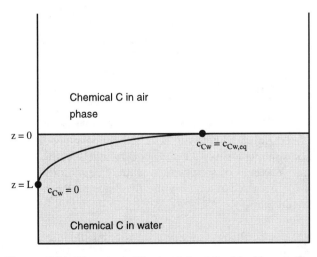

Figure 5.3 Chemical diffusion into a liquid with reaction.

This can be simplified as in previous derivations by assuming that $J_{Ww} = 0$ (stagnant fluid)

$$J_{Cw,z} = -\frac{D_{Cw}}{(1 - x_C)} \frac{dc_{Cw}}{dz} \tag{5-46}$$

Assuming a dilute solution, $(1 - x_C) = 1$. Substituting in the reaction for R_{Cw}, we have the final result that

$$D_{Cw} \frac{d^2 c_{Cw}}{dz^2} - k c_{Cw} = 0 \tag{5-47}$$

Example 5.3

Derive Eq. 5-47 starting with Eq. 4-66.

Equation 4-66 is the constitutive transport equation with diffusion and advection terms separated. Writing it in terms of the liquid phase, we have

$$\frac{\partial c_{Cw}}{\partial t} + \mathbf{u}_w \cdot \nabla c_{Cw} - D_{Cw} \nabla^2 c_{Cw} - R_{Cw} = 0 \tag{5-48}$$

For the case of no advective motion in the fluid and steady-state conditions, this simplifies to

$$D_{Cw} \nabla^2 c_{Cw} + R_{Cw} = 0 \tag{5-49}$$

For one-dimensional transport in the z direction, $\nabla^2 = \partial^2/\partial z^2$. Substituting in the reaction rate produces

$$D_{Cw} \frac{d^2 c_{Cw}}{dz^2} - k c_{Cw} = 0 \tag{5-50}$$

which is the same as Eq. 5-47.

In order to solve Eq. 5-47, we will need two boundary conditions to obtain the two constants of integration. There are several boundary conditions that we could choose, and each combination would yield a different specific solution to the differential equations. The goal of specifying boundary conditions is to obtain the simplest equation that most accurately depicts the system. We will choose here the following two conditions:

$$BC - 1: c_{Cw} = c_{Cw,eq}, \qquad z = 0 \tag{5-51}$$

$$BC - 2: c_{Cw} = 0, \qquad\quad z = L \tag{5-52}$$

The first BC specifies that equilibrium is assumed to occur at the gas–liquid interface ($z = 0$). The second BC indicates that at some depth of penetration L into the fluid, the liquid concentration of C reaches 0. This latter BC is an approximation since for a first-order reaction there will always be some finite concentration of C. However, for all practical purposes the concentration will become small enough at some depth that it will be undetectable.

To obtain the general solution for Eq. 5-47 we recognize that the equation is of the form

$$\frac{d^2x}{dy^2} - by = 0 \tag{5-53}$$

where $b = k/D_{Cw}$. Therefore, by inspection (Appendix 4) the general solution is

$$c_{Cw} = b_1 \cosh \left(\sqrt{\frac{k}{D_{Cw}}} z \right) + b_2 \sinh \left(\sqrt{\frac{k}{D_{Cw}}} z \right) \tag{5-54}$$

where b_1 and b_2 are two constants of integration that must be obtained using the two boundary conditions. Substituting the first BC into Eq. 5-54, we have

$$c_{Cw,eq} = b_1 \cosh(0) + b_2 \sinh(0) \tag{5-55}$$

Since the $\sinh(0) = 0$, and the $\cosh(0) = 1$, the first constant b_1 is

$$b_1 = c_{Cw,eq} \tag{5-56}$$

Substituting in the second boundary condition and Eq. 5-56 into Eq. 5-54 produces

$$0 = c_{Cw,eq} \cosh \left(\sqrt{\frac{k}{D_{Cw}}} L \right) + b_2 \sinh \left(\sqrt{\frac{k}{D_{Cw}}} L \right) \tag{5-57}$$

which simplifies to

$$b_2 = -c_{Cw,eq} \coth \left(\sqrt{\frac{k}{D_{Cw}}} L \right) \tag{5-58}$$

Substituting Eq. 5-58 into Eq. 5-54, and using the relationship that $\coth(x) = 1/\tanh(x)$, we have the concentration profile

$$c_{Cw} = c_{Cw,eq} \cosh \left(\sqrt{\frac{k}{D_{Cw}}} z \right) - \frac{c_{Cw,eq} \sinh \left(\sqrt{\frac{k}{D_{Cw}}} z \right)}{\tanh \left(\sqrt{\frac{k}{D_{Cw}}} L \right)} \tag{5-59}$$

To determine the flux, we start with our expression for the flux (Eq. 5-46) for a dilute solution evaluated at $z = 0$

$$J_{Cw,z}\big|_{z=0} = -D_{Cw} \frac{dc_{Cw}}{dz}\bigg|_{z=0} \tag{5-60}$$

The derivative of Eq. 5-59 is

$$\frac{dc_{Cw}}{dz} = c_{Cw,eq} \sqrt{\frac{k}{D_{Cw}}} \sinh\left(\sqrt{\frac{k}{D_{Cw}}}\, z\right) - \frac{c_{Cw,eq}}{\tanh\left(\sqrt{\frac{k}{D_{Cw}}}\, L\right)} \sqrt{\frac{k}{D_{Cw}}} \cosh\left(\sqrt{\frac{k}{D_{Cw}}}\, z\right) \tag{5-61}$$

Evaluating the derivative at $z = 0$ results in

$$\frac{dc_{Cw}}{dz}\bigg|_{z=0} = 0 - \frac{c_{Cw,eq} \sqrt{\dfrac{k}{D_{Cw}}}}{\tanh\left(\sqrt{\dfrac{k}{D_{Cw}}}\, L\right)} \tag{5-62}$$

Incorporating the result for the gradient (Eq. 5-62) into our expression for the flux (Eq. 5-60) produces the final complete solution for the flux

$$J_{Cw,z}\big|_{z=0} = \frac{D_{Cw} c_{Cw,eqw}}{L} \frac{\sqrt{\dfrac{k}{D_{Cw}}}\, L}{\tanh\left(\sqrt{\dfrac{k}{D_{Cw}}}\, L\right)} \tag{5-63}$$

In order to examine the effect of a chemical reaction on the mass transport rate, we need to compare the chemical flux with reaction to a situation that is identical but lacks reaction. In the absence of reaction we can simplify Eq. 5-47 to

$$D_{Cw} \frac{dc_{Cw}}{dz} = 0 \tag{5-64}$$

Integrating twice, and using the same boundary conditions as before (Eqs. 5-51 and 5-52)

$$J_{Cw,z}\big|_{z=0} = \frac{D_{Cw} c_{Cw,eq}}{L} \tag{5-65}$$

The effect of reaction on the flux can now be compared using a dimensionless ratio, sometimes referred to as the Hatta number, Ha, defined as

$$\text{Ha} \equiv \frac{\text{flux with reaction}}{\text{flux without reaction}} \tag{5-66}$$

Using Eqs. 5-62 and 5-65, the Ha number for this comparison is

$$\text{Ha} = \sqrt{\frac{k}{D_{Cw}}} L \coth\left(\sqrt{\frac{k}{D_{Cw}}} L\right) \tag{5-67}$$

Thus we can see that the flux increases in proportion to the Hatta number, and that the Hatta number is a ratio of the reaction rate to the diffusion constant. The ratio of these two constants will occur in many instances involving problems dealing with chemical reactions and mass transport.

If the reaction is very fast, or diffusion very slow, the tanh term will approach unity. For very fast reaction we can simplify Eq. 5-63 to

$$J_{Cw,z}\big|_{z=0} = \sqrt{D_{Cw}k}\, c_{Cw,eq} \tag{5-68}$$

For very fast reactions the flux into the liquid is a half-order function of the diffusion coefficient, but the flux is not independent of the rate constant. The flux is also proportional to the square root of the rate constant and the equilibrium concentration. The effect of reaction on the flux into a stagnant fluid is summarized in Table 5.1.

Penetration Theories

The transport of a chemical across a surface becomes mathematically more complex if the phase into which the chemical is being transported is in motion. Since truly stagnant fluids are unlikely to occur very often in natural systems (although we may wish them to be stagnant for mathematical simplicity), mass transport calculations

TABLE 5.1   Mass Transport (One Dimensional) into a Fluid at Steady State

Conditions	Local Flux
Non-steady-state diffusion to depth L, no reaction	$J_{C,z}\big|_{z=0} = \dfrac{D_C c_{C,eq}}{L}$
Steady-state diffusion, with reaction	$J_{C,z}\big|_{z=0} = \dfrac{D_C c_{C,eq}}{L} \dfrac{\sqrt{k/D_C}\,L}{\tanh(\sqrt{k/D_C}\,L)}$
Diffusion with a very fast reaction	$J_{C,z}\big|_{z=0} = \sqrt{D_C k}\, c_{C,eq}$

for these moving systems are more useful. Even though the fluid is in motion, a simplifying assumption can be made that the velocity of the fluid is constant with the depth of penetration so that the fluid *appears* to be stagnant relative to the depth of chemical penetration. Several approaches have been used to consider the transport of a chemical into a fluid that is considered to be immobile. In this section we will consider two cases: chemical transport into a falling laminar film (a wetted wall) and transport into a packet of fluid that is assumed to come to rest, for a brief period of time, at the interface separating the two phases (surface renewal). As we shall see, the resulting equations are more complex but in important ways are more representative of transport in actual systems.

Penetration Theory for Thin Films   In many engineered processes the rate of mass transfer of a chemical between a liquid and gas phase is accomplished by spreading the liquid onto media with large specific surface areas (area per volume). In air stripping, groundwaters containing volatile organic chemicals are treated by running water downward over a column of packed media while pumping air up through the media. Since the volatile chemicals rapidly partition into the air phase, little chemical remains in the water phase by the time the water reaches the end of the column. In trickling filters transport is in the reverse direction: Oxygen is transported from the air into the liquid. As a separate mass transport step, oxygen continues to diffuse through the liquid film into the biofilm. There are many different types of media used in these processes, ranging from rocks and small plastic objects to plastic sheets with flat, wavy, and contoured surfaces. The best medium for a given operation is a function of cost, space availability, and treatment levels.

The description of chemical transport in thin fluid films can be reduced in complexity to be just diffusion into or out of a falling film. The simplest case of chemical transport into a liquid is known as penetration theory since it is assumed that the transport of the chemical occurs over a short enough distance into the liquid that the velocity of the fluid is constant.

To calculate the chemical flux into a falling film, we begin by deriving the constitutive transport equation for a falling film with no reaction using a shell balance. The differential control volume has a width Δx, length Δz, and depth into the plane of width w (Fig. 5.4). After a system has been in operation for a period of time, there is a constant chemical flux into the system at any location. The system is at steady state since the system concentrations are no longer changing as a function of time. We will assume that the advective flux in the downward direction is much greater than the flux due to diffusion so that diffusion in the direction of flow can be neglected. In the absence of chemical reaction in the fluid, the system is therefore defined by transport in two directions: advection in the downward z direction, and diffusion across the plane in the transverse x direction. The rate of advection into the control volume across the plane of area $w \, \Delta x$ is

$$J_{Cw}\big|_z = u_z c_{Cw}\big|_z w \, \Delta x \qquad (5\text{-}69)$$

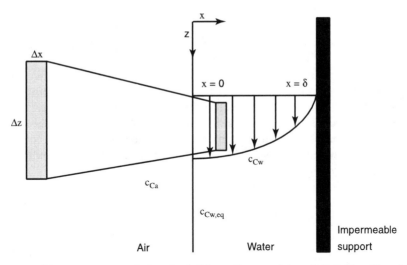

Figure 5.4   Mass transport of chemical C from the gas into a thin falling film of thickness δ. The differential control volume (expanded view) has dimensions Δx, Δz, and width w.

while the advective flux out of the control volume is

$$J_{Cw}\big|_{z+\Delta z} = u_z c_{Cw}\big|_{z+\Delta z} w\,\Delta x \tag{5-70}$$

Similarly, the diffusive flux across the plane of cross-sectional area $w\,\Delta z$ is

$$J_{Cw}\big|_x = -D_{Cw}\left.\frac{\partial c_{Cw}}{\partial x}\right|_x w\,\Delta z \tag{5-71}$$

and the flux out is

$$J_{Cw}\big|_{x+\Delta x} = -D_{Cw}\left.\frac{\partial c_{Cw}}{\partial x}\right|_{x+\Delta x} w\,\Delta z \tag{5-72}$$

Combining all four terms, dividing by $w\,\Delta x\,\Delta z$, and rearranging produces

$$u_z \frac{\left(c_{Cw}\big|_{z+\Delta z} - c_{Cw}\big|_z\right)}{\Delta z} - D_{Cw}\frac{\left(\left.\dfrac{\partial c_{Cw}}{\partial x}\right|_{x+\Delta x} - \left.\dfrac{\partial c_{Cw}}{\partial x}\right|_x\right)}{\Delta x} = 0 \tag{5-73}$$

Taking the limit as $\Delta x\,\Delta z \to 0$

$$u_z \frac{\partial c_{Cw}}{\partial z} = D_{Cw}\frac{\partial^2 c_{Cw}}{\partial x^2} \tag{5-74}$$

This result is a two-dimensional transport problem that cannot be readily integrated since u_z is a function of x. In order to specify this functionality for the advection term, we need to perform momentum balance over the control volume. We can use the same differential control volume constructed for the mass transport shell balance above, except this time we will balance momentum, $\rho_w u_w$, and not concentration. Gravity acts on the liquid to pull it down the along the wall, exerting a body force in the z direction, $\rho_w g_z$. Once the fluid has accelerated and reached a steady velocity, there is no accumulation of momentum in the fluid, and the system will operate under steady conditions. The momentum in the z direction is constant since the momentum entering the shell, $\rho_w u_{z|z} = \rho_w u_{z|z+\Delta z}$.

The only remaining force to account for is fluid shear. The fluid is held by the wall surface so that $u_z = 0$ at $x = \delta$, where δ is the fluid thickness. At the air–liquid interface there is no resistance to the falling liquid so that the fluid falls at its maximum velocity. The result of fluid falling freely at one side and being held at the other side of the fluid film is that the fluid is sheared in the x direction. We designate the shear as $\tau_{w,xz}$, where x indicates the shear direction and z the direction of momentum transfer. From Newton's Law, we can see the fluid shear is a function of the velocity gradient according to

$$\tau_{zx} = -\mu_w \frac{\partial u_z}{\partial x} \tag{5-75}$$

In combining all terms for our momentum balance, we find that there are only two remaining terms, producing the ordinary differential equation

$$\frac{d\tau_{zx}}{dx} = \rho_w g \tag{5-76}$$

where the subscript has been dropped on the gravitational constant since it is assumed that gravity acts in the z direction. Notice that the derivative of the shear in Eq. 5-76 is analogous to the derivative of the chemical flux in the constitutive transport equation.

To solve the momentum equation we use the two boundary conditions

$$BC - 1: x = 0, \qquad \tau_{zx} = 0 \tag{5-77}$$

$$BC - 2: x = \delta, \qquad u_z = 0 \tag{5-78}$$

The first boundary condition reflects our assumption that the air offers no resistance to the fluid, and therefore the fluid cannot transport any momentum (via shear) into the air. The second boundary condition is the no-slip condition at the liquid–wall interface. Integrating Eq. 5-76 and solving using Eqs. 5-77 and 5-78, we have the result that the velocity profile is

$$u_z = \frac{\rho_w g \delta^2}{2\mu_w} \left[1 - \left(\frac{x}{\delta} \right)^2 \right] \tag{5-79}$$

When $x = 0$ the fluid velocity is the maximum velocity, u_{max} calculated from Eq. 5-79 as

$$u_{max} = \frac{\rho_w g \delta^2}{2 \mu_w}$$ (5-80)

Using this definition of u_{max}, we can equivalently write Eq. 5-79 as

$$u_z = u_{max} \left[1 - \left(\frac{x}{\delta} \right)^2 \right]$$ (5-81)

The parabolic velocity profile is valid for laminar flow conditions, which occur at Reynolds numbers, $Re = u_{avg} \delta / v_w < 1$ to 6, where $u_{avg} = 2/3 u_{max}$ is the average fluid velocity. At $Re > 250$ to 500 the fluid flow is turbulent, and at intermediate Re there are some ripples in the otherwise laminar flow. Laminar flow is quickly established at a distance $L_{ent} = 0.3 \, Re \, \delta$ when fluid enters over a flat plate (Stucheli and Ozisik, 1976). Since $Re < 6$ for laminar flow, laminar conditions are established within a distance $\sim 2\delta$.

Example 5.4

Calculate the average velocity, volume flow rate, and fluid film thickness of a thin fluid film of water flowing down a flat plate.

The average velocity, u_{avg}, can be obtained by integrating the fluid velocity over the plate area, or

$$u_{avg} = \frac{\int_0^w \int_0^\delta u_z \, dx \, dy}{\int_0^w \int_0^\delta dx \, dy}$$ (5-82)

or since the integral with respect to y is just w, and the denominator is just δw

$$u_{avg} = \frac{1}{\delta} \int_0^\delta u_{w,z} \, dx$$ (5-83)

Substituting in the definition of velocity, we have the result that

$$u_{avg} = \frac{\rho_w g \delta^2}{3 \mu_w}$$ (5-84)

The volume flow rate, Q_w, is just the product of the average velocity and cross-sectional area, or

$$Q_w = u_{avg} w \delta = \frac{\rho_w g w \delta^3}{3 \mu_w} \tag{5-85}$$

The fluid thickness δ can be obtained from rearrangement of either of the above solutions as

$$\delta = \left(\frac{3 \mu_w u_{avg}}{\rho_w g} \right)^{1/2} = \left(\frac{3 \mu_w Q_w}{\rho_w g w} \right)^{1/3} \tag{5-86}$$

Combining the transport Eq. 5-74 with Eq. 5-86 produces our final differential transport equation with all values explicit in direction

$$u_{max} \left[1 - \left(\frac{x}{\delta} \right)^2 \right] \frac{\partial c_{Cw}}{\partial z} = D_{Cw} \frac{\partial^2 c_{Cw}}{\partial z^2} \tag{5-87}$$

To solve this equation we will need three boundary conditions for the chemical concentration, two as a function of the z direction and one as a function of x. We choose

$$BC - 1: x = 0, \qquad c_{Cw} = c_{Cw,eq} = c_{eq} \tag{5-88}$$

$$BC - 2: x = \delta, \qquad \frac{\partial c_{Cw}}{\partial x} = 0 \tag{5-89}$$

$$BC - 3: z = 0, \qquad c_{Cw} = c_{Cw,0} = c_0 \tag{5-90}$$

These three boundary conditions specify that: The chemical diffusing into the film is initially at its equilibrium concentration at the air–liquid interface ($x = 0$); there is no chemical flux through the wall ($x = \delta$); fluid entering the system ($z = 0$) contains the chemical at a concentration c_0.

The solution to this equation can be expressed in terms of a dimensionless concentration profile (Johnstone and Pigford, 1942) as

$$\frac{c_{Cw,L} - c_{eq}}{c_0 - c_{eq}} = 0.7857 e^{-5.1213 \Lambda^*} + 0.1001 e^{-39.318 \Lambda^*}$$

$$+ 0.03500 e^{-105.64 \Lambda^*} + 0.01811 e^{-204.75 \Lambda^*} + \cdots \tag{5-91}$$

where $\Lambda^* = D_{Cw} L / \delta^2 u_{max}$, L is the length of flow in the z direction, and $c_{Cw,L}$ the concentration in the liquid at a distance L. This solution is quite cumbersome but applicable to chemical diffusion into the fluid film over long periods of time or flow distance.

A simpler, but more restricted, solution of Eq. 5-87 can be obtained by assuming that the chemical diffusing into the fluid only diffuses (penetrates) a short distance into the fluid film. This approach, known as *penetration theory*, greatly simplifies the governing transport equation since for short penetration distances the fluid velocity is constant so that $u_z \approx u_{max}$. Thus for penetration theory the transport equation is simplified to

$$u_{max} \frac{\partial c_{Cw}}{\partial z} = D_{Cw} \frac{\partial^2 c_{Cw}}{\partial x^2} \tag{5-92}$$

Since we now assume for this model that the fluid does not penetrate very far, the boundary conditions can also be modified slightly. If the chemical penetrates only a short distance into the fluid, the chemical concentration at a distance δ (a distance essentially of $x = \infty$) is $c_{Cw} = 0$. Thus the second boundary condition becomes

$$BC - 2: x = \infty, \qquad c_{Cw} = 0 \tag{5-93}$$

The solution to this equation can be obtained using Laplace transforms (a solution technique that will not be covered here). Thus the solution without derivation is

$$\frac{c_{Cw} - c_{eq}}{c_0 - c_{eq}} = \text{erf}\left(\frac{x}{(4D_{Cw}z/u_{max})^{1/2}}\right) \tag{5-94}$$

where erf(f) is the error function. The error function is a mathematical function, in the same manner as a sin or logarithm, developed specifically for solutions to the type of problem presented here. Tables of the error function are contained in Appendix 4, although this function is also available as a built in function in most computer spreadsheet programs.

It is often common to consider the case where the influent fluid is devoid of chemical C, so that boundary condition 3 becomes

$$BC - 3: z = 0, \qquad c_{Cw} = 0 \tag{5-95}$$

In this case, the solution of the transport equation is simplified to

$$\frac{c_{Cw}}{c_{eq}} = \text{erfc}\left(\frac{x}{(4D_{Cw}z/u_{max})^{1/2}}\right) \tag{5-96}$$

where erfc(f) is the complementary error function equal to 1-erf(f).

The mass flux is obtained in the usual manner using

$$J_{Cw,0} = -D_{Cw} \left.\frac{dc_{Cw}}{dx}\right|_{x=0} \tag{5-97}$$

Evaluating the derivative of Eq. 5-96 at $x = 0$ and substituting this result into Eq. 5-97 produces the *local flux* at any point z along the flow path

$$J_{Cw,0} = c_{eq} \left(\frac{D_{Cw} u_{max}}{\pi z} \right)^{1/2}$$

(5-98)

The usual objective in making a mass transport calculation is to find out how much chemical is transported per time, or the rate of chemical transport. We can calculate the total rate of chemical C transported into the liquid film, $W_{Cw,0}$, by integrating the local flux, $J_{Cw,0}$, over the length of flow. For a fluid width w and length of flow L, the integral of Eq. 5-98 is

$$W_{Cw,0} = c_{eq} \left(\frac{D_{Cw} u_{max}}{\pi} \right)^{1/2} \int_0^w \int_0^L z^{-1/2} \, dz \, dy$$

(5-99)

Carrying out the integration produces our final result for the total flux into the fluid film

$$W_{Cw,0} = c_{eq} wL \left(\frac{4 D_{Cw} u_{max}}{\pi L} \right)^{1/2}$$

(5-100)

Surface Renewal   There are many instances of mass transport in natural and engineered environments that do not conform to the requirements of the two above theories for either a stagnant film or a very thin fluid film with little chemical penetration. In order to describe mass transport into deeper fluids when the fluid has some motion, Dankwurtz (1951) proposed the surface renewal theory. According to his model a fluid could be viewed as a series of packets of fluid that are moved into contact with the air interface for brief periods (Fig. 5.5). While in contact with the surface air a chemical can diffuse into the packet of fluid, but after the packet is

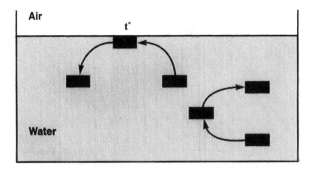

Figure 5.5 Chemical diffusion into fluid, as described by the surface renewal theory, occurs as if packets spent a time t^* in contact with the surface before being moved away from the surface. Other packets of fluid are moved about but may not come into contact with the fluid–air interface.

moved out of contact mass transport to that packet ceases. The theory derives its name from the surface–age distribution function T, defined as

$$T = se^{-st} \tag{5-101}$$

where T is the probability that a packet of fluid will be exposed to the surface, and $s[t^{-1}]$ is the fractional renewal rate. Using this renewal rate, the mean steady-state flux into the liquid across the interface is

$$J_{Cw,0} = (D_{Cw}s)^{1/2}(c_{eq} - c_{\infty}) \tag{5-102}$$

The major difficulty with applying this theory is that the function s is usually not known. However, the theory does provide an indication that transport into the fluid is proportional to the chemical gradient and the square root of the diffusion coefficient.

Boundary Layer Theory

One of the few cases where it is possible to obtain an analytical solution for two-dimensional mass transport is for steady laminar flow over a flat plate. The flow approaching the flat surface is uniform and parallel to the surface with velocity u_{∞} (Fig. 5.6). If we assume that $dP/dx = 0$ (there are no net pressure forces and the pressure does not contribute to the drag for flow), the following equations must be solved in the region above the plate

$$u_x \frac{\partial u_x}{\partial x} + u_x \frac{\partial u_y}{\partial y} = v \frac{\partial^2 u_x}{\partial y^2} \tag{5-103}$$

$$\frac{\partial u_x}{\partial x} + \frac{\partial u_y}{\partial y} = 0 \tag{5-104}$$

where we have omitted for the moment the phase subscripts. Blasius was the first to solve these two equations with the boundary conditions $u_x = u_y = 0$ at $y = 0$ and

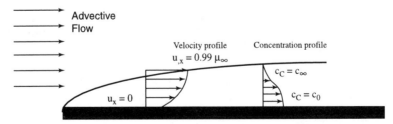

Figure 5.6 Mass transport of chemical C from a plate into a fluid according to boundary layer theory. The hydrodynamic boundary layer thickness is δ_n, while the concentration boundary layer has a thickness of δ_c.

$u_x = u_\infty$ at $y = \infty$. He introduced a stream function that allowed the set of equations to be reduced to ordinary differential equations. There were several important outcomes from his solution. First, the hydrodynamic boundary layer thickness, δ_h, defined as the point where $u_x = 0.99u_\infty$, can be calculated at any distance from the point of flow using

$$\delta_h = 5 \left(\frac{vx}{u_\infty} \right)^{1/2} \tag{5-105}$$

or in terms of the dimensionless Reynolds number defined as $\text{Re} = u_\infty x/v$ is

$$\frac{\delta_h}{x} = 5 \, \text{Re}^{-1/2} \tag{5-106}$$

Second, the velocity gradient at the surface was calculated as

$$\left. \frac{\partial u_x}{\partial y} \right|_{y=0} = 0.332 \, u_\infty \left(\frac{u_\infty}{vx} \right)^{1/2} \tag{5-107}$$

These analytical solutions made it possible to formulate the solution to the transport equation for the analogous chemical boundary conditions that $c_{Ca} = c_0$ at $y = 0$ and $c_{Aa} = c_\infty$ at $y = \infty$. However, this solution was only possible by requiring that $Sc = v_a/D_{Ca} = 1$, or that the thickness of the hydrodynamic boundary layer is equal to the thickness of the concentration boundary layer. Thus the Blasius analytical solution for mass transport from a flat plate is only valid for air, since for air $Sc \approx 1$. For water $Sc \gg 1$. For these conditions, the concentration gradient in the air was derived as

$$\left. \frac{dc_{Ca}}{dy} \right|_{y=0} = \left(\frac{0.332}{x} \, \text{Re}_x^{1/2} \right) (c_\infty - c_0) \tag{5-108}$$

where the phase subscript for air has been added to c in the concentration gradient term as a reminder that this solution is limited to $Sc = 1$, and where $\text{Re} = u_\infty x/v_w$. Although we have not solved for the concentration profile, the flux is easily obtained since it is just the diffusion coefficient of C in air times the concentration gradient obtained in Eq. 5-108, or

$$J_{Ca,y=0} = \frac{0.332 D_{Ca}}{x} \, \text{Re}_x^{1/2} (c_0 - c_\infty) \tag{5-109}$$

This solution can be used to solve for mass transport in phases other than air by using the relationship between the thickness of the hydrodynamic boundary layer,

δ_h, and the concentration boundary layer, δ_c,

$$\frac{\delta_h}{\delta_c} = Sc^{1/3} \tag{5-110}$$

In cases where the phase used for a dimensionless number is ambiguous, a phase subscript can be added to the dimensionless number. For example, we could use Sc_a here, but it will be assumed that Sc is written about the phase indicated by the variable on the left-hand side of the equation. Including the Sc into our solution for the local flux (Eq. 5-109) produces the general case applicable to a wide range of Sc

$$J_{C,y=0} = \frac{0.332D_C}{x} \, \text{Re}_x^{1/2} \, Sc^{1/3}(c_0 - c_\infty) \tag{5-111}$$

The total rate of mass transport over the whole plate can be obtained by integrating the local flux, Eq. 5-111, over the width of the plate w and the length of flow L

$$W_{C,L} = (c_0 - c_\infty) \int_0^L \int_0^w \frac{0.332D_C}{x} \, \text{Re}_x^{1/2} \, Sc^{1/3} \, dx \, dz \tag{5-112}$$

Upon integration, this yields

$$W_{C,L} = 0.664 \, wD_C \, \text{Re}_L^{1/2} \, Sc^{1/3}(c_0 - c_\infty) \tag{5-113}$$

This result indicates that the chemical flux decreases as the time (or distance from the plate entrance) increases. Thus the rate of mass transport is the highest at the plate entrance since the gradient is greatest. As the flow continues along the plate, the gradient decreases, decreasing the rate of chemical transport from the plate into the fluid.

5.3 MASS TRANSPORT IN RADIAL AND CYLINDRICAL COORDINATES USING SHELL BALANCES

As discussed in Section 4.7, it is often useful when deriving the constituent transport equation for systems with many components to begin with a shell balance in the same manner done to derive the general transport equation. In Cartesian coordinates, the analysis usually begins with a cube with differential control volume of size Δx, Δy, Δz. For cylindrical coordinates, the system consists of a ring of thickness Δr, length Δz, and angle $\Delta \theta$. In spherical coordinates, the differential control volume includes the scales Δr and angle $\Delta \theta$, but may require an additional angle in the third dimension of $\Delta \phi$. While many of the solutions for multiphase and multidimensional problems is beyond the scope of this text, we will consider here a few cases for one-dimensional transport that can easily be solved analytically.

Mass Transport to a Sphere at Rest

Spherical particles arise in a variety of systems, for example, bubbles in aerated tanks, support media in fluidized and fixed packed beds, suspended microbes and aggregates in bioreactors, etc. As a starting point in calculations of mass transport to these spherical particles, we will consider here the case of steady-state mass transport from an isolated sphere with no reaction. In order to derive an expression for the chemical flux to a sphere at rest, we start by defining our control volume as a spherical shell around the sphere of thickness Δr and surface area $4\pi r^2$ (Fig. 5.7). For this system, our mass balance equation of

$$\text{accumulation} = \text{in} - \text{out} + \text{reaction} \tag{5-114}$$

becomes only in = out because our sphere is at rest (there is no advective component to the flux) and because there is no reaction. Therefore, the rate of chemical transport into the differential control volume by diffusion is

$$\text{diffusion in} = J4\pi r^2|_r \tag{5-115}$$

or since $J = -D\, dc/dr$, we can write the flux in terms of the concentration gradient as

$$\text{diffusion in} = -D\frac{dc}{dr}4\pi r^2|_r \tag{5-116}$$

Similarly, the chemical transport out of the differential volume is

$$\text{diffusion out} = -D\frac{dc}{dr}4\pi r^2|_{r+\Delta r} \tag{5-117}$$

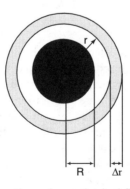

Figure 5.7   Diffusion through a spherical film of thickness Δr.

Substituting the two terms (Eqs. 5-116 and 5-117) into our mass balance (Eq. 5-114), we have

$$-D \frac{dc}{dr} 4\pi r^2 \big|_r + D \frac{dc}{dr} 4\pi r^2 \big|_{r+\Delta r} = 0 \tag{5-118}$$

Dividing by $D4\pi\Delta r$, rearranging, and taking the limit as $\Delta r \to 0$.

$$\lim_{\Delta r=0} \frac{\frac{dc}{dr} r^2 \big|_{r+\Delta r} - \frac{dc}{dr} r^2 \big|_r}{\Delta r} = 0 \tag{5-119}$$

which produces

$$\frac{d}{dr} \left(r^2 \frac{dc}{dr} \right) = 0 \tag{5-120}$$

Integrating twice results in

$$c = \frac{b_1}{r} + b_2 \tag{5-121}$$

where b_1 and b_2 are two constants of integration.

In order to solve this equation we require two boundary conditions. As an example we consider here the case of oxygen transport in water from an air bubble. The first condition should be that the oxygen concentration in the water immediately at the gas–liquid interface is the saturation concentration of oxygen, or

$$BC - 1: c = c_{Cw,eq} = c_{eq} \qquad r = R \tag{5-122}$$

For the second boundary condition we choose that at large distances from the bubble the concentration of oxygen in the fluid is a background concentration, or

$$BC - 2: c_{Cw} = c_{Cw,\infty} = c_\infty \qquad r = \infty \tag{5-123}$$

Substituting the second boundary condition into Eq. 5-121, we have

$$c_\infty = \frac{b_1}{\infty} + b_2 \tag{5-124}$$

or $b_2 = c_\infty$. From the first boundary condition and our result for b_2 we similarly obtain

$$b_1 = R(c_{eq} - c_\infty) \tag{5-125}$$

Combining, our final expression for these boundary conditions is

$$c_{Cw}(r) = \frac{R(c_{eq} - c_\infty)}{r} + c_\infty \tag{5-126}$$

The chemical flux is obtained in the usual manner, starting with

$$J_{Cw}|_{r=R} = -D_{Cw} \left. \frac{dc_{Cw}}{dr} \right|_{r=R} \tag{5-127}$$

where we have chosen to obtain the total flux into the liquid by evaluating the chemical gradient at the gas–liquid interface ($r = R$). Evaluating the gradient by taking the derivative of Eq. 5-126

$$\left. \frac{dc_{Cw}}{dr} \right|_{r=R} = R(c_{eq} - c_\infty) \left. \frac{d(1/r)}{dr} \right|_{r=R} \tag{5-128}$$

or

$$\left. \frac{dc_{Cw}}{dr} \right|_{r=R} = -\frac{(c_{eq} - c_\infty)}{R} \tag{5-129}$$

Combining with the expression for the flux, we have our final result that the flux from the sphere is

$$J_{Cw,r=R} = \frac{D_{Cw}}{R}(c_{eq} - c_\infty) \tag{5-130}$$

Transport in a Pipe

Let us consider the case of a fluid flowing laminarly through a circular pipe whose inner surface is leaking a toxic chemical that goes into solution. Let us also assume that this material reacts in solution according to first-order kinetics, and the pipe wall concentration is c_p = const. To derive a governing equation using cylindrical coordinates, we start with a shell balance. Our differential control volume is a ring having a thickness Δr and length Δz. We will assume the system is symmetrical about an angle θ.

The volume of the cylindrical control volume is obtained by multiplying the cross-sectional area of the ring in the axial direction by the thickness of the ring (Fig. 5.8). The axial cross-sectional area is equal to

$$\text{axial cross-sectional area} = \pi(r + \Delta r)^2 - \pi r^2 \tag{5-131}$$

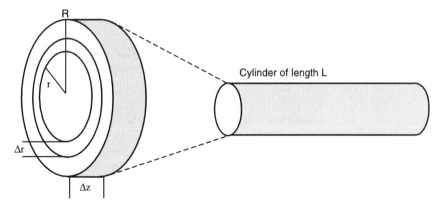

Figure 5.8 Control volume used in examining mass transport within a fluid in a pipe. The expanded portion of the pipe shows the differential volume of thickness Δr and length Δz, for an axial cross-sectional area of $2\pi r\, \Delta r$ and radial cross-sectional area $2\pi r\, \Delta z$.

Expanding the squared term and simplifying

$$\text{axial cross-sectional area} = 2\pi r\, \Delta r - \pi\, \Delta r^2 \qquad (5\text{-}132)$$

If Δr is very small, then Δr^2 is infinitesimally small and can be neglected. Accordingly, Eq. 5-132 can be simplified to

$$\text{axial cross-sectional area} = 2\pi r\, \Delta r \qquad (5\text{-}133)$$

Thus the differential control volume of the ring is $2\pi r\, \Delta r\, \Delta z$.

To derive the final equation we will develop an equation term by term, including in our mass balance about the control volume only those terms we will define as important to model mass transport in the pipe. Terms to be included at this stage include: reaction, axial convection, axial diffusion, and radial diffusion. Since the fluid flow is laminar, we do not have any advective transport in the radial direction. All terms will have units of mole/time, and since only one chemical in one phase (the water) is being modeled, we can omit both the chemical and phase subscripts from our equations.

Reaction occurs homogeneously within the differential control volume. For the first-order reaction rate $R_C = -k_1 c$ occurring in a volume $2\pi r\, \Delta r\, \Delta z$, the reaction term is the product $R\, \Delta V$, or

$$\text{reaction} = -k_1 c 2\pi r\, \Delta r\, \Delta z \qquad (5\text{-}134)$$

The net axial convection is the difference between the rate flow enters and leaves the control volume. The rate flow moves through the control volume is $u_z c A$, where A is the cross-sectional area derived above and z is the direction of flow. The net axial flow is therefore

$$\text{axial flow} = (u_z c|_z 2\pi r \, \Delta r) - (u_z c|_{z+\Delta z} 2\pi r \, \Delta r) \tag{5-135}$$

Collecting terms and simplifying

$$\text{axial flow} = u_z(c|_z - c|_{z+\Delta z})2\pi r \, \Delta r \tag{5-136}$$

Axial diffusion is obtained in the same manner as the difference between the rate a chemical diffuses into and out of the control volume. Here the rate is a production of the flux times the area for diffusion, or $(-D \, dc/dz)A$. Using the axial area derived above, this results in

$$\text{axial diffusion} = \left(-D \left.\frac{\partial c}{\partial z}\right|_z 2\pi r \, \Delta r\right) - \left(-D \left.\frac{\partial c}{\partial z}\right|_{z+\Delta z} 2\pi r \, \Delta r\right) \tag{5-137}$$

Collecting terms and simplifying

$$\text{axial diffusion} = -D \left(\left.\frac{\partial c}{\partial z}\right|_z - \left.\frac{\partial c}{\partial z}\right|_{z+\Delta z}\right) 2\pi r \, \Delta r \tag{5-138}$$

To calculate the rate a chemical moves through the control volume due to the radial flux we use the same approach as above for axial diffusion, except the flux is expressed with respect to changes in the radial direction and the cross-sectional area is $2\pi r \, \Delta z$. This produces a net radial diffusion rate of

$$\text{radial diffusion} = \left(-D \left.\frac{\partial c}{\partial r}\right|_{r+\Delta r} 2\pi r \, \Delta z\right) - \left(-D \left.\frac{\partial c}{\partial r}\right|_r 2\pi r \, \Delta z\right) \tag{5-139}$$

Simplifying results in

$$\text{radial diffusion} = -D \left(\left.\frac{\partial c}{\partial r} r\right|_{r+\Delta r} - \left.\frac{\partial c}{\partial r} r\right|_r\right) 2\pi \, \Delta z \tag{5-140}$$

where the r term must be included within the bracketed terms since c is a function of the radial direction.

The final equation is produced by combining all terms in Eqs. 5-134, 5-136, 5-138, and 5-140 into the single equation to produce

$$u_z 2\pi r \, \Delta r(c|_{z+\Delta z} - c|_z) - D2\pi r \, \Delta r\left(\left.\frac{\partial c}{\partial z}\right|_{z+\Delta z} - \left.\frac{\partial c}{\partial z}\right|_z\right)$$

$$- D2\pi \, \Delta z\left(r \left.\frac{\partial c}{\partial z}\right|_{r+\Delta r} - r \left.\frac{\partial c}{\partial r} r\right|_r\right) + k_1 c 2\pi r \, \Delta r \, \Delta z = 0 \tag{5-141}$$

Dividing by $2\pi r \Delta r \, \Delta z$

$$u_z \frac{(c|_{z+\Delta z} - c|_z)}{\Delta z} - D \frac{\left(\left.\dfrac{\partial c}{\partial z}\right|_{z+\Delta z} - \left.\dfrac{\partial c}{\partial z}\right|_z\right)}{\Delta z}$$

$$- D \frac{\left(\left.r\dfrac{\partial c}{\partial z}\right|_{r+\Delta r} - \left.r\dfrac{\partial c}{\partial r}r\right|_r\right)}{\Delta r} + kc = 0 \qquad (5\text{-}142)$$

Taking the limit as $(\Delta r \, \Delta z) \to 0$ and rearranging produces our result

$$u_z \frac{\partial c}{\partial z} - D \frac{\partial c}{\partial z} - D \frac{\partial^2 c}{\partial r^2} - \frac{D}{r}\frac{\partial c}{\partial r} + k_1 c = 0 \qquad (5\text{-}143)$$

This equation is quite complex and cannot be easily solved analytically. In order to obtain a more tractable problem, we further consider the case of chemical transport very far downstream. After the pipe has been exposed to the chemical for a long distance of flow, we can neglect the axial transport terms since the concentration profile has reached a steady condition and no longer changes with the distance of flow. Under these conditions, the transport Eq. 5-143 becomes an ordinary differential equation

$$\frac{d^2 c}{dr^2} - \frac{1}{r}\frac{dc}{dr} + \frac{k_1}{D}c = 0 \qquad (5\text{-}144)$$

This equation can be solved by inspection by recognizing that it is a zero-order modified Bessel function with a solution

$$c = b_1 I_0(\sqrt{k_1/D}\ r) + b_2 K_0(\sqrt{k_1/D}\ r) \qquad (5\text{-}145)$$

To obtain the exact solution of this equation for our system we need two boundary conditions for the concentration of A as a function of the radius. We choose

$$BC - 1: c = \text{finite}, \qquad r = 0 \qquad (5\text{-}146)$$

$$BC - 2: c = c_p, \qquad r = R \qquad (5\text{-}147)$$

The first boundary condition is chosen based on experience using Bessel functions, since $K_0(0) = \infty$. Therefore, in order for there to be a finite solution of Eq. 5-135 at $r = 0$, the constant b_2 must be equal to zero. Applying the second boundary condition that the concentration at the pipe wall is c_p, we have

$$c_p = b_1 I_0(\sqrt{k_1/D}\ R) \qquad (5\text{-}148)$$

The final solution to the concentration profile in the pipe is obtained by inserting Eq. 5-148 into Eq. 5-145, or

$$c = c_p \frac{I_0(\sqrt{k/D}\ r)}{I_0(\sqrt{k/D}\ R)} \tag{5-149}$$

Values of I_0 can be obtained from mathematical tables in standard reference books or textbooks.

PROBLEMS

5.1   In Example 5.1 we calculated the molar rate of phenol P through a 10-cm-diameter vent 20 cm deep to be 8×10^{-9} mol s^{-1} for a room concentration of 0.05% at 20°C by assuming a linear phenol gradient. Calculate the molar rate using the parabolic profile derived in Eq. 5-30, which yielded a flux of

$$J_{P,z} = -\frac{c_a D_{Pa}}{z_2 - z_1} \ln\left(\frac{1 - y_{P,1}}{1 - y_{P,2}}\right) \tag{5-150}$$

5.2   Equation 5-47 describes the diffusion of chemical C into a liquid containing chemical B where C undergoes a first-order reaction with B. Solve Eq. 5-47 using the following boundary conditions: (1) at the bottom of the container ($z = 0$), there is no flux of C; (2) at the interface ($z = \delta$), the concentration of C is equal to the equilibrium concentration of C in the liquid.

5.3   A fluid film of thickness δ_f, containing a chemical initially at a concentration c_0, flows over a biofilm of thickness δ_b, at a constant velocity u (Fig. P-5.3). Assuming only advective transport is dominant, use a shell balance to derive a differential equation for the change in concentration with distance z in the *fluid*. Assume the fluid is relatively well mixed in the vertical direction, which means the flux into the biofilm, $c(kD)^{1/2}$, can be treated as a reaction in the fluid instead of a boundary condition (watch units!). You should obtain an expression that is easily solved. Solve it.

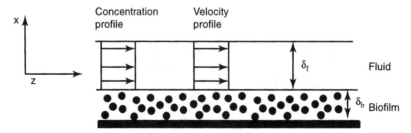

Figure P-5.3

5.4 A smart choice of boundary conditions can often lead to a simpler solution of a differential equation. Whenever possible, choose a condition that sets a variable at zero. We derived the equation $dJ_C/dz = 0$, where the equation below can be used to describe the mass flux by diffusion through a stagnant gas film

$$J_{C,z} = -\frac{c_a D_{Ca}}{1 - y_C}\frac{dy_C}{dz} \qquad (5\text{-}151)$$

Let us assume for the remainder of this problem that $y_C \ll 1$. (a) Derive an equation that is the solution of the concentration profile in the air phase given the following boundary conditions: at $z = z_1$ (the air–liquid interface), c_C is at an equilibrium concentration c_{eq}; at $z = z_2$ (the vent-free air interface), $c_C = 0$. (b) Repeat, but this time choose that at $z = 0$ (the air–liquid interface), c_C is at an equilibrium concentration c_{eq}.

$\checkmark$**5.5** Fluid enters a column 50 cm long and 5 cm in diameter at a superficial velocity (approach velocity) of 1 cm s^{-1}, filled with granular activated carbon. A chemical in solution at 150 mg L^{-1} reacts (adsorbs) according to an empirical kinetic expression, that is, reaction rate = kc^2, where k is an empirical constant equal to 0.001 L mg^{-1} s^{-1}. (a) Derive a differential expression for the concentration of chemical in the column. Neglect axial and radial diffusion. (b) Solve the expression for concentration as a function of distance for the given conditions. (c) What is the observed rate of removal? (*Hint*: look at overall removal, i.e., in minus out.)

5.6 Many times during filtration of wastewater samples a layer of microorganisms forms on the filter. This has the effect of removing particles smaller than the pore size of the filter since closely situated microbes can form pores smaller than the filter pores. If the microbes are active, however, they can also remove soluble substrate by taking substrate into the cell. Let us assume for this problem that substrate removal is half-order with respect to concentration, that is, that the removal rate is: $kc^{1/2}$. The influent concentration is 100 mg L^{-1}, the rate constant is $k = 4 \times 10^4$ (mg L^{-1} s^{-2})$^{1/2}$, $D = 10^{-5}$ cm^2 s^{-1}, and $u =$ 1 cm s^{-1}. (a) If the microbes form a layer on a filter, use a shell balance to derive a *differential* equation that describes the concentration of dissolved substrate in the microbial layer (i.e., in the microbial biofilm). Assume that there is advective and diffusive transport in the direction of flow. (b) Nondimensionalize the equation, placing all dimensionless numbers into single constants where possible. (c) Let us assume that the advective transport is much larger than diffusion. Using your result from part (a), derive and solve an equation for the concentration profile in the biofilm.

5.7 A well-mixed batch reactor contains 1 g L^{-1} of TiO$_2$. An environmental engineering student states that TCE degradation in the reactor is first order with respect to TCE concentration (c_T). His major finding is that the half-life of TCE decreases with concentration according to the graph shown in Figure

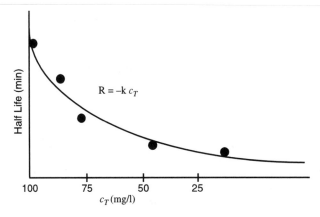

Figure P-5.7

P-5.7. Each point shown in the graph was determined in separate experiments. This finding was challenged by a professor on the thesis committee on the grounds that it made no sense. Even though you do not know anything about TiO_2 reactions, figure out who is correct. This is a thought question, and does not require a very long answer. You may wish to make some calculations, however, to defend your arguments.

5.8 A chemical generated at the tip of the tube (diameter 0.5 cm) is present at a concentration of 100 µg L^{-1} at an impermeable wall (Fig. P-5.8; not to scale). The chemical reacts with material in the tube as it diffuses ($D_C = 0.1$ cm^2 s^{-1}) through the tube according to the kinetic constant 4.8 s^{-1}. The flux through the nonreactive tip end is driven by a linear gradient through the filter with a proportionality constant of 0.44 cm s^{-1}, and the rate of chemical diffusion through the tip end is 0.0045 µg s^{-1}. Derive a differential equation, and solve for the steady-state concentration profile, within the tube for this system.

5.9 The rate of mass transfer through a porous and reactive tube was described in Problem 5.8. This question re-examines this problem from a different perspective. (a) For the system shown in Figure P-5.9 containing a porous

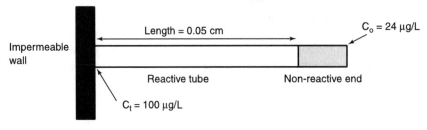

Figure P-5.8

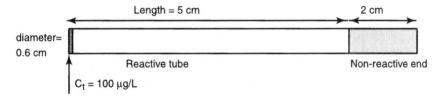

Figure P-5.9

media through which gas flows, followed by a nonreactive filter (i.e., it is a cigarette!), derive (but do not solve yet) a differential equation describing the concentration of species C in the reactive portion of the tube, neglecting radial dispersion and assuming axial advection is constant (i.e., plug flow conditions). (b) Now, assuming diffusion is negligible, derive and solve an equation for the concentration profile in the reactive tube. (c) What is the concentration of chemical C leaving the nonreactive end (i.e., being inhaled) if the air velocity in the tube is 30 cm s^{-1} and the first-order rate constant is 4.8 s^{-1}? (d) If the concentration of chemical were everywhere in the reactive tube equal to 100 μg L^{-1}, what would be the rate of chemical removal in the tube?

5.10 Chlorine gas is considered being used in a wetted wall reactor for bacterial disinfection. In water, Cl_2 dissociates to form HOCl, which can in turn form $H^+ + OCl^- \rightarrow$ HOCl is the preferred form of chlorine in water since HOCl is 40 to 80 times more effective a disinfectant than OCl^-. (a) Calculate the diffusivity of HOCl in water at 20°C using the Wilke–Change correlation given in Chapter 3. (b) In the presence of 2 mg L^{-1} of HOCl 99% of *Escherichia coli* will be killed in 0.2 min. Assuming this reaction is first order with respect to chemical concentration, calculate a first-order rate constant k_1 for disinfection of *E. coli*. (Actually, disinfection is more accurately described as being proportional to the product of time and concentration of the disinfectant, but we will assume first-order kinetics here.) (c) Let us assume that all Cl_2 in the water instantly dissociates to form HOCl. If the air–water interface concentration of HOCl is 2 mg L^{-1}, what will be the depth of HOCl penetration (defined as a 95% reduction of the surface concentration) into a wetted wall film after a distance of travel of 5 cm, assuming the maximum fluid velocity of the fluid film is 7 cm s^{-1} and no chemical reaction/consumption of chlorine? (d) In the presence of a first-order reaction, the total rate of chemical transport into the entire film after it has traveled a distance L (Bird et al., 1960, p. 553) is:

$$W_{C,L} = c_{Cw,0} u_{\max} w \left(\frac{D_{Cw}}{k_1} \right)^{1/2} \left[\left(\frac{1}{2} + u^* \right) \mathrm{erf}\, (u^{*1/2}) + \left(\frac{u^*}{\pi} \right)^{1/2} e^{-u^*} \right] \quad (5\text{-}152)$$

where $w = 15$ cm is the width of the plate, $u_{\max}$ is the maximum fluid velocity of the fluid film, and u^* is a dimensionless velocity calculated as $u^* = (k_1L/u_{max})$. Calculate $W_{C,L}$ for these conditions. (e) Show that in the absence of reaction, the Eq. 5-152 is reduced to Eq. 5-100. (f) Calculate the ratio of the rate of HOCl transported into the fluid film in the presence and absence of chemical reaction under the conditions given in this problem.

REFERENCES

Bird, R. B., W. E. Stewart, and E. N. Lightfoot. 1960. *Transport Phenomena.* John Wiley & Sons, Inc. New York, NY.

Danckwurtz, P. V. 1951. *Trans. Faraday Soc.* **47**:1014–1023.

Higbie, R. 1935. *Trans. AIChE* **31**:368–89.

Johnstone, H. F., and R. L. Pigford. 1942. *Trans. AIChE* **38**:25.

Roberts, P. V., G. D. Hopkins, C. Munz, and A. H. Riojas. 1995. *Environ. Sci. Technol.* **19**(2): 164–73.

Stucheli, A., and M. N. Ozisik. 1976. *Chem. Eng. Sci.* **31**:369.

CHAPTER 6

MASS TRANSPORT CORRELATIONS: FROM THEORY TO EMPIRICISM

6.1 DEFINITION OF A MASS TRANSPORT COEFFICIENT

While it is possible in some instances to develop an analytical solution for the flux of a chemical to or from a surface, in most instances the system is too complicated to permit such an approach. For example, while it is possible to calculate easily a concentration profile around a sphere at rest, the solution for a moving sphere is more complicated, and the case for a sphere in a packed bed of nonuniform particles has yet to be solved fully. In order to calculate chemical fluxes in more realistic environments, we start with the assumption that the flux of chemical i can be calculated using

$$\mathbf{J}_i = \mathbf{k}_{ij}(c_{ij,2} - c_{ij,1}) \tag{6-1}$$

where $\mathbf{k}_{ij}$ is a mass transport coefficient, with units $[\text{L t}^{-1}]$, that is used to calculate the transport of chemical i in phase j. The mass transport coefficient is not a vector quantity, but it appears in bold letters to distinguish it from a reaction rate constant k. The mass transport coefficient can contain a subscript if the transport correlation is specific to a chemical, but usually the chemical being referred to is clear, and it is omitted from the subscript. For example, the flux of chemical C in phase w could be written as

$$\mathbf{J}_C = \mathbf{k}_w(c_{Cw,2} - c_{Cw,1}) \tag{6-2}$$

Notice that this description of the flux no longer relies on knowing a concentration *gradient*, but a concentration *difference*. This distinction is important since specifying the gradient requires being able to quantify both a change in concentration and a distance over which this change occurs. While an overall change in concentration is easy to observe, the exact distance over which this occurs at an interface is difficult to measure.

The mass transport coefficients in Eqs. 6-1 and 6-2 provide a mechanism for introducing empirical correlations into transport rate calculations. We have already seen that dimensionless numbers are useful for scaling systems and for quantifying the relative importance of different phenomena. The Reynolds number, for example, is a ratio of the inertial force to the viscous force, and the magnitude of Re is also used to indicate when the fluid flow is laminar or turbulent.

In order to scale transport coefficients, we use the dimensionless Sherwood number, defined as

$$Sh = \frac{\mathbf{k}_{ij}x}{D_{ij}} \qquad (6\text{-}3)$$

Thus the Sherwood number is a ratio of total mass transport to diffusive transport. The length scale is defined as appropriate for the system; this characteristic length is x, L, or some length dimension for linear distances, while r (radius) or d (diameter) is used for a radial characteristic length.

In the following sections it will be shown that, in some instances, the mass transport coefficient can be derived from the analytical solution of the governing transport equation. Thus for simple systems, the mass transport coefficient is completely defined. As the system becomes more complex, empirical mass transport correlations will be required to calculate mass fluxes. Deciding which parameters to include in a correlation requires insight into the factors controlling transport. One guiding principle is often that a complex system, when reduced to more basic conditions, should display a relationship between the mass transport coefficient and the chemical diffusivity expected from the analytical solution of that simpler system. Examples of basic relationships between $\mathbf{k}_{ij}$ and D_{ij} are shown below for the "three mass transport models."

6.2   THE THREE THEORIES

In Section 5.3 three different theories were proposed to describe transport into fluids: stagnant film, penetration, and boundary layer (Fig. 6.1). These three cases are all situations where it is possible to solve analytically for the chemical flux between two liquid phases. For situations where it is not possible to solve for the chemical flux (or the chemical gradient) we start by *assuming* that a mass transfer coefficient can be specified based on an observed concentration difference. The validity of this assumption can be supported by comparing our assumption as specified in Eq. 6-2 with our observed chemical fluxes. In this manner, the three theories of mass transport shown in Figure 6.1 will be used to demonstrate how the mass transport coefficient can be analytically derived for simple cases.

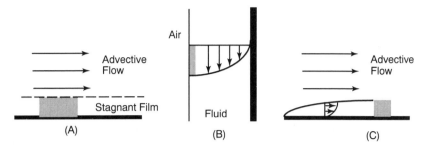

Figure 6.1   Three theories of mass transport for which analytical solutions exist: (a) transport into a stagnant fluid; (b) penetration of a chemical into a thin fluid film; (c) chemical transport from a flat plate as described by boundary layer theory.

Stagnant Film Model

For one-dimensional chemical flux out of a stagnant fluid into air, we obtained

$$J_{C,z} = -\frac{D_{Ca} c_a}{z_2 - z_1} \frac{(y_{C,eq} - y_{C,2})}{y_{A,lm}} \tag{6-4}$$

Assuming a dilute solution, $y_{A,lm} = 1$. For constant temperature and pressure, we can move the total concentration across the derivative to obtain the flux in terms of molar concentration as

$$J_{C,z} = -\frac{D_{Ca}}{z_2 - z_1} (c_{Ca,eq} - c_{Ca,2}) \tag{6-5}$$

This equation was derived to describe transport across a stagnant film of air above pure chemical C. It is also applied (incorrectly, as will be pointed out) to other systems where a stagnant film is *assumed* to exist above a surface (Fig. 6.1a). Writing this flux for this more general case of transport across a stagnant fluid, here defined as water with an interface concentration $c_{Cw,1}$, produces

$$J_{C,z} = \frac{D_{Cw}}{\delta_s} (c_{Cw,2} - c_{Cw,1}) \tag{6-6}$$

where $\delta_s = z_2 - z_1$ is the stagnant film thickness, and the minus sign has been dropped by switching the order of the chemical concentration difference. If we write our definition of the mass transport coefficient in Eq. 6-2 in terms of chemical flux only in the z direction, we have

$$J_{C,z} = \mathbf{k}_w (c_{Cw,2} - c_{Cw,1}) \tag{6-7}$$

Comparing Eqs. 6-6 and 6-7, we conclude that the mass transport coefficient for this system must be

$$\mathbf{k}_w = \frac{D_{Cw}}{\delta_s} \tag{6-8}$$

For a stagnant film, the distance over which the gradient exists has been incorporated into the mass transport coefficient. Based on the stagnant film thickness as a characteristic length scale, the mass transport coefficient can be expressed as a function of the Sherwood numbers as $Sh = \mathbf{k}_w \delta_s / D_{Cw}$ (Table 6.1). Since the thickness of the stagnant fluid is only known when it can be controlled, as in the case of the Arnold diffusion cell, the mass transport coefficient is only known a priori in this very specific case. In other cases where stagnant films are assumed, such as those shown in Figure 6.1, the mass transport is not known and must be inferred from experiments.

Penetration Theory

A similar approach can be taken to derive a mass transport coefficient for chemical penetration from air into a thin fluid film. We previously derived the instantaneous chemical flux into the thin fluid film after traveling a distance z as

$$J_C|_{x=0} = \left(\frac{D_{Aw}u_{max}}{\pi z}\right)^{1/2} c_{eq} \tag{6-9}$$

where $c_{eq} = c_{Cw,eq}$ is the equilibrium concentration of chemical C in water. This equation was derived by assuming that the entering fluid was devoid of chemical C. If the chemical in the water enters at a uniform concentration of c_0, the driving force for chemical transport is reduced, and Eq. 6-9 can be written instead as

$$J_C|_{x=0} = \left(\frac{D_{Cw}u_{max}}{\pi z}\right)^{1/2} (c_{eq} - c_0) \tag{6-10}$$

TABLE 6.1 Theoretical Mass Transport Correlations

Theory	Characteristic Length	Dimensionless Numbers	Expression
Stagnant film	$\delta_s = (z_2 - z_1)$	$Sh = \dfrac{k\delta_s}{D}$	$Sh = 1$
Penetration	z	$Sh = \dfrac{kz}{D}$, $Pe = \dfrac{u_{max}z}{D}$	$Sh = \dfrac{1}{\sqrt{\pi}} Pe^{1/2}$
Boundary layer	x	$Sh = \dfrac{kx}{D}$, $Re = \dfrac{u_\infty x}{\nu}$ $Sc = \dfrac{\nu}{D}$	$Sh = 0.332\, Re^{1/2}\, Sc^{1/3}$

Recognizing that the time of exposure of the fluid is $t = z/u_{max}$, we can write a more general expression for chemical transport over a time t into a fluid as

$$J_C|_{x=0} = \left(\frac{D_{Cw}}{\pi t}\right)^{1/2} (c_{eq} - c_0) \tag{6-11}$$

Comparing this flux to our assumed form of the mass transport coefficient, we can see that $\mathbf{k}_w$ must have the form

$$\mathbf{k}_w = \left(\frac{D_{Cw}}{\pi t}\right)^{1/2} \tag{6-12}$$

Defining a Peclet number as $Pe = Re\ Sc = uz/D$, a mass transport correlation can be expressed as a function of Pe as $Sh = (Pe/\pi)^{1/2}$ (Table 6.1).

Boundary Layer Theory

Proceeding as above, we can similarly derive the mass transport coefficient for the case of mass transport to a plate using boundary layer theory. We previously derived that the chemical flux from the plate could be expressed as

$$J_{C,y=0} = \frac{0.332 D_{Cw}}{x} Re_w^{1/2}\ Sc^{1/3} (c_{Cw,0} - c_{Cw,\infty}) \tag{6-13}$$

From comparison of our assumed relationship of the flux to some difference in chemical concentrations in the system, the mass transport coefficient for the case of mass transport from a plate into water must be

$$\mathbf{k}_w = \frac{0.332 D_{Cw}}{x} Re_x^{1/2}\ Sc^{1/3} \tag{6-14}$$

Based on Eq. 6-14, a mass transport correlation for boundary layer theory would be $Sh = 0.332\ Re^{1/2}\ Sc^{1/3}$ (Table 6.1).

Relationship Between the Mass Transport Coefficient and Diffusivity

From the formulation of the mass transport coefficients in these three systems we can see fundamentally different relationships between the transport coefficient $\mathbf{k}_w$ and the molecular diffusivity of the chemical, D_{Cw}. In the case of stagnant fluids, $\mathbf{k}_w$ $\tilde{}D_{Cw}$. However, for penetration theory, $\mathbf{k}_w$ $\tilde{}D_{Cw}^{1/2}$. Thus the two theories of transport into fluid that is otherwise undisturbed results in a fundamental difference in the response of chemical transport to changes in chemical diffusivities. For boundary layer theory we obtain a third result that $\mathbf{k}_w$ $\tilde{}D_{Cw}^{2/3}$.

These different proportionalities between $\mathbf{k}_w$ and D_{Cw} are important since they specify that in simple cases there should be a clear relationship between chemical fluxes and molecular diffusivities. By experimental examination of mass transport we can test which chemical transport theory applies to the system, that is, whether the transport coefficient is proportional to D_{Cw} raised to the $\frac{1}{2}$, $\frac{1}{3}$, or 1 power. These mass transport relationships are summarized in Table 6.2. Once a relationship between $\mathbf{k}$ and D is known, the mass transport coefficient for another chemical can be scaled using the relationship $\mathbf{k}_{Bj} = \mathbf{k}_{Cj}(D_{Bj}/D_{Cj})^n$, where $n = \frac{1}{2}$, $\frac{1}{3}$, or 1 depending on which theory (or empirical correlation) was used to describe the chemical transport of C.

Stagnant Films: Concepts versus Myth

While it is useful to consider the *possibility* of a stagnant fluid film in cases where a surface is exposed to a moving fluid (Fig. 6.1a), it must be stressed here that stagnant fluid films do not usually exist. The "film concept" or idea of a "stagnant film" is an artificial construct that helps to visualize the presence of a concentration boundary layer around an object.

In each of the cases for mass transport depicted in Figure 6.1, a slice of fluid is shown. This slice spans a depth which would be defined as a stagnant film. In the first case (Fig. 6.1a) the slice of fluid is not moving, and a chemical gradient extends across the width of the block. In the case of the falling fluid (Fig. 6.1b), the fluid is moving, but the velocity within a surface slice is constant (since the chemical does not penetrate very far) and so the surface fluid acts as if it were "stagnant." For both of these cases we can predict the relationship between $\mathbf{k}_w$ and D_{Cw}. For the defined stagnant fluid, $\mathbf{k}_w$ is proportional to D_{Cw} raised to the power 1, but for the falling film $\mathbf{k}_w$ is proportional to D_{Cw} to the $\frac{1}{2}$ power. Which case more accurately depicts the true situation? Higbie (1935) found that for the case of falling films that $\mathbf{k}_w$ was proportional to D_{Cw} to the $\frac{1}{2}$ power, thereby proving that the predictions the stagnant film theory do not apply to systems with advective or convective flow.

TABLE 6.2   Scaling Relationship between the Mass Transport Coefficient k and the Diffusion Constant D_C for Dilute Solutions

Model	Local Flux	Mass Transport Coefficient	Relationship between $\mathbf{k}$ and D_c
Stagnant film	$J_{C,z} = \dfrac{D_C}{\delta_s}(c_{C,2} - c_{C,1})$	$k = \dfrac{D_C}{\delta_s}$	$\mathbf{k} \sim D_C^1$
Penetration	$J_{C|0} = \left(\dfrac{D_C}{\pi t}\right)^{1/2}(c_{C,\text{eq}} - c_{C,0})$	$k = \left(\dfrac{D_C}{\pi t}\right)^{1/2}$	$\mathbf{k} \sim D_C^{1/2}$
Boundary layer	$J_C = \dfrac{0.332 D_C}{x}\,\text{Re}_x^{1/2}\,\text{Sc}^{1/3}(c_{C,0} - c_{C,\infty})$	$k = \dfrac{0.332 D_C}{x}\,\text{Re}_x^{1/2}\,\text{Sc}^{1/3}$	$\mathbf{k} \sim D_C^{2/3}$

Similar arguments against using a stagnant film model can be made based on comparison of the stagnant film model with the boundary layer model. The boundary layer model does not require that any portion of the fluid be stagnant (except at one point, the boundary), and for that reason the two models are quite different in construction. If the local chemical flux is predicted by the two models is compared to experimental data using chemicals with different diffusivities, it can be shown that the boundary layer model correctly predicts the chemical flux being proportional to $D_{Cw}^{1/2}$, and not $D_{Cw}^{1/2}$, as predicted by the stagnant fluid model.

Why does the stagnant fluid model fail? The main reason is that the thickness of a concentration boundary layer above the surface (perpendicular to the direction of flow) of a thin plate varies with distance from the plate edge along the direction of flow. This decrease in the concentration gradient produces a corresponding decrease in the chemical flux. In contrast, the stagnant film model predicts a constant concentration gradient at all locations. The stagnant film model therefore fails to correctly describe the local chemical flux from the surface of a flat plate.

The stagnant film model persists in the engineering literature for two main reasons. First, it is a useful conceptual model for showing concentration gradients over a transport distance. If multiple gradients exist, the size of the gradient yields insights into the factors limiting (or providing a resistance to) mass transport. The concept of a mass transport resistance is critical to understanding factors that control chemical transport in complicated systems. By using the film concept, we can easily envision layers of resistance to transport in multiphase transport problems without having to derive complicated mass transport equations at every step. Second, the film concept allows us to solve complicated problems where it is impossible to solve more accurately for chemical gradients. For example, in transport to sediments on the ocean floor there is no beginning and end to the system that we could use to apply a boundary layer solution. By considering that over reasonable distances there is some average thickness of a concentration gradient, we can simplify flux calculations and concentrate on other aspects of the problem. Therefore, a stagnant fluid model should only be used as a type of crutch when no analytical or numerical solutions are available or justified for the transport calculation.

6.3 MULTIPLE RESISTANCES DURING INTERPHASE MASS TRANSPORT

In all examples of mass transport considered until this point, we have only derived a chemical flux based on mass transport within one phase. Although several situations have involved two phases, for example, the flux of a chemical from air into water, our transport equation was written about a control volume that did not include the gas phase. For some cases it may not be possible to consider only the liquid or gas phase. There may be resistances to mass transport in both phases, as evidenced by chemical gradients in each phase at a phase interface. If there are two chemical gradients, which one is used to define the overall rate of mass transport? To answer this question, we must consider how to define the resistances to mass transport for multiple phases and multiple resistances.

For mass transport between water and air we can write the chemical flux between the two phases as either

$$\mathbf{J}_{Cw} = \mathbf{k}_w c_w (x_{C,i} - x_{C,\infty}) \tag{6-15}$$

for the flux of C from the water phase to the air phase for an overall chemical difference Δx, or as

$$\mathbf{J}_{Ca} = \mathbf{k}_a c_a (y_{C,\infty} - y_{C,i}) \tag{6-16}$$

where $\mathbf{k}_w$ and $\mathbf{k}_a$ are the water and air phase mass transport coefficients, and $x_{C,\infty}$ and $y_{C,\infty}$ are the mole fractions of C in the bulk liquid and gas phases. If there is no chemical reaction, it must be true at steady state that these fluxes are equal, or $\mathbf{J}_{Cw} = \mathbf{J}_{Ca}$.

In the case where gradients develop in both phases the interface concentrations may not simply be the equilibrium concentrations derived from bulk conditions. To see this, consider chemical transport between a liquid and a gas for the cases shown in Figure 6.2 for chemical C in a container that is initially sealed. If this two-phase system is at equilibrium, then chemical C will be distributed about the two phases, as expected based on the Henry's Law constant for the chemical with the interface concentrations equal everywhere in the phase to their equilibrium concentrations (either $x_{C,eq}$ or $y_{C,eq}$). The quantity h is the distance between the equilibrium mole fractions at the interface (Fig. 6.2a) or

$$h = y_{C,i} - x_{C,i} \tag{6-17}$$

where $x_{C,i}$ and $y_{C,i}$ are the values of the mole fractions at the interface. The Henry's

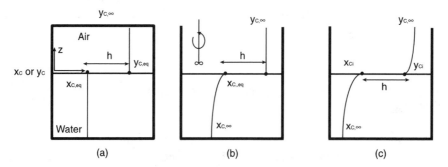

Figure 6.2   Different chemical mole fractions at the air–liquid interface: (a) system is at equilibrium; (b) top is removed from container, resulting in a concentration gradient in the liquid, but the gas is mixed, disrupting any concentration gradient; (c) without fluid mixing concentration gradients develop in both phases, but the ratio of the mole fractions at the interface is always a function of the Henry's Law constant according to the distance h.

Law constant can be related to h. Using $x_C = y_C/H_{Cyx}$, we calculate the distance h in terms of the Henry's Law constant as

$$h = y_{C,i}\left(1 - \frac{1}{H_{Cyx}}\right) \tag{6-18}$$

If we now remove the top to the container, and keep the air phase well mixed, a chemical gradient will not be able to occur in the air phase (Fig. 6.2b). However, there will be a chemical gradient in the liquid phase. Since y_C is uniform everywhere in the gas phase, we can assume that the mole fraction of C in the gas phase at the interface will be $y_{C,i} = y_{C,\infty}$, and that the mole fraction of C at the interface in the liquid phase will be $x_{C,eq}$, in equilibrium with $y_{C,\infty}$.

If we stop mixing the gas phase, a chemical gradient can now develop in the gas phase (Fig. 6.2c). Once the interface concentration of C in the air drops below that of the bulk air phase, we can no longer express the liquid-phase concentration as an equilibrium concentration related to $y_{C,\infty}$. We can still assume that Henry's Law applies here, or that $y_{C,i} = Hx_{C,i}$ but both of the interphase mole fractions are unknown.

The Overall Mass Transport Coefficient   Because we usually do not know the value of the interface concentrations, the flux of chemical i in a system between two phases (j and k) is usually calculated in terms of an overall mass transport coefficient, $\mathbf{K}_{jk}$, defined in terms of the overall driving force present in only one of the systems. For example, the overall chemical flux of C between the water and air phases could be written in terms of an overall mass transport coefficient as either

$$\mathbf{J}_{Cw} = \mathbf{K}_{wa}c_w(x_{C,eq} - x_{C,\infty}) = \mathbf{K}_{wa}(c_{Cw,eq} - c_{Cw,\infty}) \tag{6-19}$$

based on a mole fraction or concentration difference in the water phase (Δx_C, or Δc_{Cw}) or as

$$\mathbf{J}_{Ca} = \mathbf{K}_{aw}c_a(y_{C,\infty} - y_{C,eq}) = \mathbf{K}_{aw}(c_{Ca,\infty} - c_{Ca,eq}) \tag{6-20}$$

based on the air phase, where $x_{C,eq}$ is the mole fraction of C in water calculated to be in equilibrium with the bulk-phase mole fraction in the air phase, $y_{C,\infty}$, and $y_{C,eq}$ is the equilibrium condition for the corresponding bulk-phase condition of $x_{C,\infty}$. The first subscript on the mass transport coefficient is used to designate the phase used for the concentration difference.

It is possible to relate the overall mass transport coefficients to the liquid- and gas-phase mass transport coefficients when there is a linear equilibrium relationship between the gas and liquid phases. As we know from Chapter 2, this linear relationship occurs when dealing with dilute solutions. Starting with the general expression for fugacity

$$f_{Ca} = f_{Cw} \tag{6-21}$$

For a chemical C in air and water, we can simplify this equation for concentrations in terms of mole fractions as

$$y_C = x_C \frac{(\gamma_{Cw} p_C^\circ)}{P} \tag{6-22}$$

As long as the solution is dilute, the activity coefficient is constant. Since the vapor pressure of C is also a constant, the values in the parentheses are constant. Equation 6-22 is simply Henry's Law and can be rewritten in terms of mole fractions as

$$y_C = x_C H_{Cyx} \tag{6-23}$$

or in terms of concentrations as

$$c_{Ca} = c_{Cw} H_{Ccc} \tag{6-24}$$

Therefore, as long as the solution is dilute, there exists a linear equilibrium relationship for chemical C between the two phases. This linear relationship may not hold at higher concentrations as shown in Figure 6.3.

Based on Henry's Law for dilute solutions, we can now write the following equalities between the gas and liquid phases

$$c_{Ca,i} = c_{Cw,i} H_{Ccc} \tag{6-25}$$

$$c_{Ca,eq} = c_{Cw,\infty} H_{Ccc} \tag{6-26}$$

$$c_{Cw,eq} = \frac{c_{Ca,\infty}}{H_{Ccc}} \tag{6-27}$$

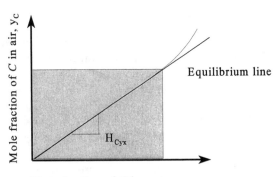

Figure 6.3 For dilute solutions (low concentrations of C in air and water) the ratio of y_C and x_C is constant and equal to the Henry's Law constant, H_{Cyx} (shaded region).

The overall mass transport coefficient can now be related to the individual mass transport coefficients by rearranging Eq. 6-19 in the form

$$\frac{1}{\mathbf{K}_{wa}} = \frac{(c_{Cw,eq} - c_{Cw,\infty})}{\mathbf{J}_{Cw}} \tag{6-28}$$

Expanding the term on the right-hand side, we can introduce the interface mole fraction to obtain

$$\frac{1}{\mathbf{K}_{wa}} = \frac{(c_{Cw,eq} - c_{Cw,i})}{\mathbf{J}_{Cw}} + \frac{(c_{Cw,i} - c_{Cw,\infty})}{\mathbf{J}_{Cw}} \tag{6-29}$$

From the assumption that Henry's Law is valid, we can substitute Eqs. 6-24 and 6-25 into Eq. 6-28, producing

$$\frac{1}{\mathbf{K}_{wa}} = \frac{(c_{Cw,eq} - c_{Cw,i})}{\mathbf{J}_{Cw}} + \frac{(c_{Ca,i} - c_{Ca,\infty})}{H_{Ccc}\mathbf{J}_{Cw}} \tag{6-30}$$

Substituting the definitions of the single-phase mass transport coefficients in Eqs. 6-15 and 6-16 for the two right-hand terms in Eq. 6-29, we have

$$\frac{1}{\mathbf{K}_{wa}} = \frac{1}{\mathbf{k}_w} + \frac{1}{H_{Ccc}\mathbf{k}_a} \tag{6-31}$$

Starting with Eq. 6-21, the following equation can also be obtained for the overall gas-phase mass transport coefficient

$$\frac{1}{\mathbf{K}_{aw}} = \frac{1}{\mathbf{k}_a} + \frac{H_{Ccc}}{\mathbf{k}_w} \tag{6-32}$$

Equations 6-31 and 6-32 demonstrate that the overall mass transport coefficient can be expressed in terms of the individual phase mass transport coefficients as long as equilibrium conditions and Henry's Law is assumed to be valid for the transport of chemical C in a dilute system. The same expression can be written in terms of mole fractions, and H_{Cyx}, but phase concentrations must be included in the derivation. For example, Eq. 6-32 can be written using H_{Cyx} as

$$\frac{1}{\mathbf{K}_{aw}} = \frac{1}{\mathbf{k}_a} + \frac{H_{Cyx}c_a}{\mathbf{k}_w c_w} \tag{6-33}$$

This derivation is left as an exercise for the reader (see Problem 6.9)

Example 6.1

In an air stripper treating groundwater contaminated with a volatile chemical, the mole fractions measured in the bulk fluid at one point in the system are $x_{C,1}$ and $y_{C,1}$. Show in a graph how this point is related to the equilibrium curve drawn in Figure 6.3.

At any point in a system the chemical flux from the liquid and gas phases must be equal, or

$$\mathbf{J}_{Cw} = \mathbf{J}_{Ca} \tag{6-34}$$

Substituting Eqs. 6-15 and 6-16 into Eq. 6-32

$$k_w c_w (x_{C,i} - x_C) = \mathbf{k}_a c_a (y_C - y_{C,i}) \tag{6-35}$$

Substituting in our defined conditions for the mole fractions at point 1 and rearranging

$$-\frac{\mathbf{k}_w c_w}{\mathbf{k}_a c_a} = \frac{(y_{C,1} - y_{C,i})}{(x_{C,1} - x_{C,i})} \tag{6-36}$$

This result indicates that our bulk and interface concentrations are related by the ratio of the two mass transport coefficients. If we plot these points along with the equilibrium line, it is easy to see that the bulk concentrations are related to the two interface concentrations, which are assumed according to Henry's Law to be in equilibrium with each other, by a line having a slope of $-\mathbf{k}_w c_w / \mathbf{k}_a c_a$ shown in Figure 6.4.

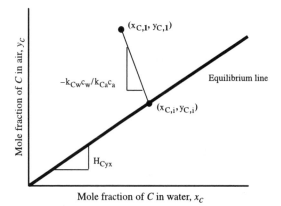

Figure 6.4   Mole fractions discussed in Example 6.1.

Mass Transport Resistances   In the analysis of a chemical reaction that occurs via a series of chemical reactions, it is often useful to identify the slowest chemical reaction since this reaction, if it is much slower than other reactions, will limit the overall rate. A similar analysis can be applied to mass transport. When the transport of a chemical occurs across an interface, the overall transport rate will be a function of the gas and liquid mass transport coefficients as shown by Eq. 6-31. If transport in one of these phases is slower than in the other phase, then the slower transport step will control the overall transport rate.

Interphase transport rates are typically analyzed in terms of the overall *resistances* to mass transport. While the transport coefficient defines the rate of chemical transport, the inverse of this coefficient yields the resistance to transport. For the overall transport coefficient, we can say that

$$\frac{1}{K_{wa}} = \text{total resistance to mass transport} \tag{6-37}$$

which according to Eq. 6-31 is the sum of the resistances to transport in the gas and liquid phases, or

$$\frac{1}{k_w} = \text{resistance in the water phase} \tag{6-38}$$

$$\frac{1}{H_{Ccc}k_a} = \text{resistance in the air phase} \tag{6-39}$$

By examining the magnitude of these resistances, it is possible to see if transport limitations in one phase controls the overall rate of mass transport. Ordinarily, the magnitude of the Henry's Law constant will define which phase limits the overall mass transport rate. For large Henry's Law constants, or as $H_{Ccc} \to \infty$, Eq. 6-31 becomes

$$\frac{1}{K_{wa}} = \frac{1}{k_w} + \frac{1}{\infty} \tag{6-40}$$

or since $1/\infty \to 0$

$$\frac{1}{K_{wa}} = \frac{1}{k_w} \tag{6-41}$$

For very large Henry's Law constants there is little resistance in the gas phase to mass transport from the liquid to the gas, and we say that the transport is liquid-phase controlled. Oxygen is an example of a chemical that is sparingly soluble in the liquid phase ($H_{Oyx} = 43,000$) and whose transport is nearly completely limited by resistance in the liquid phase. Under such conditions we expect that there will be a negligible chemical gradient in the gas phase.

For small Henry's Law constants we can make a comparison on the basis of the overall magnitude of gas- and liquid-phase resistances. As $H_{Ccc} \to 0$ the resistance in the gas phase will become increasingly large (approaching infinity at the limit) in comparison to the liquid-phase resistance, or

$$\frac{1}{H_{Ccc}\mathbf{k}_a} \gg \frac{1}{\mathbf{k}_w} \tag{6-42}$$

Therefore for very small Henry's Law constants we can approximate the overall mass transport resistance as

$$\frac{1}{\mathbf{K}_w} = \frac{1}{H_{Ccc}\mathbf{k}_a} \tag{6-43}$$

This analysis yields two useful limiting conditions that

large H_{Ccc} (water-phase controlled): $\mathbf{K}_{wa} = \mathbf{k}_w$ $\tag{6-44}$

small H_{Ccc} (gas-phase controlled): $\mathbf{K}_{wa} = H_{Ccc}\mathbf{k}_a$ $\tag{6-45}$

From an engineering perspective it is very useful to know which phase controls the overall rate of mass transport in designing systems to maximize the rate of mass transport. For example, if a system is liquid-phase controlled, mixing the gas phase will have little effect on the overall rate of mass transport unless mixing is conveyed by shear in the gas phase to the liquid phase. Similarly, if the transport is gas-phase controlled, then mixing the liquid will have little effect on the overall mass transport rate.

Example 6.2

During the transport of a chemical from air into a liquid you measure, using microprobes, interface concentrations of $x_{Ci} = 0.4$ and $y_{Ci} = 0.8$ at a time when the bulk-phase concentrations are $x_{C\infty} = 0.1$ and $y_{C\infty} = 0.9$. If the Henry's Law constant, $H_{Cyx} = 2$ (mole fraction/mole fraction), and the gas-phase mass transport coefficient, $\mathbf{k}_a c_a = 0.1$ mol cm^{-2} s^{-1}, calculate the value of h, the chemical flux across the interface, and the individual and overall liquid-phase mass transport coefficients on a mole concentration basis, $\mathbf{k}_w c_w$ and $\mathbf{K}_{wa} c_w$.

If the mole fraction at the interface in the gas phase is known, then the magnitude of h can be calculated from Eq. 6-18 as

$$h = y_{Ci}\left(1 - \frac{1}{H_{Cyx}}\right) = 0.8\,(1 - \tfrac{1}{2}) \tag{6-46}$$

$$= 0.4 \tag{6-47}$$

From our definition of the individual mass transport coefficients, the flux can be calculated as

$$\mathbf{J}_{Ca} = \mathbf{k}_a c_a (y_{C,\infty} = y_{C,i}) = 0.1\,\frac{\text{mole}}{\text{cm}^2\,\text{s}}\,(0.9 - 0.8) = 0.01\,\frac{\text{mole}}{\text{cm}^2\,\text{s}} \tag{6-48}$$

Using this flux, the liquid-phase mass transport coefficient can be calculated as

$$\mathbf{k}_w c_w = \frac{\mathbf{J}_{Cw}}{(x_{C,i} - x_{C,\infty})} = \frac{0.01 \text{ mol/cm}^2 \text{ s}}{(0.4 - 0.1)} = 0.033 \frac{\text{mol}}{\text{cm}^2 \text{ s}} \qquad (6\text{-}49)$$

From our definition of the overall mass transport coefficient in Eq. 6-31, we have

$$\frac{1}{\mathbf{K}_{wa} c_w} = \frac{1}{\mathbf{k}_w c_w} + \frac{1}{H_{cx} k_a c_a} = \frac{1}{(0.033 \text{ mol/cm}^2 \text{ s})} + \frac{1}{(2)(0.1 \text{ mol/cm}^2 \text{ s})} \qquad (6\text{-}50)$$

or after evaluation,

$$\frac{1}{\mathbf{K}_{wa} c_w} = 30 \frac{\text{cm}^2 \text{ s}}{\text{mol}} + 5 \frac{\text{cm}^2 \text{ s}}{\text{mol}} \qquad (6\text{-}51)$$

which produces the final result that

$$\mathbf{K}_{wa} c_w = 0.028 \frac{\text{mol}}{\text{cm}^2 \text{ s}} \qquad (6\text{-}52)$$

In this case 85% [(30/35) $\times$ 100] of the resistance to mass transport is due to the liquid phase. A larger amount of mass transport resistance in the liquid phase is expected since the Henry's Law constant for this chemical is greater than 1.

We can check our calculation for $\mathbf{K}_{wa} c_w$ by calculating the equilibrium condition for the mole fraction of C in the water in equilibrium with the bulk gas phase using Eq. 6-26 as

$$x_{C,eq} = \frac{y_{C,\infty}}{H_{Cyx}} = \frac{0.9}{2} = 0.45 \qquad (6\text{-}53)$$

From our definition of the overall transport coefficient written about the liquid-phase concentrations, Eq. 6-19, we calculate the flux as

$$\mathbf{J}_{Cw} = \mathbf{K}_{wa} c_w (x_{C,eq} - x_{C,\infty}) = 0.028 \frac{\text{mol}}{\text{cm}^2 \text{ s}} (0.45 - 0.1) = 0.01 \frac{\text{mol}}{\text{cm}^2 \text{ s}} \qquad (6\text{-}54)$$

which is identical to the flux calculated in Eq. 6-48. The relationships between these mole fractions is shown in Figure 6.5.

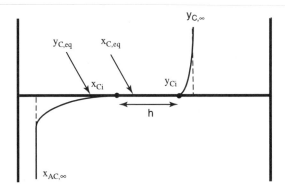

Figure 6.5   Mole fractions discussed in Example 6.2.

6.4   EMPIRICAL CORRELATIONS FOR MASS TRANSPORT COEFFICIENTS

The previous derivations of mass transport correlations have provided the theoretical framework for chemical flux calculations and are therefore a starting point for more practical systems. While it is possible to derive a concentration profile around a single spherical particle, or even a particle moving at a constant velocity in a homogeneous medium, the engineer is faced with calculating fluxes under very different conditions. For example, oxygen transport in activated sludge reactors generated by gas diffusers are a function of gas flow rates (which generate different bubble size distributions), tank geometry (particularly height), diffuser type, wastewater composition, and other factors. Thus mass transport correlations, which can be developed for specific transport situations, often play a more important role in engineering calculations than the theoretical models already developed.

In this section the rate of mass transport to spheres, thin fluid films, plates, and within pipes is reinvestigated in terms of correlations that have been reported in the literature. Since it would be impossible to review fully all the transport calculations in this area, this review is performed mainly to emphasize the nature of such mass transport correlations.

Transport to a Moving Sphere

The rate of mass transport to spheres has been extensively studied, in part, due to the ease of placing a sphere in a moving fluid, but primarily due to the large number of mass transport conditions involving packed beds. Here only transport to a sphere fixed in a convective fluid is considered. Transport to moving spheres in stagnant fluids and to spheres in packed beds is considered in later chapters.

For mass transport to a sphere at rest we derived (Eq. 5-130) that the chemical flux was

$$\mathbf{J}_{Cw}|_{r=R} = \frac{D_{Cw}}{R}(c_{Cw,0} - c_{Cw,\infty}) \qquad (6\text{-}55)$$

where we have substituted the boundary condition $c_{Cw,0}$ for the equilibrium condition $c_{Cw,eq}$. Comparing this flux with the form assumed for a mass transport coefficient (Eq. 6-2), we see that the mass transport coefficient $\mathbf{k}_w$ for this case must be related to D_{Cw} and R according to

$$\mathbf{k}_w = \frac{D_{Cw}}{R} \qquad (6\text{-}56)$$

From our definition of the Sh number for a sphere of radius R, or Sh = $\mathbf{k}R/D$, for a nonmoving sphere the theoretical prediction is that

$$Sh = \frac{\mathbf{k}_w R}{D_{Cw}} = 1 \qquad (6\text{-}57)$$

The theoretical analysis indicates that Sh = 1 for a sphere at rest. When there is advective flow past the sphere, we expect the concentration boundary layer thickness to decrease, producing an increase in the overall rate of mass transport. Experimental studies confirm this hypothesis. The best form of an empirical correlation varies, but many have the general form

$$Sh = 1 + \alpha\, Re^{\beta}\, Sc^{\gamma} \qquad (6\text{-}58)$$

where the dimensionless Re number is calculated here based on the radius R, or Re = $\mathbf{u}R/\nu$, and the Schmidt number is Sc = ν_w/D_{Cw}. The values of α, β, and γ depend on the experimental conditions. For example, for chemical transport into gases, Fröessling (Welty et al., 1984) obtained

$$Sh = 1 + 0.39\, Re^{1/2}\, Sc^{1/3}, \qquad 1 < Re < 400, \quad 0.6 < Sc < 2.7 \qquad (6\text{-}59)$$

For a slightly more restrictive range of Sc numbers ($0.6 \le Sc \le 1.85$), this correlation can be applied up to Re < 12,000 (Evnochides and Thodos, 1959).

For chemical transport in liquid streams over a modest range in Re and Sc (Garner and Suckling, 1958),

$$Sh = 1 + 0.67\, Re^{1/2}\, Sc^{1/3}, \qquad 50 < Re < 350,\ 1200 < Sc < 1525 \qquad (6\text{-}60)$$

For water Sc are generally in range of 500 to 1000, and transport correlations are usually determined in terms of a Peclet number, Pe, defined as Pe = Re Sc. At lower

Pe, the equation developed by Brian and Hales (1969)

$$Sh = (1 + 0.48\ Pe^{2/3})^{1/2}, \qquad Pe < 5000 \tag{6-61}$$

applies, while for higher Pe the Levich (1962) correlation

$$Sh = 0.63\ Pe^{2/3}, \qquad Pe > 5000 \tag{6-62}$$

is applicable. Notice that the Levich correlation does not reduce to the case of Sh = 1 for stagnant flow ($u = 0$), since this equation was developed specifically for higher advective transport rates to isolated spheres in uniform flow.

Although many of the above equations reduce to the case of simple diffusion to a sphere at rest for $u = 0$, they also are applicable to instances where advection (forced convective flow) dominates over natural convection. When

$$Re > 0.6\ Gr^{1/2}\ Sc^{-1/6} \tag{6-63}$$

natural convection will dominate, where Gr is the Grashoff number defined in terms of the particle radius as

$$Gr = \frac{R^3 \rho g\ \Delta\rho}{\mu^2} \tag{6-64}$$

For the situation of low velocities (natural convection), the mass transport coefficient can be obtained from one of the following equations (Steinberger and Treybal, 1960)

$$Sh = 1 + 0.27\ (Re\ Sc^{1/2})^{0.62}, \qquad 0.5 < Re < 1.4 \times 10^4, \quad 0.6 < Sc < 3600 \tag{6-65}$$

or

$$Sh = 1 + 0.48\ (Gr\ Sc)^{0.25}, \qquad Gr\ Sc < 10^7 \tag{6-66}$$

or

$$Sh = 1 + 0.0254\ (Gr\ Sc)^{1/3}\ Sc^{0.244}, \qquad Gr\ Sc < 10^7 \tag{6-67}$$

Empirical Correlations for Transport to and from Wetted Walls

Extensive research into mass transport to thin fluid films has produced a better understanding of the fluid flow mechanics and chemical transport to wetted walls. Fulford (1964) provides an excellent review. The flow characteristics are usually described in terms of the Re defined here for thin fluid films as $Re = u\delta/\nu$, where δ is the thickness of the fluid film, or equivalently as $Re = \Gamma/\mu$, where Γ is the flow per wetted perimeter. Re is sometimes presented as a quantity four times larger than this number, or $4\Gamma/\mu$, but there is little reason to use this form for Re other than the historical one.

Stable laminar flow is rapidly developed within a fluid film after an entrance length, L_{ent} (Stucheli and Ozisik, 1976), of

$$L_{ent} = 0.3\delta \text{ Re} \tag{6-68}$$

The laminar–turbulent transition typically occurs at Re of 250 to 400 depending on the capillary and surface forces of the fluid and support. Wavy flow can occur at even lower Re (Re from 1 to 10), although even after the onset of wavy or turbulent flow much of the fluid will still be laminar.

Based on our previous analysis for mass transport to a wetted wall when chemical penetration was limited to short distances, we derived that the mass transport coefficient at any time t (equivalent to any distance of flow by $t = z/u_{max}$) was

$$\mathbf{k}_{w|0} = \left(\frac{D_{Cw}}{\pi t}\right)^{1/2} \tag{6-69}$$

which has become known as the penetration theory mass transport coefficient. Integrating this over a length L, the mass transport coefficient can be expressed in terms of Sh, we can similarly write

$$\text{Sh} = \frac{\mathbf{k}_w L}{D_{Cw}} = \left(\frac{4u_{max}L}{\pi D_{Cw}}\right)^{1/2} \tag{6-70}$$

or using the dimensionless numbers $\text{Re}_L = u_{max}L/\nu_w$ and $\text{Sc} = \nu_w/D_{Cw}$, Eq. 6-70 can be written as

$$\text{Sh} = \frac{2}{\pi^{1/2}} \text{Re}_L^{1/2} \text{Sc}^{1/2} = \frac{2}{\pi^{1/2}} \text{Pe}_L^{1/2} \tag{6-71}$$

Vivian and Peaceman (1956) found that penetration theory overestimated the mass transport of a chemical into a liquid film, when the resistance was primarily in the gas phase, by 10 to 20%. In their studies transport was studied using transport to fluids flowing down the inside of pipes. Their correlation was

$$\text{Sh} = \frac{\mathbf{k}_w L}{D_{Cw}} = 0.754 \text{ Sc}^{1/2} \text{ Re}_\delta^{0.4} \left(\frac{gL^3}{\nu_w^2}\right)^{1/6} \tag{6-72}$$

where L is the length of contact and Re_δ is based on the fluid film thickness and average fluid velocity. Thus $\mathbf{k}_w$ is proportional to D_{Cw} to the $\frac{1}{2}$ power, as expected from penetration theory.

In early studies Gilliland and Sherwood (1934) examined the evaporation of water and eight different pure-phase organic liquids (such as chlorobenzene and propanol) into air inside a tube having a fixed diameter of 2.54 cm. Their correlation for the gas-phase mass transport coefficient, based on nearly 400 tests, was

$$\text{Sh} = \frac{\mathbf{K}_{wa}d}{D_{Ca}} = 0.023 \ \text{Re}_d^{0.83} \ \text{Sc}^{0.44} \ \frac{P}{p_{A,\text{lm}}} \tag{6-73}$$

where $\mathbf{K}_{wa}$ is the mass transport coefficient between the air and the liquid in the tube, d is the tube diameter, $\text{Re} = u_a d / v_a$ and $\text{Sc} = v_a / D_{Ca}$ are based on air properties, P is the total pressure, and $p_{A,\text{lm}}$ is the log mean air partial pressure in the tube. This correlation is valid for $2000 < \text{Re} < 35{,}000$ and $0.6 < \text{Sc} < 2.5$.

More recently, Roberts et al. (1985) examined the mass transport of six dilute organic solutes in water a countercurrent packed bed air stripping tower. A transport model developed by Onda et al. (1968) predicted transport for these chemicals more accurately than models developed by Sherwood and Holloway (1940) and Shulman et al. (1955). The Onda model incorporates separate mass transport coefficients for both the water and air phases. The water-phase mass transport coefficient was

$$\mathbf{k}_w \left(\frac{\rho_w}{\mu_w g} \right)^{1/3} = 0.0051 \ \text{Re}_{w,J}^{2/3} \ \text{Sc}_w^{-0.5} (A_p d_p)^{0.4} \tag{6-74}$$

where $\text{Re}_{w,J}$ is defined here as $\text{Re} = J_w / a_w \mu_w$, J_w is the liquid flux (kg m^{-2} s^{-1}), a_w is the specific wetted packing surface area, Sc_w is based on the liquid-phase properties ($\text{Sc}_w = v_w / D_{Cw}$), A_p is the total packing surface area, and d_p is the nominal packing size of media in the column for rings, spheres, rods, and saddles 4 to 50 mm in diameter. Notice that $\mathbf{k}_w$ is proportional to $D_{Cw}^{1/2}$, as predicted by penetration theory. The specific wetted packing surface area, a_w, can be calculated using

$$\frac{a_w}{A_p} = 1 - \exp \left[-1.45 \left(\frac{\sigma_c}{\sigma_w} \right)^{0.75} \text{Re}_{w,J}^{0.1} \ \text{Fr}^{-0.05} \ \text{We}^{0.2} \right] \tag{6-75}$$

where σ_c is the critical surface area of packing ($\sigma_c = 0.061$ kg s^{-2} for Berl saddles), σ_w is the surface tension of water, the Froude number and Weber numbers are calculated as $\text{Fr} = J_w^2 a / \rho_w^2 g$, and $\text{We} = J_w^2 / \rho_w \sigma_w A_p$. Equation 6-75 is expected to be accurate to within $\pm 20\%$ for saddles, spheres, and rods made of ceramic, glass, PVC, and wax-coated materials.

The gas-phase mass transport coefficient was

$$\frac{\mathbf{k}_a}{A_p D_{Ca}} = 5.23 \ \text{Re}_{a,J}^{0.7} \ \text{Sc}_a^{1/3} (A_p d_p)^{-2} \tag{6-76}$$

where a_a is the specific surface area of air [L^{-1}], $\text{Re}_{a,J} = J_a / a_a \mu_a$ is defined in terms of the gas flux, J_a (kg m^{-2} s^{-1}), and Sc_a is based on the gas-phase properties ($\text{Sc} = v_a / D_{Ca}$). The constant 5.23 applies only to packing larger than 15 mm, and should be 2.00 for smaller media.

Roberts et al. determined from experiments in columns packed with ceramic Berl saddles that the Onda model predicted transport rate constants of the six chemicals within an average standard deviation of 21%. The transport of two of the most volatile chemicals was completely controlled by the liquid-phase resistance (oxygen and dichlordifluormethane), while the moderately volatile chemicals (carbon tetrachloride, tetrachloroethylene, trichloroethylene, and chloroform) had resistances in both phases.

Transport to Plates and Surfaces

Based on the analysis of uniform flow field over a flat surface, Eq. 6-13 was used to describe the chemical flux from a surface as a function of travel distance x along the plate. The local mass transport coefficient (the mass transport coefficient at any point x) in Eq. 6-14 can therefore be used to specify the Sh for laminar flow over a flat plate as

$$\text{Sh} = \frac{\mathbf{k}_w x}{D_{Cw}} = 0.332 \text{ Re}_x^{1/2} \text{ Sc}^{1/3} \tag{6-77}$$

For turbulent flow ($\text{Re} > 3 \times 10^5$) it has been shown that

$$\text{Sh} = 0.0292 \text{ Re}_x^{4/5} \text{ Sc}^{1/3} \tag{6-78}$$

These equations are applicable to Schimdt numbers in the range of $0.6 < \text{Sc} < 2500$.

By integrating mass transport over a plate of length L, the overall mass transport equations can be derived from Eqs. 6-77 and 6-78 as

$$\text{Sh} = 0.664 \text{ Re}_L^{1/2} \text{ Sc}^{1/3}, \qquad \text{laminar flow} \tag{6-79}$$

$$\text{Sh} = 0.036 \text{ Re}_L^{4/5} \text{ Sc}^{1/3}, \qquad \text{turbulent flow} \tag{6-80}$$

If there is a transition in the flow from laminar to turbulent, the mass transport coefficient needs to be integrated over the flow length. This average mass transport coefficient, as $\mathbf{k}_{w,\text{avg}}$, can be calculated by using

$$\mathbf{k}_{w,\text{avg}} = \frac{\displaystyle\int_0^{L_{\text{lam}}} \mathbf{k}_w \text{ (laminar) } dx + \int_{L_{\text{lam}}}^{L} \mathbf{k}_w \text{ (turbulent) } dx}{\displaystyle\int_0^{L} dx} \tag{6-81}$$

where the laminar and mass transport coefficients can be obtained from Eqs. 6-77 and 6-78. The final solution for a transition length L_{lam} defined as the distance where

Re = 3 × 10^5 for a total length of flow L (Welty et al., 1984), is

$$\text{Sh}_{\text{avg}} = \frac{k_{w,\text{avg}} L}{D_{Cw}} = 0.664 \, \text{Re}_{\text{lam}}^{1/2} \, \text{Sc}^{1/3} + 0.036 \, \text{Sc}^{1/3} \, (\text{Re}_{\text{lam}}^{4/5} - \text{Re}_L^{4/5}) \quad (6\text{-}82)$$

This result indicates that the mass transport coefficient must be weighted by the fraction of flow that occurs within the laminar region.

6.5 MASS, ENERGY, AND MOMENTUM TRANSPORT ANALOGIES

As was shown in Section 4.2, a generalized transport equation can be developed to equally describe mass, energy (heat), or momentum transport. Dispersive transport of momentum, heat, and mass are individually described by Newton's, Fourier's, and Fick's Laws (Section 4.2), but all three laws have diffusion coefficients with units $[L^2 \, t^{-1}]$. Ratios of diffusive or advective transport rates in these different transport systems produce dimensionless numbers. Scaling transport according to these dimensionless ratios permits the development of analogies between systems and allowing greater accuracy in predictions of transport of components.

It has already been demonstrated how the dimensionless numbers such as Re, Sc, and Sh can be used to obtain mass transport coefficients. However, other dimensionless groups have been developed to scale mass transport and develop correlations for a wide range of conditions. Such groups have been developed based on analogies between different systems, while others are instead based on empirical data.

The Reynolds Analogy

The first and best-known analogy was proposed by Reynolds to relate the analogous behavior of momentum and energy transfer in the flow over a flat plate. For laminar flow over a flat plate in air, where Sc = 1, the velocity and concentration profiles are equal (see Section 5.2.3), and we can write

$$\frac{\partial}{\partial y} \left(\frac{u_x}{u_\infty} \right) \Bigg|_{y=0} = \frac{\partial}{\partial y} \left(\frac{c_{Ca} - c_{Ca,0}}{c_{Ca,\infty} - c_{Ca,0}} \right) \Bigg|_{y=0} \quad (6\text{-}83)$$

Based on the definition of the mass transport coefficient that $J_{Ca} = \mathbf{k}_a \, \Delta c_{Ca}$, we can write that at $y = 0$ (the surface of the plate)

$$J_{Ca} = -D_{Ca} \frac{\partial (c_{Ca} - c_{Ca,0})}{\partial y} \Bigg|_{y=0} = \mathbf{k}_a (c_{Ca,0} - c_{Ca,\infty}) \quad (6\text{-}84)$$

Combining the above two equations, we obtain

$$\mathbf{k}_a = \frac{D_{Ca}}{u_\infty} \frac{\partial u_x}{\partial y} \Bigg|_{y=0} \quad (6\text{-}85)$$

For air when Sc = 1, it is therefore also true that $D_{Ca} = v_a$, and Eq. 6-85 can equivalently be written as

$$\mathbf{k}_a = \frac{v_a}{u_\infty} \frac{\partial u_x}{\partial y}\bigg|_{y=0} \tag{6-86}$$

The velocity gradient at the plate surface is related to the shear rate according to Newton's Law, or

$$\mathbf{k}_a = \frac{v_a}{u_\infty} \frac{\tau_{yx}}{\mu_a} \tag{6-87}$$

The drag force due to friction at the surface of the plate set up by the motion of the fluid can be related to the quantity $\rho_a u_\infty / 2$, a quantity often referred to as the dynamic pressure, according to

$$\tau_{yx} = C_f \frac{\rho_a u_{a,\infty}}{2} \tag{6-88}$$

where C_f is a friction coefficient. Combining these last two equations, we have the final result for the Reynolds analogy for air

$$\frac{\mathbf{k}_a}{u_\infty} = \frac{C_f}{2} \tag{6-89}$$

The result is that this analogy predicts that the ratio of the mass transport coefficient to the free stream velocity in flow over a flat plate will always be equal to a constant. While this is a useful result, the real utility of the analogy is not appreciated until the same derivation is conducted for heat transport, which is shown here without derivation, to produce

$$\frac{k_{H,a}/(\rho_a c_{a,p})}{u_\infty} = \frac{C_f}{2} \tag{6-90}$$

where $k_{H,a}$ is the heat transport coefficient and $c_{a,p}$ is the heat capacity of the fluid. Since both of these equations are equal to the same constant, they are equal to each other, or

$$\frac{\mathbf{k}_a}{u_\infty} = \frac{k_{H,a}/(\rho_a c_{a,p})}{u_\infty} \tag{6-91}$$

Thus for air it is possible to obtain a mass transport coefficient directly from a heat transport coefficient.

The Chilton–Colburn Analogy

While the Reynolds analogy is limited to Sc = 1, Chilton and Colburn developed the j factor for mass transfer

$$j_D \equiv \frac{\mathbf{k}_a}{u_{a,\infty}} \, Sc^{2/3} = \frac{C_f}{2} \tag{6-92}$$

This analogy, which relates mass and momentum transport, incorporates Sc and therefore is not restricted to Sc = 1. It has been found to be applicable to $0.6 < Sc < 2500$.

For laminar flow over the plate this analogy satisfies the theoretical analysis of flow over a flat plate. Rearranging Eq. 6-77, we have

$$\frac{Sh}{Re_x \, Sc} \, Sc^{2/3} = \frac{0.332}{Re_x^{1/2}} \tag{6-93}$$

Based on the Blasius solution for flow over a flat plate (see, for example, Welty et al., 1984), the right-hand side of Eq. 6-93 is equal to $C_f/2$, or

$$\frac{Sh}{Re_x \, Sc} \, Sc^{2/3} = \frac{C_f}{2} \tag{6-94}$$

From the definitions of each of the dimensionless numbers we can expand the left-hand side

$$\frac{(\mathbf{k}_a x/D_{Ca})}{(u_x x/v_a)(v_a/D_{Ca})} \, Sc^{2/3} = \frac{\mathbf{k}_a}{u_\infty} \, Sc^{2/3} = \frac{C_f}{2} \tag{6-95}$$

obtaining the Chilton–Colburn analogy for mass transport.

The full Chilton–Colburn analogy relates all three types of mass transport, or

$$j_D = j_H = \frac{C_f}{2} \tag{6-96}$$

where the j_H factor is defined for heat transport according to

$$j_H = \frac{k_{H,a}/(\rho_a c_{a,p})}{u_\infty} \, Pr^{2/3} \tag{6-97}$$

and is applicable to $0.6 < Pr < 100$, where $Pr = \mu_a c_{p,a}/h_a$. Equation 6-94 applies to transport over flat plates and other geometries as long as there is no form drag is present, in which case $j_H = j_D$ is still true, but $C_f/2 \neq (j_H = j_D)$.

Using the definition of the j_D factor in Eq. 6-90, the mass transport correlations expressed in Eqs. 6-77 and 6-78 can be recast as

$$j_D = 0.664\,\mathrm{Re}_L^{-1/2}, \qquad \text{laminar flow} \tag{6-98}$$

$$j_D = 0.036\,\mathrm{Re}_L^{-0.2}, \qquad \text{turbulent flow} \tag{6-99}$$

The Chilton–Colburn analogy has been tested for a wide range of conditions. For example, for gas transport over a flat plat (Sherwood et al., 1975),

$$j_D = k_{a,\text{avg}}\frac{p_{A,\text{lm}}}{u_{a,\text{avg}}P}\,\mathrm{Sc}^{2/3} = j_H = 0.037\,\mathrm{Re}_L^{-0.2}, \qquad 8000 < \mathrm{Re}_L < 300{,}000$$

where the subscripts indicate that the average values are used for mass transport over the whole plate of length L. If the gas is air, and the system is dilute in concentration with respect to a chemical C, then the log mean pressure for air is $p_{A,\text{lm}} = P$.

PROBLEMS

6.1 Use Figure P-6.1 to draw profiles of the mole fractions of C in the gas and liquid phases, paying particular attention to the interface mole fractions for two cases:

$$H > 10, \text{ chemical transport from liquid to gas phase}$$

$$H < 0.01, \text{ chemical transport from gas to liquid phase}$$

Since you have not been given exact concentrations, you need to only show the shapes and relative locations of the curves.

∨ **6.2** Mass transfer coefficients were measured by Mackay and Yeun (1983) for 11 organic compounds with various Henry's Law constants in a 6 m long by 0.6 m wide wind–wave tank at 20°C. Their results confirmed the presence of both liquid- and gas-phase resistances to transport of these chemicals from

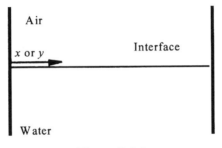

Figure P-6.1

the water to air. Their empirically derived mass transport coefficients (m s^{-1}) were

$$k_a = 1 \times 10^{-3} + 46.2 \times 10^{-3} u_{sh} \, Sc_a^{-0.67}$$

$$k_w = 1 \times 10^{-6} + 34.1 \times 10^{-4} u_{sh} \, Sc_w^{-0.5}, \qquad u_{sh} > 0.3 \text{ m s}^{-1}$$

$$k_w = 1 \times 10^{-6} + 144 \times 10^{-4} u_{sh}^{2.2} \, Sc_w^{-0.5}, \qquad u_{sh} < 0.3 \text{ m s}^{-1}$$

The shear velocity, u_{sh} [m s^{-1}] is defined in terms of the drag coefficient according to

$$C_D = \left(\frac{u_{sh}}{u_\infty} \right)^2$$

where the drag coefficient was empirically determined to be $C_D = 4 \times 10^{-4} u_\infty$, and u_∞ is the free stream velocity (m s^{-1}). For toluene, $p_T^0 = 0.029$ atm, $c_{Tw,eq} = 5.59$ mol m^{-3}, $H_{Tpc} = 5.18 \times 10^{-3}$ atm m^3 mol^{-1}, $D_{Ta} = 0.079$ cm^2 s^{-1}, and $D_{Tw} = 0.844 \times 10^{-5}$ cm^2 s^{-1}. Also, $v_a = 0.156$ cm^2 s^{-1}, and $v_w = 0.009$ cm^2 s^{-1}. (a) What are the liquid, gas, and overall mass transport coefficients for toluene transport for this system if $u_\infty = 8.6$ m s^{-1}? (*Hint:* the universal gas constant can be expressed as $\mathcal{R} = 82 \times 10^{-6}$ atm m^3 mol^{-1} K^{-1}). (b) What is the rate of toluene transport from the water to the gas (in units of mol h^{-1}), when the bulk-phase concentration of toluene is 1 mol m^{-3}?

6.3   Convert the following empirical mass transport equation developed for Sh and Pe numbers based on radius (Sh = kr/D; Pe = vr/D) to an equation based on diameter (Sh = kd/D; Pe = vd/D), collecting all constants into a compact dimensionless correlation. What would be the applicable range of Pe numbers for this new equation?

$$Sh = (1 + 0.48 \, Pe^{2/3})^{1/2}, \qquad Pe < 5000$$

6.4   Methane formed in anaerobic sediments can often be present at high concentrations in overlying waters directly above the sediment. Let us view the sediment here as analogous to flow over a smooth, flat plate. The Chilton–Colburn analogy is often used for mass transport calculations to flat plates. According to this approach mass transport is expressed in terms of a j_D factor defined as $j_D = k_j \, Sc^{2/3}/u_{j,\infty}$, where k_j is the mass transport coefficient for a chemical in phase j, and u_∞ is the approach or free stream velocity above the flat plate

$$j_D = 0.664 \, Re^{-1/2}, \qquad \text{laminar}$$

$$j_D = 0.036 \, Re^{-0.2}, \qquad \text{turbulent}$$

Calculate the flux of methane from the sediment over a total distance of 20 m, assuming that the bulk fluid velocity is 3 m min^{-1}. Properties of methane in water are: $\rho_{Mw,eq}$ = 24.1 mg L^{-1}, D_{Mw} = 1.3 × 10^{-5} cm^2 s^{-1}, γ_{Mw} = 137.5.

√ **6.5** As shown in Figure P-6.5, chemical C is diffusing from a bulk liquid into a sediment where it reacts away according to first-order kinetics with first-order kinetic constant k_1 = 6.3 × 10^{-3} s^{-1}. The diffusivity of C in both the sediment and water can be assumed here to be 700 × 10^{-8} cm^2 s^{-1}. Using microprobes (very small probes developed to measure the concentration of C), we find that the concentration of the chemical at the sediment surface is c_s = 2.0 mg/L. (a) If the first-order reaction is very fast, but the concentration at the interface is not zero (it is 2.0 mg/L) what is the maximum possible chemical flux into the sediment? (b) What is the observed liquid mass transport coefficient for this system based on a bulk concentration of chemical C of c_∞ = 8.5 mg/L? (c) Assuming that the concentration gradient is linear, what is the thickness of the concentration boundary layer, δ_c?

6.6 Equation 6-11 describes the local flux to a wetted wall at any time t. Integrate this equation over a wall of width w to obtain the total flux into the fluid after flow over a distance L. Based on this result, show that you obtain Eqs. 6-70 and 6-71 describing the mass transport coefficient and Sh for total chemical transport to a wetted wall of length L.

- **6.7** A 1 m square, thin plate of solid naphthalene is oriented parallel to a stream of air flowing at 20 m/s. The air system is at 27°C, with kinematic viscosity of 1.57 × 10^{-5} m^2 s^{-1}. The partial pressure of naphthalene is 26 Pa under these air conditions, and the diffusivity of naphthalene at 27°C is 5.90 × 10^{-6} m^2 s^{-1}. Calculate the flux from the plate at 0.3 m assuming the Sh number correlation Sh = 0.0292 Re$^{4/5}$ Sc$^{1/3}$. *interface · vapor pressure*

√ **6.8** Calculate the instantaneous rate of naphthalene dissolution (mg s^{-1}) from a sphere of pure naphthalene 100 μm in diameter falling from its own weight (ρ_n = 1.03 g cm^{-3}) in a large, undisturbed column of water. Assume D_{Nw} = 4.79 × 10^{-6} cm^2 s^{-1}.

6.9 Starting with Eqs. 6-19 and 6-20 written in terms of mole fractions, derive the expression given for the overall mass transport coefficient $\mathbf{K}_{wa}$ in terms of H_{Cyx}.

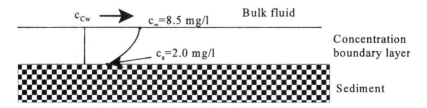

Figure P-6.5

- **6.10** Methanotrophs and propanotrophs are microorganisms that oxidize methane (M) and propane (P), respectively, but these microbes also can cometabolize certain chlorinated aliphatic compounds. You are designing an air-transport system to absorb these compounds in water being fed to a bioreactor. Chemical data available on these two compounds are: $M_M = 16$ g mol^{-1}, $D_{Mw} = 1.49 \times 10^{-5}$ cm^2 s^{-1}, $c_{Mw,eq} = 25.1$ mg L^{-1}, $M_P = 44$ g mol^{-1}, $D_{Pw} = 0.97 \times 10^{-5}$ cm^2 s^{-1}, $c_{Pw,eq} = 119$ mg L^{-1}, where equilibrium is based on pure gas. The reactor is a wetted wall system with a water film thickness of 78 μm and maximum velocity of 3 cm s^{-1} (1.5 times the average velocity). (a) Assuming penetration theory, calculate a mass transport coefficient for the total flux of methane and propane into the thin water film after flow down a flat plate 30 cm long and having a surface area of 98 m^2. (b) Calculate the total rate (mg s^{-1}) each chemical is transferred into the water assuming the entering fluid is devoid of chemical and assuming an atmosphere of 1% of each chemical is maintained in the unit. (c) Compare the rate of methane transfer calculated using penetration theory to the more empirical result of Vivian and Peaceman (Eq. 6-72)

REFERENCES

Brian, P. L. T., and H. B. Hales. 1969. *A.I.Ch.E. J.* **15**:419.

Chilton, T. H., and A. P. Colburn. 1934. *Ind. Eng. Chem.* **26**:1183.

Evnochides, S., and G. Thodos. 1959. *A.I.Ch.E. J.* **5**:178.

Fulford, G. D. 1964. In: *Adv. in Chem. Engin.*, T. B. Drewe, J. W. Hoopes, Jr., T. Vermeulen, and G. R. Cokelet (Eds.), Academic, NY, **5**:151.

Garner, F. H., and R. D. Suckling. 1958. *AIChE. J.* **4**:114–124.

Gilliland, E. R., and T. K. Sherwood. 1934. *Ind. Eng. Chem.* **26**:516.

Higbie, R. 1935. *Trans. AIChE.* **31**:368–89.

Johnstone, H. F., and R. L. Pigford. 1942. *Trans. AIChE.* **38**:25.

Levich, V. G. 1962. *Physicochemical Hydrodynamics*, Prentice Hall, NJ.

Mackay, D., and A. T. K. Yeun. 1983. *Environ. Sci. Technol.* **17**(4):211–217.

Onda, K., H. Takeuchi, and Y. J. Okumoto. 1968. *Chem. Eng. Jpn.* **1**:56–62.

Roberts, P. V., G. D. Hopkins, C. Munz, and A. H. Riojas. 1985. *Environ. Sci. Technol.* **19**(2): 164–73.

Sherwood, T. K., and F. A. L. Holloway. 1940. *Trans. Am. Inst. Chem. Eng.* **36**:39–70.

Sherwood, T. K., R. L. Pigford, and C. R. Wilke. 1975. *Mass Transfer.* McGraw-Hill, New York.

Shulman, H. L., C. F. Ulrich, A. Z. Prowx, and J. O. Zimmerman. 1955. *AIChE. J.* **1**:253–258.

Steinberger, R. L., and R. E. Treybal. 1960. *A.I.Ch.E. J.* **6**:227.

Stucheli, A., and M. N. Ozisik. 1976. *Chem. Eng. Sci.* **31**:369.

Vivian, J. E., and D. W. Peaceman. 1956. *AIChE J.* **2**(4):437–443.

Welty, J. R., C. E. Wicks, and R. E. Wilson. 1984. *Fundamentals of Momentum, Heat and Mass Transfer*, 3rd ed. John Wiley, New York.

CHAPTER 7

TRANSPORT IN SHEARED REACTORS

7.1 INTRODUCTION

Virtually all liquids exposed to the natural environment will undergo fluid mixing, and therefore fluid turbulence, even if they are not actively mixed. A column of water sitting on a lab bench can have convective currents in the column for any of several reasons. The fluid can be set in motion by being sheared by even a slight air flow over the top of the column. Evaporative cooling at the air–water interface will produce slightly cooler water at the air–water surface, which will sink, displacing warmer water from the column bottom. These two problems can be avoided by sealing the column at the top, leaving no air. There may also be temperature gradients in the room, however, which can produce gradients in the column and lead to fluid mixing. Whether these convective currents are important is dependent on the intended use of the column. If the column will be mechanically mixed or aerated, convective flow would be unimportant. In settling experiments, however, convective currents should be small enough so that they do not alter the settling velocities of particles. The presence of convective currents in water or air can produce a net rate of chemical *dispersion* that is larger than that possible solely by *molecular diffusion*. Thus chemical transport in natural and engineered systems is often controlled more by the fluid properties, in this case the mixing rate of the fluid, than by chemical properties such as the molecular diffusivity of the compound.

In this chapter the focus will be on systems that have a large energy inputs for mixing, due to paddles, mixers, stirrers, or bubbles, so that the convective currents in the system are insignificant compared to advective currents generated in the system. Under idealized conditions, a "completely mixed" reactor would be, by definition, completely homogeneous throughout. Adding a small drop of dye into a completely mixed reactor would instantaneously produce a uniform dye concentration everywhere in the reactor. While such actual mixing rates are obviously not this efficient, the time scales of fluid mixing may be rapid enough to warrant such an assumption. In making chemical transport calculations for completely mixed reactors, the importance of the bulk mixing rates must be compared to rates of chemical transport across phase interfaces (gas to water) versus transport to target surfaces (water to particle). As we shall see, such comparisons can be made on the basis of the mass transport resistances.

The modeling of mixing and turbulence in fluids is a difficult, and computationally intensive, subject, and it will only receive a cursory review here. In order to make chemical transport calculations, we shall rely upon empirical mass transport correlations to predict the outcome of the complex fluid mechanics on mass transport. Therefore, we begin with a review of the basic components of mixed and sheared fluids for subsequent calculations of chemical transport rates.

7.2   FLUID SHEAR AND TURBULENCE

In a fluid there are eddy sizes that can range in size from the size of the container to a minimum size, known as the Kolmogorov microscale, λ. This situation is independent of the type of the "container." For example, due to the spin of the earth the oceans rotate clockwise in the northern hemisphere and counterclockwise in the southern hemisphere. Constrained by the continents, or the "container," these large-scale eddies give rise to planetary-scale ocean currents known to oceanographers as *gyres*. On smaller scales, for example, in lakes, whole lake mixing waves known as *internal seiches* can span the length of the lake. Large-scale eddies, such as gyres and internal seiches, contain tremendous amounts of energy due to the momentum of such a large mass of water being set in motion.

The largest eddies dissipate into progressively smaller and smaller eddies until at last a limit is reached on the size of the smallest stable eddies that can be formed. These smallest eddies lose their energy due to viscous dissipation of their energy into heat. This progression of energy dissipation, while relatively easy to envision, is quite difficult to model mathematically. Consider a container of water being mixed by a stirring device as shown in Figure 7.1. The largest eddy is the size of the container, but there are eddies of all sorts of different sizes and different intensities in the container. The turbulence in the container is anisotropic, varying in magnitude appreciably over the container, and is particularly difficult to describe near the impeller. The description of turbulence everywhere in the container would become prohibitively complex for routine or simple calculations of mixing. Therefore, it is usually assumed that the turbulent fluid is isotropic. The relative errors in such an assumption will be discussed in sections to follow.

How can turbulence be measured? Turbulence is not in itself a fluid property, like viscosity or density, but rather a feature of the fluid flow specific to the container at some instant. If we were to measure the absolute velocity of the fluid at some point in the container in Figure 7.1a, we would likely obtain a data set such as that shown in Figure 7.1b. The instantaneous velocity in the container in the x direction, u_x, is

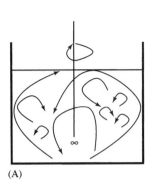 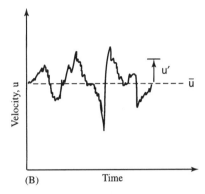

(A) (B) Time

Figure 7.1 Fluid turbulence in a mixed container. (A) Fluid mixing produces a range of eddies, with the largest eddies of size Λ transporting fluid around the container and the smallest eddies of size λ dissipating the mixing energy into heat. (B) Fluid velocity measured at a single point can be viewed as consisting of velocity fluctuations u' about some average velocity.

given as

$$u_x = \bar{u}_x + u_x' \tag{7-1}$$

where $\bar{u}_x$ is the average fluid velocity and u_x' is the velocity fluctuation. Note that the phase subscript for the bulk phase has been omitted here for convenience. The average velocity results from the larger-scale eddies, which produce the bulk flow in a given direction (in this case x). By definition, the mean of the fluctuation u_x' is zero. However, the intensity of the turbulence, $\langle u_x \rangle$, is given by the root-mean-squared velocity of the fluctuations, or

$$\langle u_x \rangle = (\overline{u_x'^2})^{1/2} \tag{7-2}$$

It can be seen from Eq. 7-2 that the intensity of the turbulence is not necessarily a function of the fluid velocity, although in many instances $\langle u_x \rangle$ increases in proportion to the average fluid velocity.

The velocity fluctuations can be incorporated into the general transport equation using Eq. 7-1. If we recognize that the velocity at any point is the sum of the two velocity components, for example, momentum transport in the x direction is

$$\frac{\partial}{\partial t}(\rho u_x) + \frac{\partial}{\partial x}(\rho u_x u_x) + \frac{\partial}{\partial y}(\rho u_y u_x) + \frac{\partial}{\partial z}(\rho u_z u_x)$$

$$= -\frac{\partial p}{\partial x} + \mu \left(\frac{\partial^2 u_x}{\partial x^2} + \frac{\partial^2 u_x}{\partial y^2} + \frac{\partial^2 u_x}{\partial z^2} \right) + \rho g_x \tag{7-3}$$

Substituting in Eq. 7-1 for each u_x term and $p = \bar{p} + p'$ for the pressure terms produces

$$\frac{\partial}{\partial t}(\rho \bar{u}_x) + \frac{\partial}{\partial x}(\rho \bar{u}_x \bar{u}_x) + \frac{\partial}{\partial_y}(\rho \bar{u}_y \bar{u}_x) + \frac{\partial}{\partial_z}(\rho \bar{u}_z \bar{u}_x) + \frac{\partial}{\partial x}(\rho \overline{u'_x u'_x}) + \frac{\partial}{\partial y}(\rho \overline{u'_y u'_x})$$

$$+ \frac{\partial}{\partial z}(\rho \overline{u'_z u'_x}) = -\frac{\partial \bar{p}}{\partial x} + \mu \left(\frac{\partial^2 \bar{u}_x}{\partial x^2} + \frac{\partial^2 \bar{u}_x}{\partial y^2} + \frac{\partial^2 \bar{u}_x}{\partial z^2} \right) + \rho g_x \qquad (7\text{-}4)$$

This is the time-smoothed form of the momentum transport equation in the x direction, with the new terms arising from the velocity fluctuations about the average values of the velocity. Similar equations could be developed for the velocity components in the y and z directions.

The Kolmogorov Microscale

Although Eq. 7-4 is useful for describing the transport of momentum during mixing or the movement of large masses of water, mass transport to particles suspended in mixed fluids can be analyzed without having to solve the momentum transport equation. The eddies produced in a turbulent fluid can be analyzed as consisting of energy-containing eddies (of scale Λ) and energy-dissipating eddies (of scale λ) that are separated widely in size (see Fig. 7.2). Kolmogorov proposed in his Universal Equilibrium Theory (using a dimensional analysis) that the smallest eddy that can exist in a fluid of viscosity ν is

$$\lambda = \left(\frac{\nu^3}{\varepsilon_d} \right)^{1/4} \qquad (7\text{-}5)$$

where ε_d is the energy dissipation rate, $\varepsilon_d = P/m_r$, where P is the power input into a reactor containing fluid of total mass m_r, and ε_d has units of $[L^2 \, T^{-3}]$. Since the kinematic viscosity has units of $[L^2 \, T^{-1}]$, λ has units of $[L]$.

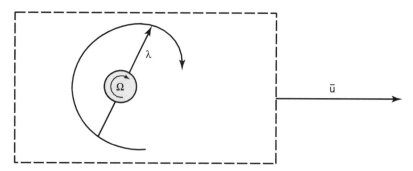

Figure 7.2 An object within a microscale eddy will rotate at a rate Ω. Note that the whole eddy (encased in a packet of fluid, shown here within the dashed box) may exist and be transported within the flow moving at some average fluid velocity.

While the largest eddies of size Λ contain most of the energy in the system and serve to transport fluid around the container, according to the Universal Equilibrium theory they are not energy dissipating eddies. These largest eddies produce a spectrum of smaller eddies, with only the smallest eddies of size λ dissipating their energy into heat. As a result, small particles (on the order of 1 μm) suspended in the fluid exist in an environment of small, energy-dissipating eddies (on the order of 100 μm for typical sheared bioreactors).

Through dimensional analysis, it is also possible to relate the energy dissipation rate and the shear rate by

$$G = \left(\frac{\varepsilon_d}{\nu}\right)^{1/2} \tag{7-6}$$

where G is defined as the mean shear rate in the vessel. The shear rate can be thought of as the velocity gradient in the eddy. Since the eddy has velocity u and size λ, $G = \Delta u/\Delta l$, or $G = u/\lambda$.

Based on these assumptions, it is possible to interrelate the shear rate and Kolmogorov microscale using Eqs. 7-5 and 7-6 as

$$\lambda = \left(\frac{\nu}{G}\right)^{1/2} \tag{7-7}$$

Because we have assumed here that all energy put into the system is dissipated by eddies of size λ, and we have assumed isotropic turbulence, the shear rates referred to in Eqs. 7-6 and 7-7 are usually called the *mean* shear rates. In practice, there are likely to be range of shear rates since the energy input into the fluid in a mixed container is not uniform throughout the container. Mean shear rates in mixed bioreactors generally range from 100 to 500 s^{-1}. Assuming a fluid viscosity of water, this implies Kolmogorov microscales in the range of 45 to 100 μm for bioreactors.

The size of the Kolmogorov microscale was deduced by a dimensional analysis based on power input and fluid viscosity. However, others have suggested equally plausible length scales, and there is experimental evidence to suggest that the issue of an appropriate size of a microscale is open to debate. Parker et al. (1970), based on earlier theoretical work by others, suggested that there were eddies in a viscous dissipation subrange, that is, that there were eddies having a size less than λ. After conducting particle velocity measurements in a laboratory turbulence tank, Hill (1992) also concluded that there were eddies smaller than the Kolmogorov microscale and that these eddies were important in considering mass transport to suspended particles. However, this subject has not been resolved.

For systems with low energy-dissipation rates, Batchelor (1952) proposed a second length scale based on dimensional arguments of

$$\lambda_D = \left(\frac{\nu D_C^2}{\varepsilon_d}\right)^{1/4} \tag{7-8}$$

where the subscript D has been added to the microscale to indicate that transport by molecular diffusion is important, and D_C is the molecular diffusivity of chemical C. Because the magnitude of chemical diffusivity is in general several orders of magnitude less than momentum diffusivity ($D_C \ll \nu$, or 10^{-5} cm^2 s^{-1} $\ll 10^{-2}$ cm^2 s^{-1}) in water, this scale describes a region where both diffusion and advection are important transport mechanisms. Equation 7-8 is therefore only useful for describing natural systems where energy dissipation rates are much lower than those for engineered vessels. For example, in an engineered system with $G = 100$ s^{-1} and $D_C = 10^{-5}$ cm^2 s^{-1}, $\lambda = 100$ μm, but λ_D would only be equal to 3 μm. In seawater with $G = 0.01$ s^{-1}, $\lambda = 1$ cm, and $\lambda_D = 300$ μm, indicating this smaller length scale would be important to the movement of chemicals and some particles. This suggests that below length scales of λ_D, diffusional processes are faster than those caused by fluid motion.

There are also arguments that support using a microscale larger than the Kolmogorov microscale. Turbulent sheared fluid is not isotropic in the sense that there are eddies larger than the microscale present in the fluid. Based on measurements of microstructure for natural waters, it was suggested that the smallest energy-containing eddies are larger by a factor of 5–10 than the Kolmogorov microscale (Lazier and Mann, 1989). Parker et al. (1970) calculated that the eddy scale most significant in imparting a relative fluid velocity on a particle through partial entrainment would be twice the size of the particle. Even if we accept that the smallest eddy that can exist is the Kolmogorov microscale, it should be recognized that there are eddies larger than the microscale that do not degenerate into smaller eddies but instead dissipate into heat. The centroid of the smallest eddy sizes that dissipate into heat was experimentally determined in a sheared vessel to be 2λ (Brodkey, 1966; Matson et al., 1972).

Kolmogorov intended from his dimensional analysis of energy dissipation rates that λ would be the size of the *smallest* eddy that could exist in a turbulent fluid; so variations by a factor of 2 or even 5 should be quite acceptable due to the order-of-magnitude nature of the estimate. However, the length scales presented above can vary by several orders of magnitude. For our discussions here we will consider λ to be a good estimate of the smallest microscale, and will use 2λ as an estimate of the average size of an eddy in a turbulent fluid.

Steady Shear Flow

Below the length scale of the microscale, chemicals and small particles exist within a local environment that is defined by an eddy of size λ for a fluid sheared at a rate **G**, where **G** is the fluid shear rate defined as the velocity gradient. Thus very small particles experience only shear flow. For constant shear flow, where $\partial \mathbf{U}/\partial t = 0$

$$\mathbf{U}(\mathbf{X}) = \mathbf{V} + \mathbf{G}\mathbf{X} \qquad (7\text{-}9)$$

where **V** is the free flow mean velocity vector, **X** is a position vector, and **G** in the most general case can be a nine-component tensor ($G_{ij} = \partial U_i/\partial X_j$) where the subscripts here indicate vector direction. Since a particle is carried along with the mean flow and sees only the microscale fluid shear, only the relative velocity influences mass transport.

The velocity gradient can be defined in terms of two components as

$$\mathbf{G} = \mathbf{\Omega} + \mathbf{E} \tag{7-10}$$

where $\mathbf{\Omega}$ is the rotation rate defined as

$$\mathbf{\Omega} = \Omega_{ij} = \frac{1}{2}\left(\frac{\partial V_i}{\partial X_j} - \frac{\partial V_j}{\partial X_i}\right) \tag{7-11}$$

and $\mathbf{E}$ is the strain rate, defined as

$$\mathbf{E} = E_{ij} = \frac{1}{2}\left(\frac{\partial V_i}{\partial X_j} - \frac{\partial V_j}{\partial X_i}\right) \tag{7-12}$$

The maximum rate of mass transport due to fluid shear occurs when $\Omega = 0$ (pure straining) or if an object does not rotate in the fluid. However, all particles suspended in sheared fluid can be expected to rotate with the fluid (Fig. 7.3). The magnitude of the strain rate is defined as

$$E = |\mathbf{E}| = (E_{ij}E_{ij})^{1/2} \tag{7-13}$$

This can be related to the energy dissipation rate using

$$\varepsilon_d = 2\nu E_{ij}E_{ij} \tag{7-14}$$

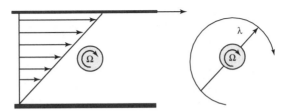

Figure 7.3   Steady shear flow can be produced by pulling a plate parallel to a second fixed plate. The shear rate is the velocity of the upper plate divided by the gap thickness, or $G = \Delta U / \Delta x$. This case of steady shear is found to be analogous to the case of a particle suspended in an eddy, since in both cases the particle exists in a laminar shear field.

By combining Eqs. 7-13 and 7-14

$$E = \left(\frac{\varepsilon_d}{2\nu}\right)^{1/2} = 0.71 \left(\frac{\varepsilon_d}{\nu}\right)^{1/2} \qquad (7\text{-}15)$$

Equation 7-15 is valid for any shearing motion (Karp-Boss and Jumars, 1996).

In order to simplify the above description of the fluid flow further, it is common to consider the case of steady shear flow with all components of $\mathbf{E}$ reduced to zero except $\mathbf{E}_{12} = \mathbf{E}_{21}$ where the magnitude of the strain rate produces a corresponding rotation rate, or

$$E_{12} = E_{21} = \Omega = \frac{G}{\sqrt{2}} \qquad (7\text{-}16)$$

where here G now becomes a scalar quantity defined as $\mathbf{V} = (Gy,0,0)$ (Leal, 1992). Substituting Eq. 7-16 into Eq. 7-13, we see the relationship that $E = G/2^{1/2}$. This same conclusion could be reached by comparing Eq. 7-7 with Eq. 7-15.

7.3 MASS TRANSPORT CORRELATIONS FOR SHEARED SYSTEMS (GENERAL)

The rate of chemical transport to particles in a sheared fluid has been investigated for spheres primarily under two conditions: with and without rotation of the sphere. Particles suspended in simple shear flow, such as that produced by a couette device, will rotate with the axis of rotation perpendicular to the flow direction. This rotation decreases the rate of mass transport compared to a particle held in the flow and unable to rotate ($\Omega = 0$). Particles suspended in a turbulent sheared fluid will also rotate with the fluid, but the direction of the shear is constantly changing as the eddies dissipate. To calculate mass transport to a sphere suspended in a sheared fluid, we will use mass transport correlations developed for two different cases: particles suspended in steady shear flow, and particles suspended in a fluid under turbulent shear conditions.

Mass Transport in a Turbulent Sheared Fluid

Mass transport in turbulent sheared fluids is usually correlated to the shear Peclet number, $Pe_G = R^2 G/D$. For isotropic turbulence the ratio between E and Ω is on the order of unity. At scales smaller than the Kolmogorov microscale the flow regime looks like steady shear flow, so for $Pe \ll 1$ (and $E/\Omega \leq 1$) mass transport can be calculated using the correlation developed for one-dimensional steady shear flow (Batchelor, 1979)

$$Sh_G = 1 + 0.29\, Pe_G^{1/2}, \quad Pe_G \ll 1 \qquad (7\text{-}17)$$

where the mass transport coefficient is obtained from the Sherwood number defined as $Sh = \mathbf{k}R/D$. For a statistically steady, but not necessarily homogeneous, turbulent flow, Batchelor (1980) proposed the equation

$$\underline{Sh_G = 0.55\ Pe_G^{1/3}, \quad Pe_G \gg 1, \text{ turbulent shear}} \qquad (7\text{-}18)$$

Although the form of the equation was based on theoretical considerations, the co-efficient 0.55 was chosen as a typical condition (a "skewness" factor of 0.6 from a range of 0.3 to 1). It is considered likely that the coefficient of 0.55 will vary in practice by 20%.

Between the limits of very small and very large Pe_G, mass transport equations are not so easily derived because it is difficult to approximate strain and shear rates. However, if we assume that these limits are accurate, then intermediate values can be easily obtained from interpolation between the two regions. By extending the upper or lower range, into the other range, Karp-Boss and Jumars (1996) proposed the two correlations

$$\underline{Sh_G = 1.014 + 0.15\ Pe_G^{1/2}, \quad 0.1 \le Pe_G \le 100} \qquad (7\text{-}19)$$

producing a lower estimate of the Sh, and

$$\underline{Sh_G = 0.955 + 0.34\ Pe_G^{1/3}, \quad 0.1 \le Pe_G \le 100} \qquad (7\text{-}20)$$

producing an upper estimate for Sh in this intermediate range of Pe.

Mass Transport in Steady Shear Flow

Mass transport in sheared fluids is also usually correlated to the shear Peclet number, $Pe_G = R^2 G/D$. At low Peclet numbers ($Pe_G \ll 1$), for one-dimensional steady shear flow, Frankel and Acrivos (1968) obtained the correlation

$$Sh_G = 1 + 0.26\ Pe_G^{1/2} + O(Pe_G^{1/2}), \quad Pe_G \ll 1 \qquad (7\text{-}21)$$

where the $O(Pe_G^{1/2})$ term is read as "on the order of $(Pe_G^{1/2})$." This last term is usually small and is dropped for calculations, producing a working equation for linear laminar shear of

$$\underline{Sh_G = 1 + 0.26\ Pe_G^{1/2} \quad Pe \ll 1, \text{ rotation, laminar shear}} \qquad (7\text{-}22)$$

Notice that this correlation reduces to Sh = 1 expected for a sphere suspended in a stagnant fluid. At high Pe the Sh asymptotes to Sh = 4.5 due to closed streamlines around the sphere when rotation is strong.

In most systems of interest, the fluid environment of small particles is a three-dimensional shear field. As the particle is shifted from dissipating eddy to eddy, the orientation of the particle continually shifts. Solutions for the more realistic case of a three-dimensional linear sheared fluid when the particle has no rotation (Batchelor,

1979) is

$$Sh_G = 1 + 0.30\ Pe_G^{1/2}, \quad Pe \ll 1,\ \text{no rotation} \qquad (7\text{-}23)$$

with an expected error on the coefficient of <3% (Karp-Boss and Jumars, 1996). For the most general case of a rotating particle in the sheared fluid, when $|\Omega|/|\mathbf{E}| < 1$

$$Sh_G = 1 + 0.29\ Pe_G^{1/2}, \quad Pe \ll 1,\ \text{rotation} \qquad (7\text{-}24)$$

with an error of <10%. Notice that the effect of rotation is to decrease the rate of mass transport to the suspended particle. Batchelor (1979) also presented a solution for the case where $|\Omega|/|\mathbf{E}| > 1$, or when the particle is turning within the sheared fluid (as might occur for a swimming bacterium) so that the only contribution to the flux due to E_ω, is

$$Sh_G = 1 + 0.19\ Pe_G^{1/2}, \quad Pe \ll 1,\ \text{strong rotation} \qquad (7\text{-}25)$$

resulting in an even lower rate of mass transport to the particle. For calculations in sheared fluids it is recommended that Eq. 7-24 be used (Table 7.1).

At very large Peclet numbers mass transport in the absence of rotation is

$$Sh_G = 0.8\ Pe_G^{1/3}, \quad Pe \gg 1,\ \text{no rotation} \qquad (7\text{-}26)$$

and when the particle is allowed to rotate

$$Sh_G = 0.60\ Pe_G^{1/3}, \quad Pe \gg 1,\ \text{rotation} \qquad (7\text{-}27)$$

These correlations do not reduce to the expected result that Sh = 1 for a stagnant fluid, but notice that the effect of rotation is to decrease the overall rate of mass transport to the suspended spherical particle. Also notice that our result in Eq. 7-27 is essentially the same result presented above for turbulent shear at very large Pe_G.

As before, these correlations also do not cover the range of Peclet numbers of most interest in our calculations, or when $0.01 \leq Pe_G \leq 100$. Correlations between the two ranges of values were developed by Karp-Boss and Jumars (1996) for both a rotating and nonrotating particle. Since only the case of a rotating particle is of general interest, it will be the only one considered here. Through extension of the

TABLE 7.1 Summary of Sherwood Correlations Used for Mass Transport to a Sphere in a Steady, Sheared Fluid

Conditions	No Rotation	Rotation
$Pe \ll 1$	$Sh_G = 1 + 0.30\ Pe_G^{1/2}$	$Sh_G = 1 + 0.26\ Pe_G^{1/2}$
$0.01 < Pe < 100$	$Sh_G = 1.004 + 0.29\ Pe_G^{1/2}$	$Sh_G = 1.002 + 0.18\ Pe_G^{1/2}$
$Pe \gg 1$	$Sh_G = 0.8\ Pe_G^{1/3}$	$Sh_G = 0.60\ Pe_G^{1/3}$

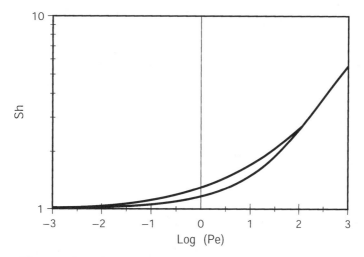

Figure 7.4   Sherwood number correlations as a function of Peclet number: Pe < 0.01 (Eq. 7-18); 0.01 < Pe < 100 (lower line, Eq. 7-24; upper line, Eq. 7-25); Pe > 100 (Eq. 7-23).

Pe ≪ 1 correlation to higher Pe numbers, they obtained

$$Sh_G = 1.002 + 0.18\ Pe_G^{1/2} \tag{7-28}$$

and using the correlation for Pe ≫ 1, they similarly obtained

$$Sh_G = 0.921 + 0.42\ Pe_G^{1/3} \tag{7-29}$$

These correlations are compared in Figure 7.4. The lower line in the range of 0.01 < Pe < 100 is based on Eq. 7-27. A comparison of these two correlations made by Karp-Boss and Jumars (1996) with their computer simulations and with data from Purcell (1978) suggests that Eq. 7-27 should be used. Recommended correlations for shear flow are summarized in Table 7.1.

Example 7.1

Calculate a range of Sherwood numbers for a 100-μm (radius) particle suspended in a sheared fluid at shear rates ≤150 s^{-1}. Assume a chemical diffusivity of 600 × 10^{-8} cm^2 s^{-1}.

For the given information, we first need to calculate the Peclet number to identify a range of Pe for choosing a mass transport correlation. At the highest shear rate, Pe is

$$\mathrm{Pe}_G = \frac{R^2 G}{D_{Cw}} = \frac{(100\ \mu\text{m})^2(150\ \text{s}^{-1})}{600\ \mu\text{m}^2\ \text{s}^{-1}} = 2500 \qquad (7\text{-}30)$$

and at a low shear rate of $1\ \text{s}^{-1}$, Pe = 17. Based on Pe $\gg$ 1, we choose Eq. 7-27 for particles allowed to rotate in a turbulent sheared fluid. The resulting behavior of the Sherwood number is shown in Figure 7.5.

7.4 MEAN SHEAR RATES IN REACTORS

In order to calculate transport rates to particles suspended in sheared fluids, the shear rate or energy dissipation rate in the system needs to be calculated. Energy can be input by paddles, mixers, and bubbles. For examining the effect of shear under controlled conditions, a laminar shear field can be generated in a rotating cylinder (couette) device. However, in systems mixed with an impeller or paddle, turbulent shear conditions will exist in the vessel, and the shear field is likely not to be homogeneous.

Calculating Mean Shear Rates in Different Systems

The mean shear rate in a stirred reactor can be approximated based on the energy input by the paddle or mixer, or by the energy dissipated by rising bubbles. As shown by Eq. 7-14, the mean shear rate is a function of the ratio of the energy dissipation

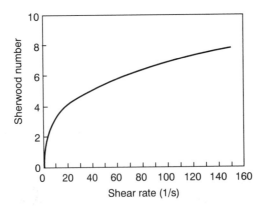

Figure 7.5 Sherwood number as a function of shear rate for $\mathrm{Sh}_G = 0.60\ \mathrm{Pe}^{1/3}$, assuming a primary particle size of 100 μm and a chemical diffusivity of 600×10^{-8} cm^2 s^{-1}.

rate, ε_d, to the momentum dissipation rate, ν. The energy dissipation rate is calculated from

$$\varepsilon_d = \frac{P}{m_{w,T}} = \frac{P}{V_T \rho_w} \qquad (7\text{-}31)$$

where $m_{w,T}$ is the total mass of water in the reactor, V_T is the total volume of the reactor, and ρ_w is the water density. Combining with Eq. 7-4, the shear rate can also be written as

$$G = \left(\frac{P}{V_T \rho_w \nu_w} \right)^{1/2} \qquad (7\text{-}32)$$

Power Input by a Paddle   The power necessary to turn the paddle in the fluid is calculated as the work per unit time, or in terms of force, is

$$P = \frac{\text{force} \times \text{distance}}{\text{time}} \qquad (7\text{-}33)$$

or more simply in terms of the velocity of the paddle relative to the fluid, U_r, as

$$P = FU_r \qquad (7\text{-}34)$$

The force exerted by the paddle, F_p, is defined in terms of the mass of fluid accelerated, or

$$F_p = b_{d,p} A_p \frac{1}{2} \rho_w U_r^2 \qquad (7\text{-}35)$$

where $b_{d,p}$ is a drag coefficient for the paddle, and A_p the cross-sectional area of the paddle normal to the fluid. Drag coefficients can be greater than unity and vary depending on the blade geometry. Based on length to width ratios of 5 and 20, the drag coefficients are 1.2 and 1.5, respectively, increasing to a maximum value of $b_{d,p}$ = 1.9. Assuming that the paddle velocity relative to the fluid is approximately 0.75 times the paddle tip velocity, or $U_r = 0.75\ U_p$, and using $U_p = \omega r$, where ω is the tip speed in radians per second and R_p is the paddle radius, the power input can be calculated as

$$P_p = \frac{0.75^3}{2^3} b_D A_p \rho_w \omega^3 R_p^3 \qquad (7\text{-}36)$$

Combining this result with our definition of the shear rate in Eq. 7-14, and using $\omega = 2\pi s$, where s is the paddle velocity in revolutions per second

$$G = \left(\frac{52.3 b_{d,p} A_p \rho_w s^3 R_p^3}{\nu_w V_T}\right)^{1/2} \tag{7-37}$$

Shear rates are often calculated directly from the torque input to the mixer. Since torque is equal to $\mathbf{F} \times \mathbf{r} \sin\theta$, for a mixer where $\sin\theta = 1$, the power can be calculated using Eq. 7-33, where U_r is defined in terms of the angular velocity as before as

$$P_p = \frac{\tau_p}{R_p} 0.75\omega R_p \tag{7-38}$$

where τ_p is defined as the net torque, or the torque used to mix the fluid minus that needed to turn the mixer in an empty vessel. Substituting this definition of power into Eq. 7-31, G can be calculated from the net torque as

$$G = \left(\frac{0.75\omega\tau_p}{\nu_w V_T \rho_w}\right)^{1/2} \tag{7-39}$$

If the torque can be measured for a system, the drag coefficient can be directly calculated. Combining Eqs. 7-36 and 7-38

$$b_{d,p} = \frac{0.0902\tau_p}{\rho_w A_p s^2 R_p^3} \tag{7-40}$$

Example 7.2

Calculate the mean shear rate in a beaker filled with 2 liters of water stirred by a paddle 17.5 cm^2 in cross-sectional area (height = 2.3 cm, width = 7.6 cm) at 100 rpm. Assume the vessel is 19 cm high, and filled to 16.3 cm.

The length-to-width ratio of the paddle is 7.6:3.3 = 3.3. The closest value that we are given for these conditions is $b_D = 1.2$ for a larger ratio. Since the drag coefficient increases with increasing length-to-width ratios, we will use 1.0 for this calculation (no drag force correction). Using the other given information in

Eq. 7-36, and assuming a fluid density of 1 g cm^{-3}, the mean shear rate is

$$G = \left(\frac{52.3 b_{d,p} A_p \rho_w s^3 R_p^3}{\nu_w V_T}\right)^{1/2}$$

$$= \left(\frac{52.3(1.0)(17.5 \text{ cm}^2)(1.0 \text{ g cm}^{-3})\left(\frac{100}{60}\text{ rps}\right)\left(\frac{7.6}{2}\text{ cm}\right)^3}{(0.01 \text{ cm}^2 \text{ s}^{-1})(2000 \text{ cm}^3)}\right)^{1/2} \qquad (7\text{-}41)$$

$$= 108 \text{ s}^{-1} \qquad (7\text{-}42)$$

Power Input by Small Bubbles   When a bubble rises in a fluid, it exerts a force on the fluid. This drag force, $F_{d,b}$, is

$$F_{d,b} = \frac{b_{d,b} A_b \rho_w U_b^2}{2} \qquad (7\text{-}43)$$

where $A_b = \pi d_b^2/4$ is the cross-sectional area of the bubble, d_b is the bubble diameter, and U_b is the rise velocity of the bubble. Multiplying this drag force by the rise velocity produces the power dissipated by a single bubble

$$P_b = \frac{\pi}{8} b_{d,b} d_b^2 \rho_w U_b^3 \qquad (7\text{-}44)$$

Drag coefficients for spheres can be expressed as a function of the bubble Reynolds number, $Re_b = U_b d_b/\nu_w$, as

$$b_{d,b} = \frac{24}{Re_b} + \frac{3}{Re_b^{1/2}} + 0.34 \qquad (7\text{-}45)$$

For a small bubble with $Re_b \ll 1, b_{d,b} \approx 24/Re_b$, Eq. 7-44 can be simplified as

$$P_b = 3\pi d_b \nu_w \rho_w U_b^2 \qquad (7\text{-}46)$$

In order to calculate the shear imparted by a swarm of bubbles, we need to calculate the number of bubbles in a vessel aerated at a rate Q_a. The number of bubbles can be calculated from the total volume of air divided by the volume of a bubble. Since the volume of air is a product of the air flow rate (Q_a) and the bubble residence time (θ_b), the total number of bubbles in the reactor is

$$N_b^* = \frac{Q_a \theta_b}{(\pi/6) d_b^3} \qquad (7\text{-}47)$$

Not all bubbles will have the same residence time in the reactor due to the convective motion set up by the rising bubbles. However, if we take the simplest case that $\theta_b = h/U_b$, where h is the height of the reactor, then Eq. 7-47 becomes

$$N_b^* = \frac{Q_a h}{(\pi/6)d_b^3 U_b} \qquad (7\text{-}48)$$

The total power input into the vessel is the product of the power per bubble and number of bubbles in the reactor. Combining Eqs. 7-46 and 7-48 with our definition of shear in Eq. 7-32, we have the mean shear rate in the reactor produced by bubbles

$$G_b = \left(\frac{18U_b Q_a h}{V_T d_b^2}\right)^{1/2} \qquad (7\text{-}49)$$

Since we have already made the assumption in calculating the bubble drag that the bubble rise velocity is small with Re $\ll$ 1, we can further simplify Eq. 7-49 using Stokes' Law for the bubble velocity to produce

$$G_b = \left(\frac{Q_a h g}{V_T v_w} \frac{(\rho_w - \rho_a)}{\rho_w}\right)^{1/2} \qquad (7\text{-}50)$$

Since $(\rho_w - \rho_a) \approx \rho_w$, this can be further simplified to

$$G_b = \left(\frac{Q_a h g}{V_T v_w}\right)^{1/2} \qquad (7\text{-}51)$$

The same result was obtained by Parker et al. (1970) using a slightly different approach. For conventional aerated activated sludge reactors they estimated G_b values in the range of 88 to 220 s^{-1}.

Example 7.3

Calculate a typical shear rate in a laboratory reactor vessel produced just by the bubbles for an air flow rate of 1 liter min^{-1}. Assume the reactor vessel is a 2-liter vessel filled with 1.5 liter of water, the height of the water is 10 cm, and bubbles are 0.01 cm in diameter.

   For this calculation we can use Eq. 7-48 if the bubbles are in the creeping flow region of Re $\ll$ 1. Using Stokes' Law, we calculate a rise velocity of 0.55 cm s^{-1}. This produces Re = (0.55 cm s^{-1} $\times$ 0.01 cm)/(0.01 cm^2 s^{-1}), or Re = 0.55. This is not really much less than unity, but we shall proceed anyway assuming the calculation is a sufficient approximation. Substituting the given data into Eq. 7-48, we have

$$G_b = \left(\frac{Q_a h g}{V_T \nu_w}\right)^{1/2} = \left(\frac{(1 \text{ L/min})(10 \text{ cm})(981 \text{ cm/s}^2)}{(1.5 \text{ L})(0.01 \text{ cm}^2/\text{s})}\right)^{1/2} \tag{7-52}$$

$$= 104 \text{ s}^{-1} \tag{7-53}$$

This is a very high shear rate for a bubbled reactor, likely due to the assumption of a very small bubble diameter for the given air flow rate. In practice, bubbles in laboratory reactors will be larger than assumed here.

Laminar Shear in Couette Devices   The shear rate produced by paddles and bubbles is not homogeneous in a vessel, making it quite difficult to study the effect of shear on mass transport. However, laminar shear can be produced in the laboratory in a carefully constructed vessel known as a couette device to study mass transport to and from suspended particles. A couette device consists of two concentric cylinders. Theoretically, both of the cylinders may rotate at the same time. The velocity distribution within the gap between the cylinders can be obtained from the Navier–Stokes equations, written in cylindrical coordinates (van Duuren, 1968), as

$$\frac{du}{dr} + \frac{u}{r} = \text{const} \tag{7-54}$$

where u is the velocity at any point in the fluid at a distance r from the center. Integrating, this becomes

$$u = b_1 r + \frac{b_2}{r} \tag{7-55}$$

Evaluating the two constants using the boundary conditions $u = 0$ at $r = R_1$ (the inner cylinder is fixed), and $u = wr$ at $r = R_2$ (outer cylinder rotates), results in

$$u = \frac{w R_2^2}{R_2^2 - R_1^2}\left(r - \frac{R_1^2}{r}\right) \tag{7-56}$$

The velocity profile in the gap is a function of the geometry of each vessel, but for an inner radius of $R_1 = 8$ cm, and a gap thickness of 0.9 cm, the velocity profile is essentially linear (Fig. 7.6).

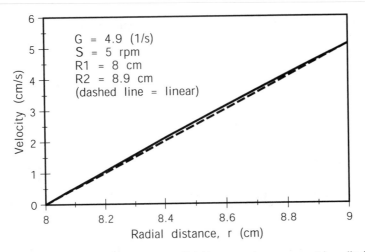

Figure 7.6 The velocity distribution in the fluid in a gap in a concentric cylinder device (solid line) is nearly linear.

The shear rate at any point can be otained from the derivative of Eq. 7-3, or

$$G(r) = \frac{du}{dr} = \frac{2wR_1^2R_2^2}{R_2^2 - R_1^2}\left(\frac{1}{r^2}\right) \tag{7-57}$$

Thus the shear rate is not a linear function of distance in the gap. The mean shear rate can be obtained from averaging the velocity distribution over the gap, or

$$G_m = \frac{1}{R_2 - R_1}\int_{R_1}^{R_2}\left(\frac{du}{dr} + \frac{u}{r}\right)dr \tag{7-58}$$

Solving, the mean shear rate is

$$G = \frac{2\pi s}{60}\left(\frac{2R_1R_2}{R_2^2 - R_1^2}\right) \tag{7-59}$$

where the rotation speed w (s^{-1}) has been replaced with the rotational speed s (in revolutions per minute, or rpm), where $w = 2\pi s/60$.

Flow within the couette device can be turbulent at higher Reynolds numbers, where the transition Re is defined as Re $= wR_2^2/v_w$. Assuming a maximum centripetal acceleration of 2 g was desirable, in a fluid with $v_w = 0.01$ cm^2 s^{-1} (20°C), van Duuren calculated the transition from laminar to turbulent flow as

$$R_2 - R_1 = 0.001365R_2^{3.084} \tag{7-60}$$

where R_2 has units of cm. The maximum recommended gap width was calculated again assuming a maximum centripetal acceleration of $2g$, using $a = w^2 R_2$, and Eq. 7-58, as

$$\frac{g}{G^2} = \frac{(R_2^2 - R_1^2)^2}{4R_1^2 R_2} \tag{7-61}$$

These results, shown in Figure 7.7, indicate the maximum gap width for laminar flow conditions. As the shear rate increases, due to higher rotational velocities, the gap width must be decreased to maintain laminar conditions.

Laminar flow in a couette device can also be obtained by fixing the outer cylinder and rotating the inner cylinder. However, the transition from laminar to turbulent conditions occurs at much lower Re, limiting the useful range of the device. For the inner cylinder rotating, the transition Re should be less than Re = 41.3 $[1 - (R_1/R_2)]^{-3/2}$. Assuming conditions of $R_1 = 8$ cm, $R_2 = 8.9$ cm, the laminar conditions are predicted up to $s = 1.5$ rpm for the inner cylinder rotating, while for the outer cylinder rotating $s < 100$ rpm according to Eq. 7-61.

Measured Shear Rates in Stirred Vessels

While the above calculations are useful for estimating the magnitude of shear rates in vessels, actual shear rates are a complex function of mixer type, impeller speed, vessel geometry, and even the location of the impeller in the vessel. A wide variation in mixing designs makes it essential to make measurements of energy dissipation rates in these vessels. The techniques to measure fluid velocities are quite common and have been in existence for a number of years, and so they will not be discussed here. Instead, we will focus on the outcome of these measurements.

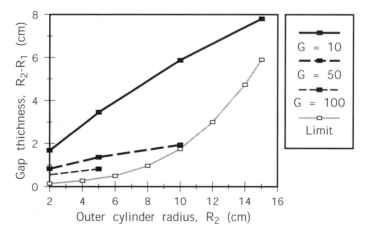

Figure 7.7 Design limits of cylinder radii for laminar shear conditions in a concentric-cylinder couette device as a function of shear rate (adapted from van Duuren, 1968).

The previous analysis for shear in a stirred vessel indicated that $G \sim s^{3/2}$ (Eq. 7-39). For an unbaffled jar test apparatus, Lai et al. (1975) found that this relationship was valid (Fig. 7.8a). For the three different propellors (A1= two-blade flat paddle, A2 = three-blade marine-type mixing propeller, and A3 = A4 = small pitched two-blade paddle) the slopes of G versus s were $\frac{3}{2}$. For example, at a mixer speed of 100 rpm the shear rate was 50 s^{-1} for paddle A1. For a mixer speed of 50 rpm, we predict a shear rate of 18 s^{-1}, in good agreement with values shown in Figure 7.8a for case A1. Changing the height above the bottom in the beaker did not affect the measured shear rate.

Introducing baffles, or stators, into the vessel substantially increases the shear rate, since the rotation of the fluid is hampered by the baffles (Fig. 7.8b). At 100 rpm Camp (1968) measured a shear rate of 180 s^{-1}, a value 2.5 times larger than without baffles present in the vessel. Camp measured somewhat higher shear rates in an unbaffled beaker than Lai et al. (1975). For example, at 100 rpm Camp obtained $G = 70$ s^{-1} compared to the 50 s^{-1} obtained by Lai et al. (1975). Temperature causes proportional increases in the fluid shear rates mainly due to the changes in fluid viscosity with shear rate (see Eq. 7-37).

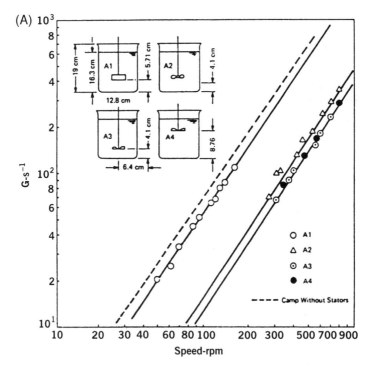

Figure 7.8 Mean shear rates in jar test vessels. (A) G in a vessel with different impeller characteristics (From Lai et al., 1975. Reprinted from Journal AWWA, Vol. 67, No. 10 (October 1975), by permission. Copyright 1975 American Water Works Association). (B) G as a function of temperature for a baffled and unbaffled vessel (Reproduced by permission of ASCE from Camp, 1968).

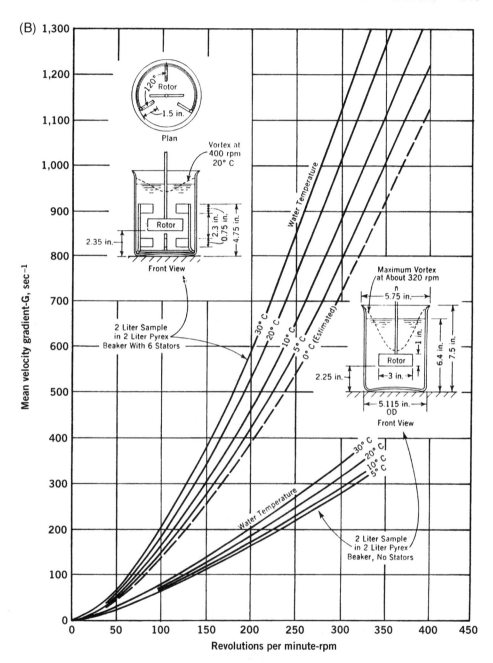

(B)

Figure 7.8 *Continued*

Example 7.4

Li (1996) measured a mean shear rate in a 1-liter beaker filled with 500 mL of water of 18.9 s^{-1} at 30 rpm. Compare this value to those of Camp (1968) and Lai et al. (1975), and that predicted from a force balance. Assume the paddle dimensions given in Example 7.2.

Measurements reported in other studies can be scaled to different conditions using Eq. 7-36. Rearranging all the constants together, we have that

$$\frac{GV_T^{1/2}}{s^{3/2}} = \left(\frac{52.3b_{d,p}A_p\rho_w R_p^3}{v_w}\right)^{1/2} = \text{const} \tag{7-62}$$

Rearranging, G at a new condition, n, can be obtained from the measured condition, m, using

$$G(n) = G(m)\left(\frac{s(n)}{s(m)}\right)^{3/2}\left(\frac{V_T(m)}{V_T(n)}\right)^{1/2} \tag{7-63}$$

From Figure 7.8a by Lai et al. (1975), the shear rate at 30 rpm for beaker A1 filled to 2 liters can be corrected to 500 mL using

$$G(n) = 10 \text{ s}^{-1}\left(\frac{2000 \text{ mL}}{500 \text{ mL}}\right)^{1/2} = 20 \text{ s}^{-1} \tag{7-64}$$

From Figure 7.8b, the data of Camp (1965) at 30 rpm for an unbaffled reactor filled to 2 liters can be corrected to 500 ml using

$$G(n) = 14 \text{ s}^{-1}\left(\frac{2000 \text{ mL}}{500 \text{ mL}}\right)^{1/2} = 28 \text{ s}^{-1} \tag{7-65}$$

The calculation in Example 7.2 can be similarly adjusted for the volume and stirring rate

$$G(n) = 107 \text{ s}^{-1}\left(\frac{30 \text{ rpm}}{100 \text{ rpm}}\right)^{3/2}\left(\frac{2000 \text{ mL}}{500 \text{ mL}}\right)^{1/2} = 35 \text{ s}^{-1} \tag{7-66}$$

Published values serve as reasonable estimates, but it would appear that actual measurements provide the most accurate values of mean shear rates for this particular jar test apparatus.

Velocity and Shear Variations in Mixed Vessels

Energy dissipation rates are not homogeneous inside stirred vessels. Velocities are much larger near the impeller than they are near the fluid surface within jar test devices used for flocculation studies (Fig. 7.9a). Flow pushed off the mixing blade is directed toward the vessel side, creating localized regions of high velocities that can affect the stability of suspended particles. Aggregates within this fluid stream could be subjected to breakup due to impaction with the vessel surface. Recirculation times in small vessels are on the order of seconds to minutes, but reaggregation times are longer (5 to 30 min) leading to aggregate sizes dominated by high-shear regions in the vessel (Stanley and Smith, 1995).

Energy dissipation rates near an impeller in a jar test device are 4 to 6 times higher in the impeller stream than calculated for the bulk average dissipation rate (Fig. 7.9b). Others have found that energy dissipation rates can vary by as much as a factor of 270 in a stirred vessel, with the bulk of the tank at an energy dissipation rate only 25% of the average rate calculated from the energy input into the tank (Tomi and Bagster, 1978). While Nagata et al. (1975) found that turbulent diffusivities were also only $\frac{1}{3}$ to $\frac{1}{4}$ as large as they are in the vicinity of the impeller, they also found that microscale eddy sizes varied from a few hundred to thousands of micrometers and were independent of impeller speed. At low impeller velocities, the larger, macroscale eddies (a few to tens of millimeters) increased with impeller speed. In all cases, it was concluded that the impeller energy was dissipated primarily by microscale eddies.

Stanley and Smith (1995) found that the energy dissipation rate should scale according to

$$\frac{\varepsilon_d}{(s\pi d_i)^3/d_i} = b_1 \qquad (7\text{-}67)$$

where s is the impeller rotation speed (s^{-1}), d_i is the impeller diameter, and b_1 is a constant found to be in the range of 0.1 to 0.12 in the impellor zone (Fig. 7.9b).

7.5 MAXIMUM SIZE OF AGGREGATES IN SHEARED FLUIDS

While smaller particles, such as bacteria, clays, and other inorganic colloids are much smaller than the Kolmogorov microscale, larger particles are also present in engineered systems used by environmental engineers. In water treatment plants the coagulation of small particles into larger, faster-settling aggregates is critical to the removal of those particles in settling tanks. In activated sludge wastewater treatment plants, rapid organic matter removal rates cannot be sustained without recirculation of bacterial biomass. In order efficiently to settle bacteria in clarifiers, they must form fast-settling flocs and aggregates.

The size distribution of particles in water and wastewater treatment plants will be addressed in greater detail in Chapter 11. However, the maximum size of aggregates that can be maintained in a sheared fluid is relevant to the current discussion of energy input and fluid shear rates. If all energy in a reactor is dissipated by eddies the size of the Kolmogorov microscale, then it can be reasoned that aggregates smaller than the microscale in an isotropic turbulent fluid should be stable. Particles

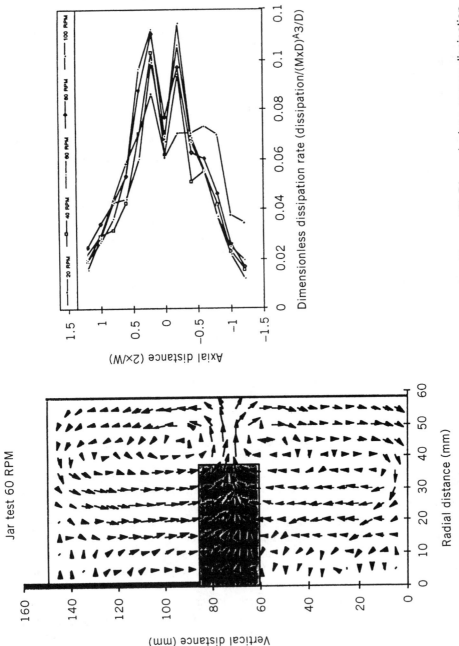

Figure 7.9 (A) General flow patterns in a test jar device mixed at 60 rpm. (B) Dimensionless energy dissipation rates in the impeller zone of the jar test device as a function of axial distance $2z/w$, where z is the distance and w is the impeller width. (Reproduced by permission of ASCE from Stanley and Smith, 1995.)

smaller than the microscale, although strained, will rotate with the eddy. Particles larger than the microscale will extend outside the eddy and be exposed to other eddies rotating in other directions. Thus particles larger than the Kolmogorov microscale will be subject to shear, breakup, and erosion rates much higher than those within the microscale.

Based on the above line of reasoning, the largest size of the aggregate should be, on average, the size of the Kolmogorov microscale. Several factors, however, complicate the argument. First, the fluid is not isotropic in the sense that there are larger eddies than the microscale present in the fluid. Parker et al. (1970) calculated that the eddy scale most significant in imparting a relative fluid velocity on an aggregate through partial entrainment would be twice the size of the aggregate diameter. More important, the Kolmogorov microscale was reasoned, using dimensional analysis, to be the size of the *smallest* eddy that could exist in a turbulent fluid. However, there are eddies larger than the microscale that do not degenerate into smaller eddies but instead dissipate into heat. The centroid of the smallest eddies that dissipate has been experimentally determined to be 2λ (Brodkey, 1966; Matson et al., 1972). Based on this experimental observation, the largest eddy size stable in an isotropic sheared fluid would be

$$d_{max} = 2\lambda \tag{7-68}$$

or based on radius

$$R_{max} = \lambda \tag{7-69}$$

as shown in Figure 7.10. Substituting in the definition of λ from Eq. 7-17 results in

$$R_{max} = \left(\frac{v_w}{G}\right)^{1/2} \tag{7-70}$$

or since $v_w \approx 0.01$ cm^2 s^{-1}

$$R_{max}[\mu m] \approx 1000 G^{-1/2} \tag{7-71}$$

There are obvious limitations on use of the microscale for estimating aggregate sizes. If the bonding between particles is stronger than the force exerted by the microscale eddy, then the floc will be able to resist the shearing force of the eddy, resulting in floc sizes larger than the microscale eddy. The addition of a polymer to aggregates is known to increase the size of aggregates produced by coagulation by strengthening the aggregate bonding forces. Median sizes of aggregates produced with coagulants such as ferric hydroxide and polymers can increase to many times their average size without the polymer (Table 7.2). The sizes of these aggregates, relative to the microscale, cannot be predicted, since it is a function of the polymer and types of particles. In the case of ferric hydroxide flocs, the average particle

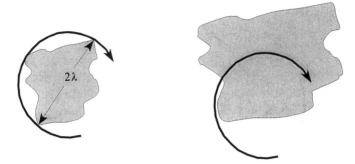

Figure 7.10 The largest sizes of aggregates, on average, is often limited by the average size of the mean eddy sizes equal to twice the size of the Kolmogorov microscale, or 2λ, as shown for the aggregate on the left. The aggregate on the right is larger due to the filaments reinforcing the aggregate to make it withstand the energy of the smaller energy-dissipating eddies.

diameters are less than the microscale, while clay aggregates flocculated with alum and a sodium alginate can be larger than the microscale.

The addition of more coagulant does not necessarily make the aggregate stronger. Coagulants such as alum and ferric chloride destabilize particles and allow them to strike each other and to stick to each other. However, polymers can bridge particles and intertwine with other polymers substantially increasing the bonding strength between particles. Increasing the doses of alum from 2 to 50 mg L^{-1}, for example, was not found to alter aggregate sizes substantially (Table 7.2). However, the addition

TABLE 7.2   Floc Sizes as a Function of Shear Rate for Different Types of Flocs

Floc Type	Shear Rate (s^{-1})	$\lambda = 1000\ G^{-1/2}$ (μm)	Median Floc Size (μm)	
			Without Polymer	With Polymer
Ferric hydroxide	50	141	17	56
+ tapwater[a]	100	100	10	48
	150	82	8	45
Ferric hydroxide	50	141	20	135
+ wastewater[a]	100	100	13	80
	150	82	10	59
Clay and alum[b]	88–98	101–107	84–120	330
	117–124	89–90	70–114	275
	151–157	80–81	51–112	208
	182–187	73–74	49–95	221
	217–222	67–68	34–97	170

[a]Leentvaar and Rebhun (1983); 10 mg L^{-1} added, anionic polymer is 3 mg L^{-1} Superfloc A100.
[b]Matsuo and Unno (1981); alum added at 2.0 to 50 mg L^{-1}, polymer is 0.5 mg L^{-1} sodium alginate with 10 mg L^{-1} alum.

of a polymer (sodium alginate) substantially increased the strength of the aggregate, allowing large increases in the average aggregate size.

Example 7.4

Using Eq. 7-71, calculate the maximum diameters of aggregates expected to occur in flocculators, mixed bioreactors, and the ocean if the size of the aggregates is limited by fluid shear. Comment on whether these aggregate sizes are reasonable.

In order to estimate the aggregate size, we need to make estimates of typical shear rates in the three systems. Biological reactors are intensely mixed in order to achieve high oxygen transport rates to flocculated microorganisms. There are wide variations in shear rates in the reactor, but mean shear rates are on the order of 100 to 1000 s^{-1}. Flocculators used in water treatment are designed to promote particle collisions and minimize breakup, and typically have much lower mean shear rates of 30 to 50 s^{-1}. The lowest shear rates of these three systems occur in the ocean. Surface waters typically have maximum shear rates of less than 0.08 to 0.44 s^{-1}, with substantially lower shear rates in deeper waters.

Using these estimates of G in Eq. 7-71, we have the results shown in Table 7.3. These estimates for particles in the ocean are fairly accurate, since highly amorphous aggregates on the order of several millimeters have been observed to occur in surface waters off the California coast (Alldredge and Gotschalk, 1987). Similarly, in water treatment plant flocculators aggregates are rarely as large as a millimeter unless the particle is part of resuspended material that may not have had sufficient time to be broken up.

The estimates for biological aggregates, however, are clearly low for bioreactors used in wastewater treatment, where aggregates can reach sizes on the order of several millimeters or more. However, when microbes are grown in pure cultures, flocs formed in the reactor are often quite smaller, and therefore much more fragile, than larger flocs found in mixed culture reactors. Why the difference? As observed by Parker et al. (1970), normal activated sludge flocs (from plants operating with good performance) include both filamentous and other types of microorganisms. If we view the aggregate as a type of concrete, these filamentous microbes act as rebar does in a cement matrix. The filaments strengthen the aggregate, allowing the aggregate to withstand shear forces exerted on the floc by

TABLE 7.3 Estimates of Aggregate Diameters Based on Fluid Shear Rates

System	$G(s^{-1})$	Maximum diameter (μm)
Bioreactors	100–1000	60–200
Flocculators	30–50	280–360
Ocean	0.08–0.44	3000–7000

smaller microscale eddies in the reactor. It is only as the aggregates reach larger sizes that they encounter eddies with sufficient energy to shear the aggregate. Thus activated sludge flocs can reach much larger sizes than our estimates, which assume the Kolmogorov microscale limits the size of the aggregates.

7.6 CHEMICAL TRANSPORT FROM BUBBLES

The generation and introduction of gas bubbles into sheared biological reactors is an efficient way to transport oxygen into the system, but there are many possible mass transport resistances. These resistances are shown in Figure 7.11 by the series of concentration gradients that would result from these resistances for the rather complicated case of oxygen transport to microorganisms within a biological aggregate. The potential resistances are: within the bulk gas phase (bulk gas-phase resistance), in the air at the air–water interface (gas-phase resistance); in the water at the air–liquid interface (liquid-phase resistance); in the bulk liquid phase (bulk liquid-phase resistance); at the water–aggregate interface; in the bulk aggregate phase (aggregate-phase resistance); at the surface of a microorganism in the aggregate (microbial mass transport resistance).

Of these resistances, we expect that the bulk gas-phase resistance will be negligible, and if the reactor is mixed either by a mechanical mixer or through the motion of the bubbles, the bulk liquid-phase resistance will also be quite small. The case of mass transport resistances around the aggregate are quite important, but for the moment we will consider the simpler case of unaggregated microorganisms. The remaining potential resistances to mass transport of oxygen to unaggregated microor-

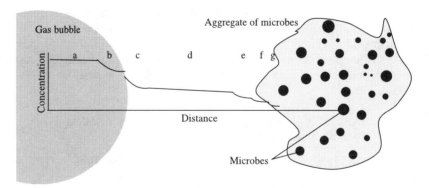

Figure 7.11   Resistances to mass transport during transport from a gas bubble to a microbe in an aggregate are due to: a, bulk gas phase; b, gas interface; c, liquid interface; d, bulk liquid; e, aggregate; f, bulk aggregate; g, microbe.

ganisms are

$$\frac{1}{\mathbf{K}_{wa}a} = \frac{1}{\mathbf{k}_w a_w} + \frac{1}{\mathbf{k}_a a_a H_{Occ}} + \frac{1}{\mathbf{k}_m a_m} \tag{7-72}$$

where $\mathbf{K}_{wa}$ is an overall mass transport coefficient [L T^{-1}] calculated from the concentration difference between oxygen saturation in the liquid, c_{eq}, and the bulk oxygen concentration in the liquid, c_∞, the three resistances on the right-hand side of Eq. 7-72 are due to the liquid phase, the gas phase, and transport of oxygen through the water to the microbes ($\mathbf{k}_m$), and H_{Occ} is the dimensionless Henry's Law constant. The areas for mass transport must be included in the resistances since they are not all equal, as they would be with only a two-phase problem. The individual areas are: water, a_w, air, a_a, and microbe, a_m. It can be seen that $a_a = a_w \neq a_m$. From a comparison of the magnitude of the mass transport resistances, it can be shown that almost all the resistance to oxygen transport is due to the liquid phase (see Problem 7.2).

The oxygen flux from a single bubble, J_{Ow}, is calculated in the usual manner from a mass transport coefficient by

$$J_{Ow} = \mathbf{K}_{wa}(c_{eq} - c_\infty) \tag{7-73}$$

where c_{eq} and c_∞ are the equilibrium and bulk concentrations of oxygen in water. Since the overall mass transport resistance is controlled by the resistance in the liquid phase, for oxygen this overall mass transport coefficient is essentially equal to the liquid-phase resistance. The rate of transport from the bubbles to the liquid, W_{Ow}, is calculated by multiplying the flux by the surface area of the bubble, A_b, or

$$W_{Ow} = \mathbf{K}_{wa}A_b(c_{eq} - c_\infty) \tag{7-74}$$

When a large number of bubbles are generated in a vessel from an aeration device, the total interfacial area is usually used to calculate the total rate of mass transport into the liquid, and Eq. 7-74 becomes

$$W_{Ow} = (\mathbf{K}_{wa}a_v)V(c_{eq} - c_\infty) \tag{7-75}$$

where a_v is the air–water interfacial area per volume of reactor [L^2 L^{-3} = L^{-1}], and V is the reactor volume. The parentheses are placed around the term $(\mathbf{K}_{wa}a_v)$ to emphasize that this term is frequently evaluated as a single constant.

Two dimensionless numbers are frequently used to describe transport from bubbles. The first is the Grashoff number, Gr, defined as

$$Gr = \frac{R_b^3 \rho_w g(\rho_w - \rho_a)}{\mu_w^2} \tag{7-76}$$

Because $\rho_w > \rho_a$ and $\nu_w = \mu_w / \rho_w$ this can be simplified to

$$\text{Gr} \approx \frac{R_b^3 g}{\nu_w^2} \qquad (7\text{-}77)$$

The second dimensionless number is the Raleigh number, Ra, defined as

$$\text{Ra} = \frac{R_b^3 g (\rho_w - \rho_a)}{\mu_w D_{Ow}} \qquad (7\text{-}78)$$

Again, this can be approximated, due to the higher density of water than air, as

$$\text{Ra} \approx \frac{R_b^3 g}{\nu_w D_{Ow}} \qquad (7\text{-}79)$$

Mass transport coefficients have been developed for a variety of cases depending on bubble size and whether the bubbles are present as single bubbles or they are present in swarms.

Single Bubbles

For small bubbles in the creeping flow region (Re $\ll$ 1) with Sc $= \nu_w / D_{Ow} \gg 1$, the mass transport coefficient can be derived from the Sh correlation

$$\text{Sh} = 0.64 \, \text{Pe}^{1/3}, \quad \text{Re} \ll 1, \text{Sc} \gg 1 \qquad (7\text{-}80)$$

where Sh $= K_{wa} R_b / D_{Ow}$ and Pe $= U_b R_b / D_{Ow}$ where U_b is the rise velocity of the bubble. The rise velocity of small bubbles obeys Stokes Law, or

$$U_b = \frac{2g}{9\nu_w} \frac{(\rho_w - \rho_a)}{\rho_w} R_b^2 \qquad (7\text{-}81)$$

Substituting the rise velocity into Eq. 7-80

$$\text{Sh} = 0.64 \left(\frac{R_b^3 2 g (\rho_w - \rho_a)}{18 \mu_w D_{Ow}} \right)^{1/3} \qquad (7\text{-}82)$$

or based in terms of dimensionless numbers, this can be computed either as

$$\text{Sh} = 0.39 \, \text{Gr}^{1/3} \, \text{Sc}^{1/3} \qquad (7\text{-}83)$$

or

$$\text{Sh} = 0.39 \, \text{Ra}^{1/3} \qquad (7\text{-}84)$$

For larger bubbles, with Re >> 1, an appropriate mass transport correlation is

$$Sh = 1 + 0.42 \, Re^{1/2} \, Sc^{1/3} \tag{7-85}$$

Swarms of Bubbles

In aerated reactors bubbles will be present as swarms of bubbles. In a swarm the average bubble rise velocity is about half that of an isolated bubble freely rising in an undisturbed fluid. A mass transport correlation for small ($R_b < 1.25$ mm) bubbles is

$$Sh = 0.31 \, Ra^{1/3} \tag{7-86}$$

while for larger bubbles

$$Sh = 0.42 \, Gr^{1/3} \, Sc^{1/2} \tag{7-87}$$

The total interfacial area of all bubbles is a function not only of the area per bubble, but also how long the bubble is in the reactor. In words

$$\text{total bubble area} = \left(\frac{\text{no. of bubbles}}{\text{time}} \right) (\text{area per bubble})$$

$$\times \, (\text{time bubble is in reactor}) \tag{7-88}$$

The first term is the gas flow rate, Q_a, divided by the volume of a single bubble, or

$$\frac{\text{no. of bubbles}}{\text{time}} = \frac{Q_a}{(4\pi/3)R_b^3} \tag{7-89}$$

Using Eq. 7-88, the area of a bubble is calculated as $4\pi R_b^2$, and defining the time the bubble is in the reactor as t_b, the total interfacial area, a_v, per volume of reactor V, is

$$a_v = \frac{3Q_o t_b}{2VR_b} \tag{7-90}$$

The time the bubbles are in the reactor should be calculated by integrating the rise velocities of the bubbles over the height of the reactor, h, or

$$t_b = \int_0^h \frac{dz}{U_b(z)} \tag{7-91}$$

Without specific reactor data, we approximate Eq. 7-91 as just $t_b = h/U_b$. For very small bubbles (Re << 1), we can calculate the rise velocity using Stokes' Law to obtain for a single bubble

$$t_b = \frac{9v_w}{2g} \frac{h\rho_w}{(\rho_w - \rho_a)R_b^2}$$

(7-92)

For larger, single bubbles

$$u_b[\text{cm s}^{-1}] = (2gR_b)^{1/2} = 44.3R_b^{1/2}$$

(7-93)

producing an average time in the reactor of

$$t_b = \frac{h}{1.41(gR_b)^{1/2}}$$

(7-94)

While the above equations are useful for predicting a_v for a bubbled vessel, once the reactor is operational a_v can be calculated from the change in reactor volume during bubbling. Defining the gas holdup, H, as

$$H = \frac{\text{volume of gas}}{\text{volume of gas + volume of liquid}}$$

(7-95)

the total interfacial area can be calculated for spherical bubbles of average diameter R_b as

$$a_v = \frac{3H}{R_b}$$

(7-96)

Determination of Mass Transport Coefficients

Oxygen transport into a vessel can occur through two routes: the air–water interface at the top of the reactor, and the air–water interface created at the bubble surfaces. Designating the separate mass transport coefficients with subscripts of surface and bubbles, a mass balance on the overall rate of air transport into the fluid in a CSTR is

$$\frac{dc_{Ow}}{dt} = \frac{Q_w c_{Ow,\text{in}}}{V} - \frac{Q_w c_{Ow,\text{out}}}{V} + [(\mathbf{K}_{wa} a_v)|_{\text{surface}}$$

$$+ (\mathbf{K}_{wa} a_v)|_{\text{bubbles}}](c_{Ow,\text{eq}} - c_{Ow,\text{out}}) - kX$$

(7-97)

where Q_w is the water flow rate into the vessel, $c_{Ow,\text{out}}$ is the bulk-phase concentration of chemical (O oxygen) being transferred from water to air, V is the reactor volume, k is a reaction rate constant, and chemical reaction is zero order with respect to c_{Ow} and first order with respect to X, the mass concentration of cells in the reactor. The interfacial area per volume term due to the liquid surface for a reactor is simply a_v = h^{-1}. Although the two routes of mass transport contribute to oxygen transport into the liquid, the much larger surface area of the bubbles usually results in $(\mathbf{K}_{aw}a_v)|_{bubbles}$ >> $(\mathbf{K}_{aw}a_v)|_{surface}$. In the analysis that follows we will only consider a single mass transport coefficient, recognizing that most of the mass transport results from oxygen transport across the bubble–water interface.

The magnitude of the terms describing the rate of oxygen production by mass transport from bubbles and oxygen utilization are usually much greater than terms describing the rate oxygen enters and leaves the reactors. Therefore, for short time periods we can simplify the above equation with $c_{Ow,out} = c$ and $c_{Ow,eq} = c_{eq}$ to

$$\frac{dc}{dt} = (\mathbf{K}_{wa}a_v)(c_{eq} - c) - kX \qquad (7\text{-}98)$$

where all the subscripts for the chemical have been omitted for simplicity.

To determine the values of the mass transport coefficient and kX, a dynamic oxygen uptake experiment can be performed. Let us assume that until a time 0 the system is at steady state so that oxygen uptake equals oxygen utilization. At time zero, we shut off the air, and in the absence of gas transport into the liquid Eq. 7-98 becomes

$$\frac{dc}{dt} = -kX \qquad (7\text{-}99)$$

From the constant rate of decrease in oxygen, we can calculate the value of the term kX as shown in Figure 7.12. If we now turn the air back on, the oxygen transport rate will rise at a nonlinear rate according to Eq. 7-88. Since we now know the

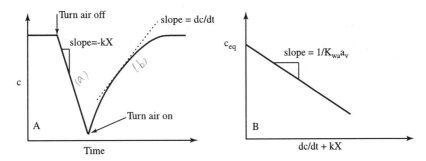

Figure 7-12 Determination of the mass transport coefficient $\mathbf{K}_{wa}a_v$. (A) A batch test is performed for a microbial suspension by turning the air off and then back on. (B) Based on data from this experiment, the mass transport coefficient can be determined as shown.

value of the term kX, we can rearrange Eq. 7-98 into a linearized form of

$$c = c_{eq} - \frac{1}{(\mathbf{K}_{wa}a_v)}\left(\frac{dc}{dt} + kX\right)$$

(7-100)

From a plot of c in the bulk phase versus time, we can obtain values of dc/dt. If we now plot the sum of $dc/dt + kX$ as a function of c, the slope of the line will be $(\mathbf{K}_{wa}a_v)^{-1}$, and the y intercept will be c_{eq} (Fig. 7.12).

This method of determining a mass transport coefficient has some limitations. First, the measurement of oxygen must be faster than the rate of change of oxygen concentration. If an oxygen probe, for example, is used to measure dissolved oxygen, the response of the probe should be fast enough that an accurate measurement of oxygen concentrations is obtained. Some probes can take minutes to reach equilibrium, and it is possible that the oxygen uptake rate can be fast enough to deplete oxygen in the reactor in only a few minutes. The second difficulty is that there will be oxygen utilization by microorganisms in the air bubble boundary layer. This means that a small fraction of the cells will see higher oxygen concentrations than that present (and measured) in the bulk liquid. If oxygen uptake is zero order with concentration, the presence of microorganisms in the bubble boundary layer should not affect the rate of oxygen utilization by the cells. However, if oxygen uptake is proportional to substrate utilization rates, and substrate uptake increases due to shear in the bubble boundary layer, then oxygen uptake will appear to increase as a result of increased substrate utilization.

A more serious limitation of this approach is that the presence of a fast reaction in a concentration gradient will accelerate oxygen transport into the liquid. We saw in Chapter 4 that a chemical reaction accelerated chemical transport (diffusion into a reactive liquid). The microbes accelerate oxygen transport in the same manner by increasing the chemical gradient around the bubbles and therefore increasing the oxygen flux into the water. Since there are no bubbles present when the oxygen utilization is measured for the cells by Eq. 7-99, the reaction rate measured in the absence of bubbles will be smaller than that measured in the presence of bubbles, producing inaccurate calculation of the mass transport coefficient.

Mass transport coefficients are measured by aerator companies using an abiotic oxygen depletion test. Oxygen is scavenged from the liquid, using a chemical such as sodium sulfite (and a cobalt catalyst). Once the chemical is fully consumed, the rate that oxygen is transferred into the liquid can be used to calculate the mass transport coefficient. In the absence of oxygen utilization of microorganisms, Eq. 7-100 can be written as

$$c = c_{eq} - \frac{1}{(\mathbf{K}_{wa}a_v)}\left(\frac{dc}{dt}\right)$$

(7-101)

Investigators have found, however, that the mass transport coefficient calculated with Eq. 7-101 is lower than that measured using Eq. 7-100. Furthermore, the mass transport coefficient has been found to increase with the biomass concentration (Albertson and DiGregorio, 1975). This has prompted some investigators incorrectly to propose that there is direct oxygen transport from the gas to the cell surface (Albertson and Digregorio, 1975). There is little scientific basis for such a hypothesis, since cells will always have some water at their surface, and therefore a mass transport resistance. There is also no biochemical basis for a pathway for direct use of oxygen from air across a cell membrane.

There are at least three factors that can account for higher oxygen transport rates in the presence of microorganisms. First, oxygen transport may increase because of decreased surface tension due to molecules produced by the microorganisms. This will decrease the liquid-phase mass transport coefficient (which is the dominant resistance to mass transport) and increase the rate of mass transport. Second, we have discussed that a reaction in the bubble boundary layer will increase the rate of oxygen transport. This has been further discussed and experimentally investigated by Sundararajan and Ju (1995). Third, fluid shear produced by the bubble boundary layer can increase the rate of substrate utilization by the microorganisms (Confer and Logan, 1991). If oxygen uptake is coupled with substrate utilization rates, then oxygen uptake can be expected to increase in conjunction with increased substrate utilization rates produced by fluid shear. This last effect is described further in the next chapter.

PROBLEMS

7.1 A culture of *Aspergillus funigatus* is subjected to a dynamic oxygen depletion test in a 14-liter fermenter. The concentration of dissolved oxygen is measured as a function of time, and the resulting data are shown in Table P-7.1. (a) Calculate the specific oxygen uptake rate for the cells. (b) What is the volumetric oxygen transfer coefficient for the fermenter? (c) Determine the concentration of oxygen in the liquid if it were in equilibrium with the gas used for aeration.

TABLE P-7.1   Oxygen Concentrations in Fermenter

Time (min)	c_{Ow} (mg L^{-1})
0	6.25
1	6.25
1.5	5.75
2	4.35
2.5	3.00
3	1.60
4	2.25
5	4.00
6	5.04
7	5.80
8	5.92

7.2 Typical values of the volumetric mass transfer coefficients for the gas and liquid phases in a well-agitated aerated system (a fermentor or activated sludge reactor) are as follows: gas phase: $k_a a_a = 3.9 \times 10^{-5}$ [g mol/atm cm^3 s]; liquid phase: $k_w a_w = 0.1$ s^{-1}. The liquid at 30°C contains 5 g DW/L (DW dry weight) of microbial cells, which may be assumed to be spheres of 2 µm diameter, containing 80% water, and having a density of 1.03 g/cm^3. Mass transfer to cells may be assumed to be equivalent to mass transfer between spheres and stagnant fluid. Pressure is 1 atm. For oxygen transfer between the gas phase and the cells, Henry's constant for oxygen may be assumed to be 10^6 cm^3 atm/g mol, and the diffusion coefficient is $D_{Ow} = 2 \times 10^{-5}$ cm^2 s^{-1}. Using Eq. 7-72, calculate for this system the overall liquid-phase volumetric mass transfer coefficient [s^{-1}], and the fraction of total mass transfer resistance attributable to each of the three mass transfer regimes.

7.3 Suspended-growth bioreactors can often be very deep. As a result, the pressures at the bottom of the tanks can be much higher than atmospheric pressure. Let us assume that we are examining oxygen transfer from a rising bubble in a tank 30 ft deep. You may wish to use some of the following data at 20°C: for oxygen, $H_{Opc} = 43,800$ atm, $D_{ow} = 1800 \times 10^{-8}$ cm^2 s^{-1}; for water, $\rho_w = 1$ g cm^{-3}, $\nu_w = 0.01$ cm^2 s^{-1}; recall that $\Delta P = -\rho g \Delta z$. (a) What would the rate of mass transfer (mg/s) be from a bubble if it was not moving, and it was at the top of the tank? Assume that the loss of oxygen from the bubble does not appreciably alter the oxygen content in the bubble relative to location. The bubble is 0.1 cm in diameter at the top of the tank, and the bulk concentration of oxygen is 4 mg L^{-1}. (b) Repeat part (a), but this time for a bubble at the bottom of the tank. (c) What would the rate of oxygen transfer be from a bubble at the top of the reactor for a shear rate of 100 s^{-1} in the bioreactor, assuming the size of the bubble is not affected by shear. (d) What would be the rate of mass transfer from the bubble at its rise velocity [$U_b = 0.77(g d_b)^{0.5}$] at the top of the reactor?

7.4 A small laboratory bioreactor is operated at a holdup of 0.1 and has a total reactor volume of 2 L at 20°C (reactor volume = gas volume + liquid volume). During operation, the microbial suspension uses oxygen so that the bulk concentration of oxygen is 6 mg/L and the saturation oxygen concentration is 8.5 mg L^{-1}. The average bubble radius is 2.5 mm, and the diffusivity of oxygen in water is 2.0×10^{-5} cm^2 s^{-1}. The following data may be useful: for air, $\rho_a = 0.0012$ g m^{-3}, $\mu_a = 0.0018$ g m^{-1} s^{-1}; for water, $\rho_w = 1.0$ g cm^{-3}, $\mu_w = 1.0$ g m^{-1} s^{-1}. (a) What are the rise velocities of a single bubble and a bubble in a swarm? (b) What is the interfacial area available for mass transfer? (c) What is the rate of oxygen transfer into the fluid under these conditions?

7.5 Wastewater treatment engineers have suggested correcting the mass transfer coefficient, k_w, by the empirical expression:

$$\frac{k_w(T)}{k_w(20)} = \theta^{T-20} \tag{7-102}$$

where $\mathbf{k}_w$ is used in the expression for the mass flux from swarms of large bubbles as $J_O = \mathbf{k}_w(c_{eq} - c)$, and $\theta = 1.024$. Water viscosities are $\mu_w(30°C) = 0.7975$ g m^{-1} s^{-1}, $\mu_w(20°C) = 1.001$ g m^{-1} s^{-1}, and $\mu_w(10°C) = 1.307$ g m^{-1} s^{-1}. Assume no reaction in the fluid. (a) What is the percent change in $\mathbf{k}_w$ between 20°C and 30°C? Between 20°C and 10°C? (b) Which theory (i.e., stagnant film, penetration, or boundary layer theory) best describes oxygen transfer from large moving bubbles to fluid? Explain why. How does the mass transfer coefficient vary with oxygen diffusivity for this theory? (c) Show how the mass transfer coefficient should be proportional to temperature from your understanding of the mass transfer theory identified in part (b). Calculate the percent change in the mass transfer coefficient with temperature as in part (a). Neglect changes in bubble size with temperature.

7.6 A 100-μm-diameter naphthalene particle is dissolving in a sheared aqueous reactor. You are interested in making a calculation on how long the dissolution might take. (a) Derive an expression for the rate of dissolution (W_{Nw}) of the naphthalene sphere in a sheared fluid, assuming that $Pe_G \ll 1$, in terms of D_{Nw}, c_{Nw}, d, etc., where d is the diameter of a sphere of pure naphthalene. (b) Recognizing that for a sphere the mass $m_N = \rho_N(\pi/6)d^3$, and that $W_{Nw} = -dm_N/dt$, rewrite this equation from part (a) in terms of mass. Simplify this equation to the case for $G = 0$ (no shear in the reactor). (c) Since you now have an expression for the rate mass is lost as a function of mass, you can integrate this equation to calculate the lifetime of the sphere. (Do this only for the case of no shear, i.e., $G = 0$.) (d) Calculate the lifetime (in days) of the naphthalene sphere assuming a sphere density of $\rho_N = 1.03$ g cm^{-3}, solubility $c_{eq} = 34.4$ mg l^{-1}, and diffusivity in water $D_{Nw} = 4.79 \times 10^{-6}$ cm^2 s^{-1}.

$\dfrac{(2.05)^3}{125}$

REFERENCES

Albertson, O. E., and D. DiGregorio. 1975. *J. Water Pollut. Control Fed.* **47**(5):976–88.

Alldredge, A. L., and C. Gotschalk. 1987. *Limnol. Oceanogr.* **33**:339–51.

Batchelor, G. K. 1952. *J. Fluid Mech.* **5**:113–33.

Batchelor, G. K. 1979. *J. Fluid Mech.* **95**:369–400.

Batchelor, G. K. 1980. *J. Fluid Mech.* **98**:609–23.

Bird, R. B., W. E. Steward, and E. N. Lightfoot. 1960. *Transport Phenomena*. Wiley, New York.

Brodkey, R. S. 1966. In: *Mixing, Theory and Practice*, Uhl and Gray, Eds., Vol. 1, Academic Press, NY.

Camp, T. R. 1968. *J. Amer. Water Works* **60**:656–73.

Confer, D. R., and B. E. Logan. 1991. *Appl. Env. Microbiol.* **57**(11):3093–100.

Frankel, A. H., and A. Acrivos. 1968. *Physics of Fluids* **11**:1913–18.

Hill, P. J. 1992. *Geophys. Res.* **97**:2295–308.

Karp-Boss, L., and P. A. Jumars. 1996. *Oceanogr. Mar. Biol. Ann. Rev.* **34**:71–107.

Lai, R. J., H. E. Hudson, Jr., and J. E. Singley. 1975. *J. Amer. Water Works Assoc.* **67**:553–57.

Lazier, J. R. N., and K. H. Mann. 1989. *Deep Sea-Res. I* **36**:1721–33.

Leal, G. 1992. *Laminar Flow and Convective Transport Processes.* Butterworth-Heinemann, Boston.

Leentvaar, J., and M. Rebhun. 1983. *Water Res.* **17**(8):895–902.

Li, X. 1996. Ph.D. Dissertation, Dept. of Chemical and Environmental Engineering, Univ. of Arizona, Tucson.

Matson, J. V., W. G. Characklis, and A. W. Busch. 1972. Proc. 27th Ind. Waste Conf., Purdue, IN. p. 894–903.

Matsuo, T., and H. Unno. 1981. *J. Environ. Engng.* **107**(EE3):527–45.

Nagata, S., M. Nishikawa, A. Inoue, and Y. Okamoto. 1975. *J. Chem. Eng. Japan* **8**(3): 243–48.

Parker, D. S., W. J. Kaufman, and D. Jenkins. 1970. Characteristics of biological flocs in turbulent regimes. SERL Rept. 70-5, College of Engineering, University of California, Berkeley, CA, 94720.

Purcell, E. M. 1978. *J. Fluid Mech.* **84**:551–59.

Stanley, R. J., and D. W. Smith. 1995. *J. Environ. Engrg.* **121**(12):902–10.

Sundararajan, A., and L.-K. Ju. 1995. *Water Env. Res.* **67**(5):848–54.

Tomi, D. T., and D. F. Bagster. 1978. *Trans. IChemE.* **56**:1–8.

van Duuren, F. A. 1968. ASCE *J. Sanit. Engrg.* **94**(SA4):671–82.

CHAPTER 8

SUSPENDED UNATTACHED AND AGGREGATED MICROORGANISMS

8.1 INTRODUCTION

In most natural environments aerobic microorganisms have typical growth rates on the order of a day. These are much less than the maximum growth rates of less than an hour obtained for nutrient-sufficient cultures. What limits the growth rates of the microorganisms? Under nutrient-sufficient conditions it is likely the time necessary to synthesize new cell material. When nutrients, or chemical substrates, limit cell growth rates, either chemical reaction rates (uptake kinetics) will limit substrate uptake and utilization or mass transport will limit substrate transport to the surface of the cell. If uptake kinetics limit substrate utilization, then concentration gradients around cells will be negligible. Conversely, under mass-transport-limited conditions we would expect large concentration gradients to persist around the cell. In order to examine under what conditions mass-transport-limited uptake is likely to arise, the rate of chemical transport to a suspended microorganism is derived below both in terms of microbial kinetics and mass transport coefficients. The question of whether concentration gradients exist around cells will be discussed separately by examining experimental evidence that under some conditions the fluid mechanical environment of the cell can be manipulated to increase chemical fluxes into microorganisms.

8.2 CHEMICAL TRANSPORT TO CELLS AT REST

Microbial growth is usually viewed from the perspective that only one chemical at a time limits growth rates, and therefore cell growth is described in terms of a kinetic model. After reviewing this kinetic model approach, it is then shown that microbial growth can also be described in terms of a limiting substrate using a mass transport model. These models are then combined in order to obtain a more general model of microbial growth when either uptake kinetics or mass transport can be limiting growth rates.

Kinetic Model

The growth rate of suspended microorganisms is proportional to the number of microorganisms, N (#/mL), or

$$\frac{dN}{dt} = \mu N \tag{8-1}$$

where μ is the growth rate constant $[t^{-1}]$. The rate of growth is a nonlinear function of the substrate concentration. While there are many expressions for microbial growth as a function of substrate concentration, the most common model is the Monod model

$$\mu = \frac{\mu_{max}c}{K_m + c} \tag{8-2}$$

where μ_{max} is the maximum rate of chemical uptake by a cell, c is the bulk concentration of a growth-limiting nutrient or substrate, and K_m is the Monod half-saturation constant defined as the chemical concentration when $\mu = \mu_{max}/2$. While the Monod model is completely empirical, it is identical in form to the Michaelis–Menten enzyme kinetic model used to describe the rate of chemical transformation by an enzyme–substrate complex. According to the Michaelis–Menten enzyme model

$$v = \frac{v_{max}c}{K_{MM} + c} \tag{8-3}$$

where v_{max} is the maximum rate of substrate reaction with enzyme, and K_{mm} is again a half-saturation constant defined as the substrate concentration when $v = v_{max}/2$. If growth is "coupled" to substrate utilization, that is, if cells grow at a rate proportional to the rate they take up substrate (and do not store substrate), then either model should describe the rate of chemical uptake by microorganisms if such uptake is limited by enzyme kinetics.

For coupled growth and substrate utilization

$$\frac{dc}{dt} = \frac{-1}{Y_{N/c}}\frac{dN}{dt} \tag{8-4}$$

Combining Eqs. 8-1, 8-2, and 8-4, we arrive at a kinetic description of the rate of substrate utilization by a single cell, W_{cell}

$$W_{cell} = \frac{-1}{N}\frac{dc}{dt} = \frac{1}{Y_{N/c}}\frac{\mu_{max}c}{(K_m + c)} \tag{8-5}$$

Mass Transport Model

The chemical flux to a spherical cell can be written in terms of a mass transport coefficient, $\mathbf{k}_w$, as

$$\mathbf{J} = \mathbf{k}_w(c_\infty - c_s) \tag{8-6}$$

where c_∞ is the substrate concentration far away from the cell surface, and c_s is the substrate concentration at the cell surface. To calculate the rate of mass transport to the cell, W_{cell}, we need to multiply the flux by the area available for mass transport. This area may not be just the surface area for the cell. While some substrate molecules can diffuse across the cell membrane, many other molecules are charged and must either enter through porins, large holes in the cell surface, or must be actively transported across the membrane. Uptake of the amino acid leucine, for example, is promoted in common Gram-negative cells such as *Escherichia coli* by special porters through porins on the cell surface. It is estimated that each *E. coli* cell contains 10^5 porins approximately $0.9~\mu m^2$ in cross-sectional area, representing roughly 13% of the surface area of a cell having a total surface area of $6.7~\mu m^2$ (Logan and Kirchman, 1991). If substrates must be actively transported across the cell surface, there will be even less surface area available for enzymes on the cell surface, resulting in a small percentage of the cell surface being available for transport. Defining the cell surface area for transport to be $4\pi R^2 f_a$, where f_a is the fraction of cell surface area available for transport, Eq. 8-6 becomes

$$W_{\text{cell}} = 4\pi R^2 f_a \mathbf{k}_w(c_\infty - c_s) \tag{8-7}$$

For spheres, the mass transport coefficient can be obtained from the Sherwood number by recalling that for a sphere at rest Sh = 1. Substituting in Sh = $k_w R/D_{Cw}$ into Eq. 8-7 produces a mass transport model

$$W_{\text{cell}} = 4\pi R~\text{Sh}~D_{Cw} f_a(c_\infty - c_s) \tag{8-8}$$

Combined Kinetic and Mass Transport Model

Since the rate of mass transport to a cell must equal the rate of chemical reaction under steady conditions, we can combine Eqs. 8-5 and 8-8 into

$$4\pi R~\text{Sh}~D_{Cw} f(c_\infty - c_s) = \frac{1}{Y_{N/c}} \frac{\mu_{\max} c}{(K_m + c)} \tag{8-9}$$

Equation 8-9 contains three different substrate concentrations. While the two definitions on the left-hand side of Eq. 8-9 are unambiguous, the concentration c on the right-hand side is not well defined. From an enzyme-kinetics point of view, the rate of uptake should be controlled by the surface concentration of the substrate. However, cell surface concentrations are not known from kinetic tests. Usually, measurements to calculate μ_{max} and K_m are based on bulk substrate concentrations. Accepting the latter situation, we will define here that $c = c_\infty$, and in subsequent calculations use only c.

Simplifying Eq. 8-9, the cell surface concentration can be calculated as a function of the bulk concentration as

$$c_s = c \left(1 - \frac{\mu_{max}}{4\pi R \ Sh \ D_{Cw} f_a (K_m + c) Y_{N/c}} \right) \qquad (8\text{-}10)$$

It is expected that mass-transport-limited conditions will only result at low substrate concentrations, or where $K_m \gg c$. For this case, Eq. 8-10 can be simplified to

$$c_s = c \left(1 - \frac{\mu_{max}}{4\pi R \ Sh \ D_{Cw} f_a K_m Y_{N/c}} \right) \qquad (8\text{-}11)$$

Example 8.1

Calculate the concentration gradient around a microorganism in undisturbed (not mixed) seawater assuming that (a) $f_a = 1.0$ and (b) $f_a = 0.001$, that the kinetics are first order, and the following constants: $\mu_{max} = 0.023 \ min^{-1}$, $D_{Cw} = 600 \times 10^{-8} \ cm^2 \ s^{-1}$, $Y_{N/c} = 10^{10}$ cells mg^{-1}, $R = 1 \ \mu m$, $K_m = 1 \ mg \ L^{-1}$.

(a) For our case $Sh = 1$ and $R = 1 \ \mu m = 10^{-4} \ cm$. Substituting our data into Eq. 8-11

$$\frac{c_s}{c} = \left(1 - \frac{0.023 \ min^{-1} \times \dfrac{1 \ min}{60 \ s} \times 10^3 \ \dfrac{cm^3}{1}}{4\pi(10^{-4} \ cm)(1)(600 \times 10^{-8} \ cm^2 \ s^{-1})(1)(1 \ mg \ L^{-1})(10^{10} \ mg^{-1})} \right) \qquad (8\text{-}12)$$

which results in

$$c_s = 0.995 \ c \qquad (8\text{-}13)$$

The small magnitude of the concentration difference suggests that mass transfer does not limit chemical uptake for these conditions, where the cell is completely available for chemical diffusion across the surface of the membrane.

(b) For the case of $f_a = 0.001$, we consider the more realistic case that a chemical must diffuse through porins on the cell surface or actively be transported into the cell. For this case, we repeat the above calculation, but with $f_a = 0.001$ we find that

$$c_s = 0.50 \ c \qquad (8\text{-}14)$$

Under these conditions, mass transport is likely to limit uptake. Fluid mixing, or fluid shear in the system that could dissipate or reduce the magnitude of the concentration boundary layer around the cell, would likely increase chemical uptake by the cell.

8.3   EFFECT OF FLUID MOTION ON MICROORGANISMS

Microbes can exist under different fluid mechanical conditions that include advective motion, simple laminar shear, and turbulent shear. If the cells are motile, swimming cells will experience a slightly different fluid environment due to the presence of both rotation about the cell's axis that drives the forward motion, and the flow along the cell in the axial direction due to that motion. Strictly advective flow past the cell's surface can occur when the cell is sinking or it is situated on a falling particle, for example, a mineral surface or a piece of detrital matter, or if the cell is part of a highly porous aggregate settling in a tank. As we have previously discussed, a microbe in a turbulent fluid will be much smaller than the Kolmogorov microscale of mixing, and will therefore see only a laminar shear field. However, in a turbulent fluid the direction of the fluid shear is constantly changing as the eddies are dissipated and reborn. Laminar shear, created in a couette device, for example, is an artificial situation in that the shear field is constant and nearly uniform. Placing a microbial suspension in a controlled environment allows us to control the magnitude of the shear field so that we can accurately determine the effect of fluid shear on cell kinetics. The effect of fluid motion on phytoplankton was first theoretically examined by Munk and Riley (1952). Experimental studies on the effect of fluid shear on microbial kinetics did not appear in the literature until several years later, but in general these studies validated Munk and Riley's calculations, as will be shown.

Advective Motion Past a Cell Fixed in a Flow Field

The first experimental evidence that fluid motion past the cell surface could increase uptake rates was provided by Canneli and Fuhs (1976). They demonstrated that phosphate uptake could be increased by 130% at fluid velocities ≤ 0.4 mm s^{-1}. At higher velocities, there was no effect of fluid motion. Extensive laboratory studies using radiolabeled compounds have also shown that the metabolic activity of heterotrophic bacteria attached to particles or surfaces can be higher, on a per-cell basis, than unattached or free-living bacteria. This has been demonstrated for a variety of substrates including glucose, gultamate, ATP, phosphate, protein hydrosylate, and amino acids (Logan and Kirchman, 1991). The reason for this difference is not well understood and has been attributed to the larger size of attached versus free-living microbes, utilization of organic matter in particles or adsorbed onto the surface, or to an increased chemical flux due to fluid flow past the particle surface.

The effect of fluid motion has been examined in the laboratory by placing cell suspensions onto porous membranes and filters, and by pulling water through the filter at controlled velocities (Logan and Dettmer, 1990). Uptake of a radiolabeled amino acid, [³H]leucine, was measured by monitoring the increase in radioactivity retained on a filter containing cells, versus that of abiotic controls. Uptake of [³H]leucine by pure cultures of *Zoogloea ramigera* held on polycarbonate membranes at very low concentrations (1 μg L⁻¹; leucine uptake was shown to be first order at this leucine concentration) increased by 55 to 65% at bulk water velocities of ~1 mm s⁻¹ compared to uptake by suspended cells (Fig. 8.1a).

A ratio of the rate of mass transport to a cell fixed in an advective flow field to one at rest is just the ratio of the Sh numbers. Since Sh = 1 for a cell in undisturbed fluid, this ratio can be simplified to the Sh number for a cell in the flow field, or

$$\frac{W_{cell} \text{ (in an advective flow field)}}{W_{cell} \text{ (undisturbed fluid)}} = \text{Sh}(u) \tag{8-15}$$

where the parenthetical material indicate the context of the calculation, and u is shown to indicate that the Sh number is calculated based on the fluid velocity. The increase in uptake with fluid motion was compared by Logan and Dettmer (1990) to that predicted by three mass transport correlations

$$\text{Sh} = 1 + 0.5 \text{ Pe} + 0.6 \text{ Pe}^2 \tag{8-16}$$

$$\text{Sh} = 1 + 45 \text{ Re}^{0.57} \tag{8-17}$$

$$\text{Sh} = (1 + 0.48 \text{ Pe}^{2/3})^{1/2} \tag{8-18}$$

based on equations proposed by Munk and Riley (1952), Logan and Alldredge

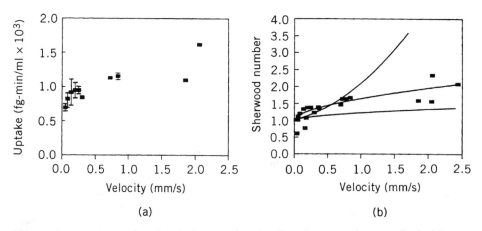

Figure 8.1 Uptake of radiolabeled leucine by *Zoogloea ramigera* cells held on a polycarbonate filter as a function velocity: (A) uptake as a function of flow; (B) Sherwood numbers calculated as the rate of uptake by a cell in the velocity flow field versus cells suspended in an undisturbed fluid (From Logan and Dettmer, © 1990, Biotechnol. Bioeng., reprinted by permission of John Wiley & Sons, Inc.).

(1989), and Brian and Hales (in Sherwood et al., 1975), respectively. It was found that an equation used by Munk and Riley (1952) to predict the effect of fluid sinking on phytoplankton provided the best agreement with the data (Fig. 8.1b).

A similar study was conducted on a natural assemblage of marine bacteria from the Delaware Bay. A water sample was filtered through a 0.8-μm filter to remove bacteriovores and autotrophs, and the cell suspension held on a 0.2-μm fibrous nylon filter. Previous experience with such seawater samples indicated that first-order uptake kinetics could be obtained by letting the samples sit overnight so that amino acids in the sample were reduced to low concentrations. When these aged samples were used in uptake experiments, uptake increased rapidly with fluid velocity when low concentrations (1 nM) of radiolabeled leucine were used (Fig. 8.2a). However, when leucine was increased to concentrations (10 nM) that were expected to saturate transporter enzymes used to facilitate leucine transport into the cell, the effect of fluid flow disappeared (Fig. 8.2b). This result is consistent with a transition from a mass-transport-limited effect when enzymes are not saturated. Increasing fluid flow decreases the thickness of the concentration boundary layer around the cell and increases uptake by surface enzymes. At high concentrations of substrate, the enzymes are saturated, and the concentration boundary layer is likely insignificant in size.

The rate of glucose uptake by marine bacteria is also thought to be enhanced by special transport enzymes embedded in the outer membrane with half-saturation uptake constants (K_{MM}) around 1 nM (Azam and Hodson, 1981). However, when the above flow experiment using aged seawater samples was repeated with a <0.8 μm sample held on a nylon filter surface, there was no effect of fluid motion on incorporation rates of [^{3}H]glucose (Logan and Kirchman, 1991). Arsenate was used to demonstrate that the transport systems for leucine and glucose are functionally different. Arsenate inhibits ATPase, the enzyme responsible for hydrolyzing ATP to provide energy for some transport systems. When 10 mM arsenate was added to the bacterial suspensions used in the flow experiments, leucine uptake was completely

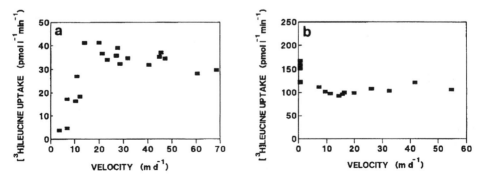

Figure 8.2 Uptake of radiolabeled leucine by marine bacteria (aged samples) held on a nylon filter as a function of fluid velocity at (A) 1 nM (first-order kinetics) and (B) 10 nM (saturated kinetics) leucine (Logan and Kirchman, 1991).

inhibited. Glucose uptake was reduced by 50%. Further experiments have shown that glucose uptake by mixed cultures can be linear with glucose concentrations over large ranges, indicating that chemical diffusion through the cell membrane can contribute to overall uptake (Logan and Fleury, 1993). It is likely that the resistance to glucose transport into the cell is primarily limited by either diffusion through the cell or slow transport enzymes, indicating that glucose mass transport to the cell is not limiting in glucose uptake by microorganisms.

Laminar Shear

Experiments with pure cultures of phytoplankton have demonstrated the positive effect of laminar fluid shear on microbial uptake kinetics. Nitrogen uptake by a pure culture of *Ditylum brightwelli* was increased by as much as 10% due to fluid shear rates of 10 s^{-1} produced in the gap of a couette device (Pasciak and Gavis, 1975). However, no conclusive results have been obtained for bacteria-sized particles using simple compounds like leucine and glucose. Leucine uptake by aged seawater bacteria was not increased by shear rates of ≤ 100 s^{-1} (Logan and Kirchman, 1991), while high shear rates have been shown actually to decrease the rate of leucine uptake by some species such as *Zoogloea ramigera* (Logan and Dettmer, 1990).

The effect of fluid shear on bacteria-sized particles predicted by mass transport correlations is predicted to be small, but should increase as the size of the particles increases. According to Eq. 7-22, leucine uptake ($D_{Lw} = 0.7 \times 10^{-5}$ cm^2 s^{-1}, estimated using the Wilke–Chang correlation) by a 1-μm particle suspended in a sheared fluid with $G = 10$ s^{-1}, would only be 3%. However, if the size of the particle is increased to 10 μm, then the increase could be as large as 31%. While a shear rate of 10 s^{-1} is quite large for natural waters (shear rates are rarely above 1 s^{-1} except in surface waters), much higher turbulent shear rates are typical of engineered systems. The effect of this turbulent shear environment on microbial kinetics is explored in the following.

Turbulent Shear

Turbulent shear rates in mixed vessels used in biological wastewater treatment reactors are typically around $G = 100$ s^{-1}. These high shear rates, coupled with a range of molecule sizes of components of dissolved organic matter, produce large predicted effects of fluid shear rates on microbial kinetics. Both Sh and Pe numbers increase with cell size and inversely with chemical diffusivity. For bacteria-sized particles mass-transfer-limited uptake will become more important as molecule size is increased, because diffusivity is inversely proportional to molecular size. It follows from Eq. 8-15 that the ratio of the rate of mass transport to a cell suspended in a sheared fluid to a cell at rest is just the Sh number, or

$$\frac{W_{\text{cell}} \text{ (in a sheared fluid)}}{W_{\text{cell}} \text{ (undisturbed fluid)}} = \frac{\text{Sh}_G}{\text{Sh} = 1} \text{Sh}_G \qquad (8\text{-}19)$$

Mass transport of a small molecule such as glucose ($D_{Gw} = 0.6 \times 10^{-5}$ cm^2 s^{-1}) to a bacteria-sized (radius = 1 μm) particle suspended in a fluid at a shear rate of 100 s^{-1} will only be increased by 11% (Eq. 7-18). However, for a macromolecule with a large molecular weight (assuming $D_{Cw} = 0.037 \times 10^{-5}$ cm^2 s^{-1}), uptake at this shear rate could be increased by 195% (Eq. 7-21).

The effect of turbulent fluid motion on bacteria degradation rates of macromolecules was examined by Confer and Logan (1991) using strains of *Zoogloea ramigera* and *Escherichia coli* able to degrade two macromolecules, bovine serum albumin (BSA) and dextran, respectively. BSA is soluble protein with a molecular weight of 68,000 dalton. Dextran is a polymer of glucose, with 95% of the molecules linked α-1,4. The dextran used in their study had an average molecular weight of 70,800, although it is commercially available with a variety of other molecular weights. Uptake of these two macromolecules was compared with uptake of two smaller molecules, leucine and glucose. The effect of fluid motion was investigated using two approaches. First, the rate of molecule incorporation was directly measured using radiolabeled molecules. Second, oxygen uptake was measured for cultures exposed to nonradiolabeled compounds. It was assumed that oxygen uptake was unaffected by shear rate but that oxygen uptake increased in proportion to substrate uptake. Measured rates were compared to Eq. 7-17 for Pe$_G$ < 0.1. For larger Pe, they used

$$\mathrm{Sh}_G = 1.54 \ \mathrm{Pe}_G^{0.153} \tag{8-20}$$

[^{3}H]BSA uptake by washed cultures of *Z. ramigera* in a BOD bottle stirred with a magnetic stir bar increased over a 24-h period, while that of an undisturbed culture was nearly constant over the same period of time (Fig. 8.3a). When the two bottles were switched, the bottle that had been stirred (and now was undisturbed) immediately exhibited decreased uptake rates, while the stirred bottle began to show increased uptake rates. Oxygen consumption of *Z. ramigera* cultures exposed to BSA was greater in stirred cultures (in BOD bottles and in a Yellow Springs Instruments DO chamber) than undisturbed cultures (BOD bottles) (Fig. 8.3b).

The overall increases in uptake due to mixing were observed to be smaller when oxygen was used to make the comparison than when radiolabeled chemicals were used (Table 8.1). Over a 24-h period there was 12.6 times as much uptake of radiolabel by stirred BSA cultures than by still samples, but in a separate experiment the oxygen uptake rate by stirred samples was only 2.3 times as large as that of still samples. A difference of 2.3 was predicted using mass transport correlations. It is likely that the comparison based on the large-molecular-weight radiolabeled compounds overestimated uptake by the stirred samples. These radiolabeled compounds were made using [^{3}H]formaldehyde in the presence of sodium cyanoborohydride. There is some evidence from other studies that the radiolabeled portion of the compound is not always incorporated into the cell after extracellular hydrolysis. If the radiolabeled isotopes were cleaved from the hydrolytic product incorporated into the cell, then uptake would not be correctly calculated by a measure of total ^{3}H associated with the cell.

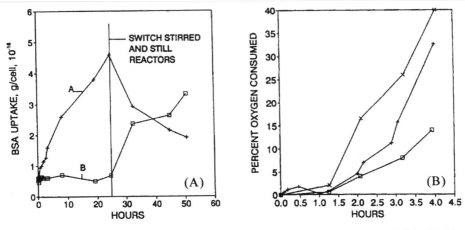

Figure 8.3 Effect of fluid motion on cultures of *Z. ramigera* degrading BSA. (A) Uptake of ³H-BSA in BOD bottles; initially A was stirred, and B still, but after 24 h the reactors were switched and B was mixed and A was still. (B) Oxygen uptake by cell suspensions: +, mixed in DO chamber; x, mixed in BOD bottle; □, still cultures in BOD bottle (Reprinted from Confer and Logan, 1991 with permission of the American Society for Microbiology).

The most conclusive evidence that the fluid environment could increase uptake by cells was provided by the oxygen uptake data. Cultures adapted to dextran and BSA had oxygen uptake rates that were 2.9 and 2.3 times as large as those of still samples. These rates were only slightly larger than those predicted using Eq. 7-17 and 8-20. When smaller-molecular-weight substrates were used, there was little effect of fluid shear. Glucose showed no effect of fluid shear, a result that was expected from previous experiments using cells fixed in a flow field. Oxygen uptake experiments using leucine were not performed, but experiments using radiolabeled leucine indicated uptake by stirred samples was greater by a factor of 1.15, a value that compared well to the predicted factor of 1.12. These experiments therefore demonstrate that turbulent fluid shear can impact overall uptake kinetics of microorganisms

TABLE 8.1 Comparison of Ratios of Uptake Predicted Using Eq. 7-17 or 8-20 and Observed for Either the Radiolabeled Substrate or Oxygen by Stirred and Still Samples (nd, not determined; Confer and Logan, 1991)

Substrate	K_C (mg l^{-1})	D_{Cw} (cm^2 s^{-1} 10^8)	Uptake (Stirred)/Uptake (Still) Predicted	Observed—³H	Observed—O$_2$
BSA	50	68	1.6	12.6	2.3
Dextran	60	37	1.8	6.2	2.9
Glucose	40	600	1.16	nd	1.0
Leucine	<100	874	1.12	1.15	nd

in a manner consistent with mass-transport-limited uptake since increasing the fluid shear rate increases uptake due to reduction of the concentration boundary layer around the cell.

8.4 TRANSPORT TO MICROBIAL AGGREGATES

The rate of chemical transport to single, free-living microorganisms is important in examining microbial kinetics, but most microorganisms present in biological reactors used for wastewater treatment are present in flocs. Substrate removal, measured in terms of COD or BOD removal, is constantly occurring at a slow rate by suspended bacteria in wastewater even during its transport to the treatment plant. Treatment is accelerated in reactors, such as activated sludge systems, by using high concentrations of biomass achieved by recycle of aggregated cells that settle in the clarifiers. Microbes must be present in flocs to be recycled in the clarifier underflow because single microbes do not settle rapidly enough to be removed and recycled in wastewater treatment clarifiers.

The analysis of the rate of substrate removal in a microbial aggregate is based on the classical analysis of mass transport to a porous catalyst suspended in a fluid. Unlike microbial aggregates, inorganic catalysts can be expensive to make and replace. In order to determine if a catalytic reactor is operating with optimum use of the catalyst, engineers have developed an approach using an effectiveness factor to quantify overall reaction efficiency. Environmental engineers have used this effectiveness factor approach to try to prove that substrate removal rates consistent with a diffusion-based model prove that there is no advective flow through microbial aggregates. As we shall see, however, a model based on either diffusion or advection can account for effectiveness factors below unity in biological reactors.

Transport by Diffusion

In order to calculate the substrate removal rate by a suspended microbial aggregate, we will assume that the aggregate is a sphere suspended in a fluid with the concentration of substrate at the fluid–aggregate interface defined as $c(R) = c_s$ (Fig. 7.13). The concentration profile in the aggregate can be developed from a shell balance approach for a spherical shell of cross-sectional area $A = 4\pi r^2$, and differential volume of $V = 4\pi r^2 \, \Delta r$. Since the reaction only occurs in the solid portion of the aggregate, these terms will need to be multiplied by the aggregate porosity, θ. For an aggregate under steady-state conditions with only diffusion into the aggregate, the flux by diffusion is

$$\text{flux in} - \text{flux out} = J|_{r+\Delta r}A - J|_r A \qquad (8\text{-}21)$$

Substituting in $J = -D_C \, dc/dr$ and the definition of the area

$$\text{net flux} = \left(-D_C \left. \frac{dc}{dr} \right|_{r+\Delta r} 4\pi r^2 \theta \right) \Bigg|_{r+\Delta r} + \left(D_C \left. \frac{dc}{dr} \right|_r 4\pi r^2 \theta \right) \Bigg|_r \qquad (8\text{-}22)$$

For the moment, the order of the reaction per unit of solid volume will be left in a general form of R_C so that the reaction in the control volume is $R_C V$, or from the definition of the control volume

$$R_C = 4\pi r^2 \, \Delta r \, \theta \tag{8-23}$$

Combining all terms, rearranging, and dividing by $4\pi \, \Delta r \, \theta$, the shell balance produces

$$D_C \left. \left(\frac{dc}{dr}\right) \right|_{r+\Delta r} - D_C \left. \left(\frac{dc}{dr}\right) \right|_r - R_C r^2 \, \Delta r = 0 \tag{8-24}$$

Dividing by Δr, and taking the limit as $\Delta r \to 0$, provides the general form of the equation

$$\frac{d^2 c}{dr^2} + \frac{2}{r}\frac{dc}{dr} = \frac{R_C}{D_C} \tag{8-25}$$

For different choices of R_c, such as zero- or first-order reactions, the particular solution to Eq. 8-25 will be different. A unique result will also be produced for each different choice of boundary conditions, as will be shown.

Zero-Order Reaction   Oxygen uptake by microorganisms can generally be considered to be zero order over a wide range of oxygen concentrations, since the half-saturation constant for oxygen, $K_m(O_2)$ is usually less than 0.1 mg L^{-1}. Let us define the rate of oxygen utilization as $R_C = k_O X$, where k_O is the specific uptake rate (g oxygen/g cell s^{-1}), and X is the cell density in the aggregate (g cell L^{-1}). For this case Eq. 8-25 becomes

$$\frac{d^2 c}{dr^2} + \frac{2}{r}\frac{dc}{dr} = \frac{k_O X}{D_O} \tag{8-26}$$

Equation 8-26 is a second-order, nonhomogeneous equation with equidimensional coefficients. The general solution for any zero-order reaction within a spherical catalyst is therefore

$$c(r) = b_1 + \frac{b_2}{r} + \frac{k_O X}{6D_O} r^2 \tag{8-27}$$

The specific solution for our system will be a function of the boundary conditions. Since we have to obtain two constants, b_1 and b_2, we will need two boundary conditions. The boundary conditions

$$r = 0, \qquad c = c_0 \tag{8-28}$$

$$r = R, \qquad c = c_s \tag{8-29}$$

indicate that at the aggregate center there is a known concentration c_0, and that at the surface of the aggregate the concentration equals the fluid bulk concentration c_s. If the surface concentration was equal to the bulk concentration, or $c_s = c_\infty$, then the resistance to mass transport at the aggregate surface would be defined as negligible. Using these two boundary conditions in Eq. 8-27, we obtain

$$c(r) = c_s - \frac{k_O X}{6D_O} r^2 \qquad (8\text{-}30)$$

but this solution is only defined for an aggregate of radius R, or

$$R = \left(\frac{k_O X}{6D_O} (c_\infty - c_0) \right)^{1/2} \qquad (8\text{-}31)$$

A more useful solution can be obtained by using the boundary conditions

$$r = 0, \qquad \frac{dc}{dr} = 0 \qquad (8\text{-}32)$$

$$r = R, \qquad c = c_\infty \qquad (8\text{-}33)$$

The first boundary condition does not require that we know the concentration at the aggregate center, and allows for this concentration to be zero or greater. For this situation, the solution to Eq. 8-27 becomes

$$c(r) = c_\infty - \frac{k_O X}{6D_O} (R^2 - r^2) \qquad (8\text{-}34)$$

First-Order Kinetics   The uptake of many nutrients and substrates at low concentrations obeys first-order kinetics. Defining the rate as $R = k_C c$, where k_C is a first-order rate constant, Eq. 8-27 becomes

$$\frac{d^2 c}{dr^2} + \frac{2}{r} \frac{dc}{dr} - \frac{k_C}{D_C} c = 0 \qquad (8\text{-}35)$$

The equation is a modified Bessel function (Appendix 4). By observation, the general solution is

$$c(r) = b_1 \frac{\sqrt{2/\pi}}{r} \sinh \left(\sqrt{\frac{k_C}{D_C}} r \right) + b_2 \frac{\sqrt{2/\pi}}{r} \cosh \left(\sqrt{\frac{k_C}{D_C}} r \right) \qquad (8\text{-}36)$$

Two boundary conditions are necessary to solve for the constants b_1 and b_2. We choose

$$r \to 0, \qquad c \text{ is finite} \qquad (8\text{-}37)$$

$$r = R_{ag}, \qquad c = c_s \qquad (8\text{-}38)$$

At $r = 0$ the sinh term is zero, but the cosh term is infinite due to r in the denominator. The only way that the concentration can be finite at the center of the aggregate is if $b_2 = 0$. Using the second boundary condition at the edge of the aggregate, when the radius $r = R_{ag}$, the final solution is

$$c(r) = \frac{c_s R_{ag}}{r} \frac{\sinh(\sqrt{k_C/D_C}\, r)}{\sinh(\sqrt{k_C/D_C}\, R_{ag})} \tag{8-39}$$

For this case of first-order kinetics we will calculate the chemical flux into the aggregate. The total flux, which is only due to diffusion, is the largest at the aggregate surface ($r = R_{ag}$) and is calculated as

$$J_C\big|_{r=R_{ag}} = -D_C \frac{dc}{dr}\bigg|_{r=R_{ag}} \tag{8-40}$$

The derivative at the aggregate surface can be obtained by taking the derivative of Eq. 8-39 and evaluating it for $r = R_{ag}$. The derivative is

$$\frac{dc}{dr} = \frac{c_s R_{ag}}{\sinh(\sqrt{k_C/D_C}\, R_{ag})} \left(\frac{1}{r} \sqrt{k_C/D_C}\, \cosh(\sqrt{k_C/D_C}\, r) + \sinh(\sqrt{k_C/D_C}\, r)\, \frac{-1}{r^2} \right) \tag{8-41}$$

and evaluating it at $r = R_{ag}$ we obtain

$$\frac{dc}{dr}\bigg|_{r=R_{ag}} = \frac{c_s R_{ag}}{\sinh(\sqrt{k_C/D_C}\, R_{ag})} \left(\frac{1}{R_{ag}} \sqrt{k_C/D_C}\, \cosh(\sqrt{k_C/D_C}\, R_{ag}) \right.$$
$$\left. + \sinh(\sqrt{k_C/D_C}\, R_{ag})\, \frac{-1}{R_{ag}^2} \right) \tag{8-42}$$

The uptake rate for a single aggregate is equal to the flux times the total surface area of the aggregate, or $W_C = J_C(4\pi R_{ag}^2)$. The rate is therefore

$$W_C\big|_{r=R_{ag}} = 4\pi R_{ag} D_C c_s [1 - \sqrt{k_C/D_C}\, R_{ag}\, \coth(\sqrt{k_C/D_C}\, R_{ag})] \tag{8-43}$$

Transport by Advection

It is a long-held view of chemical transport into a microbial aggregate that transport occurs by chemical diffusion into the aggregate interior. However, microbial aggregates are extremely loose assemblages of microorganisms containing many large holes through which fluid flow can occur. While compaction of sludges produced in secondary clarifiers may reach solids concentrations of 0.5% to 3% for suspended growth processes, most aggregates developed in suspension have significantly lower solids contents. The porosity of an aggregate increases with aggregate size, and most (>95%) of the aggregate is actually water. It seems likely that the high porosity of these aggregates will allow advective flow through the aggregate either during sedimentation or as a result of fluid shear. In the following section, we consider mass transport to microbes homogeneously distributed in aggregates. Despite recent evidence that the distribution of cells in aggregates obeys fractal scaling laws, and is not therefore homogeneous, this analysis is sufficient to demonstrate the potential for advective flow through a highly porous aggregate.

There are no fixed shapes for aggregates, since aggregates grow through collisions with other aggregates and shrink in size due to erosion and breakup. It is convenient for the mathematical analysis that follows to consider aggregates that are cubes. For transport *within* a cube, we can start with the constitutive transport equation in Cartesian coordinates with a reaction of some chemical C with microbes in the cube as

$$\mathbf{u}_{ag} \, \nabla c - D_C \, \nabla^2 c = R_C \tag{8-44}$$

where $\mathbf{u}_{ag}$ is the average intra-aggregate or intrafloc velocity. In the previous section we made the assumption that diffusive transport through the aggregate was much greater than advective transport. If we make the opposite assumption here that advective flow dominates, and we restrict our analysis to one dimension, Eq. 8-44 becomes

$$u_{ag} \frac{dc}{dx} = R_C \tag{8-45}$$

Although this equation can be solved for any order reaction, in the following analysis we consider only the case of a first-order reaction, $R_C = -k_C c$, or

$$u_{ag} \frac{dc}{dx} = -k_C c \tag{8-46}$$

Separating and integrating, the solution is

$$c(x) = b_1 e^{-k_C x / u_f} \tag{8-47}$$

For this equation we need only one boundary condition. Using the entrance condition at $x = 0$ that flow enters at a surface concentration of $c(0) = c_s$, the concentration profile in the aggregate is

$$c(x) = c_s e^{-k_C x / u_{ag}} \tag{8-48}$$

The overall rate of chemical transport to the aggregate can be calculated based on the differences in entrance and exit concentrations. The rate chemical enters is $u_f A c$, where A is the cross-sectional area of the aggregate. The rate chemical leaves the aggregate is similarly calculated using Eq. 8-48. Assuming for simplicity that the floc is a cube of length $2R_{ag}$, so that the size of the cubical aggregate is roughly that of the spherical aggregate analyzed in the previous section, the rate of chemical consumption in the aggregate is

$$W_C = u_{ag} 4R_{ag}^2 c_s (1 - e^{-k_C x/u_{ag}})$$

(8-49)

Comparing Eqs. 8-43 and 8-49, we see that it is possible to calculate the rate of chemical removal by either a diffusive or advective transport equation. In order to evaluate which equation is more appropriate for chemical removal by aggregated microorganisms, however, requires estimates of intra-aggregate velocities. In the sections to follow three cases are considered: sinking aggregates, sheared aggregates, and aggregates captured by rising bubbles.

Sinking Aggregates   Adler (1981) calculated the drainage flow rate, q, through an aggregate of radius R_{ag} as

$$q = \pi R_{d,\infty}^2 U_{s,\text{perm}}$$

(8-50)

where $U_{s,\text{perm}}$ is the settling velocity of the permeable aggregate and $R_{d,\infty}$ is the drainage radius of the aggregate defined as the equivalent radius (at infinity) through which flow would go through the aggregate. For a settling aggregate, the drainage radius can be calculated from

$$R_{d,\infty} = R_{ag} \left(1 - \frac{b_1}{\kappa^*} - \frac{b_2}{\kappa^{*3}} \right)^{1/2}$$

(8-51)

where the dimensionless constant $\kappa^* = R_{ag}/\kappa^{1/2}$, and κ is the aggregate permeability. The constants b_1 and b_2 can be determined (Adler, 1981) as

$$b_2 = -\frac{1}{b_3} \left(\kappa^{*5} + 6\kappa^{*3} - \frac{\tanh \kappa^*}{\kappa^*} (3\kappa^{*5} + 6\kappa^{*3}) \right)$$

(8-52)

$$b_1 = \frac{3\kappa^{*3}}{b_3} \left(1 - \frac{\tanh \kappa^*}{\kappa^*} \right)$$

(8-53)

$$b_3 = 2\kappa^{*2} + 3 - 3 \frac{\tanh \kappa^*}{\kappa^*}$$

(8-54)

The actual intra-aggregate velocities will be a function of r and θ in polar coordinates, but we can solve for an average velocity based on the definition of q as

$$q = \pi R_{d,\infty}^2 U_{s,\text{perm}} = \pi R_{ag}^2 u_{ag}$$

(8-55)

Solving for the intra-aggregate velocity, u_{ag}, we have

$$u_{ag} = U_{s,\text{perm}} \frac{R_{d,\infty}^2}{R_{ag}^2} \tag{8-56}$$

or substituting Eq. 8-51 into Eq. 8-56, we have

$$u_{ag} = U_{s,\text{perm}} \left(1 - \frac{b_1}{\kappa^*} - \frac{b_2}{\kappa^{*3}} \right) \tag{8-57}$$

Several correlations have been used to relate R to permeability. The most common correlation used is the Cozeny–Karmen relationship

$$\kappa = \frac{\theta^3 R_{pp}^2}{(1 - \theta)^2 45} \tag{8-58}$$

where R_{pp} is the radius of the primary particles forming the aggregate and θ is the aggregate porosity. The Kozeny–Carmen equation is only thought to be valid at much lower porosities than exist for permeable aggregates. For highly porous aggregates, the Davies correlation

$$\frac{1}{\kappa} = \frac{16}{R_{\text{cyl}}^2} (1 - \theta)^{1.5} [1 + 56(1 - \theta)^3] \tag{8-59}$$

is more frequently used (Masliyah and Polikar, 1980), where R_{cyl} is the radius of a small cylindrical filament, here assumed to be a chain of cells, composing the aggregate so that $R_{\text{cyl}} = R_{pp}$.

The settling velocity of an impermeable aggregate is predicted from Stokes' Law

$$U_{s,\text{imp}} = \frac{2g \, \Delta\rho}{9 \nu_w \rho_w} R_{ag}^2 \tag{8-60}$$

For permeable aggregates Matsumoto and Suganuma (1977) developed the correlation between impermeable and permeable spheres, such as steel wool pads, as

$$U_{s,\text{perm}} = U_{s,\text{imp}} \left(\frac{\kappa^*}{\kappa^* - \tanh \kappa^*} + \frac{3}{2\kappa^{*2}} \right) \tag{8-61}$$

The density difference between the aggregate and the fluid, $\Delta\rho$, can be expressed in terms of the primary particle or cell density, ρ_{pp}, and the aggregate porosity, θ, as

$$\frac{\Delta\rho}{\rho_w} = \frac{(\rho_{pp} - \rho_w)}{\rho_w} (1 - \theta) \tag{8-62}$$

Aggregate porosity is well known to increase with the size of an aggregate. Tambo and Watanabe (1979) developed size–porosity data for activated sludge flocs using the equation

$$(1 - \theta) = b_4 R_{ag}^{-b_5} \tag{8-63}$$

where b_5 was determined from settling velocity experiments to be 1.6. Choosing $\theta = 0.4$ for an aggregate 100 μm in radius, Logan and Hunt (1987) estimated $b_4 = 8$, resulting in the relationship

$$(1 - \theta) = 8R_{ag}^{-1.6} \tag{8-64}$$

where R_{ag} has units of μm. For a 100-μm aggregate this produces a porosity of 0.995. Microbial aggregates have a wide range of densities that vary in large part due to the methods used to estimate porosity. Mueller et al. (1968) calculated a relatively low porosity of 0.66 ± 0.10 for aggregates composed of the floc-forming microorganism *Zoogloea ramigera*, while Smith and Coakley (1984) estimated activated sludge flocs had porosities of 0.999.

Intra-aggregate flow, resulting from the high porosity of microbial aggregates composed of uniformly distributed particles, should only have a small effect on aggregate settling velocity (Fig. 8.4). However, settling velocities were calculated to be much smaller than those observed for marine particles formed from recoagulated sediments even with a high primary particle density of 2.65 g cm^{-3}. The slope in Figure 8.4

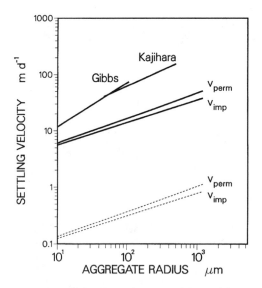

Figure 8.4 Predicted settling velocities of permeable and impermeable aggregates compared to those measured by Gibbs (1985) and Kajihara (1971). Solid lines, $\rho_p = 2.65$ g cm^{-3}, dashed lines $\rho_p = 1.06$ g cm^{-3} (reprinted from Logan and Hunt, 1987 with permission by the American Society of Limnology and Oceanography).

for impermeable particles is 0.4 but increases to only 0.44 when the particles are calculated to be permeable to the advective flow. The empirical relationships used by Gibbs (1985) and Kajihara (1971) were $bR^{0.78}$ and $bR^{0.57}$, indicating the slopes for natural particles were steeper than those predicted here. The reason for the differences between natural particles and the permeable aggregates studied here is probably that the microbes and particles are not evenly distributed in the aggregate (as assumed here). Recent studies have shown that aggregates formed by coagulation are fractal, and that their fractal structure produces settling velocities that are up to an order of magnitude larger than predicted for impermeable aggregates of the same size and density (Johnson et al., 1996). This subject is considered in further detail in Chapters 13 and 14.

For aggregate sizes of 10 to 1000 μm, intra-aggregate velocities of settling aggregate containing primary particles with a density of 2.65 g cm^{-3} increase from 2 to 100 μm s^{-1}. This advective flow increases the rate of chemical transport to cells in permeable aggregates compared to those suspended in a quiescent fluid (Fig. 8.5). The enhancement due to flow through the aggregate becomes larger as the diffusion coefficient of the chemical decreases. For a molecule with a diffusion coefficient of 10^{-5} cm^2 s^{-1}, the overall rate of mass transport to all cells in the aggregate is <10% larger than a nonsettling aggregate. However, for molecules having a large molecular weight (assuming $D_C = 10^{-7}$ cm^2 s^{-1}), the rate of transport to the settling aggregate can be 70% greater than to an aggregate at rest ($R = 1000$ μm).

Sheared Aggregates Laminar shear rotates particles and increases the rate of chemical transport to the particle's surface. Fluid shear also can produce intra-aggregate flow through highly porous aggregates. The flow rate through an aggregate of radius R_{ag} (Adler, 1981) is

$$q = \tfrac{2}{3} G R_{ag}^3 E_0^{3/2} \tag{8-65}$$

where E_0 is a tabulated function of $\kappa^* = R_{ag}/\kappa^{1/2}$. The flow rate through the aggregate is a product of the intra-aggregate velocity u_{ag}, and the collision, or drainage, cross-sectional area σ_c, or $q = u_{ag}\sigma_c$, where $\sigma_c = \pi R_{ag}^2 E_0$. The function E_0 can therefore be obtained either from tabulated values of σ_c/R_{ag}^2 or q/GR_{ag}^3 in Adler (1981), as shown in Table 8.2. For laminar shear, intra-aggregate velocities can be calculated from

$$u_{ag} = \frac{2}{3\pi} GR_{ag} E_0^{1/2} \tag{8-66}$$

When $\kappa^* > 10.89$, the aggregate becomes impermeable to shear-induced advective flow through the porous aggregate, and only diffusive transport will bring material into the aggregate.

At high shear rates there can be substantial advective flow through aggregates. Assuming a primary particle size of 1 μm, intra-aggregate flows increase to a max-

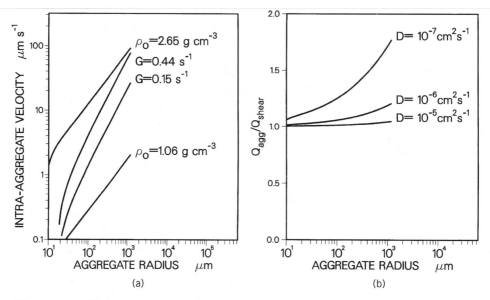

Figure 8.5 (A) Intra-aggregate fluid velocities for settling and sheared aggregates. (B) Rates of mass transport to a cell in a settling aggregate (ρ_p = 2.65 g cm^{-3}) versus a cell in sheared fluid (G = 0.44 s^{-1}) (reprinted from Logan and Hunt, © 1988 Biotechnol. Bioeng. Reprinted by permission of John Wiley & Sons, Inc.).

imum of ca. 1500 μm s^{-1} at shear rates of 1000 s^{-1} (Fig. 8.5). These calculations assume that the shear rate controls the aggregate size as specified by Eq. 7-61. As the shear rate increases, the aggregate size decreases, and according to the porosity–size relationship in Eq. 8-64 the porosity also decreases. The result is that at shear rates higher than 1000 s^{-1}, the aggregates become smaller and impermeable to advective flow despite the high shear rates.

TABLE 8.2   Numerical Values of E_0 Derived from Data Presented in Adler (1981)

κ^*	E_0	σ_c/R_{ag}^2	$q/(GR_{ag}^3)$
10.89	0	0	0
10	0.0439	0.138	0.006
8	0.1662	0.522	0.045
6	0.3326	1.045	0.128
5	0.4386	1.378	0.194
4	0.5615	1.764	0.281
3	0.7000	2.198	0.390
2	0.8403	2.640	0.514
1	0.9546	2.999	0.622

Example 8.2

Determine if a 100-μm aggregate suspended in a sheared fluid is permeable to advective flow, where the aggregate of porosity 0.90 is composed of primary particles 1 μm in radius.

Since an aggregate will only be permeable to advective flow if $\kappa^* \leq 10.89$, this question can be answered by solving for κ^* using an expression for the permeability, κ. The Kozeny–Carmen equation predicts a larger permeability for a given size aggregate; so we will use Eq. 8-58 to see if there is flow possible through the aggregate. We obtain

$$\kappa = \frac{\theta^3 R_{pp}^2}{(1 - \theta^2)45} = \frac{(0.9)^3 (1 \ \mu m)^2}{(1 - 0.9)^2 45} = 1.62 \ \mu m^2 \tag{8-67}$$

The value of κ^* is calculated from its definition as

$$\kappa^* = \frac{R_{ag}}{\kappa^{1/2}} = \frac{100 \ \mu m}{(1.62 \ \mu m^2)^{1/2}} = 78 \tag{8-68}$$

Since κ^* is not less than 10.89, this aggregate is impermeable to flow due to simple fluid shear. However, there will always be some advective flow through a porous aggregate produced by sedimentation, although this fluid velocity may be very small.

Aggregates Transported by Rising Bubbles   The presence of microorganisms in water disrupts hydrogen bonding between water molecules. A more thermodynamically favorable placement of microbes is at the air–water interface. As bubbles rise through a cell suspension, the air bubble will capture cells and transport the cells along with the rising bubble. This places the microbe in a very high shear environment. The rise velocity of a bubble with Re > 1 is

$$U_b = (2gR_b)^{1/2} \tag{8-69}$$

For a 0.1-cm bubble, $U_b = 14 \ cm \ s^{-1}$. The maximum rate of mass transfer to the microbes in the bubble boundary layer would occur if the cell experienced an advective flow equal to the rise velocity of the bubble. Therefore, an upper limit for the case of a 1-μm cell attached to a rising bubble would be a mass transport rate that is 4.4 times as large as that of a cell at rest (assumes $D_{Cw} = 0.6 \times 10^{-5} \ cm^2 \ s^{-1}$ in Eq. 7-18).

Alternately, mass transport to a cell attached to a bubble could be analyzed as a cell trapped in a sheared fluid (the bubble hydrodynamic boundary layer), with the maximum velocity gradient calculated as the difference between an immobile surface and the bubble rise velocity divided by the thickness of the bubble boundary layer (see Problem 8.4). This analysis also yields large increases in mass transport to cells captured by the rising bubble.

8.5   EFFECTIVENESS FACTORS FOR MASS TRANSPORT

As substrate diffuses into a porous reactive material, such as a porous catalyst, the substrate concentration decrease either reduces the rate of reaction or, for zero-order reactions, causes the reaction rate to be zero when the substrate is completely consumed in the aggregate. The reduction in reaction kinetics may be undesirable when the catalyst is very expensive and the catalyst center is not fully utilized, or when high reaction rates need to be maintained in the reactor. The overall efficiency of the catalyst has traditionally been evaluated in terms of an effectiveness factor, η_d, defined as

$$\eta_d = \frac{\text{observed rate}}{\text{rate if the concentration remains constant}} \qquad (8\text{-}70)$$

Microbial aggregates can also be viewed as porous catalysts, with the individual microbes creating an overall reaction rate in the aggregate. If the reaction rate remains constant everywhere in the aggregate, then the total rate can be calculated as RV, where R is the rate expression (zero order, first order, etc.) and V is the volume of the floc. For an aggregate of radius R_{ag} and assuming first-order kinetics, the denominator of Eq. 8-70 is

$$k_C c_\infty \left(\frac{4\pi}{3} R_{ag}^3 \right) \qquad (8\text{-}71)$$

Combining this result with the observed rate in the aggregate, assuming chemical transport into the aggregate by diffusion using Eq. 8-43, and assuming negligible external resistance to mass transfer (*i.e.*, that $c_s = c_\infty$), Eq. 8-70 becomes

$$\eta_d = \frac{4\pi R_{ag} D_C c_\infty [1 - \sqrt{k_C/D_C} R_{ag} \coth(\sqrt{k_C/D_C} R_{ag})]}{k_C c_\infty \left(\frac{4\pi}{3} R_{ag}^3 \right)} \qquad (8\text{-}72)$$

In calculations of mass transport with chemical reaction, the term $(k_C/D_C)^{1/2}$ frequently arises as a ratio of characteristic constants for reaction to transport by diffusion. Since diffusion occurs relative to a cross-sectional area A and reaction is a volumetric parameter, the ratio of reaction and diffusion can be nondimensionalized as

$$\phi_d = \frac{V}{A} \sqrt{\frac{k_C}{D_C}} \qquad (8\text{-}73)$$

where V is the aggregate volume, and ϕ is known as the dimensionless Thiele modulus. For a spherical aggregate the ratio of V/A is $R_{ag}/3$, or

$$\phi_d = \frac{R_{ag}}{3} \sqrt{\frac{k_C}{D_C}}, \qquad \text{sphere} \qquad (8\text{-}74)$$

Equation 8-72 can be simplified using Eq. 8-74 to

$$\eta_d = \frac{1}{3\phi_d^2} [3\phi_d \coth (3\phi_d) - 1] \qquad (8\text{-}75)$$

Similar solutions for effectiveness factors are easily obtained for other aggregate geometries such as cubes or disks, and for other reaction orders. However, the general relationship between η_d and ϕ does not vary substantially compared to the wide range of values likely to be encountered for reaction rates in comparison to diffusion rates. When $\phi < 0.1$, reaction is slow compared to diffusion and the aggregate is fully penetrated and η_d is essentially unity (Fig. 8.6). As the reaction rate becomes increasingly larger relative to the diffusive flux ($\phi > 10$), the η_d decreases directly in proportion to ϕ. At these large Thiele moduli much of the aggregate interior will be nearly devoid of chemical since the chemical will be consumed before it reaches the aggregate interior. For microbial aggregates, this can mean that microbes in the center will not receive either chemical substrates or electron acceptors such as oxygen.

The applicability of the effectiveness factor calculation for microbial aggregates has been examined in several studies using pure culture of *Zoogloea ramigera*. This

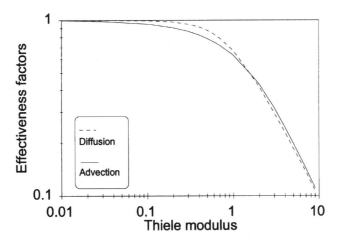

Figure 8.6 Comparison of advective and diffusive effectiveness factors (from Logan and Hunt, © 1988 Biotechnol. Bioeng. Reprinted by permission of John Wiley & Sons, Inc.).

bacterium forms large flocs and is thought to be important in floc formation in suspended growth wastewater treatment systems. Baillod and Boyle (1970) compared oxygen uptake rates of a flocculated suspension with that of a suspension partially dispersed in a blender. Glucose uptake rates were strongly correlated with temperature, but uptake increased by 10% to 35% when samples were blended. Reductions in the diffusivity of glucose due to the porous matrix of the aggregate were calculated for *Zoogloea* flocs ranging from 0.4 to 0.6 of that for glucose in water. At glucose concentrations above 3 (20°C) to 8 (30°C) mg L^{-1}, glucose uptake rates by blended and unblended samples were essentially identical. The authors explained that this lack of an effect on uptake rates at higher glucose concentrations was due to saturation of enzyme uptake systems, or zero-order kinetics. It is also possible, however, that at low glucose concentrations there was insufficient diffusive or advective transport for the substrate to penetrate into the aggregate. Thus this model did not prove diffusional transport controlled chemical transport, but it did show that low substrate concentrations uptake by aggregates could be lower than that of dispersed cells.

Similar experiments were conducted by Baillod and Boyle using activated sludge suspensions. Oxygen uptake rates increased by 50 to 65% when cultures were dispersed by blending. Glucose uptake rates at 20°C were identical in blended and unblended samples, but at 30°C glucose uptake increased by 41% when cultures were dispersed by blending. It is clear from these studies that dispersing aggregates will increase total mass transport, and as a result, total glucose chemical uptake by microorganisms. Recall from Section 8.3, however, that shear itself has no effect on glucose uptake by microorganisms.

The observation that substrate uptake obeys the relationships predicted by the effectiveness factor based on diffusion is not sufficient proof, however, that uptake by microbial aggregates is limited by *diffusion* of chemicals into the aggregate. A similar analysis can be made using assuming an effectiveness factor based on *advective* transport into the aggregate. The advective effectiveness factor, η_a, is defined in the same manner as before for η_d. We can use the previous result derived for an aggregate viewed as a cube for the overall rate of reaction observed (Eq. 8-49) for transport dominated by advective flow through the aggregate. Our previous analysis based on diffusion demonstrated that there is little effect of aggregate geometry on the effectiveness factor. For a cube of size $2R$ the surface area of a side is $4R^2$, and the volume is $8R^3$. For this case the advective effectiveness factor can be calculated assuming negligible external mass transport resistance as

$$\eta_a = \frac{u_{ag}4R_{ag}^2 c_\infty(1 - e^{-2R_{ag}k_C/u_{ag}})}{k_C c_\infty(8R_{ag}^3)} \tag{8-76}$$

Defining the Theile modulus as before as a rate of reaction to diffusion, we have an advective Thiele modulus here of $\phi_a = 2R_{ag}k_C/u_{ag}$, and Eq. 8-65 becomes simply

$$\eta_a = \frac{1 - e^{-\phi_a}}{\phi_a} \tag{8-77}$$

Comparing Eqs. 8-64 and 8-66 (Fig. 8.6), we see that there is little difference in the response of an effectiveness factor to either an advective- or diffusive-based Theile modulus. Thus the observation that microbial aggregates behave in a manner consistent with a diffusion model does not rule out the possibility of advective flow through the aggregate interior since both models provide similar predictions of overall mass transport into the aggregate.

8.6 RELATIVE UPTAKE FACTORS FOR MASS TRANSPORT

While the effectiveness factor approach is useful for determining the efficiency of impermeable catalysts, it does not incorporate several important aspects of catalysts that are unique to microbial aggregates. Microorganisms must expend energy, in the form of polymers used for adhesion, to attach to particles and other cells and remain part of the aggregate. Attachment must offer some ecological advantage for this property to have been selected in many cells over evolutionary time scales. Based on the effectiveness factor approach, a microbe in an impermeable aggregate will always have some disadvantage in nutrient uptake rates compared to those free living in solution. It is more likely that a microbe gains some nutritional advantage of growing in an aggregate compared to an unattached microbe, but this would require effectiveness factors larger than unity which is impossible. An effectiveness factor also does not incorporate external mass transport resistances into overall kinetics. If the reaction rate of a chemical is very fast at the catalyst surface, then reaction could exceed the rate at which substrate can diffuse to the aggregate's surface. We have already seen that the fluid environment of a microorganism can affect the rate of mass transport to a cell. To examine the benefits of microbial attachment, we really need to compare the rate of microbial uptake of chemical by cells attached inside impermeable and permeable aggregates to rates experienced by unattached cells in a sheared fluid.

In order to examine the benefits of microbial attachment in terms of nutrient uptake, we examine uptake kinetics using the relative uptake factor defined as

$$\gamma_a = \frac{\text{observed rate}}{\text{rate if cells are dispersed in the fluid}} \tag{8-78}$$

where both the aggregate and the dispersed cells exist in the same fluid environment. When $\gamma > 1$, aggregated cells will incorporate substrate faster than unattached cells.

The rate if all cells are dispersed in the fluid is the rate of substrate uptake by a single cell multiplied by the number of cells in the aggregate. From Eq. 8-8, the rate per cell is

$$W_{\text{cell}} = 4\pi R_c \text{ Sh } D_C f_a(c_\infty - c_s) \tag{8-79}$$

Instead of expressing the rate as a function of the concentration difference, let us instead define the unknown surface concentration as a function of the known bulk concentration as

$$c_s = f_c c_\infty \tag{8-80}$$

where f_c is defined as some fraction that relates the difference in these two substrate concentrations. The number of cells, N_c, of radius R_c, in an aggregate can be calculated from the aggregate porosity, θ, as

$$N_c = \frac{(1 - \theta)}{(4\pi/3)R_c^3} \tag{8-81}$$

Not all cells in an aggregate may be viable; so we need to reduce N_c by a fraction f_v, defined as the fraction of cells that are viable. Incorporating these calculations into Eq. 8-79 produces

$$W_f = 3 \text{ Sh } D_c f_a f_v (1 - f_c) c_\infty \frac{(1 - \theta)}{R_c^2} \tag{8-82}$$

Combining constants into a single cell efficiency, or $f = f_a f_v (1 - f_c)$, the rate calculated for all cells in the aggregate, if they are dispersed in the fluid, is

$$W_f = \frac{3 \text{ Sh } D_c f (1 - \theta)}{R_c^2} c_\infty \tag{8-83}$$

We can simplify this equation by defining a first-order rate constant k_C as

$$W_f = \frac{3 \text{ Sh } D_c f_T (1 - \theta)}{R_c^2} c_\infty = k_C c_\infty \tag{8-84}$$

To calculate the observed rate, we must solve either Eq. 8-44, assuming only diffusive transport to the aggregate, in order to solve for γ_d, or solve it assuming only advective transport and obtain γ_a. For each solution we will use a boundary condition that incorporates an external mass transport coefficient.

For mass transport by diffusion, the solution to the diffusive transport equation (Eq. 8-35) for a first-order reaction is

$$c(r) = \frac{b_1}{r} \sinh\left(\sqrt{\frac{k_C}{D_C}}\, r\right) + \frac{b_2}{r} \cosh\left(\sqrt{\frac{k_C}{D_C}}\, r\right) \tag{8-85}$$

The two boundary conditions are chosen as

$$BC1: r = 0, \qquad c = \text{finite} \tag{8-86}$$

$$BC2: r = R_{ag}, \qquad J_c = \mathbf{k}_w (c - c_\infty) \tag{8-87}$$

The first boundary condition requires that there be a finite concentration of substrate at the aggregate interior. The concentration cannot be zero since there must always be some amount of chemical (however small) at the aggregate interior for a first-order reaction. Since the sinh and cosh terms are 0 and 1, respectively, at $r = 0$, the constant b_1 must be zero, and Eq. 8-85 can be simplified to

$$c(r) = \frac{b_2}{r} \sinh \left(\sqrt{\frac{k_C}{D_C}} \, r \right) \tag{8-88}$$

If b_1 was not zero, then the term $1/r$ would become infinity and violate our requirements of boundary condition 1. The second boundary condition specifies that there will be concentration of chemical at the aggregate surface that is lower than the bulk concentration due to a chemical gradient at the aggregate surface. The flux to the aggregate is expressed in terms of a mass transport coefficient based on the aggregate radius, R_{ag}, which can also be expressed in terms of a Sh number as $Sh(R) = kR_{ag}/D_C$. The derivative of Eq. 8-88 with respect to r is

$$\frac{dc}{dr} = b_2 \left[\frac{1}{r} \sqrt{\frac{k_C}{D_C}} \cosh \left(\sqrt{\frac{k_C}{D_C}} \, r \right) - \frac{1}{r^2} \sinh \left(\sqrt{\frac{k_C}{D_C}} \, r \right) \right] \tag{8-89}$$

Using Eq. 8-89 and the second boundary condition, the constant b_2 is

$$b_2 = \frac{\mathbf{k}_w R_{ag} C_\infty}{D_C \left[\sqrt{\frac{k_C}{D_C}} \cosh \left(\sqrt{\frac{k_C}{D_C}} R_{ag} \right) - \frac{1}{R_{ag}} \sinh \left(\sqrt{\frac{k_C}{D_C}} R_{ag} \right) + \frac{k_w}{D_C} \sinh \left(\sqrt{\frac{k_C}{D_C}} R_{ag} \right) \sinh \left(\sqrt{\frac{k_C}{D_C}} R_{ag} \right) \right]} \tag{8-90}$$

Substituting Eq. 8-90 into Eq. 8-88, we obtain the solution for the concentration profile

$$c(r) = \frac{\mathbf{k}_w R_{ag} C_\infty}{r} \frac{\sinh \left(\sqrt{\frac{k_C}{D_C}} \, r \right)}{\left[\sqrt{k_C D_C} \cosh \left(\sqrt{\frac{k_C}{D_C}} R_{ag} \right) + \left(\mathbf{k}_w - \frac{D_C}{R_{ag}} \right) \sinh \left(\sqrt{\frac{k_C}{D_C}} R_{ag} \right) \right]} \tag{8-91}$$

Now that we have an expression, we can obtain a solution for the relative uptake factor based on diffusive transport. The observed rate is just the flux into the aggregate times the aggregate area. The flux is

$$\text{observed flux} = -D_C \frac{dC}{dr} \bigg|_{r=R_{ag}} \tag{8-92}$$

The gradient at the surface of the aggregate can be obtained from Eq. 8-91 evaluated at $r = R_{ag}$. The denominator of the relative uptake factor can be calculated by multiplying the rate derived in Eq. 8-84 by the volume of the aggregate, or

$$\text{rate if cells are dispersed in the fluid} = k_{C,G} c_\infty \left(\tfrac{4}{3} \pi R_{ag}^3 \right) \tag{8-93}$$

The rate constant, $k_{C,G}$, is used here to distinguish the rate constant for dispersed cells that exist in a sheared fluid, from cells in the aggregate that do not experience fluid shear. This constant is therefore defined from Eq. 8-84 as

$$k_{C,G} = \frac{3 \, Sh_G(R_c) D_c f (1 - \theta)}{R_c^2} \tag{8-94}$$

where the Sherwood number is calculated for a cell of radius R_c dispersed in a fluid with a mean shear rate G.

Before we combine these equations to obtain the relative uptake factor, we will make an additional simplification. We can use the diffusive Thiele modulus defined in Eq. 8-63 to simplify our expression. The rate constant used in the Thiele modulus is defined for a cell existing inside the aggregate. Since these cells would not experience fluid shear, Sh = 1, and we can write the Thiele modulus as

$$\phi_d = \frac{R_{ag}}{3} \sqrt{\frac{k_C(1)}{D_C}} \tag{8-95}$$

where the parentheses for the rate constant, $k(1)$, indicate that Sh = 1. We can also incorporate the external mass transport coefficient, $\mathbf{k}_w$, into a Sh number defined as $Sh_G(R_{ag}) = \mathbf{k}_w R_{ag} / D_C$, where the parentheses indicate that the Sherwood number is calculated using the aggregate radius. Combining the above results, we have finally for the overall relative uptake factor

$$\gamma_d = \frac{Sh(R_{ag})}{3 \phi_d^2 \, Sh_G(R_c)} \frac{3 \phi_d \cosh 3 \phi_d - \sinh 3 \phi_d}{3 \phi_d \cosh 3 \phi_d + [Sh(R_{ag}) - 1] \sinh 3 \phi_d} \tag{8-96}$$

The rate of mass transport to cells in the aggregate is largest when there is little external resistance to mass transport, or when

$$Sh_G \gg (3 \phi_d \coth 3 \phi_d - 1) \tag{8-97}$$

Under these conditions Eq. 8-96 can be simplified to

$$\gamma_d = \frac{1}{Sh_G(R_c)} \eta_d \tag{8-98}$$

Since $\eta_d < 1$ and $Sh_G > 1$, the relative uptake factor based on diffusion will always be smaller than unity. This result indicates that cells inside impermeable aggregates will always have a reduced rate of substrate consumption compared to unattached cells in the sheared fluid. According to this analysis, there seems to be no advantage for the cell to expend the energy to form polymers that will allow it to become firmly attached within the aggregate.

A relative uptake analysis based on advective flow produces a different result than that observed for only diffusive transport into the aggregate. The derivation proceeds in the same manner as for diffusion except this time transport into the aggregate is based on advective flow through the permeable aggregate. The advective flow inside the aggregate and past attached microbes will produce higher bacterial uptake rates compared to cells in stagnant fluid in an impermeable aggregate. The rate constant defined in Eq. 8-84 can be redefined as $k_{C,u}$, to include flow past the cells in terms of a Sh based on advective flow, defined here as Sh_u, to produce

$$k_{C,u} = \frac{3\ Sh_u(R_c)D_C f(1 - \theta)}{R_c^2} \tag{8-99}$$

where the parentheses around Sh_u indicate the basis of the calculation is the radius of a microbe. The derivation for the relative uptake factor for advection is almost identical to that calculated for diffusion except we now have two different first-order rate constants: $k_{C,u}$ for cells in the aggregate, and $k_{C,G}$ for unattached cells suspended in a sheared fluid. We can include these two rate constants in Eq. 8-78 to obtain

$$\gamma_a = \frac{u}{k_{C,G}2R}\ (1 - e^{-2Rk_{C,u}/u_{ag}}) \tag{8-100}$$

This result can be written more simply using an advective Thiele modulus defined in terms of the advective flow through the aggregate, or $\phi_a = 2R_{ag}k_{C,u}/u_{ag}$, to produce

$$\gamma_a = \frac{Sh_u}{Sh_G}\ \frac{(1 - e^{-\phi_a})}{\phi_a} \tag{8-101}$$

From this result it can be seen that the relationship between the relative uptake and effectiveness factors for advective flow through permeable aggregates is

$$\gamma_a = \frac{Sh_u}{Sh_G}\ \eta_d \tag{8-102}$$

The magnitude of γ_a can be larger than unity depending on the magnitude of Sh_u, and whether Sh_u is larger than Sh_G.

The advantages of being in a porous aggregate increase inversely with the diffusivity of the substrate. For example, when a microbe is in a porous aggregate at a shear rate of $G = 10\ s^{-1}$, the relative uptake factor is ~ 1.7 when $D_C = 10^{-7}\ cm^2\ s^{-1}$, but when $D_C = 10^{-5}\ cm^2\ s^{-1}$ $\gamma_a = 1.1$ (Fig. 8.7). The microbe degrading a large-molecular-weight compound will therefore have a substantially higher rate of chem-

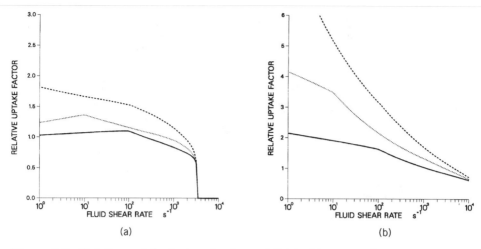

Figure 8.7 Relative uptake factors in sheared fluids ($R = 1\,\mu$m, $f = 0.01$) for $D_{Cw} = 10^{-5}$ cm^2 s^{-1} (solid line), $D_{Cw} = 10^{-6}$ cm^2 s^{-1} (dotted line), and $D_{Cw} = 10^{-7}$ cm^2 s^{-1} (dashed line) for (A) permeable aggregate and (b) permeable floc on a bubble 0.1 cm in diameter rising at 14 cm s^{-1} (From Logan and Hunt, © 1988 Biotechnol. Bioeng. Reprinted by permission of John Wiley & Sons, Inc.).

ical utilization (50% larger) than a cell suspended in the fluid. However, cells in the aggregate will only have a slightly larger increase of low-molecular-weight molecules, such as glucose, with much higher diffusivities. Thus we expect that the main advantage of microbial growth in aggregates will be to scavenge colloids and large-molecular-weight compounds from the suspending fluid.

PROBLEMS

8.1 The Damkohler number is a dimensionless number used to determine if there are mass transfer limitations in systems. It is defined as

$$Da \equiv \frac{\text{maximum rate of reaction}}{\text{maximum rate of mass transfer}} \qquad (8\text{-}103)$$

Determine whether a microbe in a marine environment can be mass transfer limited, assuming the following data: $\mu_{max} = 0.173$ h^{-1}, cell diameter = 0.5 μm, yield coefficient = 10^8 cells/nmol S, $K_s = 1$ nM, $S = 1$ nM, and $D_s = 10^{-5}$ cm^2/s. What would happen if the surface area of the cell that was available for mass transfer was only 1% of the true surface area?

8.2 Fluid shear can affect the rate of mass transfer to particles in water. This question addresses whether fluid shear in a thin fluid film of water in a wetted wall column reactor can affect microbial kinetics. Assume we have water at 20°C, a diffusion coefficient of 10^{-5} cm^2 s^{-1}, and a fluid film thickness of 100 μm. (a) What is an equation for an average fluid velocity, if the velocity profile in the fluid film is:

$$u = \frac{\rho_w g \delta_w^2}{2\mu_w} \left[1 - \left(\frac{x}{\delta_w}\right)^2 \right]$$
(8-104)

where $\rho_w = 1.0$ g cm^{-3} is the fluid density, δ_w the water film thickness, $\mu_w = 0.01$ g cm^{-1} s^{-1} the dynamic viscosity, and x the distance into the fluid film. (b) If the average shear rate is based on the average velocity from part (a) and the fluid thickness, what is the average increase in mass transfer to 1-μm spherical particles in the fluid film compared to the same particle in stagnant fluid?

8.3 The thickness of the hydrodynamic boundary layer around a bubble is $\delta_b = R_b$ $Re_b^{-1/2}$, where R_b is the bubble radius, and Re the Reynolds number based on the bubble radius, and rise velocity given from $U_b = (2R_b g)^{1/2}$, where g is the gravitational constant. Assuming a linear shear profile exists in the bubble boundary layer, calculate the maximum increase in oxygen mass transfer (D_{Ow} = 1.8 × 10^{-5} cm^2 s^{-1}) possible to a bacterium (e.g., 1 μm radius sphere) in the bubble boundary layer to a bubble at rest. The bubble is 0.1 cm in radius, and water has a kinematic viscosity of 0.01 cm^2 s^{-1}.

8.4 Calculate the minimum porosity of an aggregate 100 μm in radius that will just allow advective flow through it in a sheared reactor when the aggregate is composed of primary particles 1 μm in radius.

8.5 Calculate the intra-aggregate fluid velocities for: (a) an aggregate in sheared fluid ($G = 120$ s^{-1}), and (b) a settling aggregate assuming: $R_{ag} = 500$ μm. R_c = 10 μm, the Kozeny–Carmen equation applies, the primary particle density is 1.005 g cm^{-3}, and the aggregate porosity is 0.9854.

8.6 A microbial floc, or aggregate (ag), 500 μm in diameter is composed of microbes (M), each having a cell density $\rho_M = 1.05$ g cm^{-3} and diameter 1 μm and water having a density $\rho_w = 1.000$ g cm^{-3}. The floc is measured to have an overall, or bulk, density of $\rho_f = 1.004$ g cm^{-3} in water. (a) What percent (by volume) of the floc is water? (*Note:* you do not need this answer for other parts of this problem.) (b) Calculate a settling velocity of this floc based on the overall floc density using Stokes' Law for an impermeable sphere. (c) If a microbe on the outside of this floc experienced bulk flow conditions, how much faster would mass transport be to a *microbe* on this floc versus a microbe in stagnant fluid, assuming a chemical diffusivity of $D_{Cw} = 500 \times 10^{-8}$ cm^2 s^{-1}? (d) What would the shear rate have to be for mass transport to an unattached cell suspended in a sheared fluid in order for the mass transport to be equal to that for the microbe on the falling floc?

REFERENCES

Adler, P. M. 1981. *J. Colloid Inter. Sci.* **81**(2):531–35.

Azam, F., and R. E. Hodson. 1981. *Mar. Ecol. Prog. Ser.* **6**:213–222.

Baillod, C. R., and W. C. Boyle. 1970. *J. Sanit. Eng. Div. ASCE* **96**(SA2):525–45.

Canelli, E., and G. W. Fuhs. 1976. *J. Phycol.* **12**:93–99.

Confer, D. R., and B. E. Logan. 1991. *Appl. Environ. Microbiol.* **57**(11):3093–100.

Gibbs, R. J. 1985. *J. Geophys. Res.* **90**(C2):3249–51.

Johnson, C., X. Li, and B. E. Logan. 1996. *Environ. Sci. Technol.*, **30**(6):1911–1919.

Kajihara, M. 1971. *J. Oceanogr. Soc. Japan* **27**(4):158–62.

Logan, B. E., and A. L. Alldredge. 1989. *Mar. Biol.* **101**:443–50.

Logan, B. E., and J. W. Dettmer. 1990. *Biotechnol. Bioengin.* **25**:1135–44.

Logan, B. E., and R. C. Fleury. 1993. *Mar. Ecol. Prog. Ser.* **102**:115–24.

Logan, B. E., and J. R. Hunt. 1987. *Limnol. Oceanogr.* **32**(5):1034–48.

Logan, B. E., and J. R. Hunt. 1988. *Biotechnol. Bioeng.* **31**:91–101.

Logan, B. E., and D. L. Kirchman. 1991. *Marine Biology* **111**:175–81.

Masliyah, J. H., and M. Polikar. 1980. *Can. J. Chem. Eng.* **58**:299–302.

Matsumoto, K., and A. Suganuma. 1977. *Chem. Eng. Sci.* **32**:445–47.

Mueller, J. A., W. C. Boyle, and E. N. Lightfoot. 1968. *Biotech. Bioeng.* **10**:331–58.

Munk, W. H., and G. A. Riley. 1952. *J. Mar. Res.* **11**:215–40.

Pasciak, W. J., and J. Gavis. 1975. *Limnol. Oceanogr.* **20**:604–17.

Sherwood, T. K., R. L. Pigford, and C. R. Wilke. 1975. *Mass Transfer*, McGraw Hill. New York.

Smith, P. G., and P. Coakley. 1984. *Water Res.* **18**:117–22.

Tambo, N., and Y. Watanabe. 1979. *Water Res.* **13**:409–19.

CHAPTER 9

BIOFILMS

9.1 INTRODUCTION

Calculating the rate of chemical transport to a biofilm is usually begun in the same way as calculations to a reactive surface as described in Chapter 5. It is assumed that the biofilm completely covers the surface and that the chemical flux is determined by the diffusivities of the chemical species and the rates these chemicals are degraded by the biofilm. There are aspects of biofilms that make them different, however, from nonviable catalysts. Biofilms can increase in thickness, growing from just a few cells in depth, when a chemical diffusing into the film is not completely removed, to a thick biofilm, when cells at the bottom die due to lack of a source of chemicals for energy and growth. The former case is referred to as an incompletely penetrated or "shallow" biofilm, while the latter case is called a fully penetrated or "deep" biofilm. Depending on the bulk substrate concentration and the depth of the biofilm, it may be possible in some cases to consider the system as a zero- or first-order kinetic system. However, microbial uptake kinetics are usually described in terms of Michaelis–Menten kinetics, and may be more complicated than simple zero- or first-order reaction kinetics typically used to describe reaction kinetics for inorganic catalysts.

Most of the research in the environmental engineering field has focused on the kinetics of the biofilms, and therefore biofilm kinetic models will be carefully described in this chapter. The mass transport aspects of biofilms are perhaps equally (or more) important, and the equations used to describe the fluid above the biofilm vary between models. Different approaches will be used for biofilm-coated particles suspended in fluid or fixed to an immobile surface, with examples of how they are applied to batch and continuous-flow systems.

9.2 TRANSPORT IN THE FLUID LAYER ABOVE A BIOFILM

The use of a "stagnant film" to describe chemical transport persists in many biofilm models (Fig. 9.1). The thickness of this stagnant film is assumed to be the thickness of a concentration boundary layer above a biofilm, and it is assumed that this film thickness is constant over all portions of the biofilm. For a biofilm exposed to a turbulent fluid, for example, in a rotating biological contactor (RBC), this approach may be more practical than alternatives. For example, we could perhaps describe chemical transport in terms of boundary layer theory. This way the chemical flux would be proportional to the diffusion constant to the 1/3 power. But by using boundary layer transport calculations the flux would need to become a function of the distance from a starting point, called the plate entrance. It would be difficult to define a realistic plate entrance for an RBC biofilm in a turbulent fluid since there is no beginning and ending to the flow path across the biofilm surface. Turbulence in the fluid would shift the direction of advective flow across the biofilm and disrupt the development of concentration profiles.

The only case where advective dispersion equation in the fluid above the biofilm is solved is for the case of thin fluid films with an air interface, for example, in a tricking filter. Earlier biofilm models (e.g., Williamson and McCarty, 1980) assumed a stagnant fluid above the biofilm for this case, but as we have seen in Chapter 5, it is possible to set up the constitutive transport equation for situations involving a wetted wall. As a result, we can exactly solve for mass transport in the fluid film as a function of transport distance in a trickling filter, although the solution of these equations requires a numerical model.

Stagnant Film Models

To calculate the flux into a biofilm we assume that there is a single point or line that we can identify that separates the fluid layer above the biofilm and the biofilm

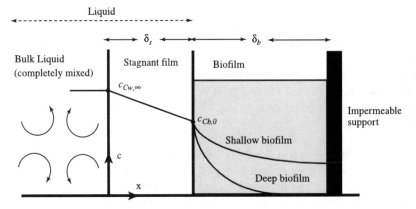

Figure 9.1 Concentration profiles in a shallow and deep biofilm assuming a stagnant fluid layer above the biofilm.

itself. There is no advective flow in the biofilm. For the case of a stagnant film, the flux into the biofilm must equal the flux across the diffusion layer. The flux across the stagnant film occurs only by diffusion, and can be described by

$$J_{C,x} = -D_{Cw} \frac{dc_{Cw}}{dx} \tag{9-1}$$

Since the concentration profile in the fluid is linear, Eq. 9-1 can be simplified for a stagnant fluid film thickness of δ_s as

$$J_{C,x} = -\frac{D_{Cw}}{\delta_s} \Delta c_{Cw} \tag{9-2}$$

The concentration difference can be obtained from our boundary conditions of $c_{Cw} = c_{Cw,\infty}$ at the interface between the well-mixed fluid and stagnant fluid interface, and $c_{Cw} = c_{Cb,0}$ at the interface between the biofilm and the stagnant fluid, to write Eq. 9-2 as

$$J_{C,x} = \frac{D_{Cw}}{\delta_s} (c_{Cw,\infty} - c_{Cb,0}) \tag{9-3}$$

The flux into the biofilm therefore requires that we calculate the thickness of the stagnant fluid film, δ_s, and that we can solve for the chemical concentration at the biofilm–fluid interface from equations describing the concentration profile in the biofilm. From Eq. 9-3 it can be easily seen that the mass transport coefficient $\mathbf{k}_w$ is

$$\mathbf{k}_w = \frac{D_{Cw}}{\delta_s} \tag{9-4}$$

Mass Transport in Packed Beds   Several correlations have been used to describe chemical transport to the surface of spheres in a packed bed. A commonly used correlation for a mass transport coefficient in a packed bed of particles of diameter d_c is based on the correlation developed by Wilson and Geankoplis (Suidan, 1986)

$$j_D = \frac{1.09}{\theta \, Re^{2/3}} \tag{9-5}$$

where θ is the bed porosity, j_D is the Chilton–Colburn j factor for mass and momentum transport, and Re the Reynolds number, which are defined here

$$j_D = \frac{Sh}{Re\ Sc^{1/3}} \tag{9-6}$$

$$Re = \frac{Ud_c}{\nu_w} \tag{9-7}$$

where U is the Darcy or superficial velocity. This correlation is applicable to $0.35 < \theta < 0.75$, $0.0016 < Re < 55$, and $165 < Sc < 70{,}600$. When $Re \ll 1$, Eq. 9-5 can be further simplified.

The fluid film thickness can be obtained from the definition of the Sh number and Eq. 9-4. From Eqs. 9-5 and 9-6 we can write

$$\frac{Sh}{Re\ Sc^{1/3}} = \frac{1.09}{\theta\ Re^{2/3}} \tag{9-8}$$

or upon rearrangement

$$Sh = \frac{1.09}{\theta} Re^{1/3}\ Sc^{1/3} \tag{9-9}$$

The Sh number is defined as $Sh = \mathbf{k}_w d_c / D_{Cw}$, but from Eq. 9.4 we can express Sh as

$$Sh = \frac{k_w d_c}{D_{Cw}} = \frac{D_{Cw}}{\delta_s}\frac{d_c}{D_{Cw}} = \frac{d_c}{\delta_s} \tag{9-10}$$

Equating Eqs. 9-9 and 9-10, the stagnant fluid film thickness can be calculated as

$$\delta_s = \frac{d_c \theta}{1.09} Re^{-1/3} Sc^{-1/3} \tag{9-11}$$

While this calculation provides a value for δ_s, it must be remembered that a stagnant film does not really exist. The calculated film thickness is not (usually!) a truly stagnant film; this is just a conceptual model with a thickness of a concentration boundary layer that results from our definition of a film according to Eq. 9-4. We could equivalently write that the flux to the biofilm as a function of the mass transport coefficient, $\mathbf{k}_w$, as

$$J_{C,x} = \mathbf{k}_w(c_{Cw,\infty} - c_{Cb,0}) \tag{9-12}$$

where the mass transport coefficient is obtained from Eq. 9-9 as

$$\mathbf{k}_w = \frac{1.09 D_{Cw}}{\theta d_c} Re^{1/3}\ Sc^{1/3} \tag{9-13}$$

This approach produces the same calculation but does not require our speculation that a stagnant fluid film exists above the biofilm. The film concept is entrenched in the biofilm literature, however, and will therefore be included in the material to follow.

Example 9.1

Calculate the stagnant fluid film thicknesses in two biofilm systems: (a) a wastewater biofilm reactor composed of rocks averaging 2.54 cm in diameter, and (b) a groundwater system containing porous media with an average grain size of 100 μm. Assume fluid properties at 20°C, a porosity of 0.38, and a fluid velocity of 0.01 cm s^{-1} ($^-$10 m d^{-1}). Comment on your results for the film thickness relative to media size.

(a) The film thicknesses can be calculated from Eq. 9-11. For the reactor system, the Re and Sc numbers are

$$Re = \frac{Ud_c}{\nu_w} = \frac{(0.01 \text{ cm s}^{-1})(2.54 \text{ cm})}{(0.01 \text{ cm}^2 \text{ s}^{-1})} = 2.54 \qquad (9\text{-}14)$$

$$Sc = \frac{\nu_w}{D_{Cw}} = \frac{0.01 \text{ cm}^2 \text{ s}^{-1}}{1000 \times 10^{-8} \text{ cm}^2 \text{ s}^{-1}} = 10^3 \qquad (9\text{-}15)$$

Using Eq. 9-11, we have for the stagnant fluid film thickness

$$\delta_s = \frac{d_c\theta}{1.09} Re^{-1/3}Sc^{-1/3} = \frac{(2.54 \text{ cm})(0.38)}{1.09}(2.54)^{-1/3}(10^3)^{-1/3} = 0.065 \text{ cm} \qquad (9\text{-}16)$$

(b) For the groundwater conditions, $Re = 0.01$ and Sc is still 10^3. The fluid film thickness is therefore

$$\delta_s = \frac{d_c\theta}{1.09} Re^{-1/3}Sc^{1/3} = \frac{(0.01 \text{ cm})(0.38)}{1.09}(0.01)^{-1/3}(10^3)^{-1/3} = 0.0016 \text{ cm} \qquad (9\text{-}17)$$

For the bioreactor case, the stagnant film thickness of 0.065 cm is relatively small compared to the particle diameter of 2.54 cm. A biofilm in a wastewater treatment bioreactor is likely to be on the order of 0.1 cm, which is larger than the calculated stagnant film thickness. For the groundwater case, however, a stagnant film of 16 μm is quite large relative to the media grain sizes of 100 μm. This suggests that many particles in the groundwater could have stagnant films that overlap, and for groundwater systems this mass transport correlation may not work very well.

An alternative approach was used by Rittmann (1982) to calculate the stagnant fluid film thickness was based on a correlation developed by Jennings (1975) for both fixed and fluidized bed reactors, of

$$\delta_s = \frac{D_{Cw}(1 - \theta)^{0.75} \, Re^{0.75} Sc^{0.67}}{9.6U} \tag{9-18}$$

where U is the Darcy or superficial velocity, and the Re and Sc numbers are defined as

$$Re = \frac{d_c U}{v_w} \tag{9-19}$$

$$Sc = \frac{v_w}{D_{Cw}} \tag{9-20}$$

where d_c is the collector or grain diameter of particles forming the packed bed, v_w the dynamic viscosity of the water, and θ the bed porosity. This correlation is valid for $0.5 \leq Re \leq 15$.

Several other correlations for chemical transport in packed beds are reported in Table 9.1, although it is not known how well these correlations apply to biofilms under these conditions.

Mass Transport to Biofilm-Coated Suspended Particles   For biofilm-coated particles that are suspended in a fluidized bed, the concentration boundary layer around the particle will be compressed by the high fluid velocities present in the bed. Re numbers in a fluidized bed are much higher than for the fixed packed bed case. Many biofilm models also describe mass transport to the biofilm in terms of a stagnant fluid (Fig. 9.2).

TABLE 9.1   Different Mass Transport Correlations Used for Packed Beds

Sh Correlation	Limits	Reference
$Sh = 1.09\theta^{-2/3} \, Re^{1/3} \, Sc^{1/3}$	$0.0016 \leq \theta \, Re \leq 55$ $950 \leq Sc \leq 70{,}000$	Wilson and Geankoplis (1966)
$Sh = \dfrac{9.6 d_c U}{D_{Cw}(1 - \theta)^{0.75} \, Re^{0.75} \, Sc^{0.67}}$	$0.5 \leq \theta \, Re \leq 15$ $Sc \approx 10^3$	Jennings (1975)
$Sh = 2.4\theta \, Re^{0.34} \, Sc^{0.42}$	$0.08 \leq Re \leq 125$ $150 \leq Sc \leq 1300$	Williamson et al. (1963)
$Sh = [2 + 0.644 \, Re^{1/2} \, Sc^{1/3}][1 + 1.5(1 - \theta)]$	$1 \leq \theta \, Re \leq 10^2$ $Sc \approx 10^3$	Gnielinski, in Shlünder (1978)
$Sh = 2 + 1.58 \, Re^{0.4} \, Sc^{1/3}$	$10^{-3} < Re < 5.8$	Ohashi et al. (1981)
$Sh = 2 + 1.21 \, Re^{0.5} \, Sc^{1/3}$	$5.8 < Re < 500$	Ohashi et al. (1981)
$Sh = 2 + 0.59 \, Re^{0.6} \, Sc^{1/3}$	$Re > 500$	Ohashi et al. (1981)

Source: Roberts et al., 1985.

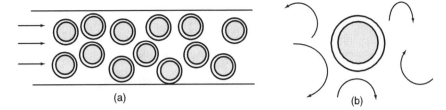

Figure 9.2 Stagnant films are assumed to exist around particles (a) fixed in a packed bed and (b) suspended in a turbulent sheared fluid.

For fluidized beds of gases or solids, the correlation developed by Gupta and Thodos (Suidan, 1986) is often used, for $1 < Re < 2140$, where j_D and Re are defined as before in Eqs. 9-6 and 9-7

$$j_D = \frac{1}{\theta}\left(0.010 + \frac{0.863}{Re^{0.58} - 0.483}\right) \tag{9-21}$$

Mass Transport to Rotating Disks   Wastewater treatment plants using rotating disks operate over a relatively narrow range of disk speeds of 1 to 2 rpm. In order to model the stagnant fluid film on a rotating disk, Rittmann (1982) assumed that

$$\delta_s = 62.5 \ \mu m \tag{9-22}$$

based on work by Gulevich et al. (1968) for flat disks 0.6 m in diameter rotating at 9.4 rpm. However, little justification was given for the use of this value.

A more widely applicable correlation for the film thickness (Grady and Lim, 1980) is

$$\delta_s = 0.00135\omega^{2/3}(R_i + R_o)^{2/3} \tag{9-23}$$

where δ_s is in cm, ω is the rotational speed in s^{-1}, and R_i and R_o are the inner and outer radii of the disks in cm. For a rotational speed of 1 rpm ($0.017 \ s^{-1}$), and $R_o = 366$ cm and $R_i = 15$ cm, $\delta_s = 47 \ \mu m$.

Wetted Wall Biofilm Models

In the case of a thin film of water falling down across a biofilm it is possible to write the complete transport equation for the chemical in the liquid phase. As shown in Figure 9.3, the fluid falling down over the biofilm is exactly the same as previously analyzed for a wetted wall (see Fig. 5.4). Equation 5-87 describes the concentration

270   BIOFILMS

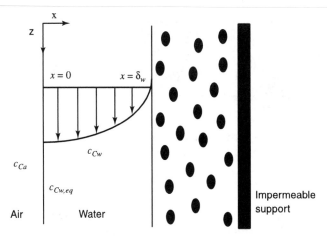

Figure 9.3 Mass transport of chemical C through a thin falling film of thickness δ_w into a biofilm of thickness δ_b.

of c_{Cw} in the liquid in one dimension as

$$u_{\max}\left[1 - \left(\frac{x}{\delta_w}\right)^2\right]\frac{\partial c_{Cw}}{\partial z} = D_{Cw}\frac{\partial^2 c_{Cw}}{\partial z^2} \tag{9-24}$$

We previously analyzed this equation for chemical transport from the gas into the liquid phase. Here we are interested in solving for the chemical flux into the biofilm. As before, we require three boundary conditions, two as a function of the z direction and one as a function of x. We choose

$$BC\ 1: z = 0, \quad c_{Cw} = c_{Cw,0} = c_0 \tag{9-25}$$

$$BC\ 2: x = 0, \quad \frac{dc_{Cw}}{dx} = 0 \tag{9-26}$$

$$BC\ 3: x = \delta_w, \quad -D_{Cw}\frac{\partial c_{Cw}}{\partial x} = J_b(c_{Cb,0}) \tag{9-27}$$

These three boundary conditions specify that: fluid entering the system ($z = 0$) contains the chemical at a concentration c_0; there is no volatilization or loss (no chemical flux) of chemical across the air–liquid interface ($x = 0$); the chemical flux across the water–biofilm interface ($x = \delta_w$) exactly equals the chemical flux into the biofilm. The flux into the biofilm J_b is a function of c_{Cw} in a manner that we must specify by solving the transport equation for chemical transport in the biofilm based upon the chemical concentration at the surface of the biofilm, $c_{Cb,0}$.

9.3 BIOFILM KINETICS

Chemical transport within a biofilm is generally considered to occur only by diffusion at a net rate balanced by chemical reaction (Fig. 9.4). From a mass balance on a differential slice of thickness Δx, the flux into the control volume of cross-sectional area A in one dimension is

$$\text{flux in} = -D_{Cb} \left. \frac{dc_{Cb}}{dx} \right|_x A \tag{9-28}$$

where the subscript b on the diffusion coefficient is used to indicate that the chemical C is in the water in the biofilm. Similarly the flux out is

$$\text{flux out} = -D_{Cb} \left. \frac{dc_{Cb}}{dx} \right|_{x+\Delta x} A \tag{9-29}$$

Defining the rate of reaction per volume of biofilm to be R_b, where the rate of reaction with respect to c_{cb} is not yet defined, the reaction in the control volume is

$$\text{reaction} = R_b A \, \Delta x \tag{9-30}$$

Combining these three terms, we have

$$-D_{Cb} \left. \frac{dc_{Cb}}{dx} \right|_x A + D_{Cb} \left. \frac{dc_{Cb}}{dx} \right|_{x+\Delta x} A + R_b A \, \Delta x = 0 \tag{9-31}$$

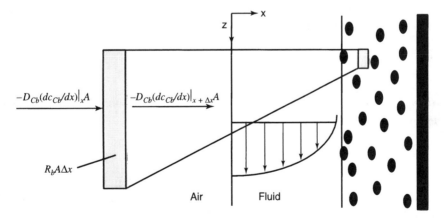

Figure 9.4 The control volume used to construct a transport equation for chemical transport within the biofilm.

Rearranging, dividing by $A \, \delta x$, and taking the limit as $\Delta x \to 0$, Eq. 9-31 becomes

$$\lim_{\Delta x \to 0} \left(\frac{dc_{Cb}}{dx} \bigg|_{x+\Delta x} - \frac{dc_{Cb}}{dx} \bigg|_{x} \right) = \frac{-R_b}{D_{Cb}}$$

(9-32)

which simplifies to

$$\frac{d^2 c_{Cb}}{dx^2} = \frac{-R_b}{D_{Cb}}$$

(9-33)

The solution of Eq. 9-33 depends on our choice of the reaction order and the boundary conditions. In the sections to follow we consider three cases: zero- and first-order kinetics, and Michaelis–Menten kinetics.

As discussed in Chapter 8, the rate of substrate utilization by a biofilm is a function of the cell density, N (# ml^{-1}), which according to Michaelis–Menten kinetics is

$$R_b = \frac{dc_{Cb}}{dt} = -\frac{N}{Y_{N/c}} \frac{\mu_{max} c_{Cb}}{(K_m + c_{Cb})}$$

(9-34)

where $Y_{N/c}$ is a yield coefficient based on cells produced per substrate consumed, μ_{max} is the maximum (saturation) uptake rate constant, and K_m is half-saturation constant or the substrate concentration that produces an uptake rate of $\mu_{max}/2$. Cell mass concentrations are more commonly used than cell number concentrations. Rewriting Eq. 9-34 in terms of cell mass concentration, X (mg L^{-1}), we have

$$R_{Cb} = \frac{dc_{Cb}}{dt} = -\frac{X}{Y_{X/c}} \frac{\mu_{max} c_{Cb}}{(K_m + c_{Cb})}$$

(9-35)

where $Y_{X/c}$ is the yield based on cell mass produced per substrate used. In many models a single constant is defined as $k_\mu = \mu_{max}/Y_{X/c}$, so that the rate is

$$R_b = -\frac{k_\mu X c_{Cb}}{(K_m + c_{Cb})}$$

(9-36)

When substrate concentrations are large ($c_{Cb} \gg K_m$), Eq. 9-36 reduces to a zero-order expression with respect to substrate concentration

$$R_b = -k_{0c} X$$

(9-37)

where $k_{0c} = \mu_{max}/Y_{X/c}$. This can be further simplified by defining a zero-order rate constant $k_0 = k_{0c} X = \mu_{max} X / Y_{X/c}$, resulting in a general expression for a zero-order rate relationship

$$R_b = -k_0$$

(9-38)

When substrate concentrations are very low ($c_{Cw} \ll K_m$), we can simplify Eq. 9-36 to a first-order rate expression

$$R_b = -k_1 c_{Cb} \tag{9-39}$$

where $k_1 = k_\mu X / K_m$.

Using these different biofilm kinetic expressions, it is possible to derive solutions of chemical transport for biofilm systems. It is common practice to nondimensionalize the transport equations so that solutions can be generalized to different values of boundary conditions. The appropriate choices for constructing these nondimensionalized expressions are shown below for the different biofilm models.

Zero-Order Biofilm Kinetics

The simplest biofilm models arise for zero-order kinetic expressions. When substrate concentrations in the biofilm remain large relative to K_m, microbial kinetics can be described in terms of Eq. 9-38, and the constitutive transport equation for chemical transport in the biofilm (Eq. 9-33) becomes

$$\frac{d^2 c_{Cb}}{dx^2} = \frac{k_0}{D_{Cb}} \tag{9-40}$$

To non-dimensionalize this system we choose the dimensionless numbers

$$c^* = \frac{c_{Cb}}{c_{Cb,0}} \tag{9-41}$$

$$x_b^* = \frac{x}{\delta_b} \tag{9-42}$$

Making the substitutions $c_{Cb} = c^* c_{C,b,0}$ and $x^2 = x_b^{*2} \delta_b^2$ into Eq. 9-40 (Harremoës, 1978), we obtain

$$\frac{d^2 c^*}{dx_b^{*2}} = \frac{k_0 \delta_b^2}{D_{Cb} c_{Cb,0}} \tag{9-43}$$

The collection of constants on the right-hand side of Eq. 9-43 is dimensionless, and can be replaced by a single constant for simplicity, to obtain the transport equation

$$\frac{d^2 c^*}{dx_b^{*2}} = \frac{2}{B_0^2} \tag{9-44}$$

where the new dimensionless constant B_0 is defined as

$$B_0 = \left(\frac{2D_{Cb}c_{Cb,0}}{k_0\delta_b^2}\right)^{1/2} \tag{9-45}$$

The constant B_0 is a dimensionless ratio of the rate chemicals diffuse into the biofilm to the rate they are reacted away, and therefore is a reciprocal of the Thiele modulus. The definition chosen for B_0 results in a useful distinction in biofilm properties for values of B_0 less than and greater than unity. The conditions that produce $B_0 > 1$ result in a fully penetrated biofilm. This means that all of the biofilm will consume substrate since substrate will be present through the biofilm at concentrations defined by Eq. 9-40. This condition of complete substrate penetration of the biofilm produces a biofilm that is referred to by most biofilm modelers as a shallow biofilm.

Shallow Biofilm   To solve Eq. 9-44 specifically for the case of a shallow biofilm, we choose the two boundary conditions

$$BC\ 1:\ x_b^* = 0,\quad c^* = 1 \tag{9-46}$$

$$BC\ 2:\ x_b^* = 1,\quad \frac{dc^*}{dx_b^*} = 0 \tag{9-47}$$

These boundary conditions specify that at the water–biofilm interface the biofilm concentration is present at its surface concentration (a dimensionless concentration of 1), and at the bottom of the biofilm there is no chemical flux through the surface (which by definition is impermeable). Using boundary conditions 9-46 and 9-47 in Eq. 9-44 and solving, we obtain the concentration profile

$$c^* = \frac{x_b^{*2}}{B_0^2} - 2\frac{x_b^*}{B_0^2} + 1, \quad \text{shallow biofilm; zero order} \tag{9-48}$$

The chemical flux into the biofilm is needed for calculations of chemical removals in reactors and other systems. From Fick's First Law, the chemical flux for the case of a shallow biofilm is

$$J_c|_{x=0} = k_0\delta_b, \quad \text{shallow biofilm; zero order} \tag{9-49}$$

Example 9.2

Calculate the flux and concentration of substrate at the biofilm surface assuming zero-order kinetics and a fully penetrated (shallow) biofilm, where the biofilm coats particles in a packed bed. Assume the following kinetic and reactor conditions: μ_{max} = 4 d^{-1}, $Y_{X/c}$ = 0.5 g/g, K_m = 10 mg L^{-1}, X = 40,000 mg L^{-1}, δ_b = 40 μm, D_{Cw} =

600×10^{-8} cm^2 s^{-1}, $c_{Cw,\infty} = 100$ mg L^{-1}, packed bed particle diameter $d_c = 1$ cm, $\theta = 0.35$, $U = 0.2$ cm s^{-1}, and fluid properties of water at 20°C.

The solution to this question requires a two-step process. First, we must solve for the flux into the biofilm, and second we must calculate the surface concentration given the expression for flux in terms of the stagnant film thickness (which in this case is really just a way of expressing a mass transport coefficient). According to Eq. 9-49, the chemical flux is

$$J_C|_{x=0} = k_0 \delta_b \tag{9-50}$$

Notice that the chemical flux is not a function of the substrate concentration for this case. The kinetic constant $k_0 = \mu_{max} X / Y_{X/c}$, or

$$k_0 = \frac{(4 \text{ d}^{-1})(40{,}000 \text{ mg L}^{-1})}{0.5 \text{ mg mg}^{-1}} = 320{,}000 \frac{\text{mg}}{\text{L d}} \tag{9-51}$$

For a biofilm thickness of 40 μm = 0.004 cm, the flux is

$$J_C = \left(320{,}000 \frac{\text{mg}}{\text{L d}}\right)(0.004 \text{ cm}) \frac{\text{L}}{10^3 \text{ cm}^3} = 1.28 \frac{\text{mg}}{\text{cm}^2 \text{ d}} \tag{9-52}$$

From Eq. 9-3, the chemical flux into the biofilm can also be expressed as

$$J_C|_x = \frac{D_{Cw}}{\delta_s} (c_{Cw,\infty} - c_{Cb,0}) \tag{9-53}$$

In order to use Eq. 9-53 to solve for $c_{Cb,0}$, we must first calculate the stagnant film thickness, δ_s. From Eq. 9-11, we have

$$\delta_s = \frac{d_c \theta}{1.09} Re^{-1/3} Sc^{-1/3} \tag{9-54}$$

The two dimensionless numbers Re and Sc are calculated from the given information as

$$Re = \frac{U d_c}{\nu_w} = \frac{(0.2 \text{ cm s}^{-1})(1 \text{ cm})}{0.01 \text{ cm}^2 \text{ s}^{-1}} = 20 \tag{9-55}$$

$$Sc = \frac{\nu_w}{D_{Cw}} = \frac{0.01 \text{ cm}^2 \text{ s}^{-1}}{600 \times 10^{-8} \text{ cm}^2 \text{ s}^{-1}} = 1666 \tag{9-56}$$

Substituting these into our expression for the film thickness, we have

$$\delta_s = \frac{(1 \text{ cm})(0.35)}{1.09} (20)^{-1/3}(1666)^{-1/3} = 0.01 \text{ cm } (= 100 \text{ μm}) \tag{9-57}$$

Rearranging Eq. 9-53 and solving

$$c_{Cb,0} = c_{Cw,\infty} - \frac{J_{Cx}\delta_s}{D_{Cw}} = \left(100 \; \frac{mg}{L}\right) - \frac{(1.28 \; mg/cm^2 \; d)(0.01 \; cm)}{600 \times 10^{-8} \; cm^2 \; s^{-1}} \quad (9\text{-}58)$$

$$= 75.3 \; mg \; L^{-1} \quad (9\text{-}59)$$

Our assumption that the biofilm is shallow should be checked using the results of our calculation. From Eq. 9-45 and the biofilm interface concentration of 75.3 mg L^{-1} we calculate that $B_0 = 3.9$, which meets our criterion for a shallow biofilm that $B_0 > 1$.

Deep Biofilm When $B_0 < 1$ the biofilm is incompletely penetrated, a condition described as a deep biofilm since the substrate is completely removed before it can diffuse to the bottom of the biofilm. For this case, we can solve Eq. 9-44 using the two boundary conditions

$$BC \; 1: x_b^* = 0, \quad c^* = 1 \quad (9\text{-}60)$$

$$BC \; 2: x_b^* = B_0, \quad c^* = 0 \quad (9\text{-}61)$$

The first boundary condition is the same as previously chosen, but the second boundary condition makes use of a result we already know: substrate will not reach the impermeable interface. While it is also true that there would be no chemical flux through the biofilm support surface, the choice of Eq. 9-61 allows for a more useful solution of the concentration profile. Using boundary conditions 9-60 and 9-61 in Eq. 9-44 and solving, we obtain the concentration profile

$$c^* = \frac{x_b^{*2}}{B_0^2} - 2\frac{x_b^*}{B_0} + 1, \quad \text{incompletely penetrated; zero order} \quad (9\text{-}62)$$

This solution is nearly identical to the first solution, and this solution is only valid for any x_b^* where $c^* > 0$.

The chemical flux into the biofilm is obtained from Fick's First Law, as

$$J_C|_{x=0} = (2D_{Cb}k_0)^{1/2}c_{Cb,0}^{1/2}, \quad \text{incompletely penetrated; zero order} \quad (9\text{-}63)$$

Example 9.3

Repeat Example 9.2 for the case when the biofilm is 1 mm thick, therefore assuming that the biofilm is an incompletely penetrated (deep) biofilm.

For this case, the flux into the biofilm is

$$J_C = (2D_{Cb}k_0)^{1/2}c_{Cb,0}^{1/2} \qquad (9\text{-}64)$$

The only problem with using this solution is that we do not know the substrate concentration at the biofilm–stagnant film interface. Substituting in our known values, we have

$$J_C = \left[2\left(600 \times 10^{-8}\ \frac{cm^2}{s}\right)\left(320{,}000\ \frac{mg}{L\ d}\right)\frac{86{,}400s}{d}\frac{1\ L^2}{10^6\ cm^6}\right]^{1/2}c_{Cb,0}^{1/2} \qquad (9\text{-}65)$$

$$J_C\ [mg\ cm^{-2}\ d^{-1}] = 0.576c_{Cb,0}^{1/2}\ [mg\ L^{-1}] \qquad (9\text{-}66)$$

where the units for the flux and concentration are as indicated. The value of $c_{Cw,b}$ can be obtained from an iterative solution using the definition of the flux as

$$J_{C,x} = \frac{D_{Cw}}{\delta_s}(c_{Cw,\infty} - c_{Cb,0}) \qquad (9\text{-}67)$$

Equating the above two equations and rearranging

$$c_{Cw,\infty} = c_{Cb,0} + \frac{0.576\delta_s}{D_{Cw}}c_{Cb,0}^{1/2} \qquad (9\text{-}68)$$

$$100\ \frac{mg}{L} = c_{Cb,0} + \left(\frac{0.576(0.01\ cm)}{600 \times 10^{-8}\ cm^2\ s^{-1}}\frac{10^3\ cm^3}{L}\frac{d}{86{,}400\ s}\right)c_{Cb,0}^{1/2} \qquad (9\text{-}69)$$

Substituting in values for $c_{Cb,0}$ until the two sides of the equation are equal, we obtain $c_{Cb,0} = 34.6$ mg/L. From our calculation of the flux in Eq. 9-66, we obtain

$$J_C = 0.576(34.6\ mg\ L^{-1})^{1/2} = 3.39\ \frac{mg}{cm^2\ d} \qquad (9\text{-}70)$$

which is only 2.6 times larger than the result for an incompletely penetrated biofilm. A check on our assumption that the biofilm is deep is made by checking the magnitude of B_0. Using Eq. 9-45, we find that $B_0 = 0.011$, which meets our criterion for a deep biofilm that $B_0 < 1$.

First-Order Biofilm Kinetics

For the case of first-order biofilm kinetics, we can substitute Eq. 9-39 into Eq. 9-33, and nondimensionalize using the parameters defined above to obtain

$$\frac{d^2c^*}{dx_b^{*2}} = \frac{k_1\delta_b^2}{D_{Cb}}c^* \qquad (9\text{-}71)$$

This can be simplified in terms of a single constant (Harremoës, 1978) to

$$\frac{d^2c^*}{dx_b^{*2}} = B_1^2 c^*$$

(9-72)

where B_1 is defined here as

$$B_1 = \left(\frac{k_1\delta_b^2}{D_{Cb}}\right)^{1/2}$$

(9-73)

Here B_1 is a Thiele modulus that indicates the importance of reaction relative to diffusion.

To solve Eq. 9-72, we choose the two boundary conditions

$$BC\ 1:\ x_b^* = 0, \quad c^* = 1$$

(9-74)

$$BC\ 2:\ x_b^* = 1, \quad \frac{dc^*}{dx_b^*} = 0$$

(9-75)

Only one solution is necessary for first-order kinetics, since there must always be a finite concentration of substrate (however small) in the biofilm. For the case of first-order kinetics (see Problem 9.14), the flux is

$$J_C|_{x=0} = k_1\delta_b c_{Cb,0}\frac{\tanh B_1}{B_1}, \quad \text{first order}$$

(9-76)

Many biofilm reactions are quite fast due to the high biomass concentration in a biofilm. It has been shown before that for fast reactions at a surface (see Section 5.2) that for large B_1, $\tanh B_1 \rightarrow 1$, and Eq. 9-76 becomes

$$J_C|_{x=0} = (k_1 D_{Cb})^{1/2}c_{Cb,0}, \quad \text{first order; fast reaction}$$

(9-77)

Example 9.4

There is considerable controversy in the research literature on the subject of whether oxygen can limit the performance of fixed film reactors used for wastewater treatment. It is relatively easy to demonstrate that oxygen, because it is sparingly soluble in water, can limit aerobic substrate removal. However, this does not address the question of anaerobic substrate removal (chemical oxidation by bacteria using alternate electron acceptors such as SO_4^{-2} and CO_2). Assuming first-order kinetics and that oxygen is present in a fluid at 8 mg L^{-1}, calculate the substrate concentrations when oxygen could begin to limit aerobic substrate removal for (a) glucose, $D_{Gw} = 600 \times 10^{-8}$ cm^2 s^{-1}, and (b) a polysaccharide, $D_{Pw} = 100 \times 10^{-8}$ cm^2 s^{-1}.

(a) To solve this problem we will assume the yield between oxygen and glucose does not result in biomass production, so that the oxidation of carbon with oxygen proceeds according to the reaction

$$C_6H_{12}O_6 + 6O_2 \rightarrow 6CO_2 + 6H_2O \qquad (9\text{-}78)$$

Thus 6 moles of oxygen will be necessary for every mole of glucose. On a mass concentration basis, this translates to 0.94 mg glucose per mg oxygen.

The rate of glucose removal by a biofilm can be considered to be a very fast reaction. Therefore, the flux into the biofilm is

$$J_{Gb}|_{x=0} = (k_1 D_{Gb})^{1/2} c_{Gb,0} \qquad (9\text{-}79)$$

Oxygen utilization at concentrations above 2 mg/L is generally zero order. We will assume here that oxygen is used in proportion to substrate until the oxygen is depleted. From our calculation above, we can write a ratio of the rate of oxygen to glucose utilization as

$$R_{\text{lim}} = \frac{J_{Ob} \times 0.94}{J_{Gb}} \qquad (9\text{-}80)$$

When $R_{\text{lim}} < 1$, oxygen can limit substrate kinetics, and when $R_{\text{lim}} > 1$, substrate removal will be limited by substrate concentrations in the fluid. Substituting in the definition of the fluxes, Eq. 9-80 becomes

$$R_{\text{lim}} = \frac{(k_1 D_{Ob})^{1/2} c_{Ob,0} \times 0.94}{(k_1 D_{Gb})^{1/2} c_{Gb,0}} \qquad (9\text{-}81)$$

or more simply

$$R_{\text{lim}} = \frac{D_{Ob}^{1/2} c_{Ob,0} \times 0.94}{D_{Gb}^{1/2} c_{Gb,0}} \qquad (9\text{-}82)$$

From the values given in the problem statement, this produces

$$R_{\text{lim}} = \frac{(1800 \times 10^{-8})^{1/2} (8 \text{ mg L}^{-1}) \times 0.94}{(600 \times 10^{-8})^{1/2} c_{Gb,0}} = \frac{13}{c_{Gb,0}} \qquad (9\text{-}83)$$

At a concentration of glucose > 13 mg L^{-1}, R_{lim} 1 and oxygen will limit substrate removal.

(b) For a high-molecular-weight polysaccharide R_{lim} is calculated as

$$R_{\text{lim}} = \frac{(1800 \times 10^{-8})^{1/2} \, (8 \text{ mg L}^{-1}) \times 0.94}{(100 \times 10^{-8})^{1/2} c_{Gb,0}} = \frac{32}{c_{Gb,0}} \tag{9-84}$$

The limiting substrate concentration is predicted to be higher for the polysaccharide than for glucose due to the lower diffusivity of the polysaccharide.

Michaelis–Menten Biofilm Kinetics

The most generally applicable case of chemical transport in a biofilm is when the biofilm kinetics are described using Michaelis–Menten expression, producing the transport equation

$$\frac{d^2 c_{Cb}}{dx^2} = \frac{-X}{D_{Cb} Y_{X/c}} \frac{\mu_{\max} c_{Cb}}{(K_m + c_{Cb})} \tag{9-85}$$

Unfortunately, no analytical solution exists for this equation. Because biofilm calculations are important in wastewater treatment processes, many analytical approximations have been suggested. A widely used analytical approximation was developed by Suidan and Wang (1985) that can be based on the dimensionless variables

$$s_\infty^* = \frac{c_{Cw,\infty}}{K_m} \tag{9-86}$$

$$s_b^* = \frac{c_{Cb}}{K_m} \tag{9-87}$$

$$s_{b0}^* = \frac{c_{Cb,0}}{K_m} \tag{9-88}$$

$$x^* = x \left(\frac{\mu_{\max} X}{Y_{X/S} D_{Cb} K_m} \right)^{1/2} \tag{9-89}$$

$$\delta_s^* = \delta_s \left(\frac{\mu_{\max} X}{Y_{X/S} D_{Cb} K_m} \right)^{1/2} \left(\frac{D_{Cw}}{D_{Cb}} \right) \tag{9-90}$$

$$\delta_b^* = \delta_b \left(\frac{\mu_{\max} X}{Y_{X/S} D_{Cb} K_m} \right)^{1/2} \tag{9-91}$$

where s_∞^*, s_b^*, and s_{b0}^* are the dimensionless substrate concentrations in the bulk fluid, in the biofilm, and at the biofilm–stagnant film interface (0 indicating the top of the biofilm), respectively, x^* is a dimensionless distance from the top of the biofilm, and δ_s^* and δ_b^* are the dimensionless thicknesses of the stagnant fluid and biofilm.

Substituting Eqs. 9-87 and 9-89 into 9-85, results in a dimensionless form of the Michaelis–Menten equation

$$\frac{d^2 s_b^*}{dx^{*2}} = \frac{s_b^*}{1 + s_b^*} \tag{9-92}$$

Nondimensionalizing the bulk substrate concentration by the half-saturation constant produces a dimensionless substrate concentration that should approach the limits described by the zero- and first-order expressions developed above. As $s_{b0}^* \to 0$ ($s_{b0}^* < 0.1$ is sufficient), Eq. 9-85 becomes a first-order expression, and as $s_{b0}^* \to 1$ ($s_{b0}^* > 0.9$), Eq. 9-85 becomes a zero-order expression. These limiting solutions are discussed further below.

Flux into the Biofilm   The solution of Eq. 9-85 (if possible) would produce a dimensionless concentration profile, but that is less important to biofilm calculations than the chemical flux into the biofilm. We can define a dimensionless flux J^* as

$$J^* = \frac{J_{C,x}}{(\mu_{max} X D_{Cb} K_m / Y_{X/S})^{1/2}} \tag{9-93}$$

Using Eq. 9-93, we can write Eq. 9-3 for the flux into the biofilm as

$$J^* = \frac{1}{\delta_s^*} (s_\infty^* - s_b^*) \tag{9-94}$$

Using the dimensionless boundary conditions

$$BC\ 1:\ x^* = 0, \quad s_b^* = s_{b0}^* \tag{9-95}$$

$$BC\ 2:\ x^* = \delta_b^*, \quad \frac{ds_b^*}{dx^*} = 0 \tag{9-96}$$

a semiempirical solution of chemical flux into a biofilm of any depth (Suidan and Wang, 1985) is

$$\delta_b^* = J^* + \tanh^{-1} \left(\frac{0.5 J^{*2} + J^*(1 + 0.23 J^{*1.19})^{-0.61}}{s_{b0}^*} \right) \tag{9-97}$$

Since this solution is not explicit for J^*, the parameters δ_b^* and s_{b0}^* must be input and J^* determined by an iterative solution technique (available on most computer spreadsheets).

In many wastewater treatment applications very deep biofilms develop so that substrate is consumed before it can diffuse to the impermeable surface. For deep biofilms, Eq. 9-85 can be directly solved. Defining the variable $\Gamma = d^2 s_b^*/dx^{*2}$ and using the chain rule, Eq. 9-92 can be transformed into

$$\Gamma \frac{d\Gamma}{ds_b^*} = \frac{s_b^*}{1 + s_b^*} \qquad (9\text{-}98)$$

This equation can be solved by separation of variables (Williamson and McCarty, 1976; Suidan, 1986) to obtain

$$\frac{J^{*2}}{2} = s_{b0}^* - \ln(1 + s_{b0}^*) \qquad (9\text{-}99)$$

While this equation produces a direct analytical solution for the flux into a very deep biofilm, it has the disadvantage of being expressed in terms of the unknown stagnant film substrate concentration, s_s^*. Using the definition of the dimensionless flux in Eq. 9-93, we can write a more convenient expression for the flux in terms of the bulk substrate concentration, s_∞^* (Suidan, 1986) as

$$\frac{J^{*2}}{2} = s_\infty^* - J^* \delta_s^* - \ln(1 + s_\infty^* - J^* \delta_s^*) \qquad (9\text{-}100)$$

This solution for a deep biofilm is implicit with respect to three parameters: s_∞^*, δ_s^*, and J^*. Thus two of the parameters must be known before the third one can be obtained. In practice, this is acceptable once the assumption of a stagnant film is made, since we usually know the bulk substrate concentration, and as we have discussed above, we can independently calculate the stagnant film thickness.

Zero- and First-Order Solutions for Shallow and Deep Biofilms

The above solutions for Michaelis–Menten kinetics can be simplified under conditions when the bulk substrate concentration is either very large or small, but different solutions arise when the biofilm is considered as either shallow (completely penetrated) or deep (incompletely penetrated). According to Suidan (1987), a biofilm can be considered deep when $\delta_b \geq 3.0$, and shallow when $\delta_b \leq 0.2$. The criteria for the kinetic expression rely on the magnitude of $s_b^* = c_{Cb}/K_m$ with first-order kinetics for $s_b^* \ll 1$, and zero-order kinetics for $s_b^* \gg 1$.

Deep Biofilm, $s_{b0}^* \ll 1$

For a deep biofilm, Eq. 9-100 can be simplified by a Taylor series expansion when $s_{b0}^* \ll 1$ since the term $[\ln(1 + s_{b0}^*)] \rightarrow (s_{b0}^* - s_{b0}^{*2}/2)$, producing

$$J^* = s_{b0}^* \qquad (9\text{-}101)$$

This result indicates that for first-order kinetics the substrate flux to deep biofilms is independent of biofilm thickness, and that the total flux is a function only of the substrate concentration at the stagnant film–biofilm interface.

Deep Biofilm, $s_{b0}^* \gg 1$ For a deep biofilm, Eq. 9-100 can again be simplified because for $s_{b0}^* \gg 1$ the log term becomes negligible compared to the other terms, resulting in

$$J^* = (2s_{b0}^*)^{1/2} \tag{9-102}$$

For zero-order kinetics chemical penetration into a deep biofilm becomes a half-order reaction.

Shallow Biofilm, $s_{b0}^* \ll 1$ Fully penetrated biofilms find less application in wastewater treatment reactors, but may arise in natural systems such as groundwater aquifers. If we assume that substrate is present everywhere in the biofilm at a concentration of s_{b0}^*, then Suidan (1986) demonstrates that the flux can be obtained from the integration

$$J^* = \frac{s_{b0}^*}{1 + s_{b0}^*} \int_0^{\delta_b^*} dz^* = \frac{s_{b0}^*}{1 + s_{b0}^*} \delta_b^* \tag{9-103}$$

For shallow biofilms, Eq. 9-103 for a first-order reaction with $s_{b0}^* \ll 1$ in the biofilm becomes

$$J^* = s_{b0}^* \delta_b^* \tag{9-104}$$

Shallow Biofilm, $s_{b0}^* \gg 1$ For a fully penetrated biofilm, from Eq. 9-103 for a zero-order reaction when $s_s^* \gg 1$, we have

$$J^* = \delta_{b0}^* \tag{9-105}$$

This result indicates that the flux into shallow biofilms is essentially dictated by the depth of the biofilm. These results for the four limiting cases are summarized in Table 9.2 along with the expressions for flux based on the bulk substrate concentrations.

TABLE 9.2   Chemical Flux into Deep and Shallow Biofilms under Conditions of Very Low and Very High Bulk Substrate Concentrations (Approximating First- and Zero-Order Kinetic Models) in Terms of Stagnant Film and Bulk Substrate Concentrations

Depth	s_{b0}^*	Flux in terms of s_{b0}^*	Flux in terms of s_∞^*
Deep	$s_{b0}^* \ll 1$	$J^* = s_{b0}^*$	$J^* = s_\infty^*/(1 + \delta_s^*)$
Deep	$s_{b0}^* \gg 1$	$J^* = (2s_{b0}^*)^{1/2}$	$J^* = \left(\dfrac{2}{s_\infty^*} + \dfrac{\delta_s^*}{s_\infty^{*2}}\right)^{-1/2} - \dfrac{\delta_s^*}{s_\infty^*}$
Shallow	$s_{b0}^* \ll 1$	$J^* = s_{b0}^* \delta_b^*$	$J^* = s_\infty^* \delta_b^*/(1 + \delta_s^* \delta_b^*)$
Shallow	$s_{b0}^* \gg 1$	$J^* = \delta_b^*$	$J^* = \delta_b^*$

Source: Suidan, 1986.

Example 9.5

Using the values given in Example 9.2, calculate the flux into the biofilm using the zero-order approximation ($s_{bo}^* \gg 1$) for biofilm kinetics for an incompletely penetrated (deep) biofilm. Compare your result to that obtained in Example 9.3.

For this case, we can obtain the flux from equations given in Table 9.2

$$J^* = \left(\frac{2}{s_\infty^*} + \frac{\delta_s^*}{s_\infty^{*2}} \right)^{-1/2} - \frac{\delta_s^*}{s_\infty^*} \tag{9-106}$$

We will first need to calculate the dimensionless parameter s_∞^* as

$$s_\infty^* = \frac{c_{Cw,\infty}}{K_m} = \frac{100 \text{ mg L}^{-1}}{10 \text{ mg L}^{-1}} = 10 \tag{9-107}$$

and assuming that the diffusion constants in the fluid and the biofilm are equal, and substituting in the definition of the variable k_0, we can write δ_s^* as

$$\delta_s^* = \delta_s \left(\frac{\mu_{max} X}{Y_{X/S} D_{Cb} K_m} \right)^{1/2} \left(\frac{D_{Cw}}{D_{Cb}} \right) = \delta_s \left(\frac{k_0}{D_{Cb} K_m} \right)^{1/2} \tag{9-108}$$

and substituting in the given values

$$\delta_s^* = (0.01 \text{ cm}) \left(\frac{320{,}000 \text{ mg/L d}}{(600 \times 10^{-8} \text{ cm}^2 \text{ s}^{-1})(10 \text{ mg/L})} \frac{d}{86{,}400 \text{ s}} \right)^{1/2} = 2.48 \tag{9-109}$$

Using these results in our expression for the dimensionless flux, we obtain

$$J^* = \left(\frac{2}{10} + \frac{2.48}{10^2} \right)^{-1/2} - \frac{2.48}{10} = 1.86 \tag{9-110}$$

From the definition of J^*, the flux is

$$J_{C,x} = J^* \left(\frac{\mu_{max} X D_{Cb} K_m}{Y_{X/S}} \right)^{1/2} = J^* (k_0 D_{Cb} K_m)^{1/2} \tag{9-111}$$

$$= (1.86) \left[\left(320{,}000 \frac{mg}{L \text{ d}} \right) \left(600 \times 10^{-8} \frac{cm^2}{s^{-1}} \right) \left(10 \frac{mg}{L} \right) \right.$$

$$\left. \cdot \frac{1 \text{ L}^2}{10^6 \text{ cm}^6} \frac{86{,}400 \text{ s}}{d} \right]^{1/2} \tag{9-112}$$

$$= 2.40 \frac{mg}{cm^2 \text{ d}} \tag{9-113}$$

This flux is only 70% of the flux calculated for the exact solution for zero-order kinetics for a deep biofilm in Example 9.3, which suggests that our calculation is *wrong*!

We began this calculation by *assuming* that the kinetics could be considered zero order and that the biofilm was deep. The requirement that we can assume zero-order kinetics is that $s_{b0}^* \gg 1$. We can check this assumption by using our result for the flux that for a deep biofilm (Eq. 9-102) that $J^* = (2s_{b0}^*)^{1/2}$. Based on our solution in Eq. 9-110, we calculate that $s_{b0}^* = 1.7$, which clearly does not meet our criterion for $s_{b0}^* \gg 1$. Thus we fail to calculate the flux in this example correctly. In order to have made our calculation correctly, K_m would have to have been much lower relative to the substrate concentration. The correct solution for this problem will be addressed using a nomograph in Example 9.6.

Other Useful Kinetic and Biofilm Relationships

So far we have considered substrate uptake kinetics, but have not mentioned biomass loss and decay. Suspended cells and biofilms undergo endogenous decay. In the case of biofilms, cells and whole patches of biomass can be sheared from the biofilm. The change in biomass thickness is a dynamic process and can be related to kinetic and biofilm parameters.

For suspended cells, the net rate of biomass formation, R_X, can be modeled as

$$R_X = Y_{X/c}R_c - b_{end}X \tag{9-114}$$

where b_{end} is the endogenous decay coefficient. The substrate flux that just supports the requirements of the cells can be obtained when $R_X = 0$, or

$$R_c = \frac{b_{end}}{Y_{X/c}} X \tag{9-115}$$

Assuming the Michaelis–Menten formulation for R_c (Eq. 9-35), Eq. 9-115 can be written as

$$b_{end} = \frac{\mu_{max}c_{Cw}}{(K_m + c_{Cw})} \tag{9-116}$$

This can be rearranged to solve for the minimum substrate concentration necessary to balance decay, defined here as c_{min}, as

$$c_{min} = \frac{b_{end}K_m}{\mu_{max} - b_{end}} \tag{9-117}$$

For aerobic systems typical values of the minimum substrate concentration are around 0.1 to 1 mg L^{-1}.

The calculation of a minimum substrate concentration also applies to substrate concentrations in biofilms except that the decay constant b_{end} needs to include losses of the biofilm due to removal processes such as fluid shear. For a biofilm, the situation is also somewhat more complicated since the chemical flux maintains the biofilm and therefore the bulk substrate concentration may be higher than the minimum substrate concentration calculated above. The thickness of a biofilm, δ_b, at steady state is

$$\delta_b = \frac{J_b Y_{X/c}}{b_{end} X} \tag{9-118}$$

Another way to view this result is that this equation indicates the chemical flux necessary to support a biofilm of thickness δ_b at steady state.

When a bioreactor is first inoculated, the biofilm will take some time to reach steady conditions. The amount of time necessary to develop the steady-state biofilm thickness was addressed by Annachhatre and Khanna (1987). Beginning with Eq. 9-114 and assuming that the biomass concentration in the biofilm is independent of thickness, the net biomass formation rate in the entire biofilm, R_X [M L^{-2} T^{-1}], is

$$R_X = \int_0^{\delta_b} r_X \, dz = Y_{X/c} \int_0^{\delta_b} r_c \, dz - bX \int_0^{\delta_b} dz \tag{9-119}$$

The rate of change of biomass per time per unit area is similarly written as

$$\frac{dB(t)}{dt} = X \frac{d\delta_b(t)}{dt} \tag{9-120}$$

where $B(t)$ is the total biomass per unit area in the biofilm at any time. Combining the above two equations

$$X \frac{d\delta_b(t)}{dt} = Y_{X/c} J_b(t) = bX\delta_b(t) \tag{9-121}$$

In dimensionless form, this becomes

$$\frac{d\delta_b^*}{dt^*} = \frac{\mu_{max}}{b} J^* - \delta_b^* \tag{9-122}$$

where t^* is a dimensionless time defined as $t^* = t/b_{end}$. The time necessary to reach steady conditions can be solved numerically by using Eq. 9-97 to relate J^* and δ_b^* to s_s^* as a function of time.

9.4 BIOREACTOR MODELING

Modeling substrate removal in a bioreactor is possible now that the hydrodynamic and biofilm models are developed. While there are many possible configurations of biofilm reactors, we will focus our discussion here on three types of reactors: fixed bed biofilm reactors, rotating biofilm contactors, and plastic media trickling filters, commonly referred to as biotowers.

Completely Mixed Biofilm Systems: Rotating Biological Contactors

The substrate concentration in the fluid in each stage of an RBC is often considered to be homogeneous, or completely mixed. This assumption makes it relatively easy to compute the effluent substrate concentration from a single-stage unit since all of the biofilm in the unit will be exposed to the the same substrate concentration. From a simple mass balance, the chemical flux into the biofilm must equal that removed in the reactor, or

$$Qc_{Cw,\infty|in} - Qc_{Cw,\infty|out} = J_{Cb}Va_b \tag{9-123}$$

where the subscripts on $c_{Cw,\infty}$ indicate the substrate concentrations entering (in) and leaving (out) the single stage reactor, Q is the flow rate into the reactor of liquid volume V, and a_b is the specific surface area of the biofilm. For a desired conversion of $c_{Cw,\infty}$ at a given a_b and Q or V, we would equate Eq. 9-123 and an expression for the substrate flux into the biofilm, and solve for the remaining unknown. Solutions for the flux such into the biofilm are in dimensionless form in Table 9.1. Converting Eq. 9-123 into dimensionless form, we obtain

$$s_{\infty|in}^* - s_{\infty|out}^* = J^*\tau^* \tag{9-124}$$

where $s_{\infty|in}$ and $s_{\infty|out}$ are the dimensionless substrate concentrations leaving and entering the reactor, J^* is the dimensionless flux into the biofilm, and τ^* is the dimensionless detention time defined as

$$\tau^* = \tau a_b D_{Cw} \left(\frac{\mu_{max}X}{Y_{X/S}D_{Cb}K_m} \right)^{1/2} \tag{9-125}$$

The solution of chemical flux into the RBC biofilm must be obtained by a trial and error approach using Eq. 9-124 and an expression for the chemical flux. As an alternative to this approach, Suidan provided a series of nomographs based on the dimensionless parameters developed above for the case of a deep biofilm (Fig. 9.5). To these nomographs you first calculate the dimensionless thickness of the stagnant film, δ_s^*, which determines which line you will use to relate the bulk effluent substrate concentration in the reactor, s_∞^*, to the flux, expressed as the ratio of J^*/δ_s^*. Typically, you then choose the reactor effluent substrate concentration and find the

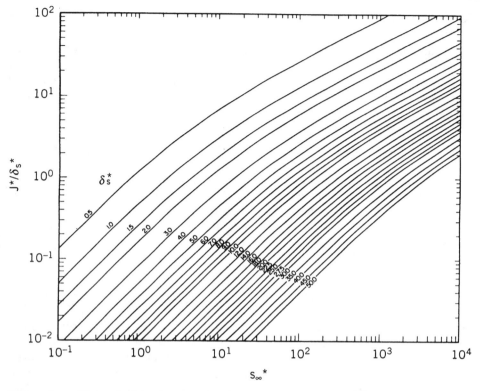

Figure 9.5 Dimensionless flux as a function of dimensionless substrate concentration and stagnant film thickness for a deep biofilm (from Suidan, 1986 by permission of ASCE).

flux from the nomograph, and then calculate the dimensionless detention time using Eq. 9-124. The final calculation of the detention time in physical space is calculated from Eq. 9-125 based on the definition of the dimensionless term.

In some instances, for example, when the removal rate is very slow relative to the detention time, the thickness of the stagnant film layer will be small enough to be ignored. For this case the calculation of the chemical flux can be calculated using the nomograph in Figure 9.6 (Suidan, 1986).

Example 9.6

Repeat Example 9.5 to calculate the chemical flux into a biofilm using the nomograph in Figure 9.5.

To calculate the flux, we will need to enter the nomograph using the following values calculated in the previous example: $\delta_s^* = 2.48$ and $s_\infty^* = 10$. Using Figure

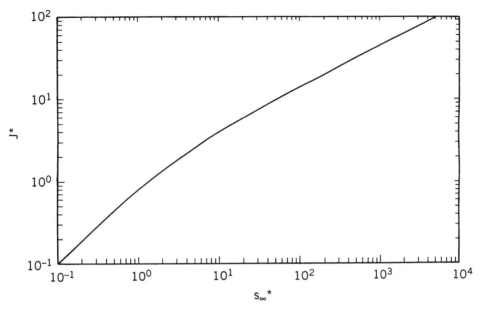

Figure 9.6 Dimensionless flux as a function of dimensionless substrate concentration for a biofilm with a stagnant film thickness equal to zero (adapted from Suidan, 1986 by permission of ASCE).

9.5, we obtain $J^*/\delta_s^* = 0.8$. Thus

$$J^* = 0.8\delta_s^* = 0.8 \times 2.48 = 1.98 \qquad (9\text{-}126)$$

This result is within 7% of the value of $J^* = 1.86$ calculated in Example 9.5 for the case where we *assumed* that the reaction could be considered zero order. We can see that our assumption of zero-order kinetics did not tremendously affect the accuracy of the calculation relative to the solution obtained in Eq. 9-126 for the given parameters and no assumption on the order of the reaction. Both of the solutions in Examples 9.5 and 9.6 do differ from the exact solution for first-order kinetics given in Example 9.3.

Example 9.7

Specify the flow rate into an RBC given a wastewater with a BOD = 500 mg/L so that the BOD is reduced to 30 mg/L. Use the following RBC design parameters and the physical constants: rotation speed, ω = 1.5 rpm; RBC disk radii, R_o = 370 cm and R_i = 15 cm; RBC total surface area, A_b = 18,000 m²; μ_{max} = 3 d⁻¹, $Y_{X/c}$ = 0.55 g/g, K_m = 20 mg L⁻¹, X = 40,000 mg L⁻¹, δ_b = 2000 μm, D_{Cw} = 400 × 10⁻⁸ cm² s⁻¹, D_{Cb} = 200 × 10⁻⁸ cm² s⁻¹, fluid properties of water at 20°C.

From Eq. 9-23, a typical film thickness for an RBC is

$$\delta_s = 0.00135\omega^{2/3}(R_i + R_o)^{2/3}$$

$$= 0.00135\left(\frac{1.5 \text{ min}}{\text{min}}\frac{\text{min}}{60 \text{ s}}\right)^{2/3}(270 + 15 \text{ cm})^{2/3}$$

$$= 0.005 \text{ cm}$$

$$= 50 \text{ μm} \tag{9-127}$$

We need to calculate the dimensionless parameters δ_s^* and s_∞^*

$$\delta_s^* = \delta_s\left(\frac{\mu_{max}X}{Y_{X/S}D_{Cb}K_m}\right)^{1/2}\left(\frac{D_{Cw}}{D_{Cb}}\right) \tag{9-128}$$

$$= (0.005 \text{ cm})\left(\frac{(0.3 \text{ d}^{-1})(d/86,400 \text{ s})(40,000 \text{ mg L}^{-1})}{(0.55)(200 \times 10^{-8} \text{ cm}^2 \text{ s}^{-1})(20 \text{ mg L}^{-1})}\right)^{1/2}$$

$$\cdot\left(\frac{400 \times 10^{-8}}{200 \times 10^{-8}}\right) = 0.79 \tag{9-129}$$

$$s_\infty^*|_{out} = \frac{c_{Cw,\infty}}{K_m} = \frac{30 \text{ mg L}^{-1}}{20 \text{ mg L}^{-1}} = 1.5 \tag{9-130}$$

At this point we do not know if the biofilm can be considered shallow or deep; so we will use Eq. 9-100 for a biofilm of any depth. The dimensionless flux J^* can therefore be calculated using

$$\frac{J^{*2}}{2} = s_\infty^* - J^*\delta_s^* - \ln(1 + s_\infty^* - J^*\delta_s^*) \tag{9-131}$$

$$0.5 = \frac{25 - 0.79J^* - \ln(26 - 0.79J^*)}{J^{*2}} \tag{9-132}$$

By trial and error J^* is calculated to be 5.88. From Eq. 9-124, we can specify the dimensionless detention time for a removal of 500 to 30 mg/L, $s_\infty^*|_{in} = 25$ and $s_\infty^*|_{out} = 1.5$, as

$$\tau^* = \frac{s_\infty^*|_{in} - s_\infty^*|_{out}}{J^*} = \frac{(25 - 1.5)}{5.88} = 4.0 \tag{9-133}$$

Converting the detention time into the dimensional domain

$$\tau = \frac{\tau^*}{a_b D_{Cw}}\left(\frac{Y_{X/S}D_{Cb}K_m}{\mu_{max}X}\right)^{1/2} \tag{9-134}$$

From the definition of $\tau = V/Q$, and $a_b = A_b/V$, where A_b is the total surface area of the RBC media (assumed to be completely covered by biofilm), we have

$$Q = \frac{A_b D_{Cw}}{\tau} \left(\frac{\mu_{max} X}{Y_{X/S} D_{Cb} K_m} \right)^{1/2} \tag{9-135}$$

$$= \frac{(18{,}000 \text{ m}^2)(200 \times 10^{-8} \text{ cm}^2 \text{ s}^{-1})}{4.0}$$

$$\cdot \left(\frac{(0.3 \text{ d}^{-1})(d/86{,}400 \text{ s})(40{,}000 \text{ mg L}^{-1})}{(0.55)(200 \times 10^{-8} \text{ cm}^2 \text{ s}^{-1})(20 \text{ mg L}^{-1})} \right)^{1/2} \frac{86{,}400 \text{ s/d}}{100 \text{ cm/m}} \tag{9-136}$$

$$= 622 \text{ m}^3 \text{ d}^{-1} \tag{9-137}$$

This flow rate will produce the desired effluent substrate concentration using a single-stage RBC. If higher wastewater flow rates must be treated, additional units will need to be added to this one. The maximum recommended loading for an RBC unit is 0.02 kg m^{-2} d^{-1} (Grady and Lim, 1980). According to the calculations above, the loading for this RBC will be

$$R = \frac{Q}{A_b} \left(s^*_{\infty|in} - s^*_{\infty|out} \right)$$

$$= \frac{622 \text{ m}^3 \text{ d}^{-1}}{18{,}000 \text{ m}^2} \frac{(500 - 30) \text{ mg}}{L} \frac{10^3 \text{ L}}{m^3} \frac{10^{-6} \text{ kg}}{mg} = 0.016 \frac{kg}{m^2 d} \tag{9-138}$$

which is less than the maximum loading rate.

Plug Flow Biofilm Systems: Packed Bed Bioreactors

Completely saturated packed beds of media are not commonly used for wastewater treatment of domestic wastewater, although packed beds are used for specialized types of treatment systems such as anaerobic packed beds and fluidized beds. Biofilms develop on most filtration beds used for water treatment and even on granular activated carbon (GAC) beds used to adsorb chemicals (primarily chlorinated and aromatic compounds) in water treatment plants. The kinetics of compound degradation are also frequently studied in the laboratory using packed beds since cultures can be retained on media. Trickling filters are unsaturated packed beds that are commonly used for wastewater treatment. In older systems rock media were used to fill cylindrical beds reaching 2 or 3 m in height. However, many of the mass transport correlations derived above and used in this section are not applicable to calculations around unsaturated media, as will be shown in the section on wetted wall bioreactors.

Microprobe studies have been used to demonstrate the presence of concentration boundary layers around biofilms in packed beds. The first such studies in the wastewater treatment area were pioneered by Bungay and co-workers (Whalen et al., 1969). Oxygen is only sparingly soluble in water. Since oxygen must diffuse from air into water, oxygen profiles often develop around biofilms even at very low substrate concentrations. An example of an oxygen concentration profile is shown in Figure 9.7 for oxygen measured above a biofilm on a rock meant to simulate biofilms in a stream. The inflection in the oxygen profile was interpreted by Whalen et al. (1969) to represent the surface of the biofilm. However, subsequent measurements by Lewandowski et al. (1990) have demonstrated that the biofilm is actually a much smaller part of the oxygen profile. That is, oxygen is rapidly depleted within the biofilm, leaving most of the concentration boundary layer outside the biofilm. Microprobe studies have now been conducted on a variety of compounds above and within biofilms, including glucose, pH, chlorine, and others (Larson and Harremoës, 1994; de Beer et al., 1994; Lens et al., 1995).

Several approaches can be used to model substrate removal in a packed bed. From a mass balance through a packed bed reactor, the constituent transport equation in one dimension is

$$\theta \frac{\partial c_{Cw}}{\partial t} = -u_x \frac{\partial c_{Cw}}{\partial x} + E_L \frac{\partial^2 c_{Cw}}{\partial x^2} - a_b J_b - \frac{\theta \mu_{max} X c_{Cw}}{Y_{X/C}(K_m + c_{Cw})} \tag{9-139}$$

where the term on the left-hand side indicates accumulation, and the right-hand terms represent advection, hydrodynamic dispersion, chemical flux into the biofilm, and uptake by suspended cells. If the packing is assumed to consist of spherical particles, the specific surface area, a_b, can be calculated from the media packing size using $a_b = 6(1 - \theta)/d_c$, where θ is the bed porosity and d_c the diameter of the packing media.

The solution of Eq. 9-139 requires an estimate of hydrodynamic dispersion in the packed bed reactor. Hydrodynamic dispersion can either be measured using suitable tracers or can be estimated from empirical equation such as the Hiby (Bear, 1972) relationship

$$\frac{E_L}{D_{Cw}} = 0.67 + \frac{0.65\ Pe}{(1 + 6.7\ Pe^{-1/2})}, \quad 10^{-2} \le Pe \le 10^2 \tag{9-140}$$

Figure 9.7 Oxygen microprobe analyses of a biofilm. (A) The original study by Whalen et al. (1969) placed the biofilm surface at the inflection point of the oxygen gradient. (Reprinted with permission, ©1969, American Chemical Society). (B) Later studies by Lewandowski et al. (1990) demonstrated, using optical density, that much of the oxygen gradient around the biofilm was contained in the fluid above the biofilm (i.e. the boundary layer) and that oxygen penetrated only a short distance into the biofilm; here, oxygen is reduced to 1.2 mg/L at the biofilm–water interface located 0.7 mm above the substratum (from Lewandowski et al., 1990, *Biotechnol. Bioeng.*, ©1990. Reprinted by permission of John Wiley & Sons, Inc.).

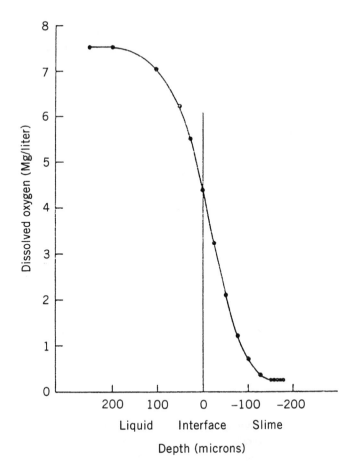

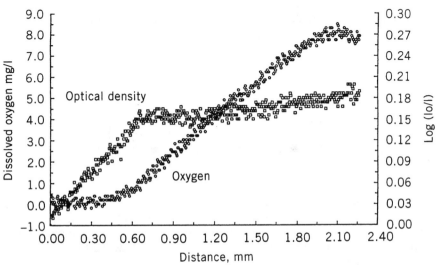

where the Peclet number is based on the chemical diffusivity and media packing diameter, or Pe $= u_w d_c / D_{Cw}$. An alternative correlation at higher Pe is

$$\frac{E_L}{D_{Cw}} = 8.8\ \mathrm{Pe}^{1.17}, \quad \mathrm{Pe} > 0.5 \tag{9-141}$$

There is no analytical solution to Eq. 9-139, and so various techniques can be used to simplify and solve this equation numerically.

Crank–Nicholson Method   One common approach to solving Eq. 9-139 is to use the Crank–Nicholson finite-difference equation. Defining $c_{Cw} = S$ as the concentration of biodegradable substrate in a reactor, Eq. 9-139 in finite-difference form becomes

$$\left(\frac{S(x,t+\Delta t) - S(x,t)}{\Delta t} \right)$$

$$= -u_x \left(\frac{S(x+\Delta x,t) + S(x+\Delta x,t+\Delta t) - S(x-\Delta x,t) - S(x-\Delta x,t+\Delta t)}{4\Delta x} \right)$$

$$+ E_L \left(\frac{S(x+\Delta x,t) + S(x+\Delta x,t+\Delta t) - 2S(x,t+\Delta t) - 2S(x,t) + S(x-\Delta x,t) + S(x-\Delta x,t+\Delta t)}{2\Delta x^2} \right)$$

$$- a_b J_b(x,t) - \frac{\theta \mu_{\max} X S(x,t)}{Y_{X/C}[K_m + S(x,t)]} \tag{9-142}$$

where $S(x,t)$ is the substrate concentration at the central space step (Fig. 9.8) and the old time step, Δt is a time increment, and Δx is a space increment. The advantage of the Crank–Nicholson method is that it is unconditionally stable and has an accuracy of order 2 in time and space. This equation must be solved implicitly by some method such as Gaussian elimination since values of S at time $t^+ \Delta t$ appear on both the left- and right-hand sides of Eq. 9-142. Boundary conditions are

$$u_x S(0,t) = u_x S(\Delta x,t) + E_L \frac{S(\Delta x,t) - S(0,t)}{\Delta x} \tag{9-143}$$

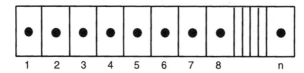

Figure 9.8   A reactor can be simulated using many different techniques. In the Crank–Nicholson and finite–difference approaches, the reactor is divided into *n* elements, each of thickness Δ*x*. Another approach is to model the system as a series of *n* completely mixed reactors, with the number of reactors defined by the extent of dispersion in the column.

at the inlet, and at the outlet of the packed bed reactor

$$S(x,t) = S(x + \Delta x,t) \qquad (9\text{-}144)$$

Finite-Difference Method   A more direct method of solving Eq. 9-139 is to use an explicit finite-difference approach. For this method, Eq. 9-139 becomes

$$\left(\frac{S(x,t + \Delta t) - S(x,t)}{\Delta t}\right) = -u_x \left(\frac{S(x + \Delta x,t) - S(x,t)}{\Delta x}\right)$$

$$+ E_L \left(\frac{S(x + \Delta x,t) + S(x - \Delta x,t) - 2S(x,t)}{\Delta x^2}\right)$$

$$- a_b J_b(x,t) - \frac{\theta \mu_{max} X S(x,t)}{Y_{X/C}[K_m + S(x,t)]} \qquad (9\text{-}145)$$

The boundary conditions for this finite-difference approach are no dispersion through the column entrance and exit. The advantage of this method is that Eq. 9-145 is explicit with respect to time so that only one iteration is necessary to solve for the substrate concentration at each time step. Unfortunately, this forward difference explicit method is not always stable.

In order for Eq. 9-145 to be stable, mass cannot be transferred out of an element faster than it is transported into that element. This is easiest to see if we first analyze the case for just dispersion (no reaction or advection). For the case of dispersion only, Eq. 9-143 becomes after some rearrangement of terms

$$S(x,t + \Delta t) = \frac{E_L \Delta t}{\Delta x^2} [S(x + \Delta x,t) + S(x - \Delta x,t)] + \left(1 - \frac{2E_L \Delta t}{\Delta x^2}\right) S(x,t) \quad (9\text{-}146)$$

Let us define Γ as

$$\Gamma = \frac{2E_L \Delta t}{\Delta x^2} \qquad (9\text{-}147)$$

This equation is stable only if $\Gamma \leq 1$. If we choose the time and distance intervals so that $\Gamma = 1$, the right-hand term will drop out of the equation and S at successive time intervals will just be the arithmetic average of the two previous time intervals.

Now let us return to the full equation. With advection and reaction, we similarly have

$$\Gamma = \frac{u_x \Delta t}{\Delta x} + \frac{2E_L \Delta t}{\Delta x^2} + \frac{a_b J_b(x,t)}{S(x,t)} + \frac{\theta \mu_{max} X}{Y_{X/C}[K_m + S(x,t)]} \qquad (9\text{-}148)$$

so that our simulations are stable with $\Gamma = 1$. With all the terms, the stability criterion is obviously much more complex. Unless the flux, J_b, can be made an explicit function of $S(x,t)$, it will be impossible to choose time and space intervals to that Γ is always equal to 1. Instead, we must make a conservative estimate of S, such as S is less than the influent concentration of S, and then choose Δx and Δt so that $\Gamma < 1$.

Complete Mix Approximation A packed bed can also be analyzed as a series of completely mixed reactors in series. In order to model dispersion, we choose the number of reactors that will produce the desired dispersion coefficient. The approach is to match the first and second moments of the residence time distribution function of flow through a completely mixed reactor. The first moment for n reactors in series is

$$\sigma = \tau \tag{9-149}$$

indicating the mean of the function is the average detention time, while the second moment for n reactors is

$$\sigma^2 = \frac{\tau^2}{n} \tag{9-150}$$

For a plug flow reactor, we can combine the reaction terms in Eq. 9-139 to produce

$$\theta \frac{\partial c_{Cw}}{\partial t} = -u_x \frac{\partial c_{Cw}}{\partial x} + E_L \frac{\partial^2 c_{Cw}}{\partial x^2} - R_C \tag{9-151}$$

For this equation we can use the three boundary conditions

$$t = 0, \quad c_{Cw} = 0$$

$$x = 0, \quad \frac{dc_{Cw}}{dx} = \frac{u_x}{E_L} c_{Cw,in}$$

$$x = L, \quad \frac{dc_{Cw}}{dx} = 0 \tag{9-152}$$

Solving, the first moment is

$$\sigma = \tau \tag{9-153}$$

which is the same as for the complete mix reactor, and the second moment is

$$\sigma^2 = \frac{2\tau^2}{Pe_L} \left(1 + \frac{(e^{-Pe_L} - 1)}{Pe_L} \right) \tag{9-154}$$

where Pe_L is the dispersion number defined as $Pe_L = u_x L/E_L$. Many texts will indicate Pe_L^{-1} as the dispersion number, d_H, or $Pe_L = 1/d_H$. When both the first and second moments are equal, we can say that the reactors are identical. Equating the two second moments, we have

$$\frac{1}{n} = \frac{2}{\text{Pe}_L}\left(1 + \frac{(e^{-\text{Pe}_L} - 1)}{\text{Pe}_L}\right)$$ (9-155)

This equation will be valid when the number of completely mixed reactors is large. The greatest disparity between the two types of reactors will occur when $2 \leq n \leq 5$. For $n \geq 12$, there will be no difference between the models. As a result of this analysis, we can simulate a plug flow reactor of length L and dispersion E_L as n completely mixed reactors in series with an overall detention time of θ.

Nomograph Method in the Absence of Dispersion   In order to derive nomographs describing chemical reaction in a packed bed, Suidan (1986) neglected uptake of chemical by suspended microbes (since without recycle they would contribute little to the overall chemical uptake) and chemical dispersion in the reactor. For these conditions, Eq. 9-139 becomes an ordinary differential equation

$$u_x \frac{dc_{Cw}}{dx} = -a_b J_b$$ (9-156)

which can also be expressed in terms of the flow rate, Q, into the reactor as

$$Q \frac{dc_{Cw}}{dx} = -a_b A_b J_b$$ (9-157)

where A_b is the total surface area of the packed bed media, and a_b is the specific surface area ($a_b = A_b/V$, where V is the reactor empty bed volume). Integrating Eq. 9-157 over the detention time in the bed, and converting to the nondimensional domain

$$\tau^* = \int_{s^*_{\infty,\text{out}}}^{s^*_{\infty,\text{in}}} \frac{ds^*_\infty}{J^*}$$ (9-158)

The flux into the biofilm is not constant in the plug flow reactor, and so Eq. 9-158 must be solved using a numerical technique. However, for the case of 99% removal of s^*_∞ in a deep-biofilm plug flow reactor, Suidan (1986) used the method of successive symmetric quadratures technique by Robinson and Healey (1984) to solve Eq. 9-158 and prepare a series of nomographs such as the one shown in Figures 9.9 and 9.10. The detention time of a reactor is calculated as the difference of the values of τ^* evaluated at the dimensionless influent and effluent substrate concentrations using the nomograph, or

$$\tau^* = \tau^*\big|_{s^*_{\infty,\text{in}}} - \tau^*\big|_{s^*_{\infty,\text{out}}}$$ (9-159)

The final value of the detention time is made by converting the dimensionless value into the physical domain.

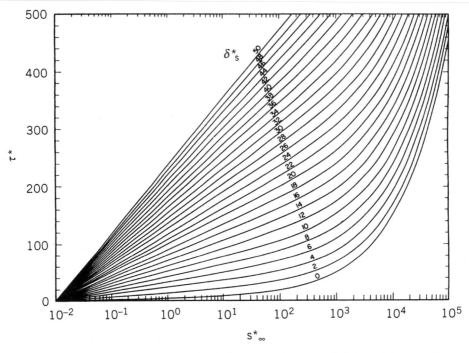

Figure 9.9 Dimensionless empty bed detention time needed to achieve a dimensionless substrate concentration of 0.01 as a function of the dimensionless film thickness for a deep biofilm (adapted from Suidan, 1986 by permission of ASCE).

Example 9.8

Specify the detention time (days) of an anaerobic packed bed reactor that can reduce a wastewater from 10,000 to 100 mg/L of COD. Assume the reactor and reaction conditions: $\mu_{max}/Y_{X/C} = 0.4$ d^{-1}, $K_m = 50$ mg L^{-1}, $X = 40,000$ mg L^{-1}, $\delta_b = 1000$ μm, $D_{Cw} = D_{Cb} = 800 \times 10^{-8}$ cm^2 s^{-1}, particle diameter $d_c = 1$ cm, $\theta = 0.38$, $u_x = 0.2$ cm s^{-1}, and fluid properties of water at 20°C.

We must first calculate the stagnant film thickness, δ_s. From Eq. 9-11, we have

$$\delta_s = \frac{d_c \theta}{1.09} Re^{-1/3} Sc^{-1/3} \tag{9-160}$$

The two dimensionless numbers Re and Sc are calculated from the given information as

$$Re = \frac{u_x d_c}{\nu_w} = \frac{(0.2 \text{ cm s}^{-1})(1 \text{ cm})}{0.01 \text{ cm}^2 \text{ s}^{-1}} = 20 \tag{9-161}$$

$$Sc = \frac{\nu_w}{D_{Cw}} = \frac{0.01 \text{ cm}^2 \text{ s}^{-1}}{800 \times 10^{-8} \text{ cm}^2 \text{ s}^{-1}} = 1250 \tag{9-162}$$

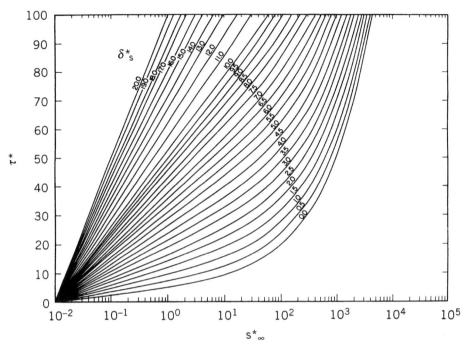

Figure 9.10 Dimensionless empty bed detention time needed to achieve a dimensionless substrate concentration of 0.01 as a function of the dimensionless film thickness for a deep biofilm (adapted from Suidan, 1986 by permission of ASCE).

Substituting these into our expression for the film thickness, we have

$$\delta_s = \frac{(1 \text{ cm})(0.38)}{1.09} (20)^{-1/3} (1250)^{-1/3} = 0.012 \text{ cm} \qquad (9\text{-}163)$$

Converting to a dimensionless distance

$$\delta_s^* = \delta_s \left(\frac{\mu_{\max} X}{Y_{X/S} D_{Cb} K_m} \right)^{1/2} \left(\frac{D_{Cw}}{D_{Cb}} \right) = \delta_s \left(\frac{\mu_{\max} X}{Y_{X/S} D_{Cb} K_m} \right)^{1/2} \qquad (9\text{-}164)$$

$$= (0.012 \text{ cm}) \left(\frac{(0.4 \text{ d}^{-1})(\text{d}/86{,}400 \text{ s})(40{,}000 \text{ mg L}^{-1})}{(800 \times 10^{-8} \text{ cm}^2 \text{ s}^{-1})(50 \text{ mg L}^{-1})} \right)^{1/2} = 0.26 \quad (9\text{-}165)$$

From the nomograph in Figure 9.10 we find the dimensionless detention times for an influent substrate concentration of $s_{\infty,\text{in}}^* = 10{,}000/50 = 200$ is $\tau^* = 21$, and for the effluent of $s_{\infty,\text{out}}^* = 100/50 = 2$ is $\tau^* = 16$. Using the difference between these values, the dimensionless detention time is $\tau^* = 5$. Converting to the dimensional domain

$$\tau = \frac{\tau^*}{a_b D_{Cw}} \left(\frac{Y_{X/S} D_{Cb} K_m}{\mu_{max} X} \right)^{1/2} \qquad (9\text{-}166)$$

The specific surface area, a_b, is calculated as

$$a_b = \frac{6(1 - \theta)}{d_c} = \frac{6(1 - 0.38)}{(1 \text{ cm})} = 3.72 \text{ cm}^{-1} \qquad (9\text{-}167)$$

Using this result in Eq. 9-166 produces

$$\tau = \frac{5}{(3.72 \text{ cm})} \left(\frac{(50 \text{ mg/L})(d/86{,}400 \text{ s})}{(0.4 \text{ d}^{-1})(800 \times 10^{-8} \text{ cm}^2 \text{ s}^{-1})(40{,}000 \text{ mg/L})} \right)^{1/2} = 0.09 \text{ d}$$
$$(9\text{-}168)$$

or about 2 h.

Thin Fluid Falling Films Covering Biofilms (Plastic Media Trickling Filters)

Modeling substrate removal by biofilms in trickling filters presents unique challenges compared to other bioreactor processes due to the complexity of chemical transport within the thin fluid film. We have already seen in Section 9.2 that the velocity profile in the fluid film is parabolic in nature and therefore cannot be considered to be plug flow. The relatively small thickness of the fluid film generally prohibits the simplifying assumptions used in other reactor models. For example, typical fluid film thicknesses over biofilms in plastic media trickling filters are 100–200 μm, which are on the order of the thicknesses of the "stagnant films" previously considered for biofilms in other types of reactors (packed beds and RBCs). If we were to apply the stagnant film concept to a trickling filter, there would be no room left for the water to flow over the biofilm (Fig. 9.11). Therefore, while stagnant film models may be used in cases where the fluid covering the biofilm is relatively deep compared to a stagnant film thickness, such models are inappropriate for trickling filters, where the fluid film is bounded by an air interface. In this section we will describe mass transport and substrate removal in plastic media trickling filters as conceptualized and modeled by Logan et al. (1987a,b).

Plastic Media Geometry   Trickling filters can be composed of several different types of media, including rocks, random plastic media, and structured plastic modules. Most new trickling filters are constructed with structured modules, and so other types of media (such as rocks) will not be further considered here. In order to model substrate removal in trickling filters, the transport equations derived in Section 9.2 will be used to solve for chemical transport in the liquid film. Wastewater flow over

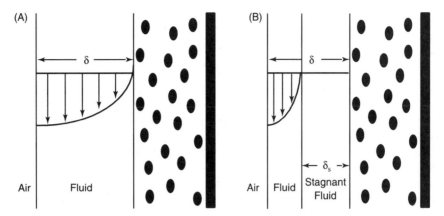

Figure 9.11 It is doubtful that a stagnant fluid film exists outside a biofilm in a trickling filter. (A) Gravity acts on all the fluid covering a biofilm, where the thickness of the fluid film can be calculated from a force balance. (B) If a stagnant film existed, this fluid would somehow have to resist the pull of gravity. The moving film could either be compressed into a thinner, but faster, moving film than expected, or the film would maintain the same thickness but somehow resting on fluid that would resist fluid shear imposed by the falling water.

the biofilm growing on the plastic media will be considered to be fully developed laminar flow at all points except where flow streams intersect or when flow leaves the end of a module. As shown in Figure 9.12, flow down a sheet of plastic media runs into adjacent sheets; the confluence of flow is considered to result in some fluid mixing (Logan et al., 1987a). Although such mixing is likely to be imperfect, without

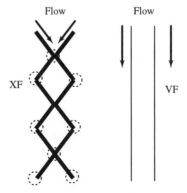

Figure 9.12 Examples of fluid films (arrows) falling down plates in cross-flow (XF) and vertical (VF) media. In XF media fluid streams intersect, generating fluid mixing at the points shown by dashed circles, while in VF media the fluid streams are separate.

additional information to specify exactly how the fluid might be mixed at every point in the module, the mixing is assumed to be complete.

Plastic modules are usually built in sizes of 0.6×1.2 m^2 and 0.6 m deep and constructed of many sheets, each bent in different ways to provide air space between adjacent sheets, and the sheets are welded together using heat or glue. Each section of uninterrupted flow within a plastic module is defined as a plate, so that each sheet may consist of a number of plates. In cross flow (XF) media, fluid will run across several plates, being mixed each time, before leaving the module. In vertical flow (VF) media, the fluid flows only along a single plate before reaching the end of the module. Variations in the number of sheets and the way the sheets are bent and welded produce a wide range of media types and surface areas (Table 9.3). For example, XF media made by manufacturer designated as "a" have plates 8.6 cm long and 5.1 cm wide, resulting in flow crossing 8 plates to reach the end of the module. Each module of XFa media has a surface area of 89 m^2 per m^3 of module volume, and consists of a total of 1560 plates in each 0.6 m of module depth per 1 m^2 of cross-sectional area.

Wastewater Composition   In most biodegradation studies, a relatively large effort is directed towards determining kinetic constants for the biodegradation of chemicals by microorganisms, but less attention is given to the sizes of molecules. As will be shown below, however, the rate of chemical diffusion into a biofilm can be just as important (and sometimes more important) than the degradation kinetics. If biofilms are "deep," and all chemicals that diffuse into the biofilm are removed, then the degradation kinetics are relatively unimportant relative to how fast the chemicals can diffuse through the liquid into the biofilm.

It was demonstrated in Chapter 3 that organic matter in water and wastewater consists of a variety of molecules with different sizes and diffusivities. In order to reduce the size distribution of molecules in wastewater to a reasonable size, Logan et al. (1987a) considered wastewater to be composed of molecules in five size ranges based on molecular size distributions by Levine et al. (1985). As shown in Table

TABLE 9.3   Plastic Trickling Filter Media Geometries (XF, cross flow; VF, vertical flow)

Media Code	Surface Area (m^2/m^3)	Plate Width (cm)	Plate Length (cm)	Plate Angle	Mixing Points	Plates (m^{-2})
XFa-89	89	5.1	8.6	60	8	1560
XFa-98	98	4.3	7.5	60	9	1960
XFa-138	138	2.8	5.0	60	14	4270
VFb-89	89	6.1	62	90	1	1410
VFb-100	100	7.0	63	90	1	1400
VFc-92	92	10.0	66	90	1	670
VFd-92	92	4.8	69	90	1	1620

Source: Logan et al., 1987a.

TABLE 9.4   Average Diffusivities of Organic Matter Having a Range of Molecular Weights

Molecular Weight Range ($M_c \times 10^{-3}$, daltons)	Average Diffusivity ($D_{Cw} \times 10^8$, cm^2 s^{-1})
3–30	112
30–50	85
50–100	65
100–500	50
500–1000	30

9.4, this characterization results in diffusivities ranging from 30 to 112 $\times$ 10^{-8} cm^2 s^{-1}.

Biofilm Model for a Multicomponent Wastewater in a Trickling Filter   In order to choose a biofilm model to include in a trickling filter model, Logan et al. (1987a) made several assumptions about substrate concentrations and trickling filter operation. Trickling filters used to treat domestic wastewater develop thick biofilms, on the order of 1 mm, that can be considered "deep" relative to substrate concentrations. Therefore, only those biofilm models that assumed the chemical was completely removed in the biofilm were considered. As described above, and in Chapter 3, wastewater consists of a variety of different types of compounds. It is unlikely that one enzyme is sufficient to degrade all organic molecules in the wastewater. Based on this assumption that the wastewater organic components were highly diverse, it was further assumed that the biodegradable organic matter (consisting of the five components identified above) would individually be at substrate concentrations that would be reacted away according to first-order kinetics.

For an incompletely penetrated, or deep, biofilm reacting according to first-order kinetics, the chemical flux into the biofilm is

$$J_b = (k_1 D_{Cb})^{1/2} c_{Cb,0} \tag{9-169}$$

Applying this model to a five-component wastewater presents a problem since this would require determining a kinetic constant, k_1, for each of the five components. There were few data available for making such an estimation, so Logan et al. (1987a) used a different approach.

If substrate removal in a biofilm is limited by mass transport to the biofilm, and microbial uptake kinetics are first order, then it was hypothesized that substrate uptake would likely occur at a rate in proportion to the rate molecules of substrate strike and combine with enzymes on the surface of the cell. Instead of viewing substrate uptake according to an enzyme–substrate model, it was instead viewed as a type of coagulation process (Fig. 9.13). Substrate molecules of radius R_S at a concentration of N_S are present in a system of cells of radius R_c at a concentration of N_c in a biofilm. According to coagulation theory (see Chapter 14), the change in

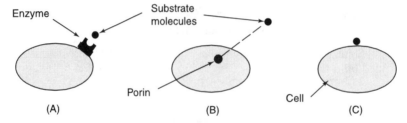

Figure 9.13 Different models of whole cell substrate uptake. (A) The conventional model is that uptake is governed by enzyme kinetics due to an enzyme binding with a specific substrate. (B) However, many molecules diffuse directly through the cell membrane, of if larger, must diffuse through pores in the membrane. (C) A more general description of whole cell uptake is to describe uptake in terms of the efficiency of attachment between the cell and a substrate molecule. By using this coagulation model, we do not need to know the mechanism of substrate uptake; instead, the rate is described as a function of the chemical's diffusivity and the efficiency of the collision. Only first-order kinetics can be described by the coagulation model.

the concentration of substrate molecules with time due to coagulation is

$$\frac{dN_S}{dt} = -\alpha\beta N_c N_S \tag{9-170}$$

where α is a sticking coefficient, defined as the rate particles strike each other to the rate that they stick, and β is a collision function describing the frequency of collisions by diffusion according to

$$\beta = 4\pi(R_c + R_S)(D_{cb} + D_{Sb}) \tag{9-171}$$

where D_{cb} is the diffusion coefficient for cells in a biofilm. Cells cannot appreciably diffuse in a biofilm, and therefore because $D_{Sb} \gg D_{cb}$, $(D_{cb} + D_{Sb}) \approx D_{Sb}$. Also, $R_c \gg R_S$, and Eq. 9-171 can be simplified to

$$\beta \approx 4\pi R_c D_{Sb} \tag{9-172}$$

Combining Eqs. 9-170 and 9-172

$$\frac{dN_S}{dt} = -4\pi\alpha R_c D_{Sb} N_c N_S \tag{9-173}$$

The concentration of cells in the biofilm can be expressed in terms of the biofilm average porosity, θ, and cell radius as

$$N_c = \frac{\text{occupied volume}}{\text{volume per cell}} = \frac{(1 - \theta)}{\frac{4}{3}\pi R_c^2} \tag{9-174}$$

Incorporating this expression into Eq. 9-173, and switching notation from number concentration, N_S, to mole concentration, we have

$$\frac{dc_{Sw}}{dt} = -\left(\frac{3\alpha D_{Sb}(1-\theta)}{R_c^2}\right) c_{Sb} \tag{9-175}$$

which is just a first-order rate expression for substrate removal in the biofilm, where terms in the large parentheses indicate the first-order rate constant, or

$$k_1 = \frac{3\alpha D_{Sb}(1-\theta)}{R_c^2} \tag{9-176}$$

The first-order rate expression can now be used in our solution for the flux into a biofilm (Eq. 9-170) to produce

$$J_b = k_b^{1/2} D_{Sb} c_{Sb,0} \tag{9-177}$$

where the rate constant for the biofilm, k_b, is defined as

$$k_b = -\frac{3\alpha(1-\theta)}{R_c^2} \tag{9-178}$$

where it is assumed for trickling filter biofilms that $\theta = 0.8$ and $R_c = 1$ μm (Logan et al., 1987a). This result produces an expression for the chemical flux into a biofilm for chemical reactions between cells and substrate molecules when the reaction rate is considered to be very fast. The most important aspect of Eq. 9-173 is that the rate constant in Eq. 9-178 is now only a function of biofilm density. This means that in order to characterize the flux of any number of different chemicals into a biofilm, we only need to know the diffusivities and concentrations of those chemicals. Since we have assumed that all the biodegradable organic matter will be taken up at a rate in proportion to the rate molecules strike cells, we no longer need kinetic data since all reactions are considered to be equally fast once the chemicals have diffused into the biofilm.

Substrate (sBOD) Removal   The components necessary to model substrate removal in a trickling filter have now been assembled and defined. Using the result for the chemical flux into a biofilm derived in Eq. 9-177, chemical transport into the biofilm can be described using Eq. 9-24 and the boundary conditions (Eqs. 9-25 to 9-27) to obtain

$$\frac{\partial c_{Cw}}{\partial t} + u_{\max}\left[1 - \left(\frac{x}{\delta}\right)^2\right]\frac{\partial c_{Cw}}{\partial z} = D_{Cw}\frac{\partial^2 c_{Cw}}{\partial x^2} \tag{9-179}$$

$$BC\ 1:\ z = 0,\quad c_{Cw} = c_{Cw,0}$$

$$BC\ 2:\ x = 0,\quad \frac{dc_{Cw}}{dx} = 0$$

$$BC\ 3:\ x = \delta,\quad J_b = k_b^{1/2} D_{Cb} c_{Cw,0} \tag{9-180}$$

These equations can be used to model substrate transport within the thin fluid film covering a biofilm on a plate so long as the fluid is not interrupted by a mixing point. Each time fluid moves from one plate (or one module) to the next in a trickling filter, the fluid will be completely mixed, and the transport equation must be solved again for the new conditions. The concentration of substrate is defined as the 5-day soluble biodegradable organic matter ($sBOD_s$) of the wastewater.

The application of Eqs. 9-179 and 9-180 to calculate organic matter removal in a trickling filter is not as difficult as it first might appear. It is true that no analytical solution exists for this equation with these boundary conditions, and therefore this system must be solved numerically. The approach used by Logan et al. (1987a) was to simulate the system indicated by Eqs. 9-179 and 9-180 until steady conditions had been obtained. Once a solution is obtained for the system at steady state for a given set of conditions of chemical diffusivities, plate length, hydraulic loading, and so forth, it can be applied to all other plates at different substrate concentrations.

The numerical solution of Eq. 9-179 was obtained using the finite-difference scheme for each component

$$S(x,z,t + \Delta t) = S(x,z,t) - \frac{u_z(x)\ \Delta t}{\Delta z} [S(z + \Delta z,t) - S(z,t)]$$

$$+ \frac{D_{Cw}\ \Delta t}{\Delta x^2} [S(x + \Delta x,t) + S(x - \Delta x,t) - 2S(x,t)] \tag{9-181}$$

where $S_j = c_{Cw}$ is used to represent the concentration of sBOD in each of j components of the sBOD, and $u_z(x)$ is indicated to be a function of the distance in the liquid film. This finite-difference equation is subject to the same limits of stability, as discussed in Section 9.4, but Δx and Δt cannot be chosen to eliminate terms since u_z is not a constant everywhere in the fluid film. Equation 9-181 is programmed in conjunction with the indicated boundary conditions, and a simulation is run until the concentration leaving the plate reaches becomes constant, indicating steady-state conditions. The fraction of sBOD remaining after flow over the plate, f_{jp}, is calculated from the concentration entering and leaving the plate as $f_j = S_{j,out}/S_{j,in}$. The total fraction of sBOD remaining in the trickling filter for each component of the sBOD is then calculated as

$$F_j = f_j^{N_p} \tag{9-182}$$

where N_p is the number of plates from the top to the bottom of the trickling filter, calculated as the product of the number of modules in the trickling filter and the number of mixing points per module. The overall removal of sBOD is calculated as the sum of the removals of each of the $n_c = 5$ components

$$\text{sBOD}_{\text{out}} = \sum_{j=1}^{n_C} F_j P_j \, \text{sBOD}_{\text{in}} \qquad (9\text{-}183)$$

where F_j is the fraction of sBOD remaining, P_j is the fraction of total sBOD in each of the n_c components, and sBOD_{in} is the total sBOD applied to the trickling filter. The solution of these equations was incorporated into a trickling filter model (TRI-FIL), now sometimes referred to as the Logan model (MOP 8, 1991; Parker et al. 1995).

Since the geometry of the media and the diffusivities of the 5-size fractions composing the wastewater are input data into the model, the only calibration parameter is the sticking coefficient, α. Using one set of sBOD removal data for XFa-30 media, α was found to be 0.0035; this result was then compared to a variety of data sets from laboratory, pilot, and field experiments (Logan et al., 1987a,b).

sBOD removal occurs more rapidly in a module of XF media than in VF media as a result of the mixing points and the slightly longer detention time of fluid in the inclined plates in the XF media. During flow over the plate organic matter near the biofilm is depleted, and molecules from the top of the fluid must diffuse across the liquid layer to reach the biofilm. This diffusion process is quite slow compared to fluid mixing. When fluid reaches the end of a plate, or the end of a module, mixing the fluid rapidly brings organic matter close to the biofilm so that it can diffuse into the biofilm and be reacted away. For example, 89% of organic matter with an average diffusivity of 112×10^{-8} cm^2 s^{-1} will be removed in a 10-module tower of XFa-98 media than in a tower of the same height composed of VFb-100 media (Table 9.5). Smaller molecules will be removed faster than larger molecules in trickling filters, since smaller molecules have a higher diffusivity and are reacted away at a faster rate than larger molecules (Table 9.5).

Using the above model, sBOD removal in a trickling filter can be predicted over a wide range of hydraulic loading rates and temperatures for any given media ge-

TABLE 9.5   Total sBOD Removals of Five Components in Wastewater as a Function of Their Diffusion Coefficients for XF and VF Media. Calculation Conditions: Ten Module (6 m) Tower, Hydraulic Loading Rate of 0.68 L m^{-2} s^{-1} With No Recycle, and a Temperature of 20°C

sBOD diffusivity ($D_{Cw} \times 10^8$, cm^2 s^{-1})	Percent Substrate Removal	
	XFa-98	VFb-100
112	89	78
80	81	69
65	76	63
50	68	55
30	54	41
Overall	73	61

ometry. The overall percent of sBOD removal for three different types of plastic media is compared in Figure 9.14. XFa medium produces the highest removals and VFc media the lowest removals for the three different media. Removal rate increases with wastewater temperature (Fig. 9.14). Higher removal rates arise for a number of reasons, but they are mainly predicted in the trickling filter model to arise from

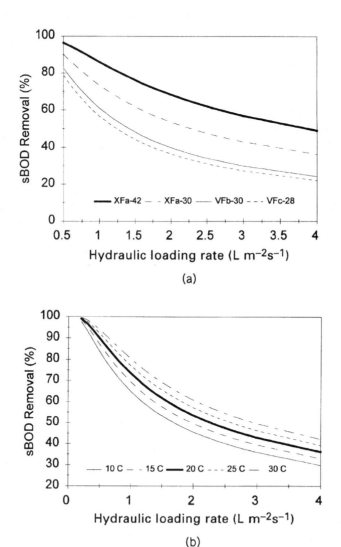

(a)

(b)

Figure 9.14 (A) sBOD removal rates in a trickling filter containing four different types of media. Simulation conditions were: 10-module tower (20 ft height), no recycle, 20°C. (B) sBOD removal in a trickling filter containing XFa-30 media at different wastewater temperatures. Simulation conditions were: 10-module tower (20 ft height), no recycle, 20°C (adapted from Logan et al., 1987b).

higher diffusion coefficients. The chemical diffusivities can be corrected using a standard temperature correction formula (Eq. 3-36). Faster-diffusing chemicals move more quickly through the liquid layer and into the biofilm, and are reacted more quickly in the biofilm as a result of increased collision frequencies with cells.

The Rate-Controlling Mechanism in sBOD Removal in Trickling Filters

The above analysis for sBOD removal has assumed that the rate of sBOD removal is primarily limited by chemical diffusivity and not microbial kinetics. Whether kinetics or mass transport controls chemical removal in a trickling filter can be evaluated using a Thiele modulus, ϕ_b, defined as

$$\phi_b = \left(\frac{\delta_w^2 k_1}{D_{Cb}}\right)^{1/2} \tag{9-184}$$

where δ_w is the thickness of the fluid film.

Swilley and co-workers (Swilley, 1966; Swilley and Atkinson, 1963; Atkinson et al., 1967) solved the transport equation with the indicated boundary conditions for two conditions: $\Phi \gg 1$ and $\Phi \ll 1$. When $\Phi \gg 1$, substrate removal is completely limited by chemical diffusion to the biofilm, and the system is said to be mass transport controlled. Two solutions are possible depending on the magnitude of a dimensionless constant, Λ, defined as

$$\Lambda = \frac{D_{Cw}L}{\delta_w^2 u_{max}} \tag{9-185}$$

where L is the length of uninterrupted flow along a plate. The two solutions for $\Phi \gg 1$ are

$$\frac{c_{Cw,out}}{c_{Cw,in}} = 0.91 e^{-2.82\Lambda}, \quad \Lambda > 0.1 \tag{9-186}$$

$$\frac{c_{Cw,out}}{c_{Cw,in}} = 1 - 1.11\Lambda^{0.656}, \quad \Lambda < 0.1 \tag{9-187}$$

When $\Phi \ll 1$, the system is kinetically controlled, and substrate removal is not affected by mass transport. Under these conditions it can be shown that the substrate removal after flow along a plate of length L is

$$\frac{c_{Cw,out}}{c_{Cw,in}} = \exp\left(-\frac{(k_1 D_{Cw})^{1/2}L}{q_w}\right) \tag{9-188}$$

where q_w is the flow per unit width of the plate.

We can use values from model calibration and operation to calculate typical values for the parameters Λ and Φ. The dimensionless constant Λ is a function of the different hydraulic loading conditions. Assuming high loading rates on XFa-98 media, and for $D_{Cw} = 112 \times 10^{-8}$ cm^2 s^{-1},

$$\Lambda = \frac{D_{Cw}L}{\delta_w^2 u_{\max}} = \frac{(112 \times 10^{-8} \text{ cm}^2 \text{ s}^{-1})(7.5 \text{ cm})}{(0.015 \text{ cm})^2(3 \text{ cm s}^{-1})} = 0.012 \tag{9-189}$$

indicating $\Lambda < 0.1$ for typical simulation conditions.

The Thiele modulus is a dimensionless ratio of diffusion, $\delta/D_{Cw}^{1/2}$, to reaction, $k_1^{1/2}$. Typical fluid thicknesses in trickling filters are 100 to 200 μm. Substituting in our definition of the first-order rate constant, assuming a relatively constant fluid film thickness of $\delta_w = 150$ μm, $R_c = 1$ μm, and $\alpha = 0.0035$, the Thiele modulus is completely defined and no longer a function of simulation conditions and is calculated to be

$$\Phi_b = \left(\frac{3\alpha\delta_w^2(1 - \theta)}{R_c^2} \right)^{1/2} = 6.9 \tag{9-190}$$

This value of $\Phi = 6.9$ is greater than unity, indicating that chemical removal is primarily mass transport controlled, but since Φ does not meet the condition of $\Phi \gg 1$, we cannot use the analytical solutions above, but instead must solve the transport equations on a case by case basis for each combination of hydraulic loading rate, chemical diffusivity, and so forth.

Maximum Substrate Removal in a Plastic Media Trickling Filter It is unlikely that all wastewaters can be described by the five components assumed above for domestic wastewater, and therefore the magnitude of the Thiele modulus could change. The analytical solutions derived by Swilley permit us to hypothesize what the maximum rate of substrate removal could be in a trickling filter in the event that sBOD removal was completely limited by mass transfer into the biofilm.

Using Eq. 9-183, we can calculate the sBOD leaving a trickling filter

$$\text{sBOD}_{\text{out}} = \sum_{j=1}^{n} (0.91e^{-2.82\Lambda_j})^{N_p} P_j \text{ sBOD}_{\text{in}}, \quad \Lambda > 0.1 \tag{9-191}$$

$$\text{sBOD}_{\text{out}} = \sum_{j=1}^{n} (1 - 1.11\Lambda_j^{0.656})^{N_p} P_j \text{ sBOD}_{\text{in}}, \quad \Lambda < 0.1 \tag{9-192}$$

where Λ must be calculated independently for each component j of a total of n_c components in the wastewater.

Kinetically Controlled Substrate Removal in a Plastic Media Trickling Filter The conventional approach to modeling substrate removal by biofilms in a trickling filter is to use the modified Velz (Parker and Merrill, 1984) equation

$$\frac{c_{Cw,\text{out}}}{c_{Cw,\text{in}}} = \left[(R_r + 1)\exp\left(\frac{k_{20}HA_s\theta^{(T-20)}}{[Q_a(R_r + 1)]^n} \right) - R_r \right]^{-1} \tag{9-193}$$

where R_r is the recycle ratio, calculated as the ratio of flow before recycle (Q_{in}) to the recycle flow (Q_r), k_{20} is a kinetic constant at 20°C, H the total media height, A_s the total specific surface area, θ a temperature correction factor usually equal to 1.035, and Q_a the flow before recycle per total cross-sectional surface area of the tower. The modified Velz equation was derived by assuming that the rate of substrate removal in the trickling filter is first order with respect to fluid detention time in the tower. Although it may not be obvious, this assumption implies that substrate removal is kinetically, and not mass transport, limited.

It can be shown that the Velz equation is really a kinetic model by deriving a model from a shell balance that has the same form as Eq. 9-193. In order to make this comparison easier, Eq. 9-193 can first be simplified for the case of no recycle and a temperature of 20°C to

$$\frac{c_{Cw,out}}{c_{Cw,in}} = \exp\left(\frac{-k_{20}HA_s}{Q_a^n}\right) \tag{9-194}$$

In order to derive an expression like this, we need to specify our control volume. Since this is a model of only removal in one dimension (length of flow), we can begin the control volume shown in Figure 9.15. For the control volume with a height of the fluid, δ_w, and a width w above a biofilm, the rates wastewater organics enter and leave the control volume are

$$\text{rate fluid enters: } u_z\delta_w w c_{Cw|z}$$

$$\text{rate fluid leaves: } u_z\delta_w w c_{Cw|z+\Delta z} \tag{9-195}$$

There is no reaction in the control volume, but we know that the chemical flux out of the biofilm occurs continuously along the edge of the fluid. The only way to derive a differential equation of the necessary form is to use the chemical flux out of the control volume as if it were a reaction in the control volume. For first-order biofilm kinetics, the flux into the biofilm is $c_{Cw}(k_1D_{Cw})^{1/2}$, where k_1 is a first-order kinetic constant. Since the flux occurs across the differential area of $w\,\Delta z$, the overall loss of chemical is the rate of removal by the biofilm, which is

$$\text{reaction rate} = c_{Cw}(k_1D_{Cw})^{1/2}w\,\Delta z \tag{9-196}$$

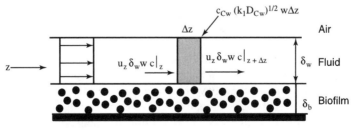

Figure 9.15 The control volume used to construct a modified Velz type of biofilm model for describing substrate removal in a trickling filter.

Combining all terms, and expressing this equation in terms of a limit as $\Delta z \rightarrow 0$, we have

$$\lim_{\Delta z \rightarrow 0} \frac{c_{Cw|z+\Delta z} - c_{Cw|z}}{\Delta z} = \frac{c_{Cw}(k_1 D_{Cw})^{1/2}}{u_z \delta_w} \tag{9-197}$$

Taking this limit produces the differential equation

$$\frac{dc_{Cw}}{dz} = -\left(\frac{(k_1 D_{Cw})^{1/2}}{u_z \delta_w}\right) c_{Cw} \tag{9-198}$$

Integrating over the length of flow, defined as the height of the tower, we obtain

$$\frac{c_{Cw,out}}{c_{Cw,in}} = \exp\left(\frac{-(k_1 D_{Cw})^{1/2} H}{u_z \delta}\right) \tag{9-199}$$

In order to transform Eq. 9-199 into 9-194, we must visualize the geometry of the system in order to relate the parameters u_z and δ_w to the hydraulic loading Q_a and specific surface area A_s. Wastewater applied at a rate Q $[L^3 \ T^{-1}]$ per cross-sectional area A_x $[L^2]$ is applied at a rate per area of Q_a $[L \ T^{-1}] = Q/A_x$ (Fig. 9.16). The specific surface area is the total cross-sectional area, A, of a module of height h, in a module of volume $A_x h$. The specific surface area is therefore related to the cross-sectional area by

$$A_s = \frac{A}{A_x h} = \frac{wh}{A_x h} = \frac{w}{A_x} \tag{9-200}$$

The rate wastewater is applied per unit area can therefore be related to fluid velocity as

$$Q_a = \frac{Q}{A_x} = \frac{QA_s}{w} = \frac{u_z w \delta_w A_s}{w} = u_z \delta_z A_s \tag{9-201}$$

Substituting in Eq. 9-201 into 9-199, we have

$$\frac{c_{Cw,out}}{c_{Cw,in}} = \exp\left(\frac{-(k_1 D_{Aw})^{1/2} H A_s}{Q_a}\right) \tag{9-202}$$

By modeling the flow of thin fluid films over angled surfaces (tennis balls) Howland (1958) demonstrated that the detention time of the fluid was proportional to the fluid flow rate raised to the 0.67 power. As a result, the flow rate Q_a in Eq. 9-202 was raised to the power 0.67. Eventually, studies demonstrated that residence times were not always proportional to $Q_a^{0.67}$, prompting a change to make the power a variable. As a result, the kinetic models generally contain the term Q_a^n, where n is

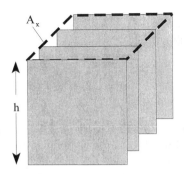

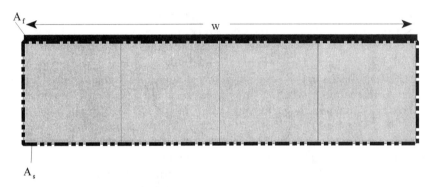

Figure 9.16 Trickling filter modules are composed of several sheets joined together to make a structurally supporting module of height h. (a) Wastewater is applied per unit of cross-sectional area, A_x. (b) Since much of the module consists of void space, the wastewater can be viewed as being applied to a single sheet of total cross-sectional area A_s. Here, that area would be four times the area of each sheet. If the fluid has a thickness of δ, then the flow is effectively applied over an area $A_f = w\delta$.

usually assumed to be $n = 0.5$ (based on experimental data). When Eq. 9-202 is modified to also include the Q_a^n term, the only remaining difference between Eqs. 9-194 and 9-202 is the difference in the rate constants. These two equations are identical when

$$k_{20} = [(k_1 D_{Cw})^{1/2}]^n \qquad (9\text{-}203)$$

The solution provided by Swilley et al. (1967) for kinetically controlled substrate removal by a biofilm, Eq. 9-188, is also identical to Eq. 9-194 derived above. The only apparent difference between these solutions is the method of defining the fluid flow velocity. Fluid applied to the plastic media is assumed to form a fluid film of thickness δ and width w over the top perimeter of the media. Therefore, a flow rate of Q applied to a module of cross section A_x can also be considered to be applied so that it is spread out over the plate to an area of $A_f = w\delta$. From the definition of q_w as the flow per unit width

$$q_w = \frac{Q}{w} = u_z \delta_w \tag{9-204}$$

Using this relationship, we see that Eqs. 9-188 and 9-194 are really the same equation if the flow rate is modeled as Q^n, where n is defined to be 1.

The above analysis has demonstrated that in order for Eq. 9-194 to be derived, it is necessary to assume that the chemical flux from the fluid is not a function of concentration gradients in the fluid, that is, that the flux to a biofilm is not mass transfer limited. The only justification for the flux to be assumed to occur without a gradient in the liquid is to assume either complete mixing or very slow biofilm kinetics. Complete mixing is not possible everywhere, since the fluid flow is laminar. Slow kinetics are not justified, since computer solutions indicate biofilm kinetics must be rapid to account for observed removal rates.

What this all means is that the modified Velz model is incorrect, since it requires conditions (either fluid mixing or slow kinetics) that are not valid for conditions in a trickling filter treating domestic wastewater. There are in fact many known problems with the Velz-type solutions. First, the kinetic "constants" appear to change as a function of media type (Parker and Merrill, 1984), a situation that should not occur if k_{20} were truly a constant. Second, the Velz equation is generally known incorrectly to predict the effect of tower height, H, on overall sBOD removal. This has led others (WEF MOP 8, 1991) to propose adjusting the kinetic constant k_{20} to remove the effect of tower height by the equation

$$k_{20}(H_2) = k_{20}(H_1) \left(\frac{H_1}{H_2}\right)^{1/2} \tag{9-205}$$

where $k_{20}(H_1)$ is the k_{20} value corrected to a standard tower height of $H_1 = 18$ ft. While such corrections help to improve trickling filter design calculations, they do not correct the problem of a fundamentally flawed model.

Oxygen Transport in a Trickling Filter   Oxygen transport into biofilms has been extensively studied using oxygen microprobes (Bungay et al., 1969; Lewandowski et al., 1991), but it is still not conclusively known whether oxygen transport into biofilms in trickling filters limits sBOD removal. Such information is important for trickling filter design. Substrate uptake by biofilms can proceed by microbes under anaerobic conditions if there are alternate electron acceptors available such as nitrate, sulfate, and carbon dioxide. Anaerobic processes are not desirable in trickling filters since volatile organic compounds and H_2S gases generated by the biofilms can be stripped into the air, causing odor problems. Many other gases such as methane can also be produced, but these may be colorless and odorless and are therefore not viewed as an operational problem. Under anaerobic conditions the treated wastewater leaving the trickling filter is of poor quality since it is usually highly turbid, smelly, and black.

In order to avoid odor generation and poor-quality wastewaters and sludges, design engineers limit the mass of sBOD applied to a trickling filter. In a study done by Richards and Reinhard (1984) it was estimated that trickling filters reached maximum sBOD removal levels of ~9 kg sBOD d^{-1} 1000-m^{-2} of XF-media surface area, and ~6.8 kg sBOD d^{-1} 1000-m^{-2} of VF-media surface area (Fig. 9.17). Oxygen-limited sBOD removal was thought to be a major factor. Richards and Reinhard's finding was based on interpolating concentration profiles taken at various depths in a trickling filter. Numerous studies at other sites have not supported the idea of maximum removal rates for wastewater treatment systems treating domestic wastewater (as shown below), but this subject will likely continue to be debated.

Oxygen transport can be modeled in a trickling filter by same approach used to model sBOD removal (Eqs. 9-179 and 9-180). The governing transport equation is

$$u_{max} \left[1 - \left(\frac{x}{\delta} \right)^2 \right] \frac{\partial c_{Ow}}{\partial z} = D_{Ow} \frac{\partial^2 c_{Ow}}{\partial z^2} \tag{9-206}$$

where O indicates the chemical being modeled is O_2. Three boundary conditions are necessary. The first two boundary conditions for oxygen transport are

$$BC\ 1: z = 0, \quad c_{Ow} = c_{Ow,0}$$

$$BC\ 2: x = 0, \quad c_{Ow} = c_{Ow,eq} \tag{9-207}$$

The first boundary condition indicates that wastewater entering the plate can contain

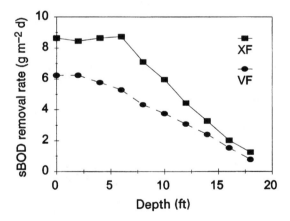

Figure 9.17 sBOD removal rate in a trickling filter normalized by modules. Each point represents the removal in a two-foot-long module of plastic media. The top of the tower is at a depth of 0 ft (from Richards and Reinhart, 1986, © WEF, reprinted with permission).

dissolved oxygen, while the second boundary condition states that oxygen at the air–water interface is at the equilibrium condition of oxygen saturation in the water at the current wastewater temperature and air pressure.

The third boundary condition is more complicated to express than the first two, since it specifies the flux of oxygen into the biofilm, and the oxygen flux is dependent on the substrate flux up to the limit of oxygen transport into the biofilm. This third boundary condition is

$$BC\ 3:\ x = \delta, \quad J_{Ob} = k_b^{1/2} D_{Cb} c_{Cw}, \quad \text{for } J_{Cb} \leq J_{Ob,\max}$$

$$J_{Ob} = J_{Ob,\max}, \quad \text{for } J_{Cb} > J_{Ob,\max} \tag{9-208}$$

where J_{Ob} and $J_{Ob,\max}$ are the actual and maximum oxygen fluxes into the biofilm. Oxygen must diffuse from the air–water interface down through the liquid to reach the biofilm surface. Substrate molecules have lower diffusivities than oxygen; at low substrate concentrations the oxygen flux will be that necessary to meet the oxygen requirements of the substrate flux. Therefore, under typical conditions the oxygen flux into the biofilm will be equal to the substrate flux into the biofilm. However, if the sBOD concentration is large, the total flux of oxygen will be insufficient to keep up with the substrate flux. For high sBOD concentrations the oxygen flux will reach a limit set by the fluid thickness and flow velocity. Since this maximum oxygen flux will depend on the hydraulic conditions, the maximum oxygen flux $J_{Ob,\max}$ must be obtained by a numerical solution of oxygen transport into the biofilm.

Oxygen transport into a biofilm is therefore a two-step process. First, if sBOD concentrations are low, the steady-state oxygen flux into the biofilm is identical to the sBOD flux. This of course assumes that the oxygen used in a BOD test by suspended microbes to degrade organic matter in the wastewater is consumed in the same proportion as oxygen used by the biofilm to degrade that substrate. Second, at large sBOD concentrations oxygen transport reaches a maximum. This maximum oxygen flux can be calculated by numerically solving Eqs. 9-206 to 9-208 using a finite-difference equation in the same manner as previously described (Eq. 9-181). The steady-state solution of these equations provides *maximum* oxygen fluxes into the biofilm for different plastic media as a function of hydraulic loading rate (Fig. 9.18). The computer program used to solve these equations called the TFO2 model (Logan et al., 1987a,b; Logan, 1993). XF media are predicted to achieve a higher rate of oxygen transport than VF media due to the effect of fluid interruptions on oxygen transport. Mixing the fluid brings oxygen-saturated wastewater down to the biofilm surface. Oxygen transport should increase with specific surface area, as demonstrated by the oxygen transport rate of XF138 media being larger than XF98 media.

Higher temperatures will increase oxygen transport rates to biofilms in trickling filters despite the decreases in solubility of oxygen (Fig. 9.19). As the fluid temperature rises, the wastewater viscosity decreases, producing thinner liquid films. This decrease produces a higher concentration gradient in the fluid film. In addition, chemical reaction rates increase due to increases in chemical diffusivities.

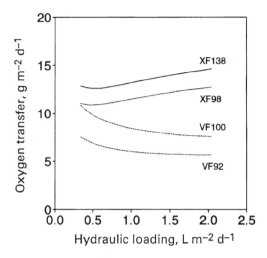

Figure 9.18 Maximum oxygen transfer rates for four different types of plastic media at steady state at 20°C (reproduced from Logan, 1993 by permission of ASCE.

Comparison of Maximum Oxygen Transport Rates with Experimental Data

The maximum oxygen transport rates in trickling filters can be compared directly to oxygen transport rates measured by Richards and Reinhart (1984). At a hydraulic loading rate of 0.9 gal/min ft^2 Richards and Reinhard measured maximum sBOD removal rates of ⁻9 g m^{-2} d^{-1} for XFa-98 media (Fig. 9.17). If sBOD removal was limited by the oxygen flux at this hydraulic loading rate, the sBOD removal

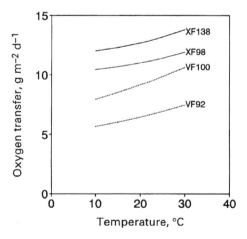

Figure 9.19 Maximum oxygen transfer rates for four different plastic media as a function of wastewater temperature at a hydraulic loading rate of 0.68 L m^{-2} s^{-1} (reproduced from Logan, 1993 by permission of ASCE).

rate would have been 11 g m^{-2} d^{-1}. Thus, if sBOD removal was limited by oxygen availability, it appears that the oxygen transport model overestimated oxygen flux into the biofilm. However, this analysis assumes that all wastewater applied to trickling filter media completely wets the surface of the media. As described below, there is good evidence that a significant portion of the biofilm is incompletely wet, resulting in a reduction of usable surface area. If oxygen did limit sBOD removal, the Richards and Reinhart data imply a surface area coverage of only 9/11 of the available area, or 82%.

Oxygen transport experiments were conducted in the laboratory by Hinton and Stensel (1995) using a synthetic wastewater and XF media. A biofilm was developed on either dextrin, with an apparent molecular weight of M_D 6,650 (D_{Ow} = 140 × 10^{-8} cm^2 s^{-1}), or sucrose (D_{Sw} = 450 × 10^{-8} cm^2 s^{-1}). No other electron acceptors except oxygen and CO_2 were available to the biofilm. They measured increasing oxygen fluxes into the biofilm in proportion to hydraulic loading rates, and may have observed a maximum oxygen transport rate at 16.3 g m^{-2} d^{-1} (Fig. 9.20). There are insufficient data at the higher loading rates, however, to justify the assumption that substrate removal was limited by oxygen transport. A comparison of the oxygen transport model predictions to the experimental conditions of Hinton and Stensel indicated that maximum oxygen transport rates of 15.6 g m^{-2} d^{-1} would have developed in the system, quite close (within 4%) of the measured rates of 16.3 g m^{-2} d^{-1}. Even conclusive evidence that oxygen limited substrate uptake by a biofilm fed a synthetic wastewater is not proof that the same thing would occur in a biotower fed wastewater. There are many alternate electron acceptors, such as NO_3^- and SO_4^{-2}, present in domestic wastewater that were not available in the synthetic wastewater. Therefore, while microbes in a biofilm fed wastewater may continue to remove sBOD using alternate electron acceptors, microbes fed a synthetic wastewater would only have CO_2 available as an electron acceptor.

The concentrations of sBOD in domestic wastewater in the United States average ⁻100 mg L^{-1}, based on a total BOD of about 300 mg L^{-1} and assuming sBOD is approximately one-third the total BOD. These conditions are not well suited to test the hypothesis of oxygen-limited substrate removal since dilution of influent wastewater with recycle flow can lower these sBODs into the range where the maximum rate of oxygen transport would still be larger than possible sBOD removal rates. Studies by Hutchinson (1975) used wastewaters with much higher sBOD concentrations, in the range of 200 to 300 mg L^{-1}. A comparison of predicted sBOD removals, based only on the chemical flux into the biofilm, and sBOD removals if oxygen was assumed to limit sBOD removals, indicated that oxygen did not limit sBOD removal (Table 9.6). While sBOD removal rates reached 21.6 g m^{-2} d^{-1}, maximum oxygen transport rates could only have been 9.3 g m^{-2} d^{-1} under these hydraulic and temperature conditions (Logan et al., 1987b; Logan, 1995).

Additional support for the calculations of maximum oxygen transport rates using the oxygen model described above come from calculated efficiencies of nitrifying trickling filters. The cell yields of nitrifying bacteria that convert ammonia to oxygen are quite low. As a result, almost all the oxygen that is used by nitrifying biofilms is required for the stoichiometric conversion of ammonia to nitrate of 4.3 mg O_2 mg

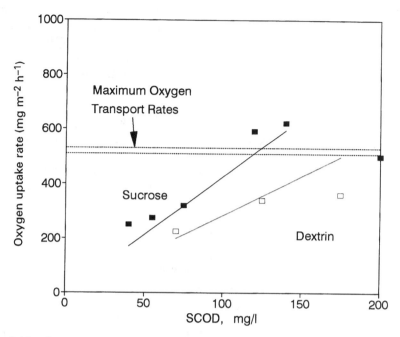

Figure 9.20 Average oxygen consumption rates measured for sucrose (■) and dextrin (□) feeds by Hinton and Stensel (1994) versus maximum substrate removal rates predicted by the TRIFIL2 model (solid lines) using the reported substrate : oxygen yield coefficients of 2.4 and 1.8 for sucrose and dextrin. The horizontal dashed lines indicate the maximum substrate removal rates if substrate removal was limited by oxygen transport into the biofilm. Relatively good agreement between the substrate and oxygen models does not permit a conclusion as to whether oxygen limited substrate removal by the bioflim (reproduced from Logan, 1994 by permission of ASCE).

TABLE 9.6   Comparison of Observed Rates of sBOD Removal (Hutchinson 1975) with Maximum Calculated Oxygen Transport Rates

Hydraulic Loading $(L\ m^{-2}\ d^{-1})$	sBOD $(mg\ L^{-1})$	sBOD Removals (%)		Observed sBOD removal rate $(g\ m^{-2}\ d^{-1})$	Maximum Aerobic sBOD removal rate $(g\ m^{-2}\ d^{-1})$
		Observed	Predicted		
0.21	233	31	46	10.4	11.5
0.41	220	27	27	16.5	10.1
0.82	256	15	15	21.6	9.3
1.11	203	12	12	18.5	9.3

Source: From Logan (1995).

NH$_3$. When ammonia concentrations are larger than a few mg L^{-1}, it is expected that oxygen flux into the biofilm will limit ammonia removal and therefore nitrification rates in the tower. Based on this assumption Parker et al. (1989) calculated the efficiency of nitrifying trickling filters by comparing observed nitrification rates in the top of the tower to maximum rates when oxygen flux would limit ammonia removal using the ratio

$$E = \frac{\text{observed nitrification rate}}{\text{maximum oxygen transfer rate}/4.3} \qquad (9\text{-}209)$$

E values ranged between 0.71 and 0.89 for XFa-138 media during normal operation of the Central Valley pilot nitrifying trickling filter (Table 9.7). Values less than unity

TABLE 9.7   Comparison of Oxygen Transfer Efficiencies (E) Calculated at Different Pilot Wastewater Treatment Plants for Nitrifying Trickling Filters

Plant (scale)	Type of Media[a]	Observed Rate (g N m^{-2} d^{-1})	Temperature (°C)	E
Central Valley, UT	XFa-140	1.4	16	(0.66)[b]
		2.6	20	0.89
		2.3	22	0.83
		3.2	18	(0.99)[c]
		1.6	15	0.71
		1.7	12	0.72
		2.1	11	0.81
Malmö, Sweden[d]	XFa-140	2.5	15	0.78
		2.8	16	0.87
		2.0	13	0.64
		2.2	13	0.71
Bloom Township, IL	VFc-89	1.2	20	0.88
		1.1	17	0.82
Midland, MI	VFc-89	1.2	13	0.86
		0.93	7	0.74
Lima, OH	VFc-89	1.2	18	0.88
		1.8	21	(1.30)[e]
		1.5	22	1.10
		1.2	22	0.76
Average at all sites[f]				0.81 ± 0.11

[a]Code for media is XF, cross flow; a through c, variable index for different manufacturers, with the number indicating specific surface area of the plastic media (m^2/m^{-3}).
[b]Value is considered out of the typical range of E since it was collected prior to steady-state operation of the plant.
[c]Plant operated in a flooding mode (BCNTF) to control biofilm predators.
[d]Plant operation was based on alternating feed between two reactors.
[e]Value is considered to be outside normal operating range.
[f]Average excludes numbers in parentheses.
Source: From Logan (1995), using data from Parker et al. (1989, 1994).

could have been produced by incomplete media wetting, predation (loss of biofilm), and competition of heterotrophic bacteria for oxygen with nitrifiers. Parker et al. (1989) demonstrated that by flooding a tower to maintain a thick biofilm at all depths in the filter, E was nearly unity ($E = 0.99$). When considering all the studies at other plants, values of E are 0.81 ± 0.11 (range of 0.74 to 1.10) when data for filters with poor biofilm development in the tower are discounted. These data therefore support the calculations of oxygen transport to biofilms by demonstrating the calculated oxygen fluxes compare favorably to those necessary to achieve ammonia removal rates in nitrifying trickling filters.

Effect of Influent Oxygen Concentration on Oxygen Transport Rates   One of the boundary conditions in the oxygen transport model is the concentration of oxygen in the wastewater entering the trickling filter. As wastewater is applied to the top of the tower, it is partially aerated as it splashes onto the media. There are no good estimates of how much the wastewater is oxygenated as a result of this process. Model simulations of oxygen transport rates in the top module of a trickling filter indicate that if oxygen is not present in the wastewater, there will be reduced oxygen transport rates in VF media compared to steady-state maximum transport rates, but oxygen transport will be relatively unaffected in XF media (Fig. 9.21). Because the wastewater is mixed several times while passing through XF media, the initial oxygen concentration affects the overall oxygen transport rates much less than in VF media. Wastewater in VF media is not mixed until it reaches the edge of the trickling filter module.

Reduced oxygen fluxes in the top module are undesirable since sBOD removal rates are highest at the top of the tower. However, in XF media mixing points quickly establish a steady condition of oxygen transport rates for modules at lower points in the tower. This can be seen by the oxygen profiles generated using the oxygen transport model (under conditions of very high sBOD loads) at several depths in a module of VF media (Fig. 9.22). As oxygen moves down through a 60-cm-long module, oxygen is able to diffuse from the surface of the liquid, at the air–water interface, to the biofilm. Within about 1.2 cm a nearly linear oxygen gradient is established in the liquid whether the wastewater entering is saturated or devoid of oxygen. By the time the wastewater leaves the module, this concentration profile is flat and is no longer a function of the influent oxygen concentration. Wastewater leaving the module is assumed to be completely mixed. For the case shown, this will result in wastewater containing oxygen at about 70% saturation being applied to the next module. Since wastewater leaving the module is independent of the concentration of oxygen in the wastewater applied to the top of the tower, the maximum oxygen transfer rates in modules after the top module are independent, are all the same, and are defined as the steady-state oxygen transport rate in the trickling filter (see Fig. 9.21).

The development of oxygen profiles can be simulated over the whole module, as shown by the three-dimensional plots of oxygen concentration in Figure 9.23. For either XF or VF media, oxygen concentrations quickly reach nearly linear profiles, indicating that oxygen diffusion is rapid. The concentrations of oxygen near the

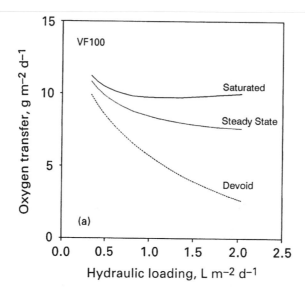

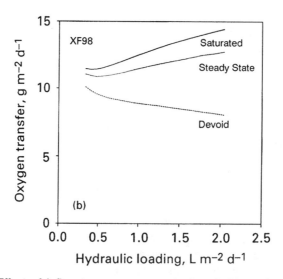

Figure 9.21   Effect of influent oxygen concentration (saturated and devoid) on the maximum rate of oxygen transport into the first module in a trickling filter as a function of hydraulic loading rate for (A) VF100 media and (B) XF-98 media. After the first module, the maximum oxygen transport rate reaches a steady condition from module to module, as shown for the two different media types (reproduced from Logan, 1993 by permission of ASCE).

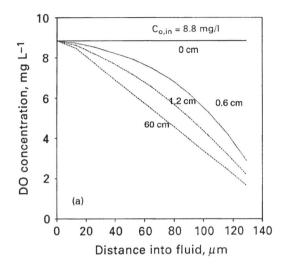

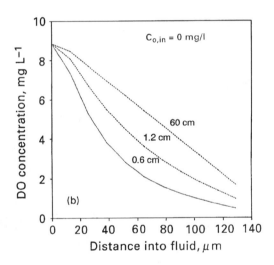

Figure 9.22 Dissolved oxygen concentrations in the fluid film as a function of distance into the fluid (distance from the air–liquid interface) at various locations within a module of vertical flow media. Applied wastewater is assumed to be (A) devoid of oxygen and (B) saturated with oxygen. Notice how the concentration profile quickly becomes nearly linear with distance into the liquid (reproduced from Logan, 1993 by permission of ASCE).

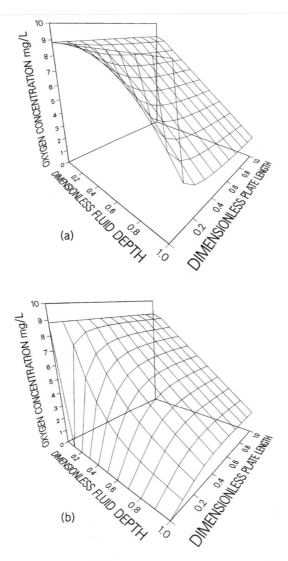

Figure 9.23   Three-dimensional dissolved oxygen profiles in wastewater flowing over a biofilm in a trickling filter simulated using the oxygen transport model. The wastewater is assumed to be either (A) fully saturated with oxygen or (B) devoid of oxygen when it is applied to the media. The dimensionless fluid depth is the distance from the air–water interface divided by the fluid film thickness; the dimensionless plate length is the distance from the plate entrance divided by the plate length. Conditions for the calculations are 20°C, VFb-100 media, and 0.68 L m^{-2} s^{-1} (reproduced from Logan 1993 by permission of ASCE).

biofilm are not zero, since there must be some oxygen in the biofilm to support microbial activities. The gradient of oxygen at the biofilm surface is a function of the biofilm kinetics since the maximum sBOD flux is a function of the chemical diffusivities of the sBOD components. Since the oxygen concentrations at the biofilm surface are quite low, however, changing the biofilm kinetic model will have little effect on the overall gradient, and therefore on the calculated maximum oxygen transport rates in the trickling filter.

The importance of oxygen transport to the aerobic removal of biodegradable organic matter in trickling filters led one research team to assume that sBOD removal could be described by calculating the oxygen transport rate into the fluid using penetration theory (Mehta et al., 1972). According to penetration theory, the absorption of a gas into a liquid in the x direction, when the liquid is falling in the z direction, can be described as

$$\frac{c_{Ow}(x,z) - c_{Ow,in}}{c_{Ow,eq} - c_{Ow,in}} = \text{erfc}\left(\frac{x}{(4D_{Ow}z/u_{max})^{1/2}}\right) \qquad (9\text{-}210)$$

where $c_{Ow,in}$ is the concentration in the liquid applied to the media, $c_{Ow,eq}$ is the equilibrium oxygen concentration, and erfc is the complimentary error function. The assumption necessary to derive this equation was that the chemical did not penetrate very far into the liquid. This assumption allowed the parabolic velocity expression to be replaced by a single constant, the maximum velocity, since the velocity would not appreciably vary if the chemical did not move very far into the fluid. According to penetration theory, the flux into a fluid film integrated over distance L (Eq. 5-100) is

$$W_{C,x=o} = c_{w,eq} wL \left(\frac{4D_{Cw}u_{max}}{\pi L}\right)^{1/2} \qquad (9\text{-}211)$$

Using this expression, it is possible to describe the overall rate of mass transport of oxygen into the fluid film (Logan, 1993) as

$$W_{Ow} = \left(\frac{6g^{1/3}}{3^{1/2}\pi v_w^{1/3}}\right)^{1/2} (D_{Ow}L)^{1/2} w^{2/3} Q^{1/3} (c_{Ow,eq} - c_{Ow,in}) \qquad (9\text{-}212)$$

where g is the gravitational constant, v_w the fluid viscosity, w the width of a plate (or the wetted perimeter), and Q the applied flow rate. The total oxygen transfer into the liquid can be calculated as the product of the number of modules, the surface area per module, and the oxygen transport rate.

Mehta et al. (1972) found good agreement with the total BOD removal in several trickling filters with the oxygen flux calculated from a penetration-theory type of model. Such agreement must be considered to be coincidental, however, since the main assumption of the penetration theory is not met for fluid flow in a trickling filter. Oxygen penetrating the fluid film penetrates quite far into the liquid and will reach the biofilm, as shown in the oxygen model simulations (Fig. 9.22). Since this result contradicts the model assumption that the velocity in the liquid is fixed, the Mehta model cannot be considered to be a valid description of oxygen transport to biofilms in trickling filters.

9.5 RECENT DEVELOPMENTS IN THE STUDY OF BIOFILMS

Biofilms continue to be studied in a variety of systems due to their importance in natural and engineered systems (Characklis and Marshall, 1990). Much of the new information we have on biofilm composition and activity is derived from microprobe studies. The use of these small probes allows investigation of processes in the biofilm on the scales of micrometers to tens of micrometers. Microprobes have evolved from just investigating oxygen profiles (Whalen et al., 1969; Lewandowski et al., 1991) to such complex studies as monitoring glucose and pH in order to observe the spatial distributions of glucose-utilizing and methangenic bacteria (Lens et al., 1995). Different types of microbes have been developed to measure a variety of compounds including sulfide (Kuhl and Jorgenson, 1992), ammonium (deBeer et al., 1993), nitrate (de Beer and Sweerts, 1989), light (Lassen et al., 1992), and chlorine (de Beer et al., 1994).

Although it is useful to assume that biofilms are homogeneous, detailed investigations of their structure indicate that the biofilm structure is quite variable with depth. Bishop et al. (1995) found that while average dry densities of biofilms were 1.17 mg TS cm^{-3}, the bulk (wet) densities increased from 1.001 to 1.003 g cm^{-3} near the top of the biofilm at the biofilm–liquid interface, to 1.01 to 1.02 g cm^{-3} near the biofilm bottom. Similarly, viable cells decreased from 72–91% at the top of the biofilm to 31–39% near the biofilm bottom.

Using confocal scanning laser microscopy (CSLM) and nuclear magnetic resonance imaging (NMRI), Lewandowski proposed a higher permeable and heterogeneous structure of a biofilm (Fig. 9.24). In this model the biofilm structure is considered to be composed of biofilm structures that are relatively less permeable than surrounding areas, so that the permeability of the overall biofilm to advective flow is quite large. Support for this view of advective flow even within the biofilm comes from observations of particles flowing within biofilms (Drury et al., 1993; Lewandowski et al., 1995). Such channel structures have been reported in a number of systems including those treating contaminated groundwaters (Massol-Deyá et al., 1995).

These observations have improved our understanding of biofilm structure and hydraulic flow on top of and through biofilms, but as yet biofilm models have not advanced any further to incorporate such new information. While such advances in modeling can be expected in the near future, it is still too early to tell how this information will modify mass transport calculations into biofilms and the design of fixed film bioreactors.

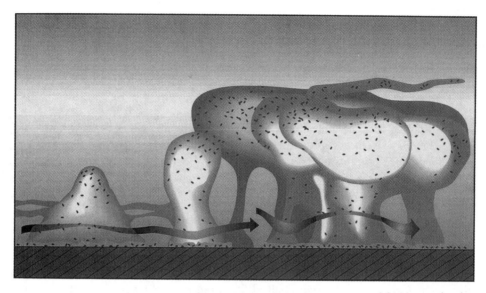

Figure 9.24 Hypothetical structure of a bacterial biofilm based on CSLM examination of pure cultures and mixed-species biofilms. It is speculated that there are channels within the biofilm matrix. (Reprinted from Lewandowski et al., 1995 with kind permission from Elsevier Science Ltd, The Boulevard, Landford Lane, Kidlington, OX5 1GB).

PROBLEMS

9.1 Calculate the glucose ($D_{Gw} = 660 \times 10^{-8}$ cm^2 s^{-1}) flux into a (a) shallow (δ_b = 10 μm) and (b) deep (δ_b = 1000 μm) biofilm assuming first-order kinetics using the kinetic constant given in Eq. 9-176 and the constants for the biofilm of α = 0.0035, θ = 0.8, a cell radius of 1 μm, and a glucose concentration of 100 mg L^{-1}. (c) How much larger is the flux into the deep biofilm than the shallow biofilm?

9.2 sBOD removal in a trickling filter at the Tolleson plant in Arizona was found to be more rapid than anticipated by design engineers. Assume the following for the filter: sBOD$_{in}$ = 27 mg/L, XFa-98 media, hydraulic loading rate of 0.46 L m^{-2} s^{-1}, wastewater temperature of 34°C. This hydraulic loading rate results in a fluid film thickness δ_w = 113 μm and a fluid velocity of 4.91 cm s^{-1}. Using the five-component molecular weight distribution given for wastewater in Table 9.4, calculate the maximum sBOD removal in the trickling filter and compare it to the observed sBODs at the following depths: 0.33 m, sBOD = 24 mg/L, 1.53 m, sBOD = 8 mg/L, and >2.85 m, sBOD < 1 mg/L. In some sewer lines much of the BOD will be solubilized anaerobically to volatile acids in the sewer line before reaching the wastewater treatment plant—especially if the travel times in the pipe are long and the temperature is warm. Therefore, repeat the calculations assuming only a single component with a diffusivity of 800×10^{-8} cm^2 s^{-1}. What can you conclude about the composition of the wastewater at this plant?

9.3 BOD removal in trickling filters was described in terms of a rate model (the modified Velz equation) and a combined mass transport/rate model. The purpose of this problem is to explore the inter-relationship of these two models. Assume for this problem that the Velz constant, $k_{20} = 0.0023$ (gpm ft^{-2})$^{0.5}$, a trickling filter contains 10 modules (20 ft tall), is loaded at 1 gpm ft^{-2}, and has no recycle. As a result, wastewater flows over each of the 9 plates in a module of Xfa-30 media at a maximum velocity of 5.7 cm s^{-1} with a fluid thickness of 143 μm. (a) Biodegradable organic matter in a wastewater has been modeled as a five-component system consisting of chemicals with diffusion coefficients of: 112, 85, 65, 50, and 30×10^{-8} cm^2 s^{-1}. Assuming just an average diffusivity, calculate a first-order rate constant for a biofilm in this trickling filter that would correspond to the Velz rate constant. (b) Based on the magnitude of the Thiele coefficient, is removal of organic matter kinetically or mass transport limited? (c) Assuming that mass transport limits sBOD removal, calculate the effluent sBOD from a trickling filter for these conditions. (d) Compare your answer in part (c) with that predicted by the Velz equation.

9.4 Compare the maximum oxygen flux predicted by the Mehta model to the maximum oxygen flux calculated using the TFO2 oxygen transport model using the results shown in Figure 9.18 for XFa-98 media assuming a hydraulic loading rate of 1.7 L m^{-2} s^{-1} and 20°C. Assume $D_{Ow} = 1800 \times 10^{-8}$ cm^2 s^{-1}, $v_w = 0.01$ cm^2 s^{-1}, $c_{Ow,eq} = 8.84$ mg L^{-1}, and that the entering wastewater is devoid of oxygen.

9.5 Starting with the expression for oxygen transport into a thin fluid film according to penetration theory (Eq. 9-210), (a) derive an expression for a mass transport coefficient k_w such that the oxygen flux into a fluid film can be described by $J_w = k_w \Delta c$, and (b) derive Eq. 9-212.

9.6 For first-order biofilm kinetics, show that for the boundary conditions of: no flux through the support at $z = 0$, and a fixed concentration of substrate (c_s) at the biofilm surface, $z = \delta_b$, that the concentration profile is:

$$c = c_s \frac{\cosh(z\sqrt{k_1/D_{Cb}})}{\cosh(\delta_b\sqrt{k_1/D_{Cb}})}$$

Using this concentration profile, show that the effectiveness factor for the biofilm is

$$\eta_d = \frac{\tanh \Phi}{\Phi}$$

where Φ is the Thiele modulus defined as $\Phi = \delta_b(k_1/D_{Cb})^{1/2}$, and δ_b is the thickness of the biofilm.

9.7   Mehta et al. (1972) report that a small (16 ft^2 top area) trickling filter treating domestic wastewater reduced the BOD from 279 to 92 mg L^{-1}. The operating conditions were 25°C, a hydraulic loading rate of 0.925 gpm ft^{-2}, a recycle ratio $R_r = 1.0$, for a tower 18 ft tall filled with 37 ft^2 ft^{-3} vertical flow media (VFb). Assuming that oxygen limited BOD removal, they calculated a removal of 21 mg BOD L^{-1} for each 2-ft-deep module, resulting in an overall BOD removal of 197 mg L^{-1} for the 18-ft height. This finding that BOD removal is constant with depth has been shown to be incorrect based on numerous measurements of BOD at different tower depths (Logan et al., 1987a). Using the modified Velz equation, calculate the BOD at depths of (a) 2 ft and (b) 18 ft for this tower. Compare these removals with those calculated and reported in the Mehta et al. (1972) study.

9.8   A trickling filter typically is loaded at a rate of 1 gpm ft^{-2}, and contains media plastic support media with a specific surface area of 30 ft^2 ft^{-3}. (a) Calculate the fluid thickness of the fluid film, assuming the plates are completely vertical, and the media "width" is 30 ft. (b) What is the maximum velocity of the fluid? (c) How far does the fluid travel down the media, before oxygen penetrates to the biofilm surface? Assume penetration theory applies, the equilibrium concentration of oxygen is 8 mg/L, and there is no oxygen in the wastewater entering the trickling filter. What do your results suggest about oxygen transfer in trickling filters?

9.9   The solution of first-order biofilm kinetics (Eq. 9-71) using the given boundary conditions was not presented in Chapter 9. In this problem, we set up the governing equation and derive solutions for the concentration profile and flux into the biofilm. (a) Starting with Eq. 9-33, obtain a differential equation (in dimensional form) describing the concentration of substrate the biofilm of any depth at steady state, with no advective motion in the biofilm, for first-order biofilm kinetics. (b) Non-dimensionalize this equation, in terms of the dimensionless constant $B_1 = [k_1\delta_b^2/D_{Cw}]^{1/2}$. (c) Solve this equation using the boundary conditions in Eqs. 9-74 and 9-75. (d) Show that the flux is Eq. 9-76.

9.10   Methanotrophic microorganisms require oxygen and methane, both of which are sparingly soluble in water. Let us assume you are analyzing mass transfer of these gases to methanotrophic biofilms in plastic media trickling filters. The diffusivities and solubility of oxygen (O) and methane (M) are: $D_{Ow} = 1800 \times 10^{-8}$ cm^2 s^{-1}; $D_{Mw} = 1940 \times 10^{-8}$ cm^2/s; $c_{Cw,eq} = 8$ mg L^{-1} (for air); $c_{Mw,eq} = 24.1$ mg L^{-1} (for pure methane gas above water). (a) Which theory (i.e., penetration, boundary layer, or stagnant film) describes the initial rate of mass transfer of these gases into the thin fluid film above the biofilm? (b) Very far down a piece of trickling filter media, considered a flat plate, assume that a constant *linear* concentration profile of both these compounds develops in the thin fluid film. If the local flux of oxygen into the biofilm is 12 g O$_2$

m^{-2} d^{-1}, and if the concentration of oxygen at the fluid–biofilm interface is 1 mg L^{-1}, what is the oxygen mass transfer coefficient (k_w[cm s^{-1}])? (c) What is the methane flux (g/m^2 d) to the biofilm assuming the same gradient as part (b)? (d) If 1.4 moles of oxygen are consumed by bacteria for every mole of methane, then what is the methane concentration (mg/L) at the biofilm–fluid surface that is required so that the methane flux is just equal the oxygen flux into the biofilm? (Assume that oxygen and methane use is exactly coupled and first order, and the reaction is very fast.) (e) What must the mole fraction of methane be in the gas phase so that the overfall flux of methane in the biofilm just equals the overall oxygen flux (12 g m^{-2} d^{-1}) into the biofilm? [Assume the biofilm–fluid interface concentration from part (d).]

9.11   Let us first look at the maximum rate of glucose consumption within a biofilm. The rate of mass transfer to a sphere can be used to analyze bacterial degradation of glucose if the surface area available for enzymes used by the cell to take up glucose is assumed to be much less than the cell surface area. Assume we have a 1-μm-diameter cell taking up glucose at a rate of 0.01% the rate of glucose transport to the cell surface. Assume $D_{Gw} = 0.6 \times 10^{-5}$ cm^2 s^{-1} for a cell in stagnant fluid. (a) What is the maximum rate (mg s^{-1}) of glucose transport to a single cell at a glucose concentration of 10 mg L^{-1}? (b) What is the total gluocse utilization rate of a 100-μm-thick biofilm (a cube) assuming no concentration gradients within the biofilm? Assume that there is 99.9% water (by volume) in the biofilm. (c) Now, let us look at the surface of a biofilm. What is the flux of glucose into the biofilm, assuming a first-order very, very fast reaction [i.e., the bulk rate that you calculated in part (b)? (d) Assuming that a microbe using a 1:1 ratio (mass basis) of oxygen to glucose, at what concentration of glucose could oxygen become a limiting factor in oxygen utilization for a single cell? Assume an oxygen concentration of 8 mg L^{-1}, and an oxygen diffusivity of 1.8×10^{-5} cm^2 s. (e) Repeat part (d), but base your comparison on diffusion into a biofilm with fast reaction.

9.12   Chemical fluxes into biofilms depend on the reaction rate in the biofilm, but the viable cell density in the biofilm is often unknown. As a result, the rate constants are often calculated from observed fluxes into the biofilm assuming negligible external mass transport resistances around the biofilm (achieved by rapidly mixing the fluid above the biofilm). (a) Calculate the flux into a "deep" biofilm assuming zero-order kinetics, a zero-order rate constant $k_0 = 25$ mg L^{-1} s^{-1}, $D_{Cw} = 600 \times 10^{-8}$ cm^2 s^{-1}, and a surface substrate concentration of 50 mg L^{-1}. (b) Assuming the flux you calculated in part (a), calculate a first-order rate constant. Can the biofilm kinetics be considered a fast reaction? (c) Using the result for a first-order concentration profile obtained in Problem 9.14, plot out the concentration profiles in the biofilm under these conditions for both zero- and first-order conditions.

9.13   A packed-bed reactor is designed to remove a toxic chemical while a biofilm is supported by anaerobic removal of acetate. What detention time of this reactor would be sufficient to reduce acetate from 400 to 4 μg/L for reactor and reaction conditions: μ_{max} = 0.56 h^{-1}, $Y_{X/c}$ = 0.12 g g^{-1}, K_m = 710 mg L^{-1}, X = 20,000 mg L^{-1}, δ_b = 1000 μm, $D_{Cw} = D_{Cb}$ = 800 $\times$ 10^{-8} cm^2 s^{-1}, bed packing diameter d_c = 0.5 cm, θ = 0.38, u_x = 0.2 cm s^{-1}, and fluid properties of water at 20°C.

9.14 For a first-order reaction in a biofilm of *any depth*, it was stated that the flux into the biofilm was $J = (k_1 \delta_b c_{CB,0}/B_1) \tanh(B_1)$ (Eq. 9-77). However, for the case of diffusion with reaction, we previously derived (see Section 5.2) that $J = (D_{Cb} c_{CB,0}/\delta_b)(B_1/\tanh B_1)$. The reason for the two different solutions is the different boundary condition. (a) Starting with Eq. 9-71, and the indicated boundary conditions (Eqs. 9-74 and 9-75), derive Eq. 9-76. (b) Show for a very fast reaction that the two equations given above reduce to the same equation.

REFERENCES

Annachhatre, A. P., and P. Khanna. 1987. *J. Environ. Eng.* **113**(2):429–33.

Atkinson, B., E. L. Swilley, A. W. Busch, and D. A. Williams. 1967. *Trans. Inst. Chem. Engin.* **45**:T257–64.

Bear, J. 1972. *Dynamics of fluids in porous media.* Dover Pub, Inc. NY.

Bishop, P. L, T. C. Zhang, and Y.-C. Fu. 1995. *Wat. Sci. Tech.* **31**(1):143–52.

Bungay, H. R., J. E. Whalen, and W. M. Sanders. 1969. *Biotechnol. Bioeng.* **11**:765–72.

Characklis, W. G., and K. C. Marshall. 1990. *Biofilms.* Wiley, NY.

de Beer, D., R. Srinivasan, and P. Stewart. 1994. *Appl. Environ. Micro.* **60**(12):4339–44.

de Beer, D. and J.-P. R. A. Sweerts. 1989. *Anal. Chim. Acta.* **219**:351–56.

de Beer, D., J. C. van den Heuvel, and S. P. P. Ottengraf. 1993. *Appl. Environ. Microbiol.* **59**:573–79.

Drury, J., P. S. Stewart, and W. G. Characklis. 1993. *Biotechnol. Bioeng.* **42**:111–17.

Grady, C. P. L. Jr., and H. C. Lim. 1980. *Biological Wastewater Treatment.* Marcel Dekker, NY.

Gulevich, W., C. E. Renn, and J. C. Liebman. 1968. *Environ. Sci. Technol.* **2**:113.

Harremoës, P. 1978. In: *Water Pollution Microbiology*, Vol. 2, R. Mitchell, Ed. John Wiley, NY, pp. 82–100.

Hinton, S. W., and S. D. Stensel. 1995. *J. Envir. Engrg. ASCE.* **120**(5):1284–97.

Howland, W. E. 1958. Proc 12th *Indus. Waste Conf. Purdue, IN*, **94**:435–465.

Hutchinson, E. G. 1975. A comparative study of biological filter media. Pres. Biotechnology Conference, May, Massey University, Palmerstown No, New Zealand.

Jennings, P. A. 1975. A mathematical model for biological activity in expanded bed adsorption columns. PhD. thesis, Dept. Civil Engin., Univ. of Illinois, Urbana.

Kuhl, M., and B. B. Jorgenson. 1992. *Appl. Environ. Microbiol.* **58**:1164–74.

Larson, T. A., and P. Harremoës. 1994. *Wat. Res.* **28**(6):1435–41.

Lassen, C., H. Plug, and B. B. Jorgensen. 1992. *Limnol. Oceanogr.* **37**:760–72.

Lens, P., D. de Beer, C. Cronenberg, S. Ottengraf, and W. Verstraete. 1995. *Wat. Sci. Technol.* **31**(1):273–80.

Levine, A. D., G. Tchobanoglous, and T. Asano. 1985. *J. Water Pollut Control Fed.* **57**(7): 805–16.

Lewandowski, Z., P. Stoodley, and S. Altobelli. 1995. *Wat. Sci. Tech.* **31**(1):153–62.

Lewandowski, Z., G. Walser, and W. G. Characklis. 1991. *Biotechnol. Bioeng.* **38**:877–82.

Lewandowski, Z., G. Walser, R. Larson, B. Peyton, and W. G. Characklis. 1990. *Proc. 1990 Specialty Conference on Environmental Engineering*, Washington, D.C., C. R. O'Melia, Ed. ASCE Publications, pp. 17–24.

Logan, B. E. 1993. *J. Environ. Engin.* **119**(6):1059–76.

Logan, B. E. 1995. *J. Environ. Engin.,* **121**(5):423–26.

Logan, B. E. 1996. *J. Environ. Engin.,* **122**(4):333–36.

Logan, B. E., S. W. Hermanowicz, and D. S. Parker. 1987a. *J. Water Pollut. Control Fed.* **59**(12):1029–42.

Logan, B. E., S. W. Hermanowicz, and D. S. Parker. 1987b. *J. Water Pollut. Control Fed.* **59**(12):1017–28.

Massol-Deyá, A. A., J. Whallon, R. F. Hickey, and J. M. Tiedje. 1995. *Appl. Environ. Microbiol.* **61**(2):769–777.

Mehta, D. S., H. H. Davis, and R. P. Kingbury. 1972. *J. Sanit. Engrg. Div., ASCE.* **98**(3): 471–89.

MOP 8, 1991. Design of municipal wastewater treatment plants. Water Environment Federation and American Society of Civil Engineers. Basic Press Inc., Brattleboro, VT.

Ohashi, H., T. Sugawara, K. I. Kikuchi, and H. Konno. 1981. *J. Chem. Eng. Japan* **14**(6): 433–38.

Parker, D., M. Lutz, B. Andersson, and H. Aspegren. 1995. *Wat. Env. Res.* **67**(7):1111–18.

Parker, D. S., and D. T. Merrill. 1984. *J. Water Pollut. Control Fed.* **56**(8):955–61.

Parker, D., M. Lutz, B. Andersson, H. Aspegren. 1984. Proc. 67th Annu. Conf. Water Env. Fed (WEF), Alexandria, VA.

Richards, T., and D. Reinhart. 1986. *J. Water Pollut. Control Fed.* **58**(7):774–83.

Rittmann, B. E. 1982. *Biotech. Bioeng.* **24**:1341–70.

Roberts, P. V., P. Cornell, and R. S. Summers. 1985. *J. Env. Eng.* **111**(6):891–905.

Robinson, A. R., and T. J. Healey. 1984. *Proc. 5th ASCE GMD Specialty Conference*, Laramie, WY.

Shülander, E. V. 1978. *Transport Phenomena in Packed Bed Reactors*. ACS Symp. Ser. 72, pp. 110–61.

Skelland, A. H. P. 1974. *Diffusional Mass Transfer*. Wiley, New York.

Suidan, M. T. 1986. *J. Env. Eng.* **112**:78–93.

Suidan, M. T., and Y.-T. Wang. 1985. *J. Environ. Eng.* **111**(5):634–46.

Swilley, E. L. 1966. Film flow models for the trickling filters. PhD dissertation, Rice University, Houston, TX.

Swilley, E. L., and Atkinson. 1963. *Proc. 18th. Ind. Waste Conf.*, Purdue, Indiana. **18**:706–32.

Whalen, W. J., H. R. Bungay III, and W. M. Sanders. *3rd. Environ. Sci. Technol.* **3**:1297.

Williamson, K., and P. L. McCarty. 1976. *J. Water Pollut. Control Fed.* **48**:9–24.

Williamson, J. E., K. E. Bazaire, and C. J. Geankoplis. 1963. *Ind. Eng. Chem. Fund.* **2**(2): 126–29.

Wilson, E. J., and C. J. Geankoplis. 1966. *Ind. Eng. Chem. Fundamentals* **5**:9–14.

CHAPTER 10

DISPERSION

10.1 INTRODUCTION

The calculation of chemical transport has centered so far on two processes: molecular diffusion and advection. Of course in real systems the situation is somewhat more complicated. On larger-than-molecular scales the properties of the fluid, the medium, or even the boundaries can affect the rate that a chemical spreads out. When examined on these macroscales the time for a chemical to distribute over a given distance is much smaller than that predicted by molecular diffusion processes. We define here *dispersion* as the spread of a chemical by any process that results in the movement of the chemical in a manner different from its average advective velocity.

The dispersion of a chemical can be thought of as being produced by the sum of several processes, including molecular diffusion, mixing, a noncontinuous flow path (mechanical mixing), and the velocity profile. As will be shown, the rate that chemical dispersion occurs can be expressed as the sum of a molecular diffusion coefficient, and dispersion coefficients for these other processes, in terms of an overall dispersion coefficient defined as

$$\mathbf{E}_C = D_C + \sum_n \mathbf{e}_{Cn} \tag{10-1}$$

where D_C is the molecular diffusivity, $\mathbf{e}_{Cn}$ dispersion coefficients for these n processes, and $\mathbf{E}$ the overall dispersion coefficient. The terms in bold can have direction, and therefore can be vectors or second-order tensors. The overall dispersion coefficient may, or may not, end up being a function of the molecular diffusion coefficient. That is, in some cases, molecular diffusion will be slow enough that chemicals having a different size will disperse at the same rate. When there is any appreciable amount of mixing in a system, for example, turbulent mixing in a lake following a chemical spill, the rate of dispersion becomes independent of molecular diffusion since the bulk properties of the fluid dictate the rate of chemical spread. In other systems, such as laminar flow in a pipe, the longitudinal spread of the chemical is a function of the chemical diffusivity, but dispersion is primarily a consequence of a nonconstant velocity in the radial direction. In multiphase systems, such as chemical transport in porous media, a chemical can be spread out over a larger distance. Due to a noncontinuous medium some molecules will be on slow-moving paths, while others will be faster-moving channels. The adsorptive behavior of individual species will also affect overall dispersion: Chemicals adsorbing to surfaces will be retarded during transport, while nonadsorbing chemicals will move with the fluid. Chemical dispersion in porous media is perhaps the most complicated situation because the flow path is not continuous and because chemical size may result in pore exclusion within particles or microaggregates based on chemical size or charge.

Dispersion will be considered here as the spread of a chemical that is larger than that produced by molecular diffusion as a result of four separate processes (Fig. 10.1). In the first case, dispersion is caused by the intrinsic motion (turbulence) of the fluid. Let us consider that a chemical is injected into a fluid as a single point source, where the fluid consists of a mixture of small (microscale) eddies. When that eddy dissipates its energy into heat, new eddies will arise that will each contain some

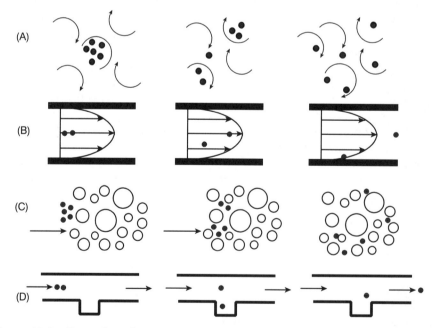

Figure 10.1 Examples of conditions that generate chemical dispersion (A) fluid eddies; (B) velocity gradients in flow in a pipe; (C) noncontinuous flow field (in porous media); (D) boundary effects.

of that chemical, resulting in the redistribution of the chemical over distances covered by these new eddies. The chemical is therefore moved within the system at a rate dependent on the size of the eddies and the rate at which larger eddies move fluid around the system.

In the second case (Fig. 10.1b) it is the fluid shear that produces chemical dispersion. Even if the fluid flow is laminar and the chemical is injected onto a single streamline, the molecules will diffuse onto other streamlines and move away from each other at a rate dependent on the chemical's molecular diffusivity and the fluid velocities at different transverse locations in the system. Molecules diffusing onto slower streamlines will move more slowly than those on a faster streamline, resulting in axial (or longitudinal) dispersion in the system. If the fluid is turbulent, average velocity gradients produce the same chemical dispersion with an important difference from the case of purely laminar flow. The rate that chemicals move from the point source will be determined by the eddy, and not molecular, diffusivity. That means that in the bulk fluid the transport of the chemical is not a function of the properties of the molecule. At the fluid boundary there will be a laminar sublayer, however, in which the transport is still governed by the molecular diffusivity.

In the third case (Fig. 10.1c), dispersion results from the fact that the medium is noncontinuous. Fluid streamlines within the porous medium must go around objects, such as soil grains, that confine the flow to specific paths. As a chemical diffuses off a streamline onto another adjacent one that may be moving to another side of the object, the chemical is spread out.

In the fourth case dispersion (Fig. 10.1d) is produced by diffusion into a pore or reversible adsorption onto a surface. One molecule diffusing into a pore will be spread out relative to another molecule that does not diffuse into the pore. Molecules can also collide with the surface, and the amount of time that a molecule spends there is not the same for all molecules. Even identical molecules will spend different times on the particle's surface, giving rise to a dispersive effect on chemical concentration over transport distances in the axial direction.

The chemical dispersion coefficient can be included into a macroscale constitutive transport equation as

$$\frac{\partial c}{\partial t} + \mathbf{u} \cdot \nabla c - \nabla(\mathbf{E} \cdot \nabla c) = 0 \qquad (10\text{-}2)$$

where the dispersion coefficient $\mathbf{E}$ has replaced the molecular diffusion coefficient D. The subscript notation will not be used in this chapter for c, D, $\mathbf{E}$ and other properties of a chemical C in the fluid unless the meaning of the subscript would be unclear. Although $\mathbf{E}$ is indicated in bold (since it can be a second-order tensor), $\mathbf{E}$ is usually either defined as a vector, with three dimensions, or treated as a scalar through solution of dispersion in the system described in one spatial dimension.

For dispersion in three spatial dimensions we can identify three different mixing coefficients, E_x, E_y, and E_z, where the subscripts indicate the direction of dispersion.

We write the transport equation in Cartesian coordinates for the case of no reaction

$$\frac{\partial c}{\partial t} + \mathbf{u} \cdot \nabla c = E_x \frac{\partial c^2}{\partial x^2} + E_y \frac{\partial c^2}{\partial y^2} + E_z \frac{\partial c^2}{\partial z^2} \qquad (10\text{-}3)$$

This equation can directly be reduced from three dimensions to one dimension, as we have done previously for molecular diffusion. For example, for a small chemical spike into an infinitely large reservoir having no net fluid motion, dispersion will be equal in all directions, and we can write in one dimension from a line source

$$\frac{\partial c}{\partial t} + u_x c = E_x \frac{\partial c^2}{\partial x^2} \qquad (10\text{-}4)$$

For a point source similar equations can be derived in cylindrical and radial coordinates.

In many systems, such as flow in a pipe or mixing in a river, it is not possible to describe dispersion in the direction of flow without considering how the velocity profile in another direction or fluid turbulence affects the chemical's dispersion. When we derive a one-dimensional dispersion coefficient, the rate of dispersion will be different depending on the fluid flow and specific system conditions. For example, for flow in a laminar tube we can describe chemical transport in the axial, x, dimension as

$$\frac{\partial c}{\partial t} + u_x c = E_L \frac{\partial c^2}{\partial x^2} \qquad (10\text{-}5)$$

where we have replaced dispersion coefficient in the x direction, E_x, with E_L, the longitudinal dispersion coefficient, since E_L has been solved specifically for the case of including all processes leading to dispersion in the direction of axial flow.

Since Eqs. 10-2 through 10-5 look, and therefore mathematically behave, in the same manner as the constitutive transport equation describing Fickian diffusion, it is not absolutely necessary for someone to understand the origin of the different mechanisms that produce chemical dispersion in order to use these equations or solutions. However, the above equations are not always successful in describing chemical dispersion in systems, and thus it would appear the current description of chemical dispersion is inadequate. Certainly, the failure of these mathematical models to describe chemical dispersion must be due to incorrect assumptions or inadequate formulation of chemical dispersion in different systems. Therefore, it is the purpose of this chapter to describe in detail how chemical dispersion coefficients are developed and applied. Some readers may wish to skip this chapter and move directly into applications of dispersion calculations found in subsequent chapters of this text.

As we shall see in the sections and chapters that follow, single dispersion coefficients are most useful in describing the rate of bulk chemical transport over large distances. This approach does result in some limitations, however, since the transport equations may not adequately describe microscale processes. In Chapter 4 we derived the constitutive dispersion coefficient by assuming that there was in a system a continuous function ξ. In a porous medium, however, the function is not continuous over the whole domain since the soil grains block the bulk fluid. The ensuing tortuous path produces dispersion that can only be described by a dispersion coefficient that is a function of fluid advection. This tie between convection and dispersion is not expected from Eq. 10-2 and will present difficulties in developing exact mathematical solutions of transport in porous media.

10.2 AVERAGING PROPERTIES TO DERIVE DISPERSION COEFFICIENTS IN TURBULENT FLUIDS

Fluid turbulence arises from the random and chaotic nature of the fluid. When a fluid is turbulent, we cannot trace the fluid paths exactly since the streamlines cross in a random manner. Fluid eddies produced by fluid turbulence move in random directions and disperse both chemicals and fluid by a convective motion. Fluid motion is a three-dimensional process, and as we have already seen, if we know the exact instantaneous fluid velocities, we can describe the concentration of a chemical at any instant and at any location using the constitutive transport equation in two dimensions in the absence of reaction as

$$\frac{\partial c}{\partial t} + u_x \frac{\partial c}{\partial x} + u_y \frac{\partial c}{\partial y} = D \frac{\partial^2 c}{\partial x^2} + D \frac{\partial^2 c}{\partial y^2} \tag{10-6}$$

This equation is a valid descriptor even of a fluid turbulence, but in order to apply this equation we would need to be able to map exactly the fluid motion at every point in our control volume. If we could do that, we would not need such equations in the first place! Therefore, what we really seek is a method to characterize chemical dispersion in a manner that allows us to predict the *average* fluid and chemical properties of the system as a function of space and time.

In order to average the properties of the system, we choose to characterize the system in terms of an average fluid velocity, which u_x is

$$u_x = \bar{u}_x + u'_x \tag{10-7}$$

where the instantaneous velocity, u_x, is the difference between the average velocity $\bar{u}_x$ over some differential cross section, and u'_x is the velocity fluctuation measured at that point (Fig. 10.2). Notice that this approach has already been introduced in Chapter 7 to describe fluid mixing in tanks. From this definition, other attributes of these velocities must follow. Integrating the velocity over some period of time must result in our average velocity, or

$$\bar{u}_x = \frac{1}{T} \int_{t_0}^{t_0 + T} u_x \, dt \tag{10-8}$$

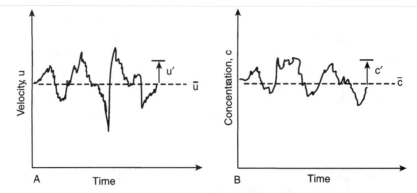

Figure 10.2 Instantaneous velocities in the x direction (A) and chemical concentrations (B) measured at a single point in a turbulent fluid. Each of these instantaneous measurements can be considered to be composed of the sum of the bulk average property and a fluctuation from that mean.

If we perform the same integration on the velocity fluctuations, we find that the integral is zero, or

$$\overline{u_x'} = \frac{1}{T} \int_{t_0}^{t_0+T} u_x' \, dt = 0 \tag{10-9}$$

Therefore, whenever we average velocities over a control volume, averages of velocity fluctuations will be zero, although the products of the fluctuations in different directions, such as $\overline{u_x' u_y'}$, will be nonzero terms.

If the fluid properties vary even at such small distances, then the concentration will also need to be described in terms of its variations at a point in space (Fig. 10.2) as

$$c = \bar{c} + c' \tag{10-10}$$

In order to see how turbulent fluctuations can be incorporated into a governing transport equation, let us examine the simplest case possible here for bulk flow in one dimension that gives rise to velocity deviations in two dimensions (Fig. 10.3). For this calculation we will use Eq. 10-6, assuming that the molecular diffusivity is a scalar, and that there is no net fluid velocity in the y direction so that $\bar{u}_y = 0$, but because there are velocity fluctuations in the y direction, $u_y' \neq 0$. Substituting in Eqs. 10-7 and 10-9 into Eq. 10-6, and averaging over time, we obtain

$$\frac{\partial \bar{c}}{\partial t} + \bar{u}_x \frac{\partial \bar{c}}{\partial x} = D \frac{\partial^2 \bar{c}}{\partial x^2} + D \frac{\partial^2 \bar{c}}{\partial y^2} + \frac{\partial(-\overline{u_x' c'})}{\partial x} + \frac{\partial(-\overline{u_y' c'})}{\partial y} \tag{10-11}$$

The two new terms on the right-hand side arise from turbulent fluctuations produced

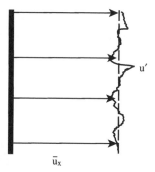

Figure 10.3 Advection in the x direction gives rise to velocity fluctuations not only in the x direction, but also in the y and z directions.

by fluid instability transported through fluid shear. These fluctuations cannot be individually predicted, since we have averaged the fluid over time, but on average it is found that these fluctuations behave in a manner analogous to Fick's Law. That is, these fluctuations give rise to random motion that spreads a chemical from a source in a way that can be described using the same equations as Fick used to describe molecular diffusion. This does not mean that a chemical spreads in proportion to a gradient; rather, it indicates that when such a gradient exists, the turbulence of the fluid will spread the chemical out in a manner consistent with changes in the concentration gradient.

From our analogy between turbulent fluctuations and diffusion, the two terms on the right-hand side of Eq. 10-11 can be replaced with mixing coefficients as

$$\overline{u_x'c'} = -e_x \frac{\partial \bar{c}}{\partial x} \tag{10-12}$$

$$\overline{u_y'c'} = -e_y \frac{\partial \bar{c}}{\partial y} \tag{10-13}$$

where e_x and e_y are mixing coefficients in the x and y directions. Rewriting Eq. 10-11 to include these new terms, we have

$$\frac{\partial \bar{c}}{\partial t} + \bar{u}_x \frac{\partial \bar{c}}{\partial x} = D \frac{\partial^2 \bar{c}}{\partial x^2} + D \frac{\partial^2 \bar{c}}{\partial y^2} + \frac{\partial}{\partial x}\left(e_x \frac{\partial \bar{c}}{\partial x}\right) + \frac{\partial}{\partial y}\left(e_y \frac{\partial \bar{c}}{\partial y}\right) \tag{10-14}$$

Assuming that both e_x and e_y are constant, we can take them outside the differential, and rearrange terms to obtain

$$\frac{\partial \bar{c}}{\partial t} + \bar{u}_x \frac{\partial \bar{c}}{\partial x} + (D + e_x) \frac{\partial^2 \bar{c}}{\partial x^2} (D + e_y) \frac{\partial^2 \bar{c}}{\partial y^2} \tag{10-15}$$

Since the diffusion and mixing coefficients have both been assumed to be constant, we can replace them with dispersion coefficients, where

$$E_x = D + e_x \tag{10-16}$$

$$E_y = D + e_y \tag{10-17}$$

Substituting these expressions into Eq. 10-15, we have more simply

$$\frac{\partial \bar{c}}{\partial t} + \bar{u}_x \frac{\partial \bar{c}}{\partial x} = E_x \frac{\partial^2 \bar{c}}{\partial x^2} + E_y \frac{\partial^2 \bar{c}}{\partial y^2} \tag{10-18}$$

At this point, it should be easy to see how our results could be generalized to three dimensions. First, we can add into our example velocity fluctuations produced in the z direction arising from fluid momentum in the x direction as

$$\frac{\partial \bar{c}}{\partial t} + \bar{u}_x \frac{\partial \bar{c}}{\partial x} = E_x \frac{\partial^2 \bar{c}}{\partial x^2} + E_y \frac{\partial^2 \bar{c}}{\partial y^2} + E_z \frac{\partial^2 \bar{c}}{\partial z^2} \tag{10-19}$$

Including advection in the y and z directions increases the terms on the left-hand side of the equation, since we now have bulk averaged transport in these directions, but does not affect the terms on the right-hand side since these are already averaged with respect to all three dimensions. As a result, we obtain the general result that

$$\frac{\partial \bar{c}}{\partial t} + \bar{\mathbf{u}} \cdot \nabla \bar{c} = \nabla \cdot \mathbf{E} \, \nabla \bar{c} \tag{10-20}$$

This result for the time-averaged constitutive transport equation indicates that we should be able to describe dispersion in such systems in exactly the same manner that we have already used to describe diffusion. There are both benefits and limitations to such an approach. The benefits are obviously that all our previously derived analytical solutions for diffusion should apply to dispersion. The limitations are that we have a model that implies chemical dispersion is forced by a concentration gradient, that is, by the properties of the chemical. In fact, in the case of a turbulent fluid, dispersion is produced as a result of the turbulent fluid and not a concentration gradient. When we change these fluid properties, for example, by changing the fluid velocity, the system may produce an entirely new pattern and intensity of velocity fluctuations, resulting in entirely new mixing coefficients e_x, e_y, and e_z. A high correlation between dispersion and advection makes it quite difficult to scale systems accurately for mixing and turbulence effects.

Longitudinal Dispersion

A special case of chemical dispersion frequently arises that is worth separately describing. In many chemical transport equations only a one-dimensional transport model is used. During flow through pipes, channels, streams, and rivers a chemical can become evenly spread across the radius or width (and depth, if appropriate) of the cross section. After this point, the chemical disperses but only in the axial, or longitudinal, direction. A one-dimensional model can be used to describe the distribution of the chemical at some distance from the point where the fluid was completely mixed across the cross section using a longitudinal transport coefficient.

We have already seen in the previous section that chemical dispersion in a two-dimensional system could be described by averaging the velocity fluctuations over a cross section. However, we considered those fluctuations to be driven by fluid turbulence. It is also possible that the velocity in the axial direction, u_x will be a function of y. If we wanted to average out the velocity due to both fluid turbulence and the velocity distribution, we would need to define

$$u_x = \bar{u}_x + [u'_x(t) + u'_x(\Delta u)] \tag{10-21}$$

where the notation in parentheses is added to indicate that these fluctuations arise from both fluid turbulence, t, and a velocity profile in the transverse or y direction (Δu). This term $u'_x(\Delta u)$ is really just a conceptual construct; a more rigorous derivation of the effect of transverse velocity gradients on dispersion will be given below. If we use Eq. 10-21 to repeat the analysis above for the case of a one-dimensional dispersion model, we obtain

$$\frac{\partial \bar{c}}{\partial t} + \bar{u}_x \frac{\partial \bar{c}}{\partial x} = (D + e_x) \frac{\partial^2 \bar{c}}{\partial x^2} + \frac{\partial}{\partial x} \overline{[-u'_x(\Delta u)c')]} \tag{10-22}$$

This last term on the right-hand side of Eq. 10-22 arises from chemical dispersion due to the velocity field as described in Figure 10.1. Following the same approach as above, we can argue that the last term can be replaced by a single coefficient, defined here as $e_{\Delta u}$, resulting in

$$\frac{\partial \bar{c}}{\partial t} + \bar{u}_x \frac{\partial \bar{c}}{\partial x} = (D + e_x + e_{\Delta u}) \frac{\partial^2 \bar{c}}{\partial x^2} \tag{10-23}$$

The separation of chemical dispersion into three terms of diffusion, turbulence, and a two-dimensional velocity field imposed on one dimension probably serves no purpose in practice, since it is difficult to separate these terms. In practice, we assign a single longitudinal mixing coefficient, E_L, to represent the sum of these three terms, or

$$E_L = D + e_x + e_{\Delta u} \tag{10-24}$$

With this definition of E_L, and now assuming that c and u_x are the average values for the system, we obtain our desired expression for a one-dimensional advection–

dispersion equation

$$\frac{\partial c}{\partial t} + u_x \frac{\partial c}{\partial x} = E_L \frac{\partial^2 c}{\partial x^2} \qquad (10\text{-}25)$$

This one-dimensional longitudinal dispersion equation applies only if the tracer initially fills the flow cross section. It can be applied to fluids under both laminar and turbulent flow conditions. Under laminar flow conditions the turbulent mixing coefficient, $e_x = 0$, but the spread of the chemical in the longitudinal direction will still be driven by velocity gradients in the transverse direction. As we shall see below for the case of chemical dispersion in a fluid flowing laminarly in a tube, the longitudinal dispersion coefficient can be derived either from an averaging approach, by averaging the velocity profile across the tube width, or by solving the equation for the concentration profile at some point downstream from the tube entrance.

10.3   DISPERSION IN NONBOUNDED TURBULENT SHEARED FLUIDS

In Chapter 3 we derived Fick's First Law for chemical diffusion based on the observation that the spread of the chemical was proportional to the concentration gradient. By analogy, dispersion can similarly be cast as

$$\mathbf{j}_C = -c_w \mathbf{E}_C \, \nabla x_C \qquad (10\text{-}26)$$

For one-dimensional transport in the x direction in a dilute system, this reduces to

$$j_{C,x} = -E_{C,x} \frac{dc_{Cw}}{dx} \qquad (10\text{-}27)$$

This formulation of dispersion implies that the concentration gradient drives dispersion, but keep in mind that this interpretation has no physical basis. The *dispersion* of a chemical in a fluid results from the dispersion of the fluid and the properties of the system. Thus chemical dispersion is a property of the system and not solely a property of the chemical. In the absence of fluid turbulence we will see that chemical dispersion can depend on the molecular diffusivity of the chemical. However, when dispersion is dominated by fluid turbulence, chemical dispersion is no longer a function of the properties of the molecule. As a result, two chemicals A and C in water under turbulent flow conditions will have the same dispersion coefficient. Accordingly, we could drop the double subscript on the dispersion coefficient, and rewrite Eq. 10-4 for turbulent dispersion as

$$j_{C,x} = -E_x \frac{dc_{Cw}}{dx} \qquad (10\text{-}28)$$

indicating that E_x is only a property of the fluid and not the chemical.

Since chemical dispersion occurs in an analogous fashion to Fickian diffusion, it is possible to make simple and direct calculations of chemical dispersion based on an estimate of $\mathbf{E}$ in a system. In the sections below we consider the case of a chemical introduced into an unbounded system, either as a spike input or as a continuous input. The chemical plume that arises in an unbounded system has properties of a normal distribution. Using equations developed to describe the properties of normal distributions, it is shown that the size of the plume can easily be calculated from the magnitude of the dispersion coefficient.

Dispersion Calculations

Spike Input (One-Dimensional Line Source)   Concentration profiles and chemical fluxes produced by chemical dispersion can be calculated in the same manner as for chemical diffusion. For chemical dispersion in an unbounded fluid, in the absence of advection and reaction

$$\frac{\partial c}{\partial t} = \mathbf{E} \; \nabla^2 c \tag{10-29}$$

Since these calculations are applicable to any phase, we will not use subscripts on concentration in this section. In one dimension, the dispersion of a chemical is

$$\frac{\partial c}{\partial t} = E_x \frac{\partial^2 c}{\partial x^2} \tag{10-30}$$

The solution of this equation requires three boundary conditions, one a function of time and two a function of distance x. We choose

$$BC\ 1:\ c(x, 0) = M/\delta \tag{10-31}$$

$$BC\ 2:\ c(\pm\infty, t) = 0 \tag{10-32}$$

$$BC\ 3:\ \frac{\partial c(0, t)}{\partial t} = 0 \tag{10-33}$$

The first boundary condition can be interpreted as a mass input m over an infinitely small area A in the y–z plane, or $M = m/A$, applied over some infinitesimal thickness δ. Concentration has units of mass per volume if we recognize that M has units of mass per unit area and δ is some small distance over which the mass is injected. This initial input of chemical decays away so that at sufficiently large distances ($x \rightarrow \infty$) the concentration is diluted out to zero (BC 2). Diffusion is symmetrical about the origin so at the origin $\partial c/\partial x = 0$. The solution of Eq. 10-30 with these three conditions is

$$c(x,\ t) = \frac{M}{(4\pi E_x t)^{1/2}} \exp\left(\frac{-x^2}{4E_x t}\right) \tag{10-34}$$

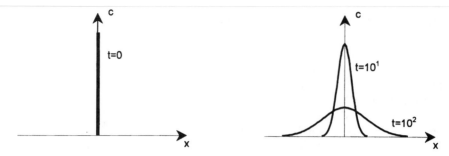

Figure 10.4 A spike input into a system at (A) time 0, and (B) at two time intervals later where the plume has spread out. Notice that as the chemical spreads out that for a spike input the concentration at the origin decreases. When mass is conserved, the areas under all curves should be equal.

A chemical spike into a system spreads out by fluid dispersion, producing a decrease in concentration at the source (Fig. 10.4).

Continuous Source When the source of the chemical is held constant, for example, from a leaking chemical drum or a buried waste, the transport equation remains the same but the boundary conditions change. In one dimension, the dispersion equation with no reaction is

$$\frac{\partial c}{\partial t} = E_x \frac{\partial^2 c}{\partial x^2} \tag{10-35}$$

The three boundary conditions chosen this time are

$$BC\ 1: t \rightarrow \infty \quad c(x) \rightarrow c_0 \tag{10-36}$$

$$BC\ 2: x = 0, \quad c = c_0 \tag{10-37}$$

$$BC\ 3: x \rightarrow \infty \quad c \rightarrow 0 \tag{10-38}$$

The chemical concentration is fixed at the source at a concentration c_0 (BC 2), and as time goes to infinity the system eventually fills up at that initial concentration (BC 1). The solution of Eq. 10-35 is

$$\frac{c}{c_0} = \text{erfc}\left(\frac{x}{(4E_x t)^{1/2}}\right) \tag{10-39}$$

where erfc is the complementary error function equal to 1-erf. The spread of the chemical under the conditions of a continuous input is shown in Figure 10.5. This solution of chemical dispersion with a constant source has applications for describing chemical dispersion in a lake, ocean, sediment, groundwater aquifer (in the absence of flow), and any case where a boundary does not affect the dispersion rate.

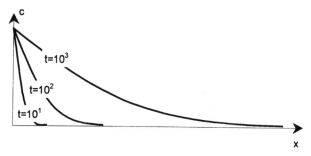

Figure 10.5 Concentration profile for a continuous input of chemical. Note that the concentration at the origin is fixed over time.

Simplifications for Dispersion Calculations

Dispersion calculations for chemicals in systems in the absence of advective flow and chemical reaction can be simplified based on the observation that the shape of a dispersion plume is the same as the shape of a normal distribution (Fig. 10.6). From the statistical properties of a normal distribution, we know that 95% of the distribution is contained within $\pm 2\sigma$, where σ is the standard deviation and σ^2 is the variance of the distribution. Therefore, 95% of the distribution is contained within a total width 4σ. Other intervals include 68% for 2σ and 99.7% for 6σ.

Since the dispersion of a chemical produces a concentration profile that has the same standard deviations and variances as a normal distribution, we can calculate the distance a chemical has spread by recognizing that 95% of the chemical plume will be contained within a distance 4σ. The variance of the chemical concentration is

$$\sigma^2 = \frac{\displaystyle\int x^2 c(x,\ t)\ dx}{\displaystyle\int c(x,\ t)\ dx} \tag{10-40}$$

where the integrals are evaluated from $-\infty$ to ∞. From Eq. 10-34, the numerator is

$$\frac{M}{(4\pi E_x t)^{1/2}} \int_{-\infty}^{\infty} x^2 \exp\left(\frac{-x^2}{4E_x t}\right) dx \tag{10-41}$$

From mathematical tables, we see that this equation has the form

$$\int_0^{\infty} y^2 e^{-ay^2}\ dy = \frac{1}{2^2 a} \sqrt{\frac{\pi}{a}} \tag{10-42}$$

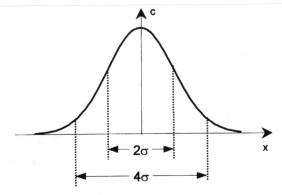

Figure 10.6   As a chemical spreads, 50% is contained within 2σ, 95% within 4σ, and 99% within 6σ.

The integration required in Eq. 10-41 is just double the integral from Eq. 10-42. Using this solution, Eq. 10-41 can be solved as

$$\frac{2M}{(4\pi E_x t)^{1/2}} \frac{1}{4^2[1/(4E_x t)]} \sqrt{\frac{\pi}{1/(4E_x t)}} \qquad (10\text{-}43)$$

which simplifies to

$$2ME_x t \qquad (10\text{-}44)$$

The denominator of Eq. 10-40 is evaluated as

$$\frac{M}{(4\pi E_x t)^{1/2}} \int_{-\infty}^{\infty} \exp\left(\frac{-x^2}{4E_x t}\right) dx \qquad (10\text{-}45)$$

which has a solution of the form

$$\int_{-\infty}^{\infty} e^{-a^2 y^2} \, dy = \frac{\sqrt{\pi}}{a} \qquad (10\text{-}46)$$

Using this solution, we can solve Eq. 10-45 as

$$\frac{c_0}{(4\pi E_x t)^{1/2}} \frac{\sqrt{\pi}}{[1/(4E_x t)]^{1/2}} = M \qquad (10\text{-}47)$$

Combining our solutions, we have the variance of the concentration profile is

$$\sigma^2 = \frac{2ME_xt}{M} \tag{10-48}$$

$$\sigma^2 = 2E_xt \tag{10-49}$$

Thus the variance of a chemical plume is a simple function of the dispersion coefficient.

If 95% of chemical is contained within 4σ, then the size of a chemical release, L, due to chemical dispersion in an unbounded system will be

$$L_E = 4\sigma = 4(2E_xt)^{1/2} \tag{10-50}$$

Example 10.1

Compare the distances that a plume of phenol will have moved after one day by diffusion ($D_{Cw} = 1.1 \times 10^{-5}$ cm^2 s^{-1}) to dispersion, assuming a dispersion coefficient of $E_x = 0.04$ m^2 s^{-1} found by Csanady (1963) for dispersion distances less than 1 km in Lake Huron.

The spread of the chemical can be described by diffusion or dispersion using Eq. 10-50. For diffusion, the distance covered by 95% of the chemical is

$$L_E = 4(2D_{Cw}t)^{1/2} = 4(2 \times 1.1 \times 10^{-5} \text{ cm}^2 \text{ s}^{-1} \times 1d \times 86{,}400 \text{ s/d})^{1/2} \tag{10-51}$$

$$= 5.5 \text{ cm}, \quad \text{diffusion} \tag{10-52}$$

For dispersion, the calculation is essentially the same

$$L_E = 4(2E_xt)^{1/2} = 4(2 \times 0.04 \text{ m}^2 \text{ s}^{-1} \times 1d \times 86{,}400 \text{ s/d})^{1/2} \tag{10-53}$$

$$= 332 \text{ m}, \quad \text{dispersion} \tag{10-54}$$

We see that a dispersion coefficient over 7 orders of magnitude (OOM) greater than the diffusion coefficient can spread a plume over distances 4–5 OOM larger than possible by diffusion due to the square root dependence of the size of the plume on the diffusion or dispersion coefficient. This calculation demonstrates that we are justified in neglecting chemical diffusion relative to dispersion in natural systems such as lakes.

Because chemical dispersion can be so large compared to diffusion, it is important to consider the dispersive velocity in comparison to diffusive velocities in such systems. Using our definition that the length of a spill spreads a distance L_E, we can calculate a dispersion velocity, u_E, as

$$u_E = \frac{L_E}{t} = \left(\frac{32E_x}{t}\right)^{1/2} \tag{10-55}$$

This dependence of velocity on time to the $-\frac{1}{2}$ power indicates that the velocity of the plume will decrease with time as the plume spreads out.

Example 10.2

Compare the velocity produced by chemical dispersion after 1 h and 1 d to a shear velocity of surface waters of 0.008 m s^{-1} produced by a wind speed of 6.3 m s^{-1} recorded 10 m above the lake surface so that $E_x = 0.04$ m^2 s^{-1}.
    From Eq. 10-55 the dispersion velocity after 1 h is

$$u_E = \left(\frac{32 \times 0.04 \text{ m}^2 \text{ s}^{-1}}{3600s}\right)^{1/2} = 0.02 \text{ m s}^{-1} \tag{10-56}$$

After one day we similarly calculate $u_E = 0.004$ m s^{-1}. This indicates that dispersion initially carries the chemical at a faster velocity than advection, although after one day the velocities due to dispersion and advection have become similar.

10.4 LONGITUDINAL DISPERSION COEFFICIENTS FOR DEFINED SYSTEMS

The longitudinal dispersion coefficient represents a special case for dispersion calculations since three different mechanisms contribute to the one-dimensional coefficient E_L: molecular diffusion, turbulent mixing, and velocity gradients in more than one dimension. It is interesting to examine mathematically which of these processes produce the largest effects on E_L. However, before we do this we should review the situation.
    In the absence of fluid mixing chemical transport should be relatively easy to calculate since the velocity flow field is well defined. Dispersion will be produced by molecules diffusing to faster or slower streamlines, and in principle we should be able to calculate the average concentration across a plane by integration. When the flow is turbulent, we have two possible cases: Turbulent eddies are either very large or eddies are very small relative to the average axial velocity. Very large eddies would destroy any velocity field in the transverse direction and make the system completely mixed across the plane perpendicular to flow. In this case, there would be no net transverse velocity gradient, and therefore no longitudinal dispersion due to dispersion in the direction perpendicular to fluid flow. In the case of small eddies,

however, the spread of the chemical by the transverse velocity field dominates and turbulence (or eddy diffusion) contributes much less to dispersion than the transverse velocity field.

In any case of significant longitudinal dispersion, we expect that the major contribution to dispersion in the axial direction will be a result of the spreading of the chemical in the transverse direction into regions with different axial velocities. In the case of dispersion under laminar flow conditions the rate of chemical dispersion should be a function of the molecular diffusivity. In the case of turbulent dispersion, the rate of spread will result from the transverse mixing. The important distinction between these situations is that in the former case dispersion will be a result of both the chemical and fluid properties, while in the latter case dispersion will not be a function of chemical properties (molecular diffusivity).

Analysis of Dispersion under Laminar Flow Conditions

In the absence of fluid mixing in the transverse direction, chemical dispersion in the axial direction, x, will result from the axial velocity being a function of the transverse direction, y, or $u_x(y)$. The relationship between the dispersion coefficient and the fluid velocities and concentrations can be derived by averaging over the cross section whether the fluid is flowing under laminar or turbulent conditions. This point is worth emphasizing, and so a dispersion coefficient for laminar flow conditions will be derived below to illustrate this process for two cases: shear flow between two plates and laminar flow in a tube. Before examining these specific systems, let us first consider the averaging process used to calculate E_L in the absence of fluid eddies.

Let us consider the case of a fluid bounded by two walls through which the fluid moves at an average velocity, $\bar{u}_x$, which can be obtained by integration of the velocity across the height of the channel h in the direction y (Fig. 10.7a) as

$$\bar{u}_x = \frac{1}{h} \int_0^h u_x \, dy \qquad (10\text{-}57)$$

Based on this definition, the velocity differs from the mean velocity as a function of y according to

$$u_x' = u_x - \bar{u}_x \qquad (10\text{-}58)$$

Since the flow is laminar across the width of the channel, the variations in velocity in this case are due only to the velocity profile across the channel width, not to fluid mixing. Similarly, the concentrations across the channel can be expressed as

$$\bar{c} = \frac{1}{h} \int_0^h c \, dy \qquad (10\text{-}59)$$

$$c' = c - \bar{c} \qquad (10\text{-}60)$$

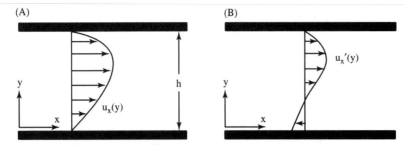

Figure 10.7 Velocity distributions between two walls under laminar flow conditions. (A) Velocity distribution as viewed from a fixed coordinate system when the velocity in the x direction is a function of y. (B) Same velocity distribution as in (A) but this time as viewed by an observer moving with the fluid at the mean velocity. Compare this situation with the one in Figure 10.3, where actual convective flow was produced in the y direction as a result of turbulence.

A transport equation describing the spread of the chemical written in terms of a fixed observer is

$$\frac{\partial}{\partial t}(\bar{c} + c') + (\bar{u}_x + u'_x)\frac{\partial}{\partial x}(\bar{c} + c') = D\left(\frac{\partial^2}{\partial x^2}(\bar{c} + c') + \frac{\partial^2 c'}{\partial y^2}\right) \qquad (10\text{-}61)$$

If we now simplify Eq. 10-61 for an observer moving with the cloud of chemical (Fig. 10.7b), we can neglect the average velocity term, since our coordinate system is moving along with the flow. Assuming that diffusion in the axial direction is smaller than the advective term, we can simplify Eq. 10-61 to

$$\frac{\partial}{\partial t}(\bar{c} + c') + u'_x\frac{\partial}{\partial x}(\bar{c} + c') = D\frac{\partial^2 c'}{\partial y^2} \qquad (10\text{-}62)$$

A comparison of the magnitude of each of these terms will show that the only two terms significant in magnitude (Fischer et al., 1979) are

$$u'_x\frac{\partial \bar{c}}{\partial x} = D\frac{\partial^2 c'}{\partial y^2} \qquad (10\text{-}63)$$

Assuming that there is no flux through the walls, or that $\partial c'/\partial y = 0$ at $y = 0$ and $y = h$, the solution of Eq. 10-63 for the chemical concentration profile in the transverse direction, or across the width of the channel, is

$$c'(y) = \frac{1}{D}\frac{\partial \bar{c}}{\partial x}\int_0^y\int_0^y u'_x \, dy \, dy + c'_0 \qquad (10\text{-}64)$$

With this description of the chemical concentration profile across the channel we can now derive an expression for the rate of chemical transport in the longitudinal, or x = axial direction. Integrating the advective flux over the width of the channel produces the net chemical rate of transport

$$W_x = w \int_0^h u_x' c' \, dy \tag{10-65}$$

where w is the channel width in the z direction. Substituting Eq. 10-64 into Eq. 10-65 produces

$$W_x = \frac{w}{D} \frac{\partial \bar{c}}{\partial x} \int_0^h u_x' \int_0^y \int_0^y u_x' \, dy \, dy \, dy \tag{10-66}$$

where the term $\int_0^h u_x' c_0' \, dy = 0$ because $\int_0^h u_x' \, dy = 0$.

The most important result of this derivation is that we have shown that the rate of chemical transport in the axial direction is proportional to the concentration gradients in the axial direction. This is exactly the same statement made by Fick's First Law, that diffusion in a direction x is proportional to the concentration gradient in the x direction. Because of the similarity of the two processes of diffusion and dispersion in the axial direction, we define a longitudinal dispersion coefficient in terms of the rate of transport as

$$W_x = -hwE_L \frac{\partial \bar{c}}{\partial x} \tag{10-67}$$

so that the flux by dispersion is

$$j_x = -E_L \frac{\partial \bar{c}}{\partial x} \tag{10-68}$$

where E_L is the longitudinal dispersion coefficient. Comparing Eqs. 10-66 and 10-67, we see that E_L can be calculated as

$$E_L = -\frac{1}{hD} \int_0^h u_x' \int_0^y \int_0^y u_x' \, dy \, dy \, dy \tag{10-69}$$

Since the longitudinal dispersion coefficient can be used for transport calculations in the same manner as molecular diffusion, we can also use it in Fick's Second Law, as

$$\frac{\partial \bar{c}}{\partial t} = E_L \frac{\partial^2 \bar{c}}{\partial x^2} \tag{10-70}$$

These equations describe dispersion from a moving coordinate system. If we use a fixed coordinate system, the average velocity of the bulk fluid must be incorporated back into the governing transport equation as

$$\frac{\partial \bar{c}}{\partial t} + \bar{u}\,\frac{\partial \bar{c}}{\partial x} = E_L\,\frac{\partial^2 \bar{c}}{\partial x^2} \tag{10-71}$$

Because we have dropped some of the terms in Eq. 10-61 in deriving Eq. 10-71, the longitudinal dispersion equation is only valid after a period where the fluid has traveled far enough, since the addition of the chemical so that c' becomes small because cross-sectional diffusion evens out the cross-sectional concentration gradients. After a period of time $t = 0.4h^2/D$ the cloud of chemical will be normally distributed across the channel of width h. The variance of the cloud will then grow linearly with time so that the cloud will move with mean velocity $\bar{u}_x$ and spread at a rate in the longitudinal direction described by the variance $\sigma^2 = 2E_L t$.

Simple Shear Flow between Two Plates   One of the simplest cases of longitudinal dispersion is the spread of a chemical between two parallel plates of infinite length and width when the top plate is moving at a velocity $u_p/2$ relative to the bottom plate moving at the same velocity in the other direction. The two plates are separated by a distance h. The velocity profile between the two plates is laminar and can be described as

$$u_x(y) = u_p\,\frac{y}{h} \tag{10-72}$$

where u_p is the maximum velocity of the fluid achieved by a no-slip condition at the fluid wall occurring at a distance $h/2$ from the centerline between the plates at the origin of our system (Fig. 10.8).

At a time larger than $0.4h^2/D$ a chemical injected into the fluid will have spread out so that the dispersion in the longitudinal direction can be calculated from

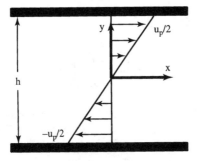

Figure 10.8   Shear flow between two plates with each plate moving at a velocity of $u_p/2$ in opposite directions under laminar flow conditions. Therefore, the velocity is in the x direction but the magnitude is a function of y.

Eq. 10-69 as

$$E_L = -\frac{1}{hD} \int_{-h/2}^{h/2} u_p \frac{y}{h} \int_{-h/2}^{y} \int_{-h/2}^{y} u_p \frac{y}{h} \, dy \, dy \, dy \qquad (10\text{-}73)$$

Solving, we have

$$E_L = \frac{u_p^2 h^2}{120D}, \quad \text{laminar shear between two plates} \qquad (10\text{-}74)$$

This result indicates that *dispersion* in the axial direction is inversely proportional to the *molecular diffusion* coefficient of a chemical in the fluid. When the diffusion coefficient is very large, the chemical will undergo little dispersion, since molecular diffusion will allow the molecules to move across the width of the channel over the same period of time as advection in the axial direction. Conversely, compounds with low diffusivities will be highly dispersed since some chemical on slowly moving streamlines, or streamlines in the $-u_p/2$ direction, will move much more slowly in the $+u_p$ axial direction compared to the average flow and will therefore be highly dispersed in the fluid.

Laminar Flow in a Pipe   The analysis of longitudinal dispersion of laminar flow in a pipe is based on the classic work of Taylor (1953). The velocity distribution in the radial direction in a pipe is

$$u_r = u_{\max}(1 - r^2/R^2) \qquad (10\text{-}75)$$

where u_r is the velocity in the axial direction that is a function of the radial distance from the center, $u_{\max}$ the maximum velocity at the pipe center, and R the pipe radius (Fig. 10.9). The convective dispersion equation in two dimensions in cylindrical

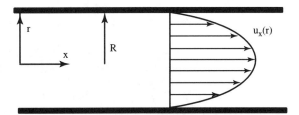

Figure 10.9   Velocity profile in a pipe of radius R under laminar flow conditions. Fluid flow in the axial (x) direction produces a velocity profile with pipe radius that is parabolic in shape. Dispersion in the axial direction is produced by different fluid velocities in the radial direction even in the absence of fluid convection in the radial direction.

coordinates is

$$\frac{\partial c}{\partial t} + u_{max}\left(1 - \frac{r^2}{R^2}\right)\frac{\partial c}{\partial x} = D\left(\frac{\partial^2 c}{\partial r^2} + \frac{1}{r}\frac{\partial c}{\partial r} + \frac{\partial^2 c}{\partial x^2}\right) \qquad (10\text{-}76)$$

Through comparison of the magnitudes of the terms we determine that we can neglect $\partial^2 c/\partial x^2$ and $\partial c/\partial t$. Setting $r^* = r/R$, and using a moving coordinate system, Eq. 10-76 becomes

$$\frac{u_{max}R^2}{D}\left(\tfrac{1}{2} - r^{*2}\right)\frac{\partial \bar{c}}{\partial x} = \frac{\partial^2 c'}{\partial r^{*2}} + \frac{1}{r^*}\frac{\partial c'}{\partial r^*} \qquad (10\text{-}77)$$

Using the boundary condition that $\partial c/\partial r^* = 0$ at $r^* = 1$, integrating twice, and substituting in the average velocity across the pipe, $\bar{u} = u_{max}/2$, we have

$$c' = \frac{\bar{u}R^2}{4D}\left(r^{*2} - \frac{r^{*4}}{2}\right)\frac{\partial \bar{c}}{\partial x} + b_1 \qquad (10\text{-}78)$$

From this definition of concentration in Eq. 10-78, we can calculate the longitudinal dispersion coefficient as before as

$$E_L = \frac{\bar{u}^2 R^2}{48D}, \quad \text{laminar flow in a pipe} \qquad (10\text{-}79)$$

Once again we see that the longitudinal dispersion coefficient is inversely proportional to the molecular diffusion coefficient of a chemical in water. Taylor demonstrated that his analysis was valid for E_L in the range $(4L/R) < \text{Pe} < 7$, where $\text{Pe} = \bar{u}L/D$.

Aris (1956) hypothesized that as fluid velocity decreases in a pipe, the longitudinal dispersion coefficient should approach the case of pure diffusion since $E_L \rightarrow 0$ as $(D + e_{\Delta u}) \rightarrow D$. By examining a lower range of $\text{Pe} \ll 4L/D$, Aris extended Taylor's analysis for flow in a pipe to obtain

$$E_L = D + \frac{\bar{u}^2 R^2}{48D}, \quad \text{laminar flow in a pipe} \qquad (10\text{-}80)$$

This result demonstrated that dispersion was additive to diffusion in a pipe. In many cases, however, since D is a very small number, $E_L \gg D_{Cw}$. For example, in a tube of radius $R = 2$ mm with $\bar{u} = 1$ cm s^{-1}, $E_L = 20$ cm^2 s^{-1}, assuming a typical diffusivity for a chemical in water of $D_{Cw} = 10^{-5}$ cm^2 s^{-1} (Fischer et al., 1979). This analysis does require, however, that the chemical has been in the tube at time $t > 0.4R^2/D_{Cw}$ so that the chemical is dispersed normally in a tube. For our example, this would require a distance of 800 cm, or 4000 tube radii. Additional information on the relationship between diffusion and dispersion for flow in a laminar tube was presented in Section 3.4.

Analysis of Dispersion under Turbulent Flow Conditions

The analysis of turbulent shear flow in a pipe (Fig. 10.10) is somewhat more complex than that of longitudinal dispersion under laminar flow conditions due to assumptions that must be made about the nature of the fluid turbulence. However, the same averaging procedure used above to obtain a single longitudinal dispersion coefficient can be used to derive a longitudinal dispersion coefficient under turbulent flow conditions. In the two sections below longitudinal dispersion coefficients are derived for the simple cases of turbulent flow in a pipe and turbulent flow down an infinitely wide plane.

Turbulent Flow in a Pipe   For turbulent flow in a pipe, the analysis is conducted in the same manner as for laminar flow conditions except that the turbulent diffusivity e_r is used to describe the dispersion of the chemical across the channel width instead of the molecular diffusivity, D, since $e_r \gg D$ for turbulent flow conditions. Therefore, we can begin the analysis for turbulent flow with Eq. 10-63 using e_r as

$$ u_x' \frac{\partial \bar{c}}{\partial x} = \frac{\partial}{\partial r} \left(e_r \frac{\partial c'}{\partial r} \right) \qquad (10\text{-}81) $$

The turbulent mixing coefficient is a function of r and therefore remains inside the derivative. Similarly, the dispersion coefficient is obtained from Eq. 10-69 for turbulent diffusion in a pipe as

$$ E_L = -\frac{1}{R} \int_0^h u_r' \int_0^r \frac{1}{e_r} \int_0^r u_r' \; dr \; dr \; dr \qquad (10\text{-}82) $$

At this point, something must be known about the nature of the fluid dispersion term e_r as a function of r. Taylor (1954) recognized that all turbulent velocity profiles in pipes can be described by the same equation in terms of the shear velocity, u_{sh},

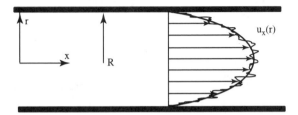

Figure 10.10 Velocity profile in a pipe of radius R under turbulent flow conditions. Fluid turbulence now produces instantaneous velocity fluctuations in both the radial and axial directions, but only those that arise in the radial direction significantly contribute to axial dispersion. Therefore, dispersion in the axial direction is a combination of radial dispersion and the velocity profile in the radial direction.

according to

$$u_x(y) = 2\bar{u} - u_{sh} f(r^*) \tag{10-83}$$

where $\bar{u}$ is the velocity averaged across the pipe cross section, u_{sh} is the shear velocity defined as $u_{sh} = (\tau_0/\rho_w)^{1/2}$, where τ_0 is the shear stress at the pipe wall, ρ_w is the density of water, and $f(r^*)$ is an empirical function that describes the shape of velocity profiles in pipes under turbulent flow conditions as a function of the dimensionless pipe radius r/R. Taylor reasoned that a description of momentum dispersion across the pipe would also describe chemical dispersion across the pipe, as known from the "Reynolds Analogy" (see Chapter 5).

This analogy between the transport equations under turbulent conditions is an important point and worth further discussion. It was shown in Chapter 4 that there was an analogy between the three laws describing dispersion of heat, mass, and momentum for air since the diffusion coefficients for these quantities were identical for air. In water, however, this equality failed since the diffusion coefficients of heat, mass, and momentum for water were not of the same magnitude. However, under turbulent flow conditions we have already seen that chemical dispersion is much larger than molecular diffusion. In fact, chemical dispersion occurs at exactly the same rate as momentum diffusion. This means that the momentum dispersion coefficient for water, which we could designate as $e_{w,r}$, would be identical to the mass dispersion coefficient, $e_{C,r}$, which we have written here in abbreviated notation as e_r. Based on this condition, we can write for turbulent flow conditions that

$$e_r = \frac{J_{Cw}}{-\partial \bar{c}/\partial r} = \frac{\tau}{-\partial(\rho u_x)/\partial r} \tag{10-84}$$

To continue our analysis, Taylor next used this equality to relate the fluid shear velocity to e_r, making use of the fluid shear term in Eq. 10-84. As already discussed, the shear stress at the wall is related to the shear velocity by $u_{sh} = (\tau_0/\rho_w)^{1/2}$. The shear at any distance is a function of radial distance according to $\tau = \tau_0(r/R) = \tau_0 r^*$. Combining these expressions, the shear can be expressed as a function of radial distance and shear velocity as

$$\tau = \rho u_{sh}^2 r^* \tag{10-85}$$

Substituting Eq. 10-84 for τ in and Eq. 10-83 for u_x into Eq. 10-85, we have

$$e_r = \frac{\tau}{-\partial(\rho u_x)/\partial r} = \frac{\rho u_{sh}^2 r^*}{-\partial(\rho[2\bar{u} - u_{sh} f])/\partial r} = \frac{R r^*}{df/dr^*} u_{sh} \tag{10-86}$$

From experimental results of velocity profiles in pipes, Taylor was able to tabulate values of f as a function of r, resulting in a tabulated expression for $u' = u_x(r) - \bar{u}$

and also for e_r. The governing transport equation in radial coordinates becomes

$$u'_r \frac{\partial \bar{c}}{\partial r} = e_r \left(\frac{\partial c'^2}{\partial r^2} + \frac{1}{r} \frac{\partial c'}{\partial r} \right)$$

(10-87)

This expression was then integrated using the experimental description of the velocity profile in a pipe, to obtain a longitudinal dispersion coefficient of

$$E_L = 10.1 R u_{sh}, \quad \text{turbulent flow in a pipe}$$

(10-88)

Notice that for a turbulent fluid E_L is no longer a function of the molecular diffusivity of the compound and is only a property of the fluid velocity and system configuration (pipe radius).

Turbulent Flow down an Infinitely Wide Plane   To analyze longitudinal dispersion in channels Elder (1959) assumed that the channel was an infinitely wide plane and that the flow could be described according to the von Karmen logarithmic velocity profile

$$u'_x = \frac{u_{sh}}{\kappa} [1 + \ln(y/h)]$$

(10-89)

where κ is the von Karmen constant, which is assumed to be equal to 0.40. A force balance over the depth of flow h (Fischer et al., 1979) yields

$$\tau = \rho e_y \frac{du_x}{dy} = \tau_0 [1 - (y/h)]$$

(10-90)

where τ_0 is the shear stress at the bottom, which we have seen before is calculated from the shear velocity as

$$u_{sh} = (\tau_0/\rho)^{1/2}$$

(10-91)

Combining the above two results, the transverse mixing coefficient can be calculated from

$$e_y = \kappa y u_{sh} [1 - (y/h)]$$

(10-92)

The longitudinal dispersion, E_L, can be calculated by assuming that the e_y equally describes mass and momentum transport, and by substituting Eqs. 10-91 and 10-92 into Eq. 10-82 and integrating. The first two integrations are quite difficult (Fischer et al., 1979) but produce

$$c' = \frac{\partial c'}{\partial x} \frac{h}{\kappa^2} \left(\sum_{n=1}^{\infty} \frac{1}{n^2} \left(\frac{h-y}{y} \right)^n - 0.648 \right)$$

(10-93)

The third integration produces the longitudinal dispersion coefficient

$$E_L = \frac{0.404 u_{sh} h}{\kappa^3}$$ (10-94)

Assuming $\kappa = 0.40$, we have the simple result (Fischer et al., 1979)

$$E_L = 0.593 u_{sh} h, \quad \text{turbulent flow down a plane}$$ (10-95)

where once again we see that for turbulent flow conditions the longitudinal dispersion coefficient is independent of the molecular diffusivity because the spread of the chemical occurs at a rate proportional to the fluid properties and not the properties of the chemical.

10.5 DISPERSION IN POROUS MEDIA

A description of chemical transport in porous media must include at least two phases and therefore is more difficult than chemical transport in just one-phase systems such as water or air. The very nature of the flow resulting from fluid pushed through a collection of impermeable particles is not just complex; a critical assumption made in deriving the general transport equation is not strictly valid for porous media. In setting up the control volume for the GTE we assumed that there was some arbitrary function ξ that was *continuous* over the domain that would be described by our resulting equation. If we examine the two-dimensional control volume shown in Figure 10.11a, we can see that our mass balance will be different depending on whether our differential element falls far from the solid particle, near, or within the solid particle. Our difficulty in defining a representative microscopic element for the

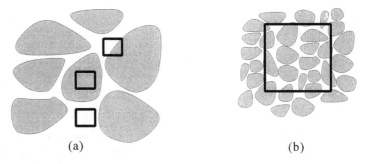

(a)   (b)

Figure 10.11 Different possibilities for choosing a differential element in porous media shown in two dimensions. (a) A microscale element that defines flow in the system will vary depending on whether it is drawn outside, inside, or overlapping a porous medium particle. (b) A more macroscopic differential element is chosen to include many particles leading to a conventional advection–dispersion transport equation.

system means that equations developed to describe the macroscopic properties of the system will, to some extent, be incomplete.

The situation relative to deriving a governing transport equation is of course not hopeless. Instead of defining our control element as some truly microscopic property of the system, as implied by a differential element, we instead can consider a differential element that will preserve the macroscopic properties of the system. Our differential element therefore is defined to contain many porous medium particles (Fig. 10.11b). This definition of the control volume has the effect of averaging the fluid properties out over many particles in the system. In Chapter 12 we will show that this control volume leads to the constitutive transport equation for a conservative, nonadsorbing chemical

$$\frac{\partial c}{\partial t} + \mathbf{u} \cdot \nabla c = \nabla(\mathbf{E} \cdot \nabla c) \tag{10-96}$$

which is exactly the same form as we previously derived for transport in one-phase flow such as a river. Thus our choice of a control volume that includes the porous medium particles produces a governing equation that looks and behaves as a typical advection–dispersion equation. However, this approach does not give us any insight into how to specify important macroscopic parameters such as the transverse and longitudinal dispersion coefficient. Paradoxically, the microscopic properties of the system must be used to develop terms in the macroscopic governing transport equation.

Porosity   The porosity of the porous medium affects fluid hydraulics and chemical transport. Medium porosity, θ, is defined as

$$\theta = \frac{\text{volume voids}}{\text{total volume}} \tag{10-97}$$

The porosity of a porous medium cannot be predicted but instead must be measured. In laboratory columns this is usually done by weighing packed columns dry and saturated. The porosity of a medium is a function of the soil grain size distribution and packing arrangement of particles. Generally, repacked soils have a higher porosity and a larger fraction of mobile small particles than undisturbed soils.

Even when a porous medium is composed of perfectly homogeneous sized spheres, the porosity of the media may not be homogeneous. For example, consider the arrangement of spherical particles shown in two dimensions in Figure 10.12. When the spheres are arranged in a linear or cubic orientation, the porosity is 0.4764. If particles are offset in a rhombohedral orientation, $\theta = 0.2596$ (Fig. 10.11). When columns are packed with spherical particles, both cubic and rhombohedral packing arrangements will contribute to the overall porosity, but other, more random particle arrangements will also occur. The final orientation of the spheres in the porous medium will be some mixture of cubical, rhombohedral, and random packing, re-

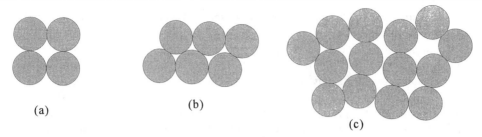

(a)

(b)

(c)

Figure 10.12 Packing of even identical-sized spheres in different orientations can result in different porosities, making it nearly impossible to produce a truly homogeneous medium. (a) Perfect cubic arrangement (θ = 0.4764), (b) perfect rhombohedral arrangement (θ = 0.2596), and (c) random arrangement.

sulting in medium porosities in the range of 0.32 to 0.40 even for porous medium composed only of spherical particles.

Diffusion versus Dispersion in a Porous Medium We have already discussed that dispersion can be described as a sum of different terms as

$$E_L = D + e_x + e_{\Delta u} \tag{10-98}$$

where D is the molecular diffusivity, e_x a mixing coefficient due to fluid turbulence, and $e_{\Delta u}$ a term arising from a nonuniform velocity profile. For a porous medium, we can add an additional term, e_{mech}, defined here as a mechanical dispersion. Thus we could now write an overall longitudinal dispersion coefficient as

$$E_L = D_s + e_x + e_{\Delta u} + e_{mech} \tag{10-99}$$

where we have also indicated the molecular diffusivity must now take into account the porous structure of the medium. In the absence of fluid motion, it has already been shown (Chapter 3) that chemical diffusivity in a porous medium is

$$D_{Cw,s} \equiv D_s = D_{Cw} \frac{\theta}{\tau_f} \tag{10-100}$$

where τ_f is a tortuosity factor representing the tortuous paths in the medium. We have simplified the notation for the effective diffusivity of chemical C in water in the soil as D_s, the molecular diffusivity of a chemical in a porous medium. A typical value is $D_s = 0.67 D_{Cw}$ for a chemical in a sandy material.

Let us consider only the case of laminar flow in the porous medium, or that $e_x = 0$. Therefore, there are really three basic mechanisms that contribute to chemical dispersion in porous media: molecular diffusion, hydrodynamic dispersion, and mechanical dispersion. Porous media can be viewed as a series of channels or tubes moving through a solid structure, with each tube having a variable diameter and direction (Fig. 10.13). Hydrodynamic dispersion results from the parabolic velocity profile of the fluid in each tube. Thus within each tube there is dispersion as a result of pore diffusion and Taylor–Aris dispersion ($D_s + e_{\Delta u}$). The velocity profile results in dispersion that is inversely proportional to the molecular diffusivity of the chemical in the presence of any appreciable advective velocity. Taylor–Aris dispersion was previously presented for tubes of uniform diameter, but other tube geometries and configurations are discussed below.

Mechanical dispersion is produced by the noncontinuous nature of the medium: Fluid must flow around particles, leading to different path lengths and different mean velocities. As a consequence of the heterogeneity of media, the average tube diameter, and therefore the average fluid velocity, is not the same in all tubes. Fluid moving from point A within two different tubes will arrive at point B at two different times, resulting in chemical dispersion of fluid within a medium. The average velocity of the fluid is the sum of all the different fluid velocities. The velocity at any point, u, can be described in terms of the sum of the average and instantaneous velocity, or $u = \bar{u} + u'$, as was done previously for turbulent flow (see Eq. 10-58), except here the deviations from the velocity are due to flow in different channels (Fig. 10.14). The same mathematical approach used previously for fluid turbulence can also be applied here to produce a mechanical dispersion coefficient, e_{mech}. Thus longitudinal dispersion for laminar flow in a porous medium can be described as the sum of three factors as

$$E_L = D_s + e_{\Delta u} + e_{\text{mech}} \tag{10-101}$$

where the first two terms of the right-hand side of Eq. 10-101 arise from Taylor–Aris dispersion, and the last term is due to mechanical dispersion. The combination of mechanical and hydrodynamic dispersion produces an overall dispersion rate of a chemical that is many times larger than that of molecular diffusion alone.

Figure 10.13 The porous medium can be analyzed as a series of tubes through which water flows. Within each tube dissolved chemicals will spread in the direction of flow according to Taylor–Aris dispersion. Since the average velocity is not the same in all tubes, mechanical dispersion results from chemicals moving from one tube to another tube.

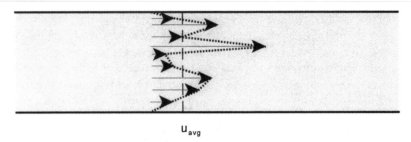

$$u_{avg}$$

Figure 10.14 Fluid velocities through a porous medium can be viewed on a macro-scale as an average velocity (shown by the vertical dashed line) plus deviations from the average, u' (each shown by an arrow) shown, or $u(y) = \bar{u} + u'$ (shown as the dashed line). As a result, the overall longitudinal dispersion for transport in a porous medium can be calculated in terms of an overall dispersion coefficient, E_L, in the same manner as previously described for turbulent flow or laminar flow in a pipe. The slower-moving paths lead to "negative velocities" relative to the average velocity of all flow velocities in the porous medium.

Longitudinal Dispersion in Channels of Different Geometries

Dispersion can be calculated from the solution of a transport equation if the geometry of a tube, or more generally of a channel, is well defined. For example, if a porous medium consisted of equal-sized grains packed in a perfectly homogeneous manner, then we might choose to define the medium as a series of concentric tubes, all of uniform diameter (Fig. 10.15). Dispersion in each tube would occur according to the Taylor–Aris model, or

$$E_L = D_s + \frac{\bar{u}^2 R^2}{48 D_s} \tag{10-102}$$

where D_s is the molecular diffusivity corrected for the tortuosity of the porous medium, and E_L the longitudinal dispersion coefficient. Curved tubes introduce com-

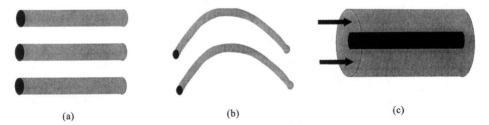

(a)   (b)   (c)

Figure 10.15 Different type of tube arrangements assuming flow in the porous media occurs as flow through a series of capillary tubes: (a) flow through circular and parallel tubes, (b) flow through curved tubes, (c) flow through an annulus where the porous media particles form the inner annulus.

plications beyond that implied by the above equation because secondary circulatory motions can develop in the presence of sharp curves (Jones, 1968).

The tube could instead be considered to be formed around a series of particles, resulting in a model of fluid flow in a concentric annulus. For narrow gaps, dispersion in a concentric annulus (Nunge and Gill, 1970) is

$$E_L = D + \frac{8}{495} \frac{u_{max}^2 R_o^2 (1 - w_g)}{4D} \tag{10-103}$$

where R_o is the radius of the outer cylinder, w_g the width of the gap between the cylinders, and u_{max} the maximum velocity of the flow between the cylinders. As $1/w_g \rightarrow 100$, Eq. 10-103 approaches the same expression for dispersion in a tube. For a wide range of radius ratios and eccentricities the dispersion coefficient may vary by over two orders of magnitude (Sankarasubramanian, 1969).

Even if the porous medium consisted of homogeneous-sized particles, a tube would not be perfectly uniform in diameter or straight. For example, if the medium is considered to be a series of repeating jagged plates, such as that shown in Figure 10.16a, enlargements and contractions in the flow will increase dispersion. Turner (1958) considered a porous medium to consist of a series of deadend pockets (Fig. 10.16b), although for creeping flow such pockets may not truly be stagnant, since there can be flow through some portion of the pocket. For Pe > 100, where Pe = ud/D, the mean concentration can be modeled (Gill and Ananthakrishnan, 1966) as

$$\frac{c}{c_0} = \frac{1}{2} \left[1 - \mathrm{erf} \left(\frac{x - u_{max}t/2R^{*2}}{(4Dt/R^{*2})^{1/2}} \right) \right] \tag{10-104}$$

where $R^* = a/b$, R_o is the outer tube radius, or the tube radius without rectangular side zones, and R_i the inner radius or radius with side zones (Fig. 10.16b). The overall dispersion coefficient (Aris, 1959) is

$$E_L = \frac{u_{max}R^2}{192D} \left(\frac{8R^{*2} + 12R^{*4} \ln R^{*2} - 7R^{*4}}{R^{*4}} \right) \tag{10-105}$$

These deadend structures are expected to produce an increase in the dispersion co-

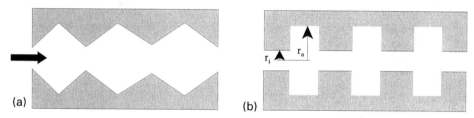

(a)   (b)

Figure 10.16 Different configurations analyzed for noncylindrical tubes: (a) sharp-edged sides, (b) rectangular with deadend pockets.

efficient 1–11 times the value for a straight capillary tube (Aris, 1959). The analysis begun by Turner on the effect of side pores has been expanded to a view of the porous medium consisting of a series of macro- and micropores. A more complete analysis of the effect of these pores on the overall transport of a chemical in soils will be conducted in Chapter 12.

Plume Size   In each of these cases for different channel geometries we always have the result that the size of a pulse of chemical introduced into the medium spreads out in the longtitudinal direction in a manner consistent with a normal distribution. Therefore, the size of the chemical plume is

$$L = (32E_L t)^{1/2} \tag{10-106}$$

For example, for the case of spreading by Taylor–Aris dispersion, where dispersion is much larger than diffusion, the plume size can be related to pore diffusivity and mean velocity by using Eq. 10-99 in Eq. 10-106 to obtain

$$L = 0.82 R\bar{u} \left(\frac{t}{D_s} \right)^{1/2} \tag{10-107}$$

where we see that the size of the plume is inversely proportional to the molecular diffusivity. For cases where mechanical dispersion dominates, which is more typical for groundwater flow, E_L must be determined for the medium or calculated from a dispersion correlation.

Theories of Dispersion in Porous Media

Models of chemical dispersion in porous media based only on the analysis of motion in a single tube, no matter what the tube geometry is assumed to be, are incomplete unless mechanical dispersion is included in the analysis. One method of modeling mechanical dispersion is to use a statistical approach. Although it is impossible to predict the exact path of any chemical tracer in a system, the rules of probability can be used to predict the average spatial distribution of a cloud of chemical tracers at different times. With time, the characteristics of the cloud of tracers fits a normal distribution, as we have already discussed. In the previous section we considered only dispersion in the direction of flow, or longitudinal dispersion. In theory the dispersion coefficient $\mathbf{E}$ can be a second-order tensor with nine components, or

$$\mathbf{E} = \begin{bmatrix} E_{xx} & E_{xy} & E_{xz} \\ E_{yx} & E_{yy} & E_{yz} \\ E_{zx} & E_{zy} & E_{zz} \end{bmatrix} \tag{10-108}$$

In a homogeneous and isotropic medium dispersion is constant throughout the medium. If the local coordinate axes are rotated to align with the principal axes so that the off-diagonal terms are zero (Bear, 1972), then the dispersion tensor is reduced to only three terms

$$\mathbf{E} = \begin{bmatrix} E_{xx} & 0 & 0 \\ 0 & E_{yy} & 0 \\ 0 & 0 & E_{zz} \end{bmatrix} \tag{10-109}$$

For advective flow in the x direction, it is usually assumed that dispersion perpendicular to the flow in the y and z directions occurs at the same rate. This further reduces the complexity of dispersion in the system to only two components: longitudinal dispersion, where $E_{xx} \equiv E_L$, and transverse dispersion where $E_{yy} = E_{zz} \equiv E_t$. Using this notation, Eq. 10-109 can be written simply as

$$\mathbf{E} = \begin{bmatrix} E_L & 0 & 0 \\ 0 & E_t & 0 \\ 0 & 0 & E_t \end{bmatrix} \tag{10-110}$$

The importance of mechanical dispersion to the overall rate of dispersion becomes evident if a statistical analysis of dispersion in a porous medium is conducted without including molecular processes (Saffman, 1959). In this case dispersion in the transverse and longitudinal direction would become equal, but in fact longitudinal dispersion is typically greater than transverse dispersion, or $E_L > E_t$, by a factor of 3 to 10, although Blackwell (1959) reports a maximum ratio of 24.

Recognizing the importance of molecular diffusivity to dispersion in porous media, Saffman (1960) introduced molecular processes into a statistical model of dispersion in porous media. His model, which examined an assembly of randomly oriented and distributed straight circular tubes comparable to the sizes of pores, was meant to model a homogeneous and isotropic porous medium. The velocity within a tube was dependent on the angle of the tube relative to the direction of flow and the radial location in the tube. The analysis also assumed that molecular diffusion and macroscopic mixing are the same order of magnitude, and that sufficient time has passed for concentration variations to be smoothed out across the cross section of the capillary tube, or that $l/\bar{u} \gg R^2/8D$, a condition necessary to incorporate Taylor–Aris dispersion into his model. For large times, he calculated the covariance of velocity components perpendicular and in the direction of flow, integrated these covariances, and obtained for Pe $\ll$ 1

$$\frac{E_L}{D} \approx m + \frac{\text{Pe}^2}{15}$$

$$\frac{E_t}{D} = m + \frac{\text{Pe}^2}{40} \tag{10-111}$$

where m is a constant and l, the length of the capillary tube, was assumed to be equal to the soil grain diameter or $l = d$. From comparison to our expressions for diffusion in terms of a tortuosity factor, τ_f, we can see that $m = \theta/t_f$. Saffman predicted from his analysis that $m = 1/3$, but empirical studies in the absence of velocity suggest that $m = 2/3$. This suggested a model of porous media consisting of a series of capillary tubes was not quite correct, particularly at low Pe. For larger Pe, in the range $1 << \text{Pe} << 8R^2/l^2$, Saffman obtained

$$\frac{E_L}{D} \approx \frac{\text{Pe}}{6} \ln \frac{3 \text{ Pe}}{2} - \frac{17 \text{ Pe}}{72} - \frac{R^2 \text{ Pe}^2}{48l^2} + \left(m + \frac{4}{9} \right) + O(\text{Pe}^{-1})$$

$$\frac{E_t}{D} \approx \frac{3 \text{ Pe}}{10} - \frac{R^2 \text{ Pe}^2}{40l^2} + \left(m - \frac{1}{3} \right) + O(\text{Pe}^{-1}) \qquad (10\text{-}112)$$

Saffman's results are compared to several experimental data sets in Figure 10.17. The agreement with the data is acceptable considering the assumptions necessary to derive the predictive model. Bear (1972) suggests that from analysis of this figure dispersion can be considered to scale differently with Pe depending on the magnitude of Pe. For Pe $<< 1$ molecular diffusion is dominant and $E_L/D = $ constant. Dispersion

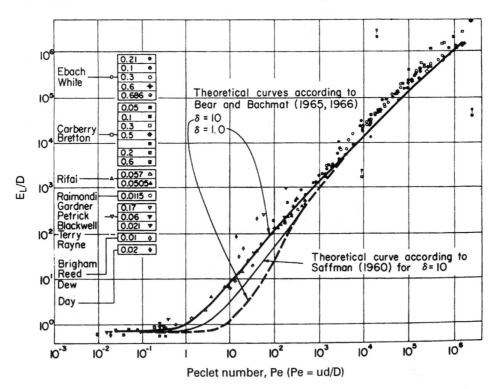

Figure 10.17 Longitudinal dispersion as a function of the Peclet number comparing experimental results with the model of Saffman (1960) (Reprinted from Bear, 1972, with kind permission of Dover Publications, Inc.).

at this scale is only a function of the medium tortuosity because the porous medium limits free diffusion of the molecules. For $0.4 < \text{Pe} < 5$, molecular diffusion is of the same order of magnitude as mechanical dispersion. At larger Pe the mechanisms of mechanical and hydrodynamic dispersion are no longer additive and actually work against each other. We see that $E_L/D = 0.5 \, \text{Pe}^n$, where n is in the range of 1 to 1.2. When molecular diffusion is large, it tends to reduce longitudinal dispersion, and therefore theories that neglect molecular diffusion will fail to predict overall chemical dispersion in this range of Pe. For very large Pe ($\text{Pe} > 10^3$), mechanical dispersion predominates so long as the flow remains laminar (Darcy's Law applies) and $E_L/D = 1.8 \, \text{Pe}^n$, where $n = 1$. When inertia and fluid turbulence may no longer be neglected, E_L/D scales with Pe^n, where $n < 1$.

Packed Bed Analysis of Dispersion   A statistical approach was also used by Koch and Brady (1985) to describe dispersion in porous media except that they modeled the flow within a packed bed of spheres and not within capillary tubes. They found at high Pe numbers ($\text{Pe} > 1$) that convective dispersion was nearly all due to mechanical dispersion, and as a result dispersion was independent of molecular diffusion and increased linearly with Pe. However, they also considered the effect of diffusion into porous medium particles on dispersion. Zero-velocity regions near and in the particles produced nonmechanical dispersion effects that eventually dominated at very high Pe, resulting in dispersion coefficients that increased as Pe^2. The effect of just diffusive boundary layers near the solid surfaces, analyzed by defining the particles to be impermeable, resulted in nonmechanical contributions to dispersion that increased as $\text{Pe} \ln \text{Pe}$ and Pe^2.

Dispersion coefficients derived by Koch and Brady (1985) are summarized in Table 10.1 for both longitudinal and transverse dispersion in terms of the $\text{Pe} = ud/D$, where u is the pore velocity, d the grain diameter, D the molecular diffusivity, and κ the bed permeability. Their solutions consider cases where there can be chemical diffusion into the particle, and therefore m is a ratio of the equilibrium concentrations of a chemical in the fluid and in the particle. Presumably, the chemical concentration in a solid particle is obtained from calculating the adsorbed mass on

TABLE 10.1   Longitudinal (K_L) and Transverse (E_t) Dispersion Coefficients in Porous Medium Analyzed as a Packed Bed[a]

Flow conditions	K_L	E_t
$\text{Pe} \ll \phi^{-1} \ll 1$	$1 + \dfrac{3(\alpha - 1)\phi}{\alpha + 2} + \dfrac{2^{1/2}\text{Pe}^2}{15\phi^{1/2}}$	$1 + \dfrac{3(\alpha - 1)\phi}{\alpha + 2} + \dfrac{2^{1/2}\text{Pe}^2}{60\phi^{1/2}}$
$\text{Pe} \gg 1$	$1 + \dfrac{3\lvert\text{Pe}\rvert}{4} + \dfrac{\pi^2 \phi \lvert\text{Pe}\lvert\ln\lvert\text{Pe}\rvert}{6} + \dfrac{(1 + \gamma)^2 D\phi\text{Pe}^2}{15 m D_p}$	$1 + \dfrac{2^{1/2} 63 \phi^{1/2} \lvert\text{Pe}\rvert}{320}$

[a]See the text for definitions of α, ϕ, m and Pe. D is molecular diffusivity in water while D_p is the diffusivity in the particle.
Source: Koch and Brady 1985.

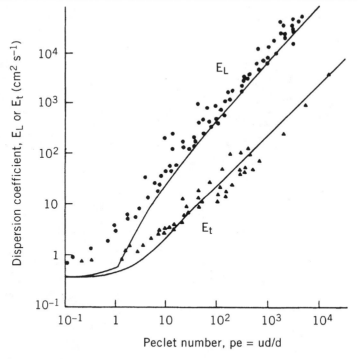

Figure 10.18   Comparison of experimental results of Fried and Comarnous (1971) with equations proposed by Koch and Brady (1985) (Reprinted from Koch and Brady, 1985, with permission of Cambridge University Press).

the particle and dividing by the particle volume. If $m = 0$, their solutions can be used for nonadsorbing impermeable particles. The factor γ is calculated as

$$\gamma = \frac{\phi(1 - m^{-1})}{1 - \phi(1 - m^{-1})} \qquad (10\text{-}113)$$

where ϕ is defined as the solid volume fraction of the bed, or $\phi = (1 - \theta)$. The term D^* is defined $D^* = D_p/Dm$, where D is the molecular diffusivity of the chemical in the fluid, and D_p is the chemical diffusivity in the porous medium. It was demonstrated by Koch and Brady that there was excellent agreement between their equations and experiments by Fried and Combarnous (1971) for dispersion coefficients in a packed bed (Fig. 10.18).

Example 10.3

Compare the theoretical predictions of longitudinal dispersion in a porous medium derived by Saffman and Koch and Brady.

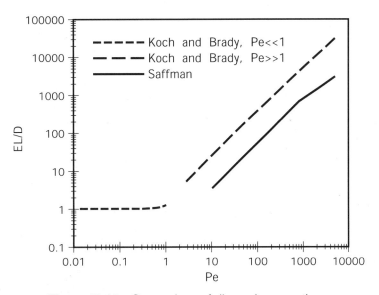

Figure 10.19   Comparison of dispersion equations.

Using Eq. 10-110, longitudinal dispersion predicted by Saffman for Pe $\gg$ 1 is plotted in Figure 10.19 assuming $m = 0.67$ and that the term $O(\text{Pe}^{-1}) = \text{Pe}^{-1}$. Values of E_L/D calculated using the results of Koch and Brady assume the porous medium is impermeable, or that there is no chemical concentration within the porous medium. Values calculated using Saffman's equations are about an order of magnitude lower than the dispersion predicted by Koch and Brady. Since both models have sucessfully been compared to experimental data, the reason for the discrepancies between the two theoretical models is unknown.

Dispersion Correlations

There have been several laboratory studies to determine empirical correlations for chemical dispersion in packed beds. All these correlations to describe chemical dispersion were developed by varying the fluid velocity and porous medium grain size. The relating dispersion coefficients were either cast as a function of the Peclet number defined as Pe = ud/D, where u is the interstitial or pore velocity, or as a function of the Reynolds number. Despite the inclusion of the molecular diffusivity in the Pe number, a result of the expected importance of molecule size on dispersion, none of these correlations for dispersion was developed by varying the diffusivities of molecules used in the studies. Correlations should only be used for the same conditions they were developed under, and therefore these correlations really only apply to the product ud and not ud/D. Because of their empirical nature, these correlations should be used with caution when examining the effect of molecular diffusivity on dispersion.

One of the best-known correlations for predicting chemical dispersion in packed beds is the Hiby correlation (Hiby, 1959, in Bear, 1972),

$$\frac{E_L}{D} = 0.67 + \frac{0.65\ Pe}{1 + 6.7\ Pe^{-1/2}}, \quad 10^{-2} < Pe < 10^2 \tag{10-114}$$

For larger Pe numbers, it is better to use the Blackwell correlation (Blackwell et al., 1959)

$$\frac{E_L}{D} = 8.8\ Pe^{1.17}, \quad Pe > 0.5 \tag{10-115}$$

Example 10.4

Compare the experimental correlations for longitudinal dispersion in porous media obtained by Hiby (1962) and Blackwell et al. (1959) with the predictions of Koch and Brady (1985).

The two correlations are compared in Figure 10.20. Notice how much larger dispersion is predicted to be using the Blackwell correlation than by either the Hiby or Koch and Brady equations. Although not shown in the figure, the equation developed by Hiby compares very well to Saffman's equation over the valid range of Pe for the Hiby equation.

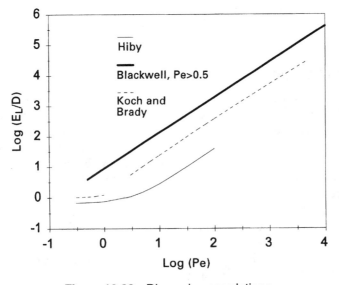

Figure 10.20   Dispersion correlations.

Harleman et al. (1963) conducted a series of experiments on dispersion in packed beds of sand and spheres. Instead of correlating E_L/D with Pe, they instead chose to examine E_L/v as a function of two different Reynolds numbers. Typically, Re is based on the soil grain diameter. Harleman et al. used a mean soil grain diameter, d_{50}, so that Re is defined as $\text{Re}_{50} = ud_{50}/v$, and obtained

$$\frac{E_L}{v} = 0.9\ \text{Re}_{50}^{1.2}, \quad \text{sand} \tag{10-116}$$

where the coefficient 0.9 becomes 0.66 for a spherical porous medium. Agreement between this correlation and experimental data was quite good, as shown in Figure 10.21. Re can also be expressed in terms of the medium permeability, κ, as $\text{Re}_{\kappa} = u\kappa^{1/2}/v$. Using this definition of Re, they obtained

$$\frac{E_L}{v} = 83\ \text{Re}_{\kappa}^{1/2}, \quad \text{sand} \tag{10-117}$$

where the coefficient 83 would be equal to 54 for a spherical porous medium.

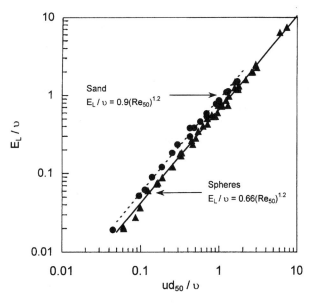

Figure 10.21 Longitudinal dispersion as a function of the Reynolds number based on a mean porous median particle diameter. (From Harleman et al. 1963, in Dullien 1992).

Example 10.5

Many molecules and colloids present in groundwaters have molecular diffusivities much smaller than those used to develop dispersion correlations. Estimate longitudinal dispersion coefficients for large molecules and colloids using the Blackwell correlation (Eq. 10-113). Assume a uniform porous medium with a grain size of 120 μm and a pore velocity of 1 m d^{-1}.

As examples of typical large molecules, we can calculate the diffusivities of dextrans and proteins using the Frigon and Polsen correlations (Eqs. 3-33 and 3-34). From these diffusion coefficients, we can estimate the "size" of the molecule, defined here as a hydrodynamic diameter, using the Stokes–Einstein equation (Eq. 3-25).

Longitudinal dispersion coefficients for dextrans and proteins having molecular weights of 1, 10, 100, and 1000×10^3 amu are shown in Figure 10.22 as a function of molecule diameter along with other particles ranging in size from 0.01 to 3 μm. Dispersion coefficients or bacteria-sized particles are predicted to reach 0.1 cm^2 s^{-1}, while those for the dextrans and proteins would range from 0.03 to 0.07 cm^2 s^{-1}.

Dispersivity In the dispersion equations above we have seen that dispersion is often some function of the fluid velocity, or $E_L \sim u^n$ where n is some power. The one-dimensional advection–dispersion equation can be rearranged to produce

$$\frac{1}{u_x}\frac{\partial c}{\partial t} = -\frac{\partial c}{\partial x} + \frac{E_L}{u_x}\frac{\partial c^2}{\partial x^2} \qquad (10\text{-}118)$$

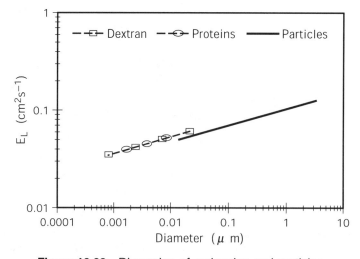

Figure 10.22 Dispersion of molecules and particles.

Defining the dispersivity, α_D, as $\alpha_D = E_L/u_x$, this equation becomes

$$\frac{1}{u_x}\frac{\partial c}{\partial t} = -\frac{\partial c}{\partial x} + \alpha_D \frac{\partial c^2}{\partial x^2} \tag{10-119}$$

where the dispersivity has units of length.

If E_L was always proportional to fluid velocity, dispersivity would be a constant. However, it has been found that dispersivity increases with the scale of the dispersion measurement (Fig. 10.23). Chemical dispersivity at a distance of 100 m can be orders of magnitude larger than dispersivity at a scale of 1 m. This phenomenon of dispersivity increasing with transport distance may be somewhat analogous to the case of horizontal dispersion in the ocean, where it was found that the horizontal dispersion coefficient increased with the plume size raised to the 4/3 power. The increase in dispersivity with distance shown in Figure 10.23 is perhaps larger than a 1:1 increase, although the large variability in the dispersivity at any given scale precludes any direct correlation between the dispersion and distance.

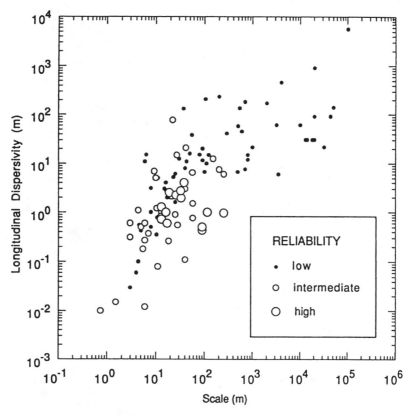

Figure 10.23 Longitudinal dispersivity measured in field tests has been found to be a function of the scale of the field measurements (from Gelhar et al., 1992).

PROBLEMS

10.1   Let us derive Eq. 10-74 for laminar shear flow between two plates moving in opposite directions, each at a velocity v_p, separated by a distance h, by a different approach. (a) Starting with Eq. 10-64, derive an expression for c' assuming that $u' = u_p y/h$. Position the origin at a point midway between the plates. (b) Show that the dispersion coefficient can also be written as

$$E_L = -\frac{1}{h(\partial \bar{c}/\partial x)} \int_{-h/2}^{h/2} \frac{u_p y}{h} c' \, dy \qquad (10\text{-}120)$$

(c) Using the above results, derive Eq. 10-74.

(handwritten annotations in left margin)
$E_t = 6.4 E \times 10^{-5}$
$H_L = 3.6 E \times 10^{-3}$
$E_L = 6.5 \times 10^{-3}$ cm/s
0.23 cm

10.2   A field experiment is conducted to determine the transverse and longitudinal dispersion coefficients in porous media. A chemical tracer ($D = 800 \times 10^{-8}$ cm^2 s^{-1}) is injected into the well, as shown in Figure P-10.2. The chemical is measured at several heights at the detection well, situated 30 m downgradient. It is found that the chemical has spread 0.2 m in the vertical direction and that the plume takes 2.7 h to pass the detection well. Soil analysis indicates an average soil grain diameter of 250 μm and soil porosity of 0.38, and pumping tests reveal an average groundwater (Darcy) velocity of 5 m d^{-1}. (a) What is the transverse dispersion coefficient, E_t(cm^2 s^{-1}) for this aquifer? (b) What is the longitudinal dispersion coefficient, E_L(cm^2 s^{-1})? (c) How does the experimentally measured longitudinal dispersion coefficient compare with that predicted by the Blackwell correlation? (d) What is the chemical dispersivity for this site?

10.3   A disinfectant is injected into water in a 2-cm-diameter pipe flowing at a velocity of 3 cm s^{-1}. (a) Although Taylor's analysis of dispersion is not valid under these conditions (why?), calculate the dispersion coefficient for the chemical assuming a very slow (negligible) reaction rate and a diffusivity of 1.22×10^{-5} cm^2 s^{-1}. (b) Compare the velocity of the chemical produced by fluid dispersion to the average fluid velocity in the pipe after 10 min.

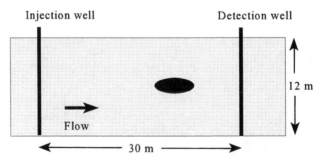

Injection well

Detection well

12 m

Flow

30 m

Figure P-10.2

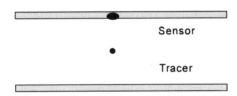

Figure P-10.6

$\sqrt{}$**10.4** Water flow down an inclined plate 40 cm long is turbulent. The water is 17 cm high, and flows at an average velocity of 45 cm s^{-1} and a shear velocity of 5 cm s^{-1}. Calculate a dispersion coefficient and the size of a chemical plume at the end of the plate. $10-95$

10.5 Fluid flow down a vertical wall can be described by the parabolic velocity profile

$$u_x = \frac{3\bar{u}y}{2h}\left(2 - \frac{y}{h}\right) \tag{10-121}$$

where h is the fluid thickness, the origin ($y = 0$) is at the wall, and the maximum and average fluid velocities are related by $\bar{u} = (3/2)u_{\max}$. Show that the longitudinal dispersion coefficient is given by

$$E_L = \frac{2h^2\bar{u}^2}{105D_C} \tag{10-122}$$

$\sqrt{}$**10.6** A small mass of chemical tracer is injected into the fluid halfway between two plates spaced 1 cm apart, as shown in Figure P-10.6. After 1000 s, the sensor records the presence of the tracer. If the experiment is repeated, but the walls are 4 cm apart, use a simple property of diffusion to calculate how long it will be before the sensor detects the chemical tracer. βk?

$\sqrt{}$**10.7** A capillary tube 0.5 mm in diameter and 1 m in length is filled with water. A salt tracer (mass 10^{-4} g; $D_{Tw} = 10^{-5}$ cm^2 s^{-1}) is injected into the tube center. Find the maximum concentration of the salt and the length of the salt diffusion cloud 10 min after the salt is injected.

REFERENCES

Aris, R. 1956. *Proc. Roy. Soc. A* **235**:67–77.

Aris, R. 1959. *Chem. Eng. Sci.* **11**:194.

Bear, J. 1972. *Dynamics of Fluid in Porous Media*. American Elsevier, N.Y.

Blackwell, J. 1959. Amer. Inst. Chem. Eng. And Soc. Petrol. Eng. 52nd Annual Meeting, San Francisco, Preprint No. 29.

Blackwell, J., R. Rayne, and W. M. Terry. 1959. *Trans. A.I.M.E.* **217**:1–8.

Csanady, G. T. 1963. *J. Geophys. Res.* **71**:5389–99.

Dullien. F. A. 1992. *Porous Media: Fluid Transport and Pore Structure*, 2nd ed. Academic Press, San Diego, CA.

Elder, J. W. 1959. *J. Fluid Mech.* **5**:544–60.

Fischer, H. B., E. J. List, R. C. Y. Koh, J. Imberger, and N. Brooks. 1979. *Mixing in Inland and Coastal Waters*. Academic Press, New York.

Fried, J. J., and M. A. Combarnous. 1971. *Adv. Hydrosci.* **7**:169–282.

Gelhar, L. W., C. Welty, and K. R. Rehfeldt. 1992. *Water Resour. Res.* **28**(7):1955–74.

Gill, W. N., and V. Ananthakrishnan. 1966. *AIChE J* **12**:906.

Harleman, D. R. F., P. F. Melhorn, and R. R. Rumer. 1963. *J. Hydraul. Div. Proc. ASCE* **89**(HY2):67.

Hiby, J. W. 1962. In: *The Interactions between Fluid and Particles*. P. A. Rottenberg, Ed., Inst. of Chem. Engin., London.

Jones, W. M. 1968. *Brit. J. Appl. Phys. Ser.* 2 **1**:1559.

Koch, D. L., and J. F. Brady. 1985. *J. Fluid Mech.* **154**:399–427.

Nunge, R. J., and W. N. Gill. 1970. In: *Flow through Porous Media*, pp. 179–95, American Chemical Society, Washington, DC.

Saffman, P. G. 1959. *J. Fluid Mech.* **6**(3):321–49.

Saffman, P. G. 1960. *J. Fluid Mech.* **7**(2):194–208.

Sankarasubramanian, P. 1969. M.S. Thesis, Clarkson College of Technology, Potsdam, N.Y.

Turner, G. A. 1958. *Chem. Eng. Sci.* **10**:14.

Taylor, G. I. 1953. Proc. Royal Acad. *Soc. London Ser. A.* **219**:186–203.

Taylor, G. I. 1954. Proc. Royal Acad *Soc. London Ser. A.* **223**:446–68.

CHAPTER 11

RIVERS, LAKES, AND OCEANS

11.1 INTRODUCTION

Engineers are concerned about many aspects of chemical transport in natural waters since they must often answer practical questions about chemical mobility in such systems. For example, if an industrial wastewater is discharged from the side of a river, how long will it take for the plume to spread out or mix over the width of the river? What will be the maximum concentration of the chemical 500 m down river? If there is a spill of the chemical, what will be the chemical concentrations 1 mile down the river where a town takes in river water to be used for drinking water? The rate of chemical mixing in rivers, particularly when many towns and cities are situated on the same river, makes it critical that we understand the fluid hydraulics and chemical dispersion in surface waters. Although many undergraduate engineering programs include courses in hydraulics, the topic of chemical dispersion receives relatively little attention.

A great deal has been learned about chemical transport in lakes, rivers, and oceans by carefully measuring the dispersion of conservative tracers in these systems. The use of fluorescent dyes for tracers has been particularly helpful in studying dispersion rates. Fluorescent dyes can be measured over many orders of magnitude in concentration (down to ppb) with fluorometers. High sensitivity in fluorometers results from the emission of light by the dyes (fluorescence) at a wavelength different from the excitation wavelength. In a fluorometer the light impinging on a sample is filtered out to remove all light below the wavelength of the fluorescent signal. The sample is therefore illuminated at a wavelength higher than the fluorescent light wavelength; when the light is detected, all wavelengths above the critical detector wavelength are filtered out. As a result, the fluorescent light is detected against a completely dark background. This high contrast provides great sensitivity of the signal. Fluorometers are relatively nonsophisticated and can easily be taken to the field for real-time data acquisition. The main interfering substances are the many components in dissolved organic matter, but the use of blanks (or different dyes) can help mitigate these problems. Variations in temperature must also be compensated for in relating the intensity of the fluorescence signal to chemical concentration.

Field measurements of chemical dispersion are interpreted using transport equations. In many instances, it is possible to simplify the system by examining dispersion from a moving coordinate system. This simplification reduces the complexity of the transport equation to a one-dimensional system that can be analytically solved. Perhaps more important, this focuses the analysis on a distinguishing feature of the system: a single chemical dispersion coefficient. We have already seen in the previous chapter that it is possible to simplify dispersion calculations by answering a basic question: How far has the chemical spread? If we give our answer as the distance that encompasses 95% of the chemical spread, the calculation of this distance L is just a simple analytical expression $L = (4Et)^{1/2}$. Thus the system can be characterized by a simple analytical model using a single dispersion coefficient for the system.

In this chapter we examine the result of the effect of fluid turbulence on the mixing rates of chemicals in different surface waters. For the most part the emphasis will be on using measured dispersion coefficients to predict the rate at which chemicals move away from a source in a system. Such calculations will have use in estimating maximum chemical concentrations over the lifetime of a spill, or steady chemical concentrations in systems receiving domestic and industrial wastewaters. In the final section of this chapter we will consider cases where chemicals that are spilled into the system are not infinitely soluble. When sparingly soluble chemicals enter into a system, they may persist as a pure phase and continue to provide a source of chemical contamination in the system for long periods of time relative to the hydraulic detention times of the system.

11.2 CHEMICAL TRANSPORT IN RIVERS

Fluid flow in rivers is always turbulent. Even in slowly meandering rivers there are always eddies that produce chemical mixing over distances that approach the size of the "container," or the depth or width of the river. The driving forces of gravity and pressure are counterbalanced by shear stresses exerted by the walls of the river. Differences in river contours, directions, depths, and average velocities can result in a variety of mixing patterns too complex to model completely. Thus it is inevitable that we simplify the system and model only the average properties of the river.

The mean velocity distribution within a river follows a logarithmic profile with depth. The mean velocity can be related to the mean wall shear stress according to

$$\tau_o = \tfrac{1}{2} f \rho \bar{u} \qquad (11\text{-}1)$$

where τ_o is the shear stress at the wall or river bank, ρ is the water density, $\bar{u}$ is the average river velocity, and f is the Darcy–Weisbach friction factor. The shear velocity is calculated from the wall shear stress as

$$u_{sh} = (\tau_o/\rho)^{1/2} \tag{11-2}$$

Therefore, the average and shear velocities in the river can be related using

$$\bar{u} = u_{sh}(8/f)^{1/2} \tag{11-3}$$

For flow in circular pipes the friction factor can be easily obtained from pipe friction diagrams as a function of the Reynolds number and the roughness of the pipe. For noncircular pipes we use the characteristic length scale $4R_h$, where R_h is the hydraulic radius calculated as the ratio of the cross sectional area to the wetted perimeter.

Shear stress in an open channel such as a river can be found from a force balance to be

$$\tau_o = \rho g S R_h \tag{11-4}$$

where S is the slope of the energy grade line, or the channel slope for uniform flow. The shear velocity in the open channel is

$$u_{sh} = (gSR_h)^{1/2} \tag{11-5}$$

For uniform open channel flow in a river of depth h, since the width of the river w is much larger than h, and therefore $R_h = wh/(w + 2h) \approx h$, the shear velocity can more simply be calculated as

$$u_{sh} = (gSh)^{1/2} \tag{11-6}$$

The slope of the river can be related to flow using the Manning equation

$$Q = \frac{1.49}{n} A R_h^{2/3} S^{1/2} \tag{11-7}$$

where Q is the flow rate (ft^3 s^{-1}), S the dimensionless slope (L L^{-1}), A the cross-sectional area (ft^2), and n the roughness coefficient, typically assumed to be $n = 0.030$.

For the case of two-dimensional flow in very wide channels, the logarithmic velocity distribution as a function of the channel depth and width is

$$u = \bar{u} + \frac{u_{sh}}{\kappa} + \frac{2.30}{\kappa} u_{sh} \log \frac{y}{h} \tag{11-8}$$

where y is the distance from the wall or river bank, h the channel depth, and κ the von Karman constant, usually assumed to be 0.40 for nonstratified flow and with no suspended sediment, as shown in Figure 11.1 (Fisher et al., 1979). High concentrations of sediments lowers κ from 0.4 to 0.21. From Eq. 11-8, this results in $u = \bar{u}$ at $y/h = 0.368$.

Based on these fluid-mechanical relationships, we can now explore dispersion in rivers. From our derivation of the constitutive transport equation we expect that a total of nine dispersion coefficients could arise from fluid turbulence in a system. In practice, however, only three different mixing coefficients are used to describe mixing in a river (Fig. 11.2). If we write the transport equation in Cartesian coordinates for the case of no reaction, we have

$$\frac{\partial c}{\partial t} + \mathbf{u} \cdot \nabla c = E_x \frac{\partial c^2}{\partial x^2} + E_y \frac{\partial c^2}{\partial y^2} + E_z \frac{\partial^2 c}{\partial z^2} \tag{11-9}$$

In our coordinate system for rivers, we will define vertical mixing as the z direction, transverse mixing across the river as the y direction, and longitudinal mixing along the river length as the x direction. Mixing is usually analyzed separately in each direction, resulting in separate, one-dimensional equations for each spatial dimension. In the vertical and transverse directions, $u_z = u_y = 0$. For vertical mixing we replace E_z with E_v and use the equation

$$\frac{\partial c}{\partial t} = E_v \frac{\partial^2 c}{\partial z^2} \tag{11-10}$$

and for transverse mixing across the width of the river, E_y is replaced by E_t, resulting in

$$\frac{\partial c}{\partial t} = E_t \frac{\partial^2 c}{\partial y^2} \tag{11-11}$$

The situation for one-dimensional mixing in the longitudinal direction is somewhat more complex. It will be shown that the longitudinal mixing coefficient e_x can be neglected in comparison to longitudinal dispersion resulting from fluid gradients in the direction perpendicular to the direction of flow, $e_{\Delta u}$. We will analyze longitudinal dispersion using E_L, a longitudinal dispersion coefficient from a fixed reference as

$$\frac{\partial c}{\partial t} + u_x \frac{\partial c}{\partial x} = E_L \frac{\partial^2 c}{\partial x^2} \tag{11-12}$$

where u_x is the average velocity in the direction of flow. This use of specialized subscript notation on the various dispersion coefficients should serve to remind us about the coordinate system being used in dispersion calculations and the nature of the fluid that drives dispersion in the system.

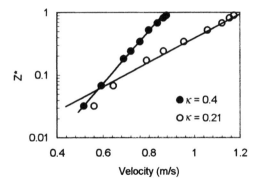

Figure 11.1 Velocity profiles in a channel 0.090 m deep and 0.85 m wide in the presence and absence of suspended sediments (0.1 mm sand). When sand is present, the von Karmen constant is reduced from κ = 0.40 to 0.21. (Data from Vanconi and Brooks, 1957; redrawn from Fischer et al., 1979.)

Vertical Mixing Coefficient

Mixing in the vertical direction in a river, for example, with depth from the bottom to the top of the river, is based upon an analysis by Elder (1959) for flow down an infinitely wide inclined plane. The velocity profile for flow down this plane is

$$u' = \frac{u_{sh}}{\kappa} [1 + \ln(z/h)] \qquad (11\text{-}13)$$

where κ = 0.4 is the von Karmen constant. A force balance over the depth, h, of the flow (Fischer et al., 1979) yields

$$\tau = \rho E_v \frac{du}{dy} = \tau_o [1 - (z/h)] \qquad (11\text{-}14)$$

where τ_o is the shear stress at the bottom defined in Eq. 11-2 as $u_{sh} = (\tau/\rho)^{1/2}$. From these relationships, Elder obtained the vertical mixing coefficient for momentum, which is the same here for mass concentration, as

$$E_v = \kappa u_{sh} z [1 - (z/h)] \qquad (11\text{-}15)$$

Averaging over the depth of flow, and assuming κ = 0.40, we have the simple result (Fischer et al., 1979)

$$\bar{E}_v = 0.067 u_{sh} h \qquad (11\text{-}16)$$

Vertical dispersion coefficients measured by others for a wide range of systems compare well to Eq. 11-16. For example, from measurements in an unstratified atmospheric boundary layer, Csanady (1976) found that $\bar{E}_v = 0.05 u_{sh} h$.

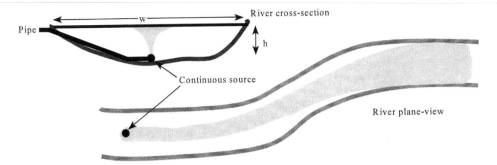

Figure 11.2 Mixing in a river from a continuous point source (pipe discharge) shown from both the cross-section and plane views. The river has a height *h* (*z* direction), width *w* (*y* direction), and mixes over a length *L* (*x* direction) based on the distance measured along the river centerline.

Transverse Mixing Coefficient

Rivers do not have well-characterized velocity profiles across their width. Therefore, transverse mixing coefficients have been developed from numerous experiments. From laboratory experiments on rectangular channels varying in width from 0.36 to 2.83 m, it has been found that the transverse mixing coefficient is a function of shear velocity and depth, according to

$$E_t \cong 0.15 u_{sh} h, \qquad \text{straight channels} \qquad (11\text{-}17)$$

where the coefficient ranges from 0.1 to 0.2 (Table 11.1). Field experiments by Fischer (1967b) on regularly shaped irrigation canals produced higher transverse mixing coefficients of 0.24–0.25. There is some controversy on the extent to which channel width plays a role in the transverse mixing coefficient, but the conclusion of Fischer et al. (1979) was that the coefficient in Eq. 11-17 is likely to be correct to within only ±50%.

Natural streams and rivers differ from the uniform rectangular channels examined by researchers in laboratory studies in at least three ways: The depth may vary, natural channels curve, and there are large irregularities in the channel sides. Variations in depth are not thought to affect vertical mixing coefficients, but the latter two factors produce significantly larger mixing coefficients in natural versus laboratory channels. The gently sloping and irregular sides also provide substantial ecological benefits, which engineers did not originally anticipate in their design of channels and water intakes at industrial sites. Long, narrow channels constructed for cooling water intake at power plants were originally designed and constructed to be highly regular channels nearly rectangular in shape. Unfortunately, fish that swam into the channel, particularly during cooler months, were unable to swim at sufficient speeds against the channel fluid velocities long enough to escape the channel. As the fish tired they were eventually drawn with the water and impinged on screens, covering the intake pipes to the power plant. Fish impingement rates exceeded 1

TABLE 11.1 Experimental Measurements of Transverse Mixing in Rectangular Open Channels with Smooth Sides

Type of Bottom Roughness and Channel[a]	Channel Width w (cm)	Mean Flow Depth h (cm)	Mean Velocity $\bar{u}$ (cm s^{-1})	Shear Velocity u_{sh} (cm s^{-1})	E_t (cm^2 s^{-1})	$\dfrac{E_t}{u_{sh}d}$
Wooden cleats	283	14.8–37.1	23.5–37.1	3.81–6.04	9.6–36.9	0.16–0.18
Smooth	85	1.5–17.3	27.1–42.8	1.6–2.2	0.64–2.9	0.09–0.20
Smooth	110	1.7–22.0	30.0–50.4	1.4–2.6	0.79–3.3	0.11–0.24
Stones	110	6.8–17.1	35.3–42.8	3.6–5.2	4.8–7.5	0.11–0.14
Metal lath	110	3.9–6.4	37.3–45.9	3.7–4.0	2.0–3.5	0.14
Rectangular blocks	59.7	12.5–13.2	30.5–81.4	3.0–16.3	3.7–36.3	0.10–0.18
0.4-mm sand	45–60	1.4–4.0	19.7–20.3	1.6–2.1	0.34–0.88	0.11–0.14
2.0-mm sand	30	1.6–3.4	20.0–20.4	1.9–2.4	0.74–0.92	0.14–0.20
2.7-mm sand	45–60	1.3–3.9	19.5–20.4	1.8–2.8	0.59–1.16	0.13–0.26
Sand dunes (IC)	18–30	66.7–68.3	63–66	6.1–6.3	102	0.24–0.25

[a]All experiments in the laboratory except for the last table entry, which was an irrigation canal (IC).

Source: Fischer et al., 1979.

TABLE 11.2 Experimental Measurements of Transverse Mixing in Open Channels with Curves and Irregular Sides

Channel	Channel Geometry	Channel Width w (m)	Mean Flow Depth h (m)	Mean Velocity $\bar{u}$ (m s^{-1})	Shear Velocity u_{sh} (m s^{-1})	E_t (m^2 s^{-1})	$\dfrac{E_t}{u_{sh}d}$
Missouri River near Blair, Nebraska	Meandering river	200	2.7	1.75	0.074	0.12	0.6
Laboratory model of the IJssel River	Groins on sides and gentle curvature	1.22	0.9	0.13	0.0078	—	0.45–0.77
IJssel River	Groins on sides and gentle curvature	69.5	4.0	0.96	0.075	—	0.51
Mackenzie River from Fort Simpson to Norman Wells	Generally straight or slight curvature; numerous island and sand bars	1240	6.7	1.77	0.152	0.67	0.66
Missouri River downstream of Cooper Nuclear Station, Nebraska	Reach includes one 90° and one 180° bend	210–270	4	5.4	0.08	1.1	3.4
Potomac River; 29 km reach below the Dickerson Power Plant	Gently meandering river with up to 60° bends	350	0.73–1.74	0.29–0.58	0.033–0.051	—	0.52–0.65

Source: Fischer et al., 1979.

million fish per day at some sites (Kleinstreuer and Logan, 1980). Such monumental catastrophes were averted at other sites by providing wider, more slowly moving waters with many nearly stagnant side pools where fish could rest.

Curves and stream side irregularities can therefore be expected to be a component of natural streams, rivers, and many new engineered channels. These two factors result in transverse chemical dispersion coefficients of

$$E_t = 0.6(\pm 50\%)u_{sh}h, \qquad \text{natural rivers and streams} \qquad (11\text{-}18)$$

For slowly meandering streams with moderate sidewall irregularities the range of the transverse mixing coefficient is 0.4–0.8, with few systems ever having this coefficient less than 0.4 (Table 11.2). While it is difficult to specify exactly what is meant by "slowly meandering," it is suggested that for such streams $w\bar{u}/Ru_{sh} < 2$, where w is the stream width and R is the radius of the curve.

From our analysis of chemical dispersion in Chapter 10 it follows from Eq. 10-27 that the size of a chemical plume in a river resulting from a spike input into the river is

$$L_t = 4(2E_t t)^{1/2} \qquad (11\text{-}19)$$

Substituting in our estimate for a slowly meandering river (Eq. 11-18) indicates that mixing will occur over a length L_t after time t of

$$L_t = 0.44(u_{sh}th)^{1/2} \qquad (11\text{-}20)$$

Analytical Solutions for Vertical and Transverse Mixing with Advection

There are several approaches to describe chemical dispersion across a stream or river, but analysis of downstream chemical concentrations can be simplified by assuming that vertical mixing can be neglected. Comparing the magnitude of the transverse and vertical mixing coefficients given above, we see that $E_t \approx 2\text{--}12E_v$, but most rivers are many times wider than they are deep. For example, a typical channel 1 m deep might be 30 m wide or more. Thus, while vertical mixing rates can be calculated, it is usually assumed for downstream calculations of chemical concentrations in the river that the river is completely mixed in the vertical direction.

For a river moving downstream in the x direction, the concentration of a chemical introduced at a rate $\dot{m}$, resulting in an initial concentration c_o for a stream with a flow rate Q (Fischer et al., 1979), is

$$c(x, y) = \frac{Qc_o}{\bar{u}h(4\pi E_t t)^{1/2}} \exp\left(-\frac{y^2}{4E_t t}\right) \qquad (11\text{-}21)$$

where $x = 0$ is the point of discharge and y is the distance across the stream at time t. Based on the average flow velocity, with $t = x/\bar{u}$, we can simplify this result to

$$c(x, y) = \frac{Qc_o}{h(4\pi E_t x\bar{u})^{1/2}} \exp \left(-\frac{y^2\bar{u}}{4E_t x} \right) \tag{11-22}$$

These two equations apply only to infinitely wide streams. For a channel of width w, the boundary effects can be accounted for by assuming no flux at the stream edges ($\partial c/\partial y = 0$ at $y = 0$ and w) and the method of superposition. In this method the flux at a boundary $y = -w$ is balanced by a hypothetical source at $y = 2w$, and the flux at the boundary $y = w$ is balanced by a source at $y = -2w$, and so forth (Fischer et al., 1979). The solution is most easily presented in terms of the three dimensionless variables

$$x^* = x\frac{E_t}{\bar{u}w} \tag{11-23}$$

$$y^* = \frac{y}{w} \tag{11-24}$$

$$c^* = \frac{c}{c_o} \tag{11-25}$$

and recognizing that the initial concentration at the source c_o results from the rate chemical is added into the stream with $Q = \bar{u}wh$ according to

$$c_o = \frac{\dot{m}}{\bar{u}wh} \tag{11-26}$$

the dimensionless concentration profile in the stream is

$$c^* = (4\pi x^*)^{-1/2} \sum_{n=-\infty}^{\infty} \left[\exp \left(\frac{-(y^* - 2n - y_o^*)^2}{4x^*} \right) + \exp \left(\frac{-(y^* - 2n + y_o^*)^2}{4x^*} \right) \right] \tag{11-27}$$

In practice, only three iterations ($n = 3$) are necessary to specify c^*.

The application of Eq. 11-27 to streams results in c^* within 5% of its mean value across the stream for $x^* > 0.1$. This allows us easily to calculate the distance that the stream will travel before there is complete mixing across the stream width (i.e., less than a 5% variation in concentration) as

$$L_{CM} = \frac{0.1\bar{u}w^2}{E_t}, \qquad \text{center discharge} \tag{11-28}$$

A similar analysis can be performed for a side discharge. In this case the width of the stream is twice the value used above for the centerline discharge, resulting in the w term being replaced by $2w$, and this term's being squared. Complete mixing over a length of flow of a stream from the point of injection when the input is the side of a stream is

$$L_{CM} = \frac{0.4\bar{u}w^2}{E_t}, \qquad \text{side discharge} \qquad (11\text{-}29)$$

Using these relationships, it is possible to evaluate whether a wastewater discharge into a receiving water body will partially or fully migrate across a stream or river.

Longitudinal Mixing and Longitudinal Dispersion Coefficients

Longitudinal dispersion produced only by turbulent mixing in the direction of flow, e_x, is about the same magnitude as the mixing produced in the transverse direction, e_y, in an infinitely wide stream because in both cases there would be no boundaries to inhibit fluid motion. If the longitudinal mixing coefficient were defined only in terms of mixing due to turbulence, $E_L(e_x)$, by assuming $e_x = e_y$, we would have

$$E_L(e_x) = 0.6(\pm 50\%)u_{sh}h, \qquad \text{natural rivers and streams} \qquad (11\text{-}30)$$

There is little interest in the magnitude of this longitudinal mixing coefficient, however, since longitudinal dispersion produced by the velocity gradients, $E_L(e_{\Delta u})$, is usually much larger than longitudinal mixing produced by turbulence. For example, we saw in Chapter 10 that the longitudinal dispersion coefficient for a logarithmic velocity profile was $E_L = 5.93u_{sh}h$; this dispersion coefficient was determined primarily by the velocity profile, or $E_L(e_{\Delta u})$. The ratio of this longitudinal dispersion coefficient, $E_L(e_{\Delta u})$, to the longitudinal mixing coefficient, $E_L(e_x)$, for a straight channel (Eq. 11-17) is therefore

$$\frac{E_L(e_{\Delta u})}{E_L(e_x)} = \frac{5.93u_{sh}h}{0.15u_{sh}h} = 40 \qquad (11\text{-}31)$$

Since the net dispersions due to shear and mixing are additive, turbulent mixing can be neglected in comparison to that produced by the logarithmic profile for chemical dispersion calculations in streams and rivers in the direction of flow.

Velocity profiles in natural streams do follow approximately the logarithmic velocity profile used by Elder that $u = \bar{u} + (u_{sh}/\kappa)[1 + \ln(z/h)]$, but actual measurements of longitudinal dispersion in natural streams have shown that observed values of $E_L/u_{sh}h$ are much larger than expected from his analysis. Values of $E_L/u_{sh}h$ typically range from 140 to 500, although they can be as large as 7500 in the Missouri River (Yotsukura et al., 1970) or as little as 8.6 (Yuma Mesa canal in Arizona; Schuster, 1965), as shown in Table 11.3. The reason that Elder's analysis for an infinitely wide plane does not apply to natural streams is that the transverse velocity is influenced by the river geometry, and as a result the velocity across the river width varies too much (Fischer, 1967a).

TABLE 11.3 Experimental Longitudinal Dispersion in Open Channels

Channel	Depth h (m)	Width w (m)	Mean Velocity $\bar{u}$ (m s^{-1})	Shear Velocity u_{sh} (m s^{-1})	E_t (m^2 s^{-1})	$\dfrac{E_L}{u_{sh}h}$
Chicago Ship Canal	8.07	48.8	0.27	0.0191	3.0	20
Clinch River, Tennessee	0.85	47	0.32	0.067	14	235
	2.10	60	0.94	0.104	54	245
Copper Creek, Virginia	0.49	16	0.27	0.080	20	500
	0.85	18	0.60	0.100	21	250
	0.49	16	0.26	0.080	9.5	245
River Derwent	0.25		0.38	0.14	4.6	131
Green-Duwamish River, WA	1.10	20		0.049	6.5–8.5	120–160
Missouri River	2.70	200	155	0.074	1500	7500
Sacramento River	4.00		0.53	0.051	15	74
South Platte River	0.46		0.66	0.069	16.2	510
Yuma Mesa A Canal	3.45		0.68	0.345	0.76	8.6

Source: Fischer et al., 1979.

A Simplified Approach for Calculating Dispersion in Real Streams Real streams contain bridge piers, partially and fully submerged wrecks, sandbars, bends, pools, and riffles. While the average dispersion rate of chemicals in such channels cannot be predicted perfectly in every case, Fischer et al. (1979) demonstrate that it is possible to estimate dispersion coefficients within a factor of four using Taylor's analysis of longitudinal dispersion in a channel. It must be assumed that the chemical has already been fully mixed across the river, since Taylor's analysis cannot be applied until after an initial mixing period. After the chemical has fully mixed across the river, which as shown above can be calculated as L_{CM}, the plume distribution across the channel will be nearly Gaussian.

Based on work by Fischer (1966, 1967a, 1975), a longitudinal dispersion coefficient was developed by neglecting the vertical profile in the stream and applying Taylor's analysis to the transverse velocity profile. He began with the definition of the longitudinal dispersion coefficient in Eq. 10-69, or

$$E_L = -\frac{1}{hD} \int_0^h u' \int_0^y \int_0^y u' \, dy \, dy \, dy \qquad (11\text{-}32)$$

Fischer used the dimensionless variables $y^* = y/h$, $u_{sh} = u'/\sqrt{\overline{u'^2}}$, and $E^* = E_t/\bar{E}_t$, where $\bar{E}_t$ is the average transverse mixing coefficient for the river. The term $\sqrt{\overline{u'^2}}$ is not the intensity of the velocity deviation, but a measure of the extent that the turbulent averaged velocity deviates throughout the cross section from its cross-sectional mean. With these variables, Fischer proposed that

$$E_L = \frac{h^2 \overline{u'^2}}{\bar{E}_t} I$$ (11-33)

where I is a dimensionless integral defined as

$$I = -\int_0^1 u_{sh} \int_0^{y^*} \frac{1}{E^*} \int_0^{y^*} u_{sh} \, dy^* \, dy^* \, dy^*$$ (11-34)

Fischer selected a dimensionless integral value of $I = 0.07$ based on his analysis of real streams. The characteristic length was chosen as $h = 0.7w$, where w is the full width of the stream, and he used an average transverse mixing coefficient of $\bar{E}_t = E_t = 0.6u_{sh}d$. Fischer found from laboratory experiments that the ratio $\sqrt{\overline{u'^2}}/\bar{u}^2$ ranged from 0.17 to 0.25, with a mean of 0.2, and indicated that values were similar in streams. Using these results, he obtained

$$E_L = \frac{0.011\bar{u}^2 w^2}{u_{sh}h}$$ (11-35)

While the above analysis can be used to understand the dispersion of dissolved chemicals in rivers, it should be noted that not all material in water consists of neutrally buoyant material. Rivers contain suspended particles with densities nearly equal to and much larger than that of water, and chemicals can be adsorbed to these particles. In an elegant experiment Granata and Horne (1985) demonstrated that the dispersion of some species of phytoplankton was larger than that of a passive tracer (Rhodamine B dye). The longitudinal dispersion rate of a fast-sinking culture of *Odontella* was 280 cm^2 s^{-1} in a large seawater flume compared to that of 60 to 80 cm^2 s^{-1} for the dye and a slower-sinking species of *Skeletonema*. This study demonstrates that even naturally occurring particles, such as single-celled phytoplankton, have the potential for greater dispersion in natural systems than neutrally buoyant particles and dissolved chemicals.

Analytical Solution for Advection with Longitudinal Mixing

For a continuous input of chemical into a river, where the chemical is dispersed over the width of the river, the solution to the advection–dispersion equation (Eq. 11-12) for longitudinal dispersion for the three boundary conditions

$$BC \ 1: \quad t = 0, \quad\quad c = 0$$

$$BC \ 2: \quad x = 0, \quad\quad c = c_0$$ (11-36)

$$BC \ 3: \quad x \to \infty, \quad\quad c \to 0$$

in the x direction as a function of time is

$$c(x,\, t) = \frac{c_0}{2} \left[erfc \left(\frac{x - u_x t}{(4E_L t)^{1/2}} \right) + erfc \left(\frac{x + u_x t}{(4E_L t)^{1/2}} \right) \exp \left(\frac{u_x x}{E_L} \right) \right] \quad (11\text{-}37)$$

This could occur, for example, where a wastewater discharge enters a river and is well dispersed across the width of the river.

If chemical reaction is included in the governing transport equation, at steady state the one-dimensional transport equation for a first-order reaction becomes

$$u_x \frac{dc}{dx} - E_L \frac{d^2 c}{dx^2} + k_1 c = 0 \quad (11\text{-}38)$$

where k_1 is a first-order rate constant. For the boundary conditions

$$BC\ 1: \quad x = 0, \qquad uc_0 = uc - E_L \frac{dc}{dx}$$
$$BC\ 2: \quad x = L, \qquad \frac{dc}{dx} = 0 \quad (11\text{-}39)$$

These two boundary conditions satisfy a mass balance at the source and at a distance L downstream. The solution of this equation is credited to Danckwurtz (1953)

$$\frac{c}{c_0} = \frac{4\beta}{(1 + \beta)^2 e^{-(Pe/2)(1-\beta)} - (1 - \beta)^2 e^{-(Pe/2)(1+\beta)}} \quad (11\text{-}40)$$

where $Pe_L = u_x L / E_L$ is a Peclet number based on L, and β is a dimensionless number calculated as

$$\beta = \left(1 + \frac{4 k_1 E_L}{u_x^2} \right)^{1/2} \quad (11\text{-}41)$$

For the case of no longitudinal dispersion, or $E_L = 0$, this result becomes that of simple first-order decay, or $c/c_0 = \exp(-k_1 L / u_x)$.

11.3 MIXING IN LAKES

The transport of chemicals within lakes is more complicated than that of dispersion in rivers due to the lack of a defined advective velocity and the complexity of mixing patterns that typically vary seasonally. The typical structure of a deep dimictic lake consists of a well-mixed epilimnion at the surface of a lake, a relatively stable hypolimnion at the bottom of the lake, and a separating layer called the thermocline or metalimnion (Fig. 11.3). Mixing in a lake is driven by a number of different forces, including entering and exiting streams, thermal gradients, wind, and coriolis effects (gyres set up from the earth's rotation). Lakes mixed by these forces can result in whole-lake currents or whole-lake motion (i.e., the whole lake actually moves like water in a bucket) to the smallest mixing patterns of surface waves set up by the wind. The overall direction of different eddies can therefore range from random motion produced by eddies the size of the Kolmogorov microscale to the lake diameter or length depending on the lake topography. These whole-lake flows are referred to as whole-lake seiches, while surface waves set up mixing patterns that are often called surface seiches.

One major factor in the distribution of chemicals in a lake is the frequency of whole-lake mixing. Monomictic lakes mix once a year and remain mixed during the winter. Dimictic lakes mix twice a year with ice formation in the winter that prevents wind shear and mixing. Polymictic lakes are shallow lakes that mix frequently from top to bottom due to wind shear. Some deep lakes do not regularly mix each season, and may mix only once every few years. Meromictic lakes are very deep lakes that cannot be mixed from top to bottom. The typical progression of mixing for a dimictic lake begins with whole-lake mixing (surface to bottom) in the springtime as the lake warms. The increasing photo period of the spring season heats up surface waters, causing density differences in the water in the lake. Since water has its highest density at 4°C, heating water at 0°C to 4°C will increase the water density, causing it to fall. Thus heating produces density gradients in the water, resulting in lake mixing. As the water continues to heat above 4°C, colder water will stabilize on the bottom, while warmer, less dense water will begin to stabilize at the surface. Continued heating and wind mixing at the surface eventually can result in a well-mixed surface layer of warmer water that floats on top of a colder layer. Since wind shear cannot extend to very great depths, the surface waters remain relatively uniform in temperature and separated by a thermal gradient from the lower hypolimnion. Although density gradients set up the differences in the fluid separation, viscosity differences contribute to the stability of a thermocline. For example, the viscosity of water is 23% less at 20°C than at 10°C, but the water density changes by only 0.15% (Fig. 11.4).

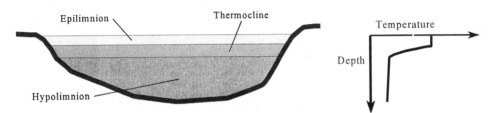

Figure 11.3 Lake structure and temperature profile showing that temperatures are relatively constant in the epilimnion and hypolimnion; they are separated by a thin zone of rapid temperature change called the metalimnion or thermocline.

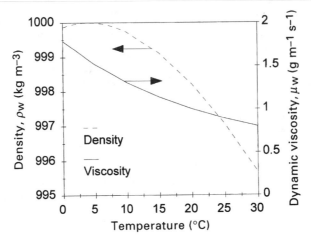

Figure 11.4 Properties of water. Notice the rapid change in fluid viscosity in comparison to the percent change in density.

During lake stratification in the summer there is relatively little dispersive transport between the different lake layers, although the depth to the thermocline can vary daily or not at all, depending on heating, mixing intensity, and other factors. Chemicals introduced into the epilimnion will mix relatively quickly and be dispersed throughout the mixed layer but will be transported only at very reduced rates to the waters in the thermocline and hypolimnion. In the fall lake cooling will again result in a second period of whole-lake mixing due to the sinking of cooler water from the lake surface to the lake bottom. By the winter time the lake can become stratified again, this time with colder, less dense water on the surface of the lake reaching 0°C, and warmer, denser water accumulating on the lake bottom. Not all lakes mix with the same frequency or intensity, resulting in a wide range of mixing patterns. Mixing in stratified systems is reviewed in more detail by Fernando (1991).

Chemicals introduced into these lakes can have different detention times depending on whether the lake is stratified or not, and depending on where the contaminant enters the lake. The detention time, τ, for a lake can be calculated in the usual manner of $\tau = Q/V$, where Q is the net flow rate in and V is the lake volume. The detention time is often referred to as the turnover time or holding time of the lake. Detention times can vary from days to years or longer; some saline lakes such as Mono Lake in California have no outflow and therefore have an infinite detention time. Chemicals already present in a warm stream entering a stratified lake could disperse throughout the epilimnion and have a much shorter detention time than chemicals in a cold water that entered into the hypolimnion. In the latter case, the chemical would be trapped in the lake until the lake mixed in the spring or fall. Thus lake structure is an important factor in chemical transport in lakes.

Wind produces mixing in surface waters by exerting a drag force on the lake surface waters. Many factors contribute to the effectiveness of the wind in shearing the water, such as wind strength and wind speed, stability of the wind, length of uninterrupted wind flow over the lake, and the degree of wave development (Fisher et al., 1979). Wind stress on the water can be calculated as

$$\tau = C_D \rho_a u_{10}^2 \tag{11-42}$$

where C_D is an empirical drag coefficient that incorporates all the factors affecting the wind stress on the lake, ρ_a is the density of the air, and u_{10} the wind velocity 10 m above the lake surface. For wind speeds up to 5 m s^{-1}, $C_D = 1.0 \times 10^{-3}$ and rises linearly to $C_D = 1.5 \times 10^{-3}$ for speeds up to 15 m s^{-1} (Hicks, 1972).

The stability of a vertically stratified water column in a lake is usually described in terms of the stability frequency, $F[t^{-1}]$, sometimes referred to as the Brunt–Väisälä frequency, defined as

$$F = \left(\frac{g}{\rho} \frac{d\rho}{dz} \right)^{1/2} \tag{11-43}$$

where ρ is the density of the water, g the gravitational constant, and $d\rho/dz$ the density gradient. When water at one density is displaced into water of a different density, there will be either a positive or negative buoyant force that will act to move the water back to a condition of density equilibrium. This mixing drives an oscillation of water about the center with a frequency given by Eq. 11-43. Large stability frequencies will result in stronger stratification, since mixing in the system is confined to a thin layer bounded by a sharp concentration gradient.

Fluid mixing produced by wind shear results in fluid turbulence, which is usually quantified in terms of the mean shear rate, G, or the energy dissipation rate, ε_d. Recall that these two quantities are related by $G = (\varepsilon_d/\nu)^{1/4}$, where ν is the kinematic fluid viscosity. Dillon et al. (1981) measured energy dissipation rates in the Green Peter Reservoir near Corvallis, Oregon. They found that the energy dissipation rates in the reservoir were nearly identical to that predicted based on the wind shear of the water. According to atmospheric boundary layer theory for unstratified flow, the variation in ε_d a distance z from the boundary is

$$\varepsilon_d = \frac{(\tau/\rho_w)^{3/2}}{\kappa z} \tag{11-44}$$

where ρ_w is the water density and κ is the von Karmen constant, usually taken to be 0.40. The shear rate τ was calculated using Eq. 11-42 with $C_D = 1.3 \times 10^{-3}$, and u was evaluated at 5 m above the water surface. They found that

$$\varepsilon_d = 0.51(\pm 0.09 \text{ cm}^3 \text{ s}^{-3}) z^{-1.03(\pm 0.05)} \tag{11-45}$$

where the values in parentheses indicate standard errors (Fig. 11.5). Within experimental error, this result predicts that ε_d is inversely proportional to z. The coefficient of 0.51 was within the range of the value predicted from wind shear measurements

Dissipation (cm^2 s^{-3})

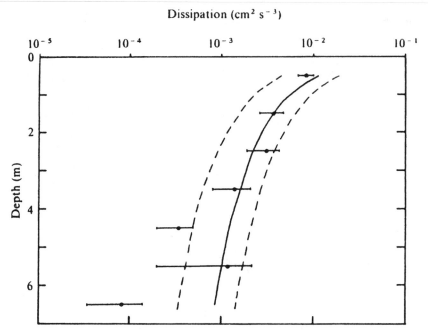

Figure 11.5 Energy dissipation rates in the upper mixed layer of a lake are produced primarily by wind shear and decrease with distance from the surface layer. The thermocline is located at 7 m. Solid line is calculated from the wind shear while the error bars are ±SE. Dotted lines calculated from highest and lowest winds. (Reprinted with permission from *Nature*, Dillon et al., 1981, © 1981 Macmillan Magazines Ltd.).

of $(\tau/\rho_w)^{3/2}\kappa^{-1}$ = 0.53 cm^3 s^{-3}. Energy dissipation rates during a period where wind speeds averaged 4.8 m s^{-1} varied from ~10^{-2} cm^2 s^{-3} at 1 m below the surface (which had waves 1 to 20 cm in height) to ~10^{-3} cm^2 s^{-3} at a depth of 6 m. Fluid shear dropped off by an order of magnitude in the thermocline located at 7 m. Thus it appears that wind shear can be used to predict fluid shear, and therefore fluid mixing rates in lakes and reservoirs.

Vertical Mixing Rates in Lakes Vertical mixing in the epilimnion is very rapid and can be neglected in comparison to the mixing rates in other directions. Mixing in the hypolimnion ranges from that of molecular diffusivities up to values of 10^{-4} m^2 s^{-1}, and is higher during periods of strong winds and high inflow and outflow. The vertical mixing coefficient e_v, and therefore the vertical dispersion coefficient E_v, varies with the stability frequency F according to

$$e_v = aF^{-2n} \tag{11-46}$$

where n ranges widely from n = 0.2 to 2.0 (Fischer et al., 1979).

Horizontal Mixing Rates in Lakes   Horizontal mixing rates in the epilimnion are of course much greater than those in the hypolimnion. The mixing coefficient e_x in the direction of mean motion was calculated by Murthy (1976) for large lakes (Lake Ontario) as

$$e_x = 1.2 \times 10^{-2} \sigma_x^{1.07} \qquad (11\text{-}47)$$

where σ_x is the standard deviation in the x direction. The mixing coefficient in the direction perpendicular to the mean flow, $e_y = e_h$, is smaller by a factor of 10. Mixing coefficients are quite variable in lake systems. For example, in Lake Huron under similar wind conditions Csanady (1963) found that $e_x = 0.04$ m^2 s^{-1} for cloud sizes of 1 km, a value about 100 times smaller than that predicted by Eq. 11-47.

In the study of mixing in Lake Ontario Murthy (1976) found that mixing in the hypolimnion could be described by

$$e_x = 1.8 \times 10^{-4} \sigma_x^{1.33} \qquad (11\text{-}48)$$

This mixing coefficient is approximately two orders of magnitude smaller than that for mixing in the epilimnion. The scatter of data was quite large, with values of σ_x ranging from 10 to 1000 m. This suggests that detailed information on the velocity structure in a lake is very important to predict vertical and horizontal mixing, and that mixing coefficients are best determined by measurements of specific systems under conditions of interest.

11.4 MIXING IN ESTUARIES

Estuaries occur whenever a river opens into an ocean, but for practical purposes, they are also considered to require a semienclosed area. The major factors that determine the extent of the estuary are the distances over which tides influence water flows and areas containing salinity gradients. Chemical mixing in estuaries is quite complicated since the motion of the water is influenced not only by temperature, advection, wind, topography, and so forth, but also salinity and tidal pumping. Readers interested in a more thorough description of mixing in estuaries and oceans are referred to Fischer et al. (1979). We will only be able to review the basics of mixing here.

Vertical Mixing in Estuaries

In the case of pure river flow through an estuary we would expect that vertical mixing could be described according to our result for rivers, $e_v = 0.067 u_{sh} d$. In an estuary, however, the water flow velocity varies from zero at the end of slack tide to a maximum midway between low and high tides. Typical values of vertical mixing are around 50 to 70 cm^2 s^{-1}. A formula for a vertical mixing coefficient, suggested by Bowden (1967), is

$$e_v = 0.0025 U_a h \tag{11-49}$$

where U_a is the depth mean amplitude of the current, and h is the average depth.

Transverse Mixing in Estuaries

Currents produced in estuaries during tidal pumping make it very difficult to predict "typical" transverse mixing coefficients in estuaries. Water flow can move toward one shore at the water surface and simultaneously be moving to the opposite shore at the sediment surface. Flows can be caught in side pools, and tides can join with air flows to set up gyres. For steady flow in a river, we expect that $e_t = 0.15 u_{sh} h$, or that $e_t/u_{sh} h = 0.15$. There are few measurements of transverse mixing coefficients in estuaries, but then again in the United States there are not that many large estuaries to study. Mixing coefficients reported by Ward (1974) in various estuaries around the world were larger than those for rivers, and were $e_t/u_{sh} h = 1.0$, 0.42, and 1.03 for the San Francisco Bay, California; Cordova Bay, British Columbia; and Gironde estuary, France. A separate study by Fischer (1974) in the Delaware Bay found $e_t/u_{sh} h = 1.2$. Therefore, mixing in an estuary is on average about ten times as rapid as transverse mixing in a river, although local variations in mixing coefficients should be expected in any area.

Longitudinal Mixing in Estuaries

The forward dispersion of chemicals due to longitudinal dispersion in estuaries has been studied by a number of investigators throughout the world. The detailed analysis of dispersion in these systems is less useful to engineers than some understanding of the magnitude of these terms. Most of the values for longitudinal dispersion range from $E_L = 100$ to 300 m^2 s^{-1}, indicating that substantial dispersion occurs in these systems. Specific values for several estuaries are given in Table 11.4. Interestingly, these values can be smaller than dispersion coefficients listed for longitudinal dispersion in some moderately sized nontidal rivers. For example, in a 200-m-wide region in the Missouri E_L was reported to be 1500 m^2 s^{-1}.

11.5   MIXING IN THE OCEAN

While the largest ocean currents are produced by the earth's rotation through Coriolis forces that produce large gyres, energy dissipation rates that contribute to mixing are primarily driven by the wind. The model of wind-induced shear described above to describe energy dissipation rates in lakes applies equally well to fluid shear rates in the ocean. Energy dissipation rates are therefore the highest at the ocean surface and decrease geometrically with distance from the surface. Estimates of maximum shear rates at the ocean surface vary of course in proportion to wind velocity. Soloviev et

TABLE 11.4 Observed Longitudinal Dispersion Coefficients in Estuaries around the World

Estuary	E_L (m² s⁻¹)
Delaware	100
	500–1500
Hudson	160
Mersey	160–360
Potomac	55
	20–100
Rio Quayas, Equador	760
Rotterdam Waterway	280
San Francisco Bay	200
Severn (summer)	54–122
(winter)	124–535
Thames (low flow)	53–84
(high flow)	338

Source: Fischer et al., 1979.

al. (1988) found that wind speeds of 7.5 to 8 m s⁻¹ measured 10 m above the sea surface, which had wave heights of 1.5 to 2 m, produced maximum shear rates of $2 < G < 7$ s⁻¹ in the top 5 m of the water, with much lower values ($G = 0.25$ s⁻¹) at depths near 10 m. Estimates during "stormy" conditions are similar in magnitude, with maximum of 6.6 s⁻¹ (Monin and Ozmidov, 1985). Shear rates in deeper (10 to 110 m) waters have been found to average 0.15 to 0.44 s⁻¹ with wind speeds of ~8 m s⁻¹ (Moum and Caldwell, 1985), likely decreasing to a background level of 0.001 s⁻¹ in deep ocean waters (Shay and Gregg, 1984).

MacKenzie and Leggett (1993) examined data from 15 studies spanning 14 geographic regions (1088 dissipation rate estimates), with all measurements but two (from a reservoir) made in marine system. Based on the boundary layer model, they found for 818 literature-derived estimates of turbulent dissipation rates measured at 11 geographic sites that 58% of the data variability could be explained by the boundary layer model equation

$$\varepsilon_d = \frac{10^{-3}(C_D \rho_a / \rho_w)^{3/2} w^3}{\kappa z} = 5.82 \times 10^{-6} \frac{w^3}{z} \qquad (11\text{-}50)$$

where w is the wind speed (m s⁻¹), z the sampling depth (m), and ε_d the energy dissipation rate (W m⁻³, where 1 W m⁻³ = 10⁻³ m² s⁻³). Using the full data set for more complex mixing environments where turbulence was known to be a factor of free convection, breaking waves in the upper 1.5 m of the water column, current shear, and upwelling, they obtained the correlation

$$\log \varepsilon_d = 2.688 \log w - 1.322 \log_{10} z - 4.812 \qquad (11\text{-}51)$$

which explained 54% of the variance in surface layer dissipation rates.

The 4/3 Law

Fluid turbulence in the upper mixed layer in the ocean leads to predictable mixing coefficients, but the mixing coefficient has been found to be a function of the size of the plume. Chemical dispersion in the open ocean is unaffected by boundaries, and therefore the rate that a chemical spreads out is proportional to the size of the cloud to the 4/3 power (Fig. 11.6). Okubo (1974) found that the mixing coefficient in the transverse, or horizontal, direction, E_t, is

$$E_t = b_E L^{4/3} \tag{11-52}$$

where b_E is a coefficient that ranges from 0.002 to 0.01 cm$^{2/3}$ s^{-1} as measured in surface, upper mixed, layer, thermocline, and a 300-m depth patch (Fischer et al., 1979). This relationhip is known as the 4/3 law, since it predicts dispersion of the cloud constantly increases with its size. The 4/3 law applies only after the chemical has spread over a distance that is relatively large compared to the microscale of

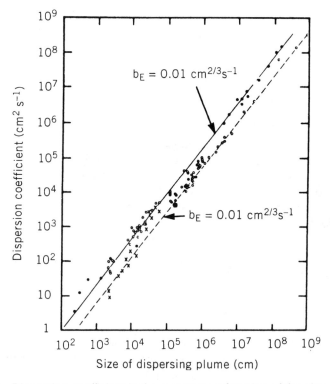

Figure 11.6   Dispersion coefficient in the ocean as a function of the size of the chemical plume: • surface float, ○ upper mixed layer, × coastal thermocline, ■ 300-m-deep patch. (Data from Okubo 1974; reprinted from Fischer et al., 1979, with kind permission of Academic Press, Inc.)

turbulence but smaller than the largest scales of turbulence. The chemical cloud cannot be affected by boundaries.

Comparison of Dispersion Coefficients in Different Systems

Based on our analysis of dispersion in natural systems, we can now summarize typical dispersion coefficients for these systems. As shown in Table 11.5, dispersion potentially spans a range of 13 orders of magnitude, from 10^{-5} cm^2 s^{-1} for simple molecular diffusion to 10^8 cm^2 s^{-1} for dispersion on the scales of thousands of kilometers in the open ocean. Even in one system, such as a river, dispersion can range in orders of magnitude from 10^3 cm^2 s^{-1} for transverse mixing to 10^7 cm^2 s^{-1} for longitudinal dispersion. But in all cases we must remember that dispersion rates in natural systems are always much larger than those for molecular processes.

11.6 TRANSPORT OF CHEMICALS PRESENT AS PURE PHASES

So far we have considered the spread of chemicals that enter into natural water bodies when the chemical is completely dissolved in water. Accidental spills of pure chemicals into rivers and lakes that are only sparingly soluble, or immiscible, means that the chemical may not be immediately dissolved. In this section we consider the rate of dissolution of these chemicals in rivers. Chemicals more dense than water will quickly sink, forming pools of the pure chemical on the bottom of a river. These pools can rest freely on the top of the sediments or may sink down into the sediment (Fig. 11.7). During the spill of the chemical tiny droplets freely suspended in the water can form. In both cases, since these chemicals are not very soluble in water, they will dissolve very slowly and persist for long periods of time.

Chemicals less dense than water, such as crude oil, will float on the water surface, rapidly forming surface slicks. Many volatile chemicals will be immediately lost to the air, but some chemicals, although sparingly soluble, will dissolve into the water due to the high surface area per volume ratio of the slick. The geometry of the slicks, combined with mass transport calculations, provides a mechanism for estimating the rates of chemical transport into the various phases.

TABLE 11.5 Typical Dispersion Coefficients in Rivers, Lakes, and the Ocean

System	Location	Dispersion Coefficient (cm^2 s^{-1})
Rivers	Vertical	$0.5 \times (10^3 - 10^4)$
	Transverse	$10^3 - 10^4$
	Longitudinal	$10^3 - 10^7$
Lakes (epilimnion)	Vertical	$10^{-1} - 10^2$
	Transverse	$10^1 - 10^4$
Ocean (scale dependent)	Horizontal	$10^1 - 10^8$

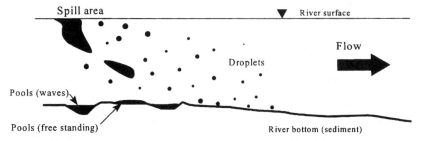

Figure 11.7   Different forms of immiscible fluids that remain as pure phases following a spill into a river: large pools, both free standing and those that fill in sediment waves, and droplets that can be suspended or resting on the sediment surface. (Adapted from Thibodeaux, 1996).

Chemical Transport from Droplets and Chemical Pools in Rivers

The rate of a chemical C dissolution can be calculated in terms of a mass transport coefficient between the water w and pure chemical phase c, using the mass transport coefficient $\mathbf{K}_{wc}$, as

$$W_C = \mathbf{K}_{wc}A(c_{eq} - c_{\infty}) \tag{11-53}$$

where W_C is the dissolution rate ($M\,T^{-1}$), c the pure chemical phase, A the interfacial area between the water and chemical phases, $c_{Cw,eq} = c_{eq}$ the equilibrium concentration of C in water (the solubility of C in water), and $c_{Cw,\infty} = c_{\infty}$ the background concentration of chemical in the water, assumed here to be zero. The rate the chemical leaves the pure chemical phase can also be expressed in terms of the total mass of the chemical, m_C, as

$$W_C = -\frac{d(m_C)}{dt} \tag{11-54}$$

Combining these two results, the mass of chemical C is lost at a rate

$$\frac{dm_C}{dt} = -\mathbf{K}_{wc}c_{eq}A(m_C) \tag{11-55}$$

where the parentheses around $A(m_C)$ indicate that the surface area of the immiscible liquid is a function of the total mass of pure chemical C. Separating and integrating, the total time for the chemical to dissolve in the water, t_C, starting with an initial mass of m_0, is

$$\int_{m_0}^{0} \frac{dm_C}{A(m_C)} = -\mathbf{K}_{wc}c_{eq}\int_{0}^{t_C} dt \tag{11-56}$$

Evaluating the integral on the right-hand side, the time for complete chemical dissolution, defined as the lifetime of the chemical, t_C, is

$$t_C = \frac{1}{\mathbf{K}_{wc} c_{eq}} \int_0^{m_0} \frac{dm_C}{A(m_c)} \qquad (11\text{-}57)$$

From this result the total lifetime of the chemical can be seen to be a function of the chemical mass, total interfacial area, mass transport coefficient, and the solubility of the chemical in water. In order to calculate the lifetime of the chemical, we must specify the geometry of the pure phase in the system.

If the chemical is well dispersed across the width of the river, then the total source strength, or concentration of the chemical past the spill area, can be calculated assuming complete vertical mixing as

$$c_{Cw} = \frac{A(m_C) \mathbf{K}_{wc} c_{eq}}{Q} \qquad (11\text{-}58)$$

If the chemical is incompletely mixed, the spread of the chemical from the point source can be estimated using techniques described earlier in this chapter.

Mass Transport from Droplets   A chemical spill will result in the formation of droplets of pure chemical of various diameters. Based on the surface tension of the liquid, a maximum droplet radius can be calculated for a stable suspended droplet (Thibodeaux, 1996) as

$$R_d = 1.90 \left(\frac{\sigma_{Cw}}{\rho_C - \rho_w} \right)^{1/2} \qquad (11\text{-}59)$$

where R_d is the drop radius (cm), σ_{Cw} the surface tension of pure chemical C in water (N/m), ρ_C the mass density of pure chemical C (g cm^{-3}), ρ_w the mass density of water (g cm^{-3}), and 1.90 is a constant that includes unit conversions.

In order to calculate the lifetime of the droplet in water, we need to specify the surface area for mass transport as a function of total mass. The mass of a spherical drop, $m_{C,d}$, can be expressed as a function of the density of the pure phase and the volume as

$$m_{C,d} = \rho_C \frac{4\pi}{3} R_d^3 \qquad (11\text{-}60)$$

The total surface area of the drop, A_d, is $4\pi R_d^2$. Combining these relationships for mass and area, we have an expression for the interfacial area of a drop, A_d

$$A_d(m_C) = 4\pi \left(\frac{3m_{C,d}}{4\pi\rho_C} \right)^{2/3} \qquad (11\text{-}61)$$

The lifetime of a drop can now be calculated by substituting Eq. 11-61 into Eq. 11-57, to produce

$$t_{C,d} = \frac{1}{K_{wc}c_{eq}4\pi} \left(\frac{4\pi\rho_C}{3}\right)^{2/3} \int_0^{m_0} \frac{dm_C}{m_C^{2/3}}$$ (11-62)

Upon integration, this produces

$$t_{C,d} = \frac{3m_0^{1/3}}{K_{wc}c_{eq}4\pi} \left(\frac{4\pi\rho_C}{3}\right)^{2/3}$$ (11-63)

The mass transport coefficient for a drop can be derived from mass transport correlations for the Sherwood number previously given for spheres. For rapidly settling droplets, Sh is calculated for particle of radius R_d settling at a velocity of U given by Stokes' Law (as long as $Re \ll 1$) using correlations for Sh_u. Alternatively, the drop can be assumed to be suspended in fluid sheared at a rate G, and the Sh calculated using equations for Sh_G. Once the drop reaches the bottom of the river, $Sh = 1$ based on the particle radius, and the drop will likely have a reduced surface area for mass transport since some of the surface area will be blocked by the sediment. As sediment continues to fall on the drop, it will become buried, resulting in a greatly reduced surface area for mass transport. The lifetime of this buried drop is then a function of the surface area for mass transport in a manner further described for immiscible liquid dissolution in Chapter 12.

Mass Transport from Pools   If pure-phase chemical C is present in larger volumes than drops in the river, it will either form a free-standing pool or it will spread out until it is contained within ripples, sandwaves, mounds, or other sediment formations at the bottom of the river. Free-standing pools must be nonwetting to the sediment and therefore will spread out at a height controlled by the surface tension of the chemical in water (Thibodeaux, 1996), according to

$$h_p = \left(\frac{2\sigma_{Cw}}{g(\rho_C - \rho_w)}\right)^{1/2}$$ (11-64)

where h_p is the height of a free-standing pool, σ_{Cw} is the surface tension between the pure chemical C phase and water, and g is the gravitational constant. If the pool is spread out over a distance such that the length of the pool is large compared to the pool height (Fig. 11.8), then the interfacial area for mass transport, $A(m_C)$, is approximately that of the top of the pool. This area can be expressed in terms of the total chemical mass as

$$A_p(m_C) = \frac{m_{C,p}}{\rho_C h_p}$$ (11-65)

Figure 11.8 A free-standing pool of chemical can reach heights h_p limited by the surface tension of the chemical in water.

Substituting this area into Eq. 11-57, the lifetime of the free-standing pool, $t_{C,p}$, is

$$t_{C,p} = \frac{\rho_C h_p}{\mathbf{K}_{wc} c_{eq}} \int_0^{m_0} \frac{dm_C}{m_{C,p}} \qquad (11\text{-}66)$$

In this case the integration limits must be reconsidered since $\ln(0)$ is not defined. It is unlikely that the pool will persist as free-standing pure chemical over its whole lifetime. Therefore, we allow some fraction of chemical to remain at the end of the "lifetime" of the pool so that the lower limit in Eq. 11-59 becomes $f_p m_0$, where f_p is defined as the fraction of the mass of the pool remaining. Integrating Eq. 11-59 between these new limits produces

$$t_{C,p} = -\frac{\rho_C h_p \, \ln(f_p)}{\mathbf{K}_{wc} c_{eq}} \qquad (11\text{-}67)$$

where a typical value of f_p can be considered to be 0.05.

The bottom of a river is usually not perfectly flat and contains ripples and dunes of fine material, as shown in Figure 11.9. If the pure phase of chemical rests between sediment ripples or waves, the height calculated above cannot be used. Instead, the pool geometry will be constrained by the sediment surface geometry. Ripples and dunes have similar geometry but differ in height. Ripples have dune crests separated by a Λ of ~30 cm and heights (H_v) of 0.6 to 6 m. Higher levels of fluid shear and turbulence produce larger dunes, which can range in length from 60 cm to several meters and in height from 6 cm to a few meters. Ripples may exist separately or as a part of dunes. In the Mississippi River dunes have been reported to reach hundreds of meters and heights of 10 m (Thibodeaux, 1996).

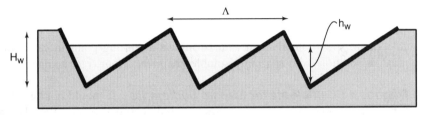

Figure 11.9 A pool of immiscible chemical may spread out to fill bottom waves formed in the sediment. The wave height is H_w, while the chemical height is h_w; the waves are separated by Λ.

If we consider the ripples and dunes to be a series of waves, the interfacial area for each pool of chemical remaining within the wave is

$$A_w(m_C) = \frac{2m_{C,w}}{\rho_C h_w} \tag{11-68}$$

where $m_{C,w}$ is the total mass of chemical in the wave, ρ_C the density of pure-phase C, h_w is the height of the pool within the wave where $h_w \leq H_w$ and H_w is the height or amplitude of the wave crest. The lifetime of the chemical pool in the wave is

$$t_{C,w} = \frac{\rho_C h_w}{\mathbf{K}_{wc} c_{eq}} \tag{11-69}$$

Combined Dissolution from Drops and Pools   When all types of droplets and pools are present in a river, the total mass of chemical at the site is

$$m_C = m_{C,d} + m_{C,p} + m_{C,w} \tag{11-70}$$

and it also follows that if there is complete mixing of the chemical leaving the contaminated river area, the total concentration of the chemical in the river water is

$$c_{Cw} = c_{Cw,d} + c_{Cw,p} + c_{Cw,w} \tag{11-71}$$

The concentration of C in the water cannot exceed the solubility limit of c_{eq}. At concentrations less than this the total concentration of C in the water can be obtained from Eq. 11-58 for each of the different immiscible forms of the chemical. For all forms, Eqs. 11-58 and 11-71 result in a total concentration mixed over the river height of

$$c_{Cw} = \frac{\mathbf{K}_{wc} c_{eq}}{Q \rho_c} \left[\frac{3m_{C,d}}{R_d} \left(1 - \frac{\mathbf{K}_{wc} c_{eq} t}{\rho_C R_d} \right) + \frac{m_{C,p}}{h_p} \exp \left(- \frac{\mathbf{K}_{wc} c_{eq} t}{\rho_C h_p} \right) \right.$$

$$\left. + \frac{2m_{C,w}}{h_w} \left(1 - \frac{\mathbf{K}_{wc} c_{eq} t}{\rho_C h_w} \right) \right] \tag{11-72}$$

While this approach is obviously a simplification of a very complex process of immiscible liquid spills into river, it at least allows a starting point in calculations of chemical dissolution from sparingly soluble chemicals from river sediments.

Mass Transport Coefficients for Stream Sediments   It should not be surprising that mass transport from a stream bottom is similar to that from a flat plate and that mass transport coefficients for stream bottoms are similar in form to those developed from boundary layer theory. However, sediments are not perfectly flat plates that have a completely uniform flow that approaches the plate. Christy and Thibo-

TABLE 11.6   Mass Transport Coefficients Developed to Predict Chemical Transport from Sediments

Mass Transport Coefficient	Conditions	Reference
$K_{ws} = 0.449 \left(\dfrac{(g\,Re)^{2/3}}{v^{1/3}l} \right)^{1/3} Sc^{-2/3}$	$Re > 590$, where $Re = \overline{u}h/v_w$	Kramers and Kreyger (1956)
$K_{ws} = \dfrac{0.114u_{sh}}{[1 + 9.6(H_v - h_v)^{1/2}]} Sc^{-2/3}$	Laminar and turbulent conditions	Christy and Thibodeaux (1982)
$K_{ws} = \left(\dfrac{u_{sh}v_w}{y_0} \right)^{1/2} Sc^{-2/3}$	Turbulent flow conditions	Levich (1962), Novotny (1969)

deaux (1982) examined published data on furfural and chloroform dissolution from shallow circular pans embedded in the sandy bottom of laboratory-scale models of flowing streams. They developed a mass transport coefficient for a forced-convection bottom-water system applicable to both flat and wavy bedforms as

$$K_{ws} = \frac{0.114u_{sh}}{Sc^{2/3}[1 + 9.6(H_w - h_w)^{1/2}]} \tag{11-73}$$

where u_{sh} is the shear velocity, $Sc = v_w/D_{Cw}$, H_w is the height of the sediment wave, and h_w is the height of the immiscible liquid in the wave form. It is assumed that the mass transport coefficient for the water and pure-phase chemical is the same as that for the water–sediment experiment.

Mass transport coefficients have been developed by others to describe chemical transport in streams and rivers. Kramers and Kreyger (1956) examined the dissolution of benzoic acid over short distances from flat surfaces in laboratory experiments under both laminar and turbulent flow conditions. Their results, shown in Table 11.6, are applicable to $Re > 590$, where the $Re = \overline{u}h/v_w$, where $\overline{u}$ is the average velocity in the river and h is the river height. Novotny (1969) was interested in the transport of oxygen, calcium, and chemicals of geological interest from river sediment surfaces. He considered the stream bottom to be a rough surface with protrusions that extended through the laminar sublayer into the turbulent core of the river flow. The mass transport coefficient he derived was therefore a function of the stream shear velocity and roughness coefficient, y_o, as shown in Table 11.6. In all these cases for chemical transport from sediment surfaces it can be seen that, as expected from boundary layer theory, the mass transport coefficient is proportional to $D_{Cw}^{2/3}$.

Mass Transport from Oil Spills in the Ocean

Crude oils are complex mixtures that contain hundreds to thousands of different components. Most are relatively immiscible. They range in size and properties from waxes and high-molecular-weight asphaltenes with low vapor pressures to low-molecular-weight aromatic compounds such as benzene, xylene, toluene, and poly-nuclear aromatic hydrocarbons (PAHs). Straight-chain hydrocarbons range in molec-ular weight from 4 to 35 or more.

When oil is spilled onto water, it immediately begins to spread out, quickly form-ing a thin oil slick. Several processes occur simultaneously. Low-molecular-weight aliphatics, volatile aromatics (such as xylene), and other hydrocarbons volatilize into the air and dissolve in the seawater (approximately 15 to 20% lost), leaving only the higher-molecular-weight hydrocarbons in the oil. Within a few hours a water-in-oil emulsion forms from the mixing produced by surface waves, with the water content of the oil slick reaching 70 to 85%, resulting in nearly a threefold increase in the thickness of the slick and therefore the volume of oil requiring removal or treatment. However, formation of this emulsified oil layer greatly increases the oil viscosity and adhesion to surfaces (such as shoreline areas), reducing the spreading, evapo-ration, and dissolution of oil components. Oil drop attachment to inorganic particles with a high density, such as resuspended sediment, can also cause the otherwise lighter-than-water oil to sink. Photochemical and biogeochemical reactions then act upon the oil to alter its properties further.

While most media attention is paid to the visible effects of the oil on organisms, such as birds, otters, and seals, there are perhaps more dangerous and less visible effects of the low-oil components on virtually all organisms that come into contact with the water. Many of the aromatic and aliphatic hydrocarbons in oil can have sublethal effects on marine organisms, particularly during developmental stages. While concentrations of 1 to 100 ppm may be lethal to adult forms, 0.01 to 0.1 ppm may produce sublethal effects, and 0.1 to 1 ppm lethal effects during larval forms of the same organisms. In general, those compounds with lower boilng points, and therefore those chemicals that leave the oil phase most quickly, are the most toxic.

The rate that chemicals are transported from oil slicks can be quantified from a mass transport analysis of the different low-vapor-pressure chemicals in an oil phase. As a rough estimate on transport rates the oil phase is treated as a single phase o, containing many different components A, B, etc., as shown in Figure 11.10. Within the oil phase all chemicals are completely mixed, and the oil phase is assumed to provide the mass transport resistance. As a result, concentration gradients exist in the air and water but not the oil phase. Since we are concerned with the highly volatile components, we will assume that $y_i > x_i$ and therefore that the Henry's Law constant is much greater than unity. The chemical concentrations at the oil–water and air–oil interfaces are in equilibrium with the concentrations of the pure phase. At the air–oil interface, the mole concentration of C in air, $c_{Ca,i}$, can be related to the mole fraction of the chemical in the oil slick, x_{Co}, using fugacity relationships as

$$c_{Ca,i} = \frac{p_{C,i}}{RT} = \frac{x_{Co}\gamma_{Co}p_c^o}{RT} \tag{11-74}$$

where γ_{Co} is the activity coefficient of the chemical C in oil, and p_C^o is the vapor pressure of pure chemical C in a liquid phase. The concentration of a chemical in

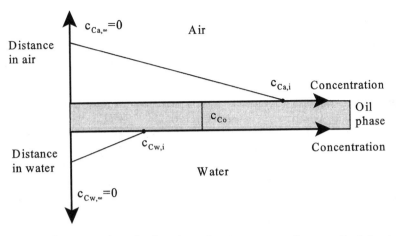

Figure 11.10 Concentrations in the air and water surrounding an oil slick when the concentration of the chemical C in the oil slick is c_{Co}. (Adapted from Thibodeaux, 1996).

the water phase at the water–oil interface is

$$c_{Cw,i} = x_{Co}\gamma_{Co}c_{Cw,eq} \tag{11-75}$$

where $c_{Cw,eq}$ is the mole concentration of C in the water in equilibrium with oil phase containing C at a mole fraction of x_{Co}.

The rate of chemical transport from the oil slick to the air and water is the sum of rates of chemical evaporation and dissolution, and can be expressed in terms of two mass transport coefficients as

$$J_{C,i} = -\mathbf{k}_a(c_{Ca,i} - c_{Ca,\infty}) - \mathbf{k}_w(c_{Cw,i} - c_{Cw,\infty}) \tag{11-76}$$

where $\mathbf{k}_a$ and $\mathbf{k}_w$ are the air- and water-phase mass transport coefficients. Assuming a negligible background concentration of C in the air and water, the bulk-phase concentrations can be neglected, or $c_{Ca,\infty} = c_{Cw,\infty} = 0$, resulting in

$$J_{C,i} = -\mathbf{k}_a c_{Ca,i} - \mathbf{k}_w c_{Cw,i} \tag{11-77}$$

The flux can be expressed in terms of the total moles in the oil phase, n_o, and the cross-sectional area of the oil slick, A, as

$$J_{C,i} = \frac{d(x_{Co}n_o/A)}{dt} \tag{11-78}$$

where the terms are left within the derivative since they may be a function of time. Combining the above expressions for mass transport from the oil slick, the overall rate of chemical transport from the slick is

$$\frac{d(x_{Co}n_o/A)}{dt} = -\mathbf{k}_a \left(\frac{x_{Co}\gamma_{Co}p_C^o}{RT} \right) - \mathbf{k}_w(x_{Co}\gamma_{Co}c_{Cw,\text{eq}}) \qquad (11\text{-}79)$$

Simplifications for Evaporation from a Multicomponent System   In order to use Eq. 11-79 to estimate the relative lifetime of chemicals in the oil slick, we can make some simplifying assumptions. First, we will assume that the oil spill area is large and that over the time period for evaporation of high-volatility components the cross-sectional area A is constant. Second, it is also assumed that n_o is constant (even though it will likely change by 15 to 20%). With these simplifications we can now move n_o and A outside the derivative, and solve for x_{Co} as a function of time as

$$x_{Co} = x_{Co,0}e^{-bt} \qquad (11\text{-}80)$$

where $x_{Co,0}$ is the concentration of C in the oil slick at time 0, and b is the rate constant calculated as

$$b = \frac{A}{n_o} \left[\mathbf{k}_a \left(\frac{\gamma_{Co}p_C^o}{RT} \right) + \mathbf{k}_w(\gamma_{Co}c_{Cw,\text{eq}}) \right] \qquad (11\text{-}81)$$

The rate of chemical loss from the oil slick from evaporation is much faster than that of dissolution, and so the kinetics are predominantly controlled by evaporation. As a result, b can be approximated as

$$b \approx b_e = \frac{\mathbf{k}_a\gamma_{Co}p_C^oA}{n_oRT} \qquad (11\text{-}82)$$

Regnier and Scott (1975) examined the evaporation of different n-alkane components of Arctic diesel 40, a fuel oil, of 3-mm-thick samples in the laboratory. They found that the evaporation of ten of the n-alkanes (C_9 to C_{18}) could be described by a single first-order constant, b_e, defined as

$$p_{C,\text{eq}} = 1.44b_e^{1.25} \qquad (11\text{-}83)$$

where $p_{C,\text{eq}}$ is the vapor pressure of the pure compound (atm) and b_e has units of min^{-1}. Although this result is only for 3-mm-thick films, correlations can be made to other film thicknesses. From Eq. 11-75, b should scale as $A/n_o = 1/h_o$, where h_o is the thickness of the oil film. The half-life of a chemical undergoing first-order decay, defined as the time for the chemical to decrease to one-half of its original concentration, or $0.5x_{Co,0}$, can be calculated as

$$t_{1/2} = -\frac{\ln[0.5x_{Co,0}/x_{Co,0}]}{b_e} = \frac{\ln 2}{b_e} \qquad (11\text{-}84)$$

Combining this result with Eq. 11-83, the half-life of an *n*-alkane in an oil slick h_o (mm) thick can be estimated as

$$t_{1/2} = 0.92 h_o p_{C,eq}^{-0.8} \tag{11-85}$$

where $p_{C,eq}$ (atm), h_o (mm), and $t_{1/2}$ (min). This equation predicts a rapid volatilization of *n*-alkanes. For example, the vapor pressure of *n*-nonane is 5.64×10^{-3} atm, resulting in a half-life of 3 h for a 3-mm-thick oil slick.

How does the evaporation of chemicals compare to the mass of chemicals lost by other routes? The Alaskan oil spill in Prince William's Sound in 1989, which consisted of approximately 10 million gallons of crude oil, provides a good example of the fate of an oil spill. It was estimated by Exxon officials that during the first 4 weeks 70% of the oil was lost to evaporation (35%), recovery (17%), incineration (8%), biodegradation (5%), and dispersal (5%). Oil slicks accounted for 10% and shorelines another 18% (Abelson, 1989). It was hoped that the remaining oil slick, micrometers in thickness, would be degraded by photooxidation and microorganisms in seawater. Approximately 4000 workers tried to reduce the visible effects of the oil on the shoreline, but the ultimate effects of the spill on the aquatic and terrestrial ecosystems will be difficult to quantify.

PROBLEMS

11.1 The chemical sulfur hexafluoride (SF_6) has been used by numerous researchers instead of fluorescent dyes to study dispersion, since it is inexpensive and easily sampled on a GC. Clark et al. (1996, 1997) used SF_6 as a conservative tracer in an experiment in the Hudson River and found that the longitudinal dispersion coefficient increased from 23 ± 3 to 78 ± 17 m^2 s^{-1} as the chemical plume increased in size. Typical transverse mixing coefficients for this river are 0.2 m^2 s^{-1}. Assuming average river conditions of a cross-sectional area of 3000 m^2, flow of 210 m^3 s^{-1}, and depth of 5.5 m, calculate a transverse mixing (E_t) and longitudinal dispersion coefficient (E_L) for this river.

11.2 There is a side discharge of wastewater into a river 100 ft wide, 5 ft deep, moving at an average velocity of 2 ft s^{-1}, with a slope of 0.0002, moving in a relatively straight path. (a) Find the time of travel for complete mixing across the width of the channel. (b) If a free standing pool of chemical *F* was present on the bottom of the river, how long would it last before it was fully ($\geq 95\%$) dissolved? You have the following information available on this chemical: D_{Fw} = 0.7×10^{-5} cm^2 s^{-1}, $K_w c_w = 1.5 \times 10^{-5}$ mol cm^{-2} s^{-1}, $\rho_F = 1.1$ g cm^{-3}, ρ_w = 1.0 g cm^{-3}, $\sigma_{Fw} = 40 \times 10^{-3}$ kg s^{-2}, $c_{Fw,eq} = 10^{-5}$ mol l^{-1}, and $M_F = 200$ g mol^{-1}. Comment on your answer.

11.3 Dichloroethylene (M_D = 96 g mol^{-1}) is spilled into a river and forms droplets, free-standing pools, and filled pools (waves). You decide to estimate how long it could persist if removal is only due to dissolution. You find for this chemical that the density is 1.22 g cm^{-3}, solubility of 2000 mg L^{-1}, viscosity of 0.33 cm^2 s^{-1}, and diffusivity in water of 2×10^{-5} cm^2 s^{-1}. The river has a shear velocity of 0.1 m s^{-1}, a roughness coefficient (y_o) of 2 cm, and a height of 10 m. Assume water has a density of 1.0 g cm^{-3}, and a viscosity of 0.01 cm^2 s^{-1} at ambient temperature. (a) How long till a 5-cm droplet dissolves in the river? (b) How long would a 6-cm high pool last? Use Novotny's mass transfer analysis.

11.4 In Section 11.6 the expression for the lifetime of a drop of a chemical in a river was derived by assuming that the mass transfer coefficient was constant. (a) Derive a similar expression for the lifetime of the drop assuming that the mass transfer coefficient is a function of the size. In order to simplify your calculations, assume a Sherwood number for a particle in a stagnant fluid. (b) Use the results from part (a) to calculate the lifetime (in days) of a chemical droplet with a density of 1.7 g cm^{-3} and a surface tension of 0.03 N m^{-1} in water (density of 1 g cm^{-3}) at 20°C. Other chemical properties are $D_{Cw} = 0.9 \times 10^{-5}$ cm^2 s^{-1}, molecular weight of 142, and solubility of 5 ppm.

11.5 The concentrations of chemicals in rivers may be affected by dispersion and chemical reaction. Assume a river with conditions of the Sacramento and continuous release of a chemical with a half-life of one hour. (a) Assuming no chemical reaction, calculate the concentrations of a chemical at a distance one-half hour downstream of the source over the next 50 m. (b) At steady state, calculate the chemical concentrations at these same locations. (c) Compare the results from part (b) with those for the case of no dispersion at the same locations.

REFERENCES

Abelson, P. H. 1989. *Science* **244**:629.

Bowden, K. F. 1967. *Phys. Fluids Suppl.* **10**:S278–80.

Christy, P. S., and L. J. Thibodeaux. 1982. *Environ. Prog.* **1**:126–29.

Clark, J. F., P. Schlosser, M. Stute, and H. J. Simpson. 1996. *Environ. Sci. Technol.* **30**:1527–532.

Clark, J. F., P. Schlosser, M. Stute, and H. J. Simpson. 1997. *Environ. Sci. Technol.* **31**(1): 308.

Csanady, G. T. 1963. *J. Fluid Mech.* **17**:360–84.

Csanady, G. T. 1976. *J. Geophys. Res.* **71**:5389–399.

Danckwurtz, P. V. 1953. *Chem. Eng. Sci.* **2**(1):1–13.

Dillon, T. M., J. G. Richman, C. G. Hansen, and M. D. Pearson. 1981. *Nature* **290**:390–92.

Elder, J. W. 1959. *J. Fluid Mech.* **5**:544–60.

Fernando, H. J. S. 1991. *Annu. Rev. Fluid Mech.* **23**:455–93.

Fischer, H. B. 1966. Technical Rep. KH-R-12, California Inst. Technol., Pasadena, CA.

Fischer, H. B. 1967a. *J. Hydraul. Div. Proc. ASCE* **93**:187–216.

Fischer, H. B. 1967b. J.S. Geological Survey Paper 575-D, D267–72.

Fischer, H. B. 1974. Int. Symp. Discharge Sewage from Sea Outfalls. London. Paper No. 37, pp. 1–8.

Fischer, H. B. 1975. *J. Environ. Eng. Div. Proc. ASCE* **101**:453–55.

Fischer, H. B., E. J. List, R. C. Y. Koh, J. Imberger, and N. Brooks. 1979. *Mixing in Inland and Coastal Waters*. Academic Press, New York.

Granata, T. C., and A. J. Horne. 1985. *J. Plankton Res.* **7**:947–53.

Hicks, B. B. 1972. *Boundary-Layer Meteorol.* **3**:201–13.

Kleinstreuer, C., and B. E. Logan. 1980. *Water Res.* **14**:1047–54.

Kramers, H., and P. J. Kreyger. 1956. *Chem. Eng. Sci.* **6**:42.

Levich, V. G. 1962. *Physicochemical Hydrodynamics*, Prentice Hall, Englewood Cliffs, N.J., p. 170.

MacKenzie, B. R., and W. C. Leggett. 1993. *Mar. Ecol. Prog. Ser.* **94**:207–16.

Monin, A. S., and R. V. Ozmidov. 1985. *Turbulence in the Ocean*, translated by H. Tennekes. D. Reidel Pub. Co., Dordrecht.

Moum, J. N., and D. R. Caldwell. 1985. *Science* **230**:315–16.

Murthy, C. R. 1976. *J. Phys. Oceanogr.* **6**:76–84.

Novotny, V. 1969. In *Advances in Water Pollution Research*, S. H. Jenkins, Ed., Pergamon Press, Elmsford, NY, pp. 39–50.

Okubo, A. 1974. Rapp. P.-v. *Réun. Cons. Int. Explor. Mer.* **167**:77–85.

Regnier, Z. R., and B. F. Scott. 1975. *Environ. Sci. Technol.* **9**(5):469–72.

Schuster, J. C. 1965. *J. Hydraul. Div. Proc. ASCE* **91**:101–24.

Shay, T. J., and M. C. Gregg. 1984. *Nature* **310**:282–85.

Soloviev, A. V., N. V. Vershinsky, and V. A. Bezverchnii. 1988. *Deep-Sea Res.* **35**(12A):1859–79.

Thibodeaux, L. J. 1996. Environmental Chemodynamics, 2nd ed. Wiley, NY.

Vanconi, V. A., and N. H. Brooks. 1957. Laboratory studies of the roughness and suspended load of alluvial streams. Rep. No. E-68, Sedimentation Laboratory, California Institute of Technology, Pasadena, CA.

Ward, P. R. B. 1974. *J. Hydraul. Div. Proc. ASCE* **100**:755–72.

Yotsukura, N., H. B. Fischer, and W. W. Sayre. 1970. U.S. Geological Survey Water-Supply Paper 1899-G.

CHAPTER 12

CHEMICAL TRANSPORT IN POROUS MEDIA

12.1 INTRODUCTION

A mathematical description of chemical transport in porous media is quite challenging due to the number of phases, complex hydraulics, and chemical interactions with dissolved and adsorbed organic matter. In order to make any calculations on the mobility and fate of chemicals introduced into groundwater aquifers, we need to make simplifying assumptions about the nature of the system. In most cases we will not be limited by our ability to formulate and solve a mathematical transport equation, but instead by an inability to specify the coefficients used in transport equations. The most integral components of these equations, such as dispersion coefficients or adsorption coefficients, may vary over distances of centimeters to meters, making it difficult to predict the exact chemical concentration at any point. By choosing constants that are spatially and temporally averaged over large parts of the system, it may be possible to predict the average behavior of chemical transport in the system.

12.2 POROUS MEDIA HYDRAULICS

The hydraulics of water flow in the subsurface environment will be simplified to one dimension. In the saturated zone water motion is essentially horizontal, while in the unsaturated, or vadose, zone water motion occurs in the vertical direction from the surface to the saturated zone. Water velocity in the saturated zone is described in terms of Darcy's Law

$$U = -K \frac{\Delta h}{\Delta l} \tag{12-1}$$

where h is the hydraulic head, $\Delta h/\Delta l$ the hydraulic gradient, and K the hydraulic conductivity. Since the gradient is dimensionless, the hydraulic conductivity has units of velocity. The Darcy velocity, U, also called a superficial velocity, is related to the average velocity of the water by the soil porosity, θ, or

$$U = u\theta \qquad (12\text{-}2)$$

where u is the actual velocity, or pore velocity, of water in the saturated porous media. The Darcy velocity is useful for calculating the total water flow rate, Q, through a given cross section of area A, as $Q = UA$. The flow rate is therefore

$$Q = -KA\frac{\Delta h}{\Delta l} \qquad (12\text{-}3)$$

Typical Darcy velocities range from <0.3 m yr^{-1} to 1.5 m d^{-1}, with hydraulic conductivities ranging from $<10^{-8}$ m s^{-1} for clay to $>10^{-2}$ m s^{-1} for gravel (Table 12.1).

Darcy's Law can also be written in terms of the soil permeability, κ, as

$$U = -\frac{\kappa}{\mu}\frac{\Delta p}{\Delta l} \qquad (12\text{-}4)$$

where μ is the dynamic viscosity of water. This form is preferred by reservoir engineers and is convenient for describing flow in packed columns since the fluid velocity is related to a pressure drop over a length (usually the length of the column). Because $\Delta p = \rho g \Delta h$, the hydraulic conductivity is related to the soil perme-

TABLE 12.1   Typical Values of Soil Porosities and Hydraulic Conductivities

Soil Type	Porosity, θ	Hydraulic conductivity K (m s^{-1})
Clay	0.40–0.70	$<10^{-8}$
Peat		10^{-8}–10^{-7}
Silt	0.35–0.50	10^{-7}–10^{-6}
Loam		10^{-7}–10^{-5}
Sand	0.25–0.50	
Very fine sand		10^{-6}–10^{-5}
Fine sand		10^{-5}–10^{-4}
Coarse sand		10^{-4}–10^{-3}
Sand with gravel		10^{-2}–10^{-3}
Gravel	0.25–0.40	$>10^{-2}$

Sources: Freeze and Cherry (1979); Tschobanuglous and Schroeder (1985).

ability by

$$K = \kappa \, \frac{\rho g}{\mu} \tag{12-5}$$

where ρ is the density, μ the dynamic viscosity of water, and g the gravitational constant. Permeability has units of $[L^{-2}]$ and is often expressed in units of darcy, where 1.01×10^{-8} darcy = 1 cm^2. This relationship evolves from the awkward definition that a saturated porous medium has a permeability of one darcy if a single-phase fluid of 1 cp viscosity flows through it at a rate of 1 cm^3 s^{-1} cm^{-2} under a pressure or equivalent hydraulic gradient of 1 atm cm^{-1}, or from Eq. 12-4

$$\kappa = \frac{Q\mu}{A} \frac{\Delta h}{\Delta p} = \frac{(1 \text{ cm}^2 \text{ s}^{-1})(1 \text{ cp})}{(1 \text{ cm}^2)} \frac{1}{(1 \text{ atm cm}^{-1})} \equiv 1 \text{ darcy} \tag{12-6}$$

The permeability of soils can be related to the soil porosity and soil grain diameter, d_g, by the semiempirical Kozeny–Carmen equation

$$\kappa = \frac{d_g^2 \theta^3}{180(1 - \theta)^2} \tag{12-7}$$

From these relationships we can see that $K \propto \kappa \propto d_g^2$, although the hydraulic conductivity is additionally a function of the grain size distribution.

The hydraulic conductivity of an aquifer is a function of the drawdown of the groundwater table, or the piezometric head, and varies in response to pumping rates at a well. The height of the groundwater table in a *confined* aquifer is m, the height of an impermeable confining layer, while in an *unconfined* aquifer the height is set by the hydrological and soil characteristics (Fig. 12.1). If a well is placed in a confined aquifer, the water will rise to some height above the confining layer; if that height reaches the surface, the well is called an artesian well. Pumping the aquifer in a confined aquifer at a rate $Q = UA_w$ lowers the piezometric head measured in an observation well at a distance R_1 from the center of the well to a height h_1, and at a distance R_2 to a height h_2. The flow into the well occurs in the horizontal direction radially around the well in the $-r$ direction with $A_w = m2\pi r$. Using Darcy's Law to specify U, we have

$$Q = K(A_w = m2\pi r) \frac{\Delta h}{\Delta r} \tag{12-8}$$

Separating terms, taking the limit as $\Delta h/\Delta r \rightarrow \partial h/\partial r$, and integrating, we have for steady conditions that

$$K = \frac{Q \ln(R_2/R_1)}{2\pi m(h_2 - h_1)}, \quad \text{confined aquifer} \tag{12-9}$$

When a well is placed in an unconfined aquifer, the water height in the observation

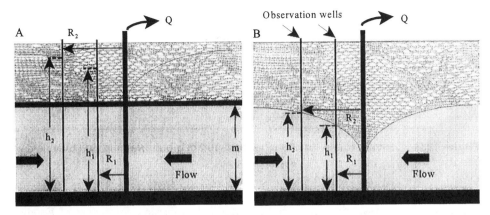

Figure 12.1 Characteristics of (A) confined and (B) unconfined aquifers used to calculate the hydraulic conductivity of the soil by pumping through a well at a rate Q, while observing the drawdown of the groundwater table.

wells is the same as that piezometric head. A mass balance over the flow into the well measured at two distances R_1 and R_2 at steady state produces for an unconfined aquifer

$$K_t = \frac{Q \ln(R_2/R_1)}{\pi(h_2^2 - h_1^2)}, \quad \text{unconfined aquifer} \tag{12-10}$$

12.3   CONTAMINANT TRANSPORT OF CONSERVATIVE TRACERS

The transport of a chemically unreactive, nonadsorbing chemical tracer in porous media in one dimension has already been introduced in Section 10.5. In that section we derived the governing advection–dispersion equation and presented different analytical solutions. The control volume used to derive a one-dimensional transport equation is a differential slice that contains many porous particles so that it represents the average hydraulic properties of the porous medium. The flow occurs through a control volume $V = \theta A_x \, \Delta x$, where θ is the porosity, and A_x the cross-sectional area of the control volume of thickness Δx (Fig. 12.2). Accumulation within the control volume is

$$\text{accumulation} = \theta A_x \, \Delta x \, \frac{\partial c}{\partial t} \tag{12-11}$$

Advection into and out of the control volume at a velocity u in the x direction is

$$\text{advection in} - \text{advection out} = \theta u A_x c|_x - \theta u A_x c|_{x+\Delta x} \tag{12-12}$$

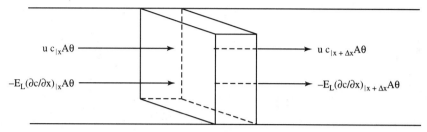

Figure 12.2   Control volume for chemical transport in groundwater in one dimension.

Chemical dispersion occurs in the direction of flow due to hydrodynamic and mechanical factors. Assuming an overall longitudinal dispersion coefficient of E_L, dispersion into and out of the differential slice is

$$\text{dispersion in} - \text{dispersion out} = \theta A_x \left(-E_L \frac{\partial c}{\partial x} \right) \Bigg|_x - \theta A_x \left(-E_L \frac{\partial c}{\partial x} \right) \Bigg|_{x+\Delta x} \quad (12\text{-}13)$$

Collecting terms, dividing by $\theta A_x \, \Delta x$, and taking the limit as $\Delta x \to 0$ produces

$$\frac{\partial c}{\partial t} = E_L \frac{\partial^2 c}{\partial x^2} - u \frac{\partial c}{\partial x} \quad (12\text{-}14)$$

Notice that since we have defined the flow to occur only through the pores at a velocity of u, the soil porosity θ drops out of the governing transport equation. We have encountered equations of this form before, and therefore the solution of Eq. 12-14, subject to the boundary conditions for a continuous source of concentration c_0 in a system initially devoid of the tracer, is given by

$$BC\ 1: t = 0, \quad c = 0$$
$$BC\ 2: x = 0, \quad c = c_0 \quad (12\text{-}15)$$
$$BC\ 3: x \to \infty, \quad c \to 0$$

is

$$c(x,\ t) = \frac{c_0}{2} \left[\operatorname{erfc} \left(\frac{x - ut}{(4E_L t)^{1/2}} \right) + \operatorname{erfc} \left(\frac{x + ut}{(4E_L t)^{1/2}} \right) \exp \left(\frac{ux}{E_L} \right) \right] \quad (12\text{-}16)$$

For flow in porous media, the second term on the right-hand side of Eq. 12-16 is negligible since values of E_L/ux produce a maximum error of 3% in $c(x,\ t)$ (Ogata and Banks, 1961), and the solution becomes

$$\frac{c(x,\ t)}{c_0} = \tfrac{1}{2} \operatorname{erfc} \left(\frac{x - ut}{(4E_L t)^{1/2}} \right) \quad (12\text{-}17)$$

Replacing the distance by $\xi = x - ut$, or the distance from the center of the dispersing plume, produces

$$\frac{c(\xi, t)}{c_0} = \frac{1}{2} \, \mathrm{erfc} \left(\frac{\xi}{(4E_L t)^{1/2}} \right) \qquad (12\text{-}18)$$

which is a familiar form of the error function solution for diffusion into an infinitely deep medium.

Example 12.1

A chemical spill occurs in a pond that is the water source of a second pond located at a distance of 1500 m from the first pond, as shown in Figure 12.3. Calculate the time for the nonadsorbing chemical to travel through the soil assuming (a) the chemical travels at the same average velocity of the water, and (b) the chemical undergoes dispersion, with $E_L = 10^{-6}$ m^2 s^{-1}, and the chemical is defined as reaching the second pond when the concentration is $0.05c_0$. Assume a soil porosity of 0.35, a hydraulic conductivity of 6×10^{-6} m s^{-1}, and an elevation difference of 25 m.

(a) The average water velocity, u, can be obtained using Darcy's Law (Eq.12-1) and $u = U/\theta$ (Eq. 12-2), or

$$u = -\frac{K}{\theta} \frac{\Delta h}{\Delta l} = -\frac{6 \times 10^{-4} \text{ m s}^{-1}}{0.35} \frac{-25 \text{ m}}{1500 \text{ m}} = 2.86 \times 10^{-5} \text{ m s}^{-1} \quad (12\text{-}19)$$

Assuming the chemical moves at the same velocity as the water, the travel time, t, is therefore

$$t = \frac{L}{u} = \frac{1500 \text{ m}}{2.86 \times 10^{-5} \text{ m s}^{-1}} \frac{d}{86400 \text{ s}} = 607 \text{ d} \qquad (12\text{-}20)$$

(b) In order to allow for dispersion, we can use Eq. 12-16, but first it will be

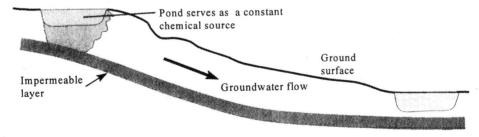

Figure 12.3 Chemical transport between two ponds by groundwater flow.

shown that for the conditions given in the problem only the first term is important. This can be seen by evaluating the right-hand terms at a time close to that calculated in part (a). Assuming a time of 607 d, the arguments of the two terms on the right-hand side of Eq. 12-16, the complementary error function and the exponential term are

$$\frac{x + ut}{(4E_L t)^{1/2}} = \frac{(1500 \text{ m}) + (2.86 \times 10^{-5} \text{ m s}^{-1})(607 \text{ d} \times 86400 \text{ s d}^{-1})}{[4(10^{-6} \text{ m s}^{-1})(607 \text{ d} \times 86400 \text{ s d}^{-1})]^{1/2}}$$

$$= 209 \qquad (12\text{-}21)$$

$$\frac{ux}{E_L} = \frac{(2.86 \times 10^{-5} \text{ m s}^{-1})(1500 \text{ m})}{(1 \times 10^{-6} \text{ m}^2 \text{ s}^{-1})} = 43{,}000 \qquad (12\text{-}22)$$

These two terms are erfc(209) = 0, and exp(43,000) → ∞, so that their product is zero. Thus Eq. 12-16 can be simplified to Eq. 12-17, or

$$c(x, t) = \frac{c_0}{2} \left[\text{erfc} \left(\frac{x - ut}{(4E_L t)^{1/2}} \right) \right] \qquad (12\text{-}23)$$

Taking the value of $c/c_0 = 0.05$, and recognizing that erfc(ϕ) = 1 − erf(ϕ) where ϕ is some argument, we can write Eq. 12-21 as

$$\text{erf} \left(\frac{x - ut}{(4E_L t)^{1/2}} \right) = 1 - (2 \times 0.05) = 0.90 \qquad (12\text{-}24)$$

Since erf(ϕ) = 0.9, then ϕ = 1.163087, and t can be solved by trial and error in the equation

$$\frac{x - ut}{(4E_L t)^{1/2}} = \frac{(1500 \text{ m}) - (2.86 \times 10^{-5} \text{ m s}^{-1})(t[d] \times 86400 \text{ s d}^{-1})}{[4(10^{-6} \text{ m s}^{-1})(t[d] \times 86400 \text{ s d}^{-1})]^{1/2}}$$

$$= 1.163087 \qquad (12\text{-}25)$$

which results in a time t = 601 d. Thus dispersion results in the arrival of the plume 6 days earlier than expected based on the groundwater velocity.

12.4 TRANSPORT WITH REACTION

If chemical reaction is included in the governing transport equation, at steady state the one-dimensional transport equation for first-order decay of the chemical becomes

$$u \frac{dc}{dx} - E_L \frac{dc^2}{dx^2} + k_1 c = 0 \qquad (12\text{-}26)$$

where k_1 is a first-order rate constant. For the boundary conditions

$$BC\ 1{:}\ x = 0, \quad uc_0 = uc - E_L \frac{dc}{dx}$$

(12-27)

$$BC\ 2{:}\ x = L, \quad \frac{dc}{dx} = 0$$

These two boundary conditions satisfy a mass balance at the source and at a distance L downstream. This equation was previously solved for the same boundary conditions in Chapter 11. The solution (Danckwurtz, 1953) is

$$\frac{c}{c_0} = \frac{4\beta}{(1 + \beta)^2 e^{-(Pe/2)(1-\beta)} - (1 - \beta)^2 e^{-(Pe/2)(1+\beta)}}$$

(12-28)

where $Pe_L = uL/E_L$ is the column Peclet number based on a column of length L, and β is a dimensionless number calculated as

$$\beta = \left(1 + \frac{4kE_L}{u^2} \right)^{1/2}$$

(12-29)

This solution has the limitation that the travel distance is assumed to be finite and there is no dispersion at the column exit (BC 2). This assumption that dispersion ends at a distance L is reasonable in a column with either a high porosity or a completely liquid-filled reactor ($\theta = 1$) since the end of the column will prevent a dispersion profile. However, mechanical dispersion in porous media is probably not seriously affected by the column end (the flow does not "see" the column end).

Two limiting cases arise for the above case. For large dispersion coefficients, the column becomes completely mixed. Recall that in Chapter 9 we found that a column reactor with a known dispersion coefficient could be analyzed as a series of completely mixed reactors. For the case of no longitudinal dispersion, or $E_L = 0$ in Eq. 12-26, the chemical concentration profile becomes that of simple first-order decay over distance x, or

$$\frac{c}{c_0} = \exp\left(-\frac{k_1 x}{u} \right)$$

(12-30)

Non-Steady-State Solution for Dispersion in an Infinite Medium: Continuous Source

A more useful solution of chemical transport in a porous medium is a non-steady-state solution where the boundaries are not assumed to affect chemical transport. The nonsteady form of Eq. 12-26 is the partial differential equation

$$\frac{\partial c}{\partial t} + u \frac{\partial c}{\partial x} - E_L \frac{\partial c^2}{\partial x^2} + k_1 c = 0$$

(12-31)

For the boundary conditions

$$c(x = 0, t \geq 0) = c_0$$

$$c(x > 0, t = 0) = 0 \qquad \text{(12-32)}$$

$$c(x = \infty, t \geq 0) = 0$$

the solution (Ogata, 1970) is

$$\frac{c(x, t)}{c_0} = \tfrac{1}{2} \, \text{erfc} \left(\frac{x - ut\Gamma}{(4E_L t)^{1/2}} \right) \exp \left(\frac{ux(1 - \Gamma)}{2E_L} \right)$$

$$+ \tfrac{1}{2} \, \text{erfc} \left(\frac{x + ut\Gamma}{(4E_L t)^{1/2}} \right) \exp \left(\frac{ux(1 + \Gamma)}{2E_L} \right) \qquad \text{(12-33)}$$

where

$$\Gamma = (1 + 2\gamma)^{1/2} \qquad \text{(12-34)}$$

$$\gamma = \frac{2k_1 E_L}{u^2} \qquad \text{(12-35)}$$

Applying L'Hospital's theorem, it can be shown that terms containing $x + ut$ are small relative to the other terms (Runkel, 1996) and Eq. 12-33 can be simplified to

$$\frac{c(x, t)}{c_0} = \tfrac{1}{2} \, \text{erfc} \left(\frac{x - ut\Gamma}{(4E_L t)^{1/2}} \right) \exp \left(\frac{ux(1 - \Gamma)}{2E_L} \right) \qquad \text{(12-36)}$$

This expression may be further simplified as $\Gamma \rightarrow 1$. When $\gamma \leq 0.0025$, Γ can be approximated by the first two terms in a binomial expansion of Eq. 12-34, so that $\Gamma \approx 1 + \gamma$, and Eq. 12-36 becomes

$$\frac{c(x, t)}{c_0} = \tfrac{1}{2} \, \text{erfc} \left(\frac{x - ut(1 + \gamma)}{(4E_L t)^{1/2}} \right) \exp \left(\frac{-k_1 x}{u} \right) \qquad \text{(12-37)}$$

Dispersion in an Infinite Medium: Source of Finite Duration   In many instances a source of chemical may be present only for a finite period of time, t_f. For the boundary conditions

$$c(x = 0, t_f \geq t \geq 0) = c_0$$

$$c(x = 0, t > t_f) = 0 \qquad \text{(12-38)}$$

$$c(x > 0, t = 0) = 0$$

$$c(x = \infty, t \geq 0) = 0$$

and for times $t \leq t_f$ the above solutions can be used. For $t > t_f$, the exact analytical solution (Runkel, 1996) is

$$
\frac{c(x, t)}{c_0} = \frac{1}{2}\left[\text{erfc}\left(\frac{x - ut\Gamma}{(4E_L t)^{1/2}}\right) - \text{erfc}\left(\frac{x - u(t - t_f)\Gamma}{(4E_L t)^{1/2}}\right) \right] \exp\left(\frac{ux(1 - \Gamma)}{2E_L}\right)
$$

$$
+ \frac{1}{2}\left[\text{erfc}\left(\frac{x + ut\Gamma}{(4E_L t)^{1/2}}\right) - \text{erfc}\left(\frac{x + u(t - t_f)\Gamma}{(4E_L t)^{1/2}}\right) \right] \exp\left(\frac{ux(1 + \Gamma)}{2E_L}\right) \quad (12\text{-}39)
$$

A common approximate solution was developed by Rose (1977)

$$
\frac{c(x, t)}{c_0} = \frac{1}{2}\left[\text{erfc}\left(\frac{x - ut(1 + \gamma)}{(4E_L t)^{1/2}}\right) \right.
$$

$$
\left. - \text{erfc}\left(\frac{x - u(t - t_f)(1 + \gamma)}{(4E_L t)^{1/2}}\right) \right] \exp\left(\frac{-k_1 x}{u}\right) \quad (12\text{-}40)
$$

Capillary Tube Model A somewhat different approach was taken by Brenner (1990) to model chemical transport in porous media. In this case chemical transport is assumed to occur within capillary tubes so that chemicals undergo Taylor–Aris dispersion within the tube. According to Brenner's model of the system, this microscale surface reaction manifests itself at a macroscale as a first-order volumetric chemical reaction rate, k^+. For a first-order irreversible reaction, the Taylor–Aris boundary condition is replaced by

$$
BC: r = R, \quad -D\frac{\partial c}{\partial r} = k_1 c \quad (12\text{-}41)
$$

where k_1 is a first-order rate constant describing the reaction at the wall. The governing equation must also be written in a slightly different form, as

$$
u_c \frac{dc}{dx} - E_L \frac{dc^2}{dx^2} + k^+ c = 0 \quad (12\text{-}42)
$$

where u_c is the mean velocity of the chemical and E_L is the Taylor–Aris dispersion coefficient. This macroscopic form of the transport equation is identical to that above except for the velocity term and the definition of k^+. Since the chemical reacts away at the wall, the mean velocity of the chemical is actually larger than that of water, and $u_c > u$. Chemicals that make their way through the tube are only those that remain on or near the center streamline. Those molecules that diffuse on slower streamlines reach the wall and are removed. As a consequence, the chemical molecules move faster than the bulk phase, which must sample all streamlines.

The difference between u_c and u depends on the rate of reaction, which can be specified in terms of the dimensionless Damkohler number, Da, defined as the ratio of the reaction rate to the rate of transport. For this case of diffusion in pores, the

Damkohler number for transport in a porous medium of particles of radius R is

$$Da = \frac{k_1 R}{D} \qquad (12\text{-}43)$$

The ratio of u_c to u is

$$\frac{u_c}{u} = 1 + \frac{1 + (\beta_0 \, Da^{-1} - 2\beta_0^{-1})^2}{3[(\beta_0 \, Da^{-1})^2 + 1]} \qquad (12\text{-}44)$$

where $\beta_0 \equiv \beta_0(Da)$ is the smallest positive root of the transcendental equation

$$\beta J_1(\beta) - Da \, J_0(\beta) = 0 \qquad (12\text{-}45)$$

where J_n is a Bessel function of order n. In general, the reaction rate constant k^+ can be obtained from

$$k^+ = \frac{\beta_0^2 D}{R^2} \qquad (12\text{-}46)$$

Brenner's equations have useful applications for very slow or very fast reactions relative to diffusion. For very slow reactions, or $Da \ll 1$

$$k^+ = \frac{2k_1}{R}, \quad Da \ll 1$$

$$\beta_0 = 2 \, Da \qquad (12\text{-}47)$$

For very fast reactions relative to diffusion, or $Da \gg 1$, the rate of diffusion to the capillary tube wall controls the overall rate of reaction, resulting in

$$k^+ = \frac{5.783 D}{R^2}, \quad Da \gg 1$$

$$\beta_0 = 2.4048 \qquad (12\text{-}48)$$

where 2.4048 is the smallest possible root of the equation $J_0(\beta) = 0$. An additional simplification can be made for very fast reactions. If the reaction at the wall is instantaneous, the boundary condition at the wall can be replaced by

$$BC: r = R, \quad c = 0 \qquad (12\text{-}49)$$

The reaction rate constant k^+ is still given as in Eq. 12-48, but because $\beta_0 = 2.4048$ and $Da \to \infty$, Eq. 12-44 becomes

$$\frac{u_c}{u} = 1.564 \qquad (12\text{-}50)$$

indicating that the chemical moves at a velocity 1.564 times as fast as that of water when it reacts instantaneously at the wall surface.

12.5 TRANSPORT WITH CHEMICAL ADSORPTION

In most cases chemicals transported in a porous medium undergo very slow chemical reactions in the soil but can rapidly adsorb to organic matter or mineral surfaces. A common form of the one-dimensional transport equation when chemical adsorption is included is

$$\frac{\partial c}{\partial t} + u \frac{\partial c}{\partial x} - E_L \frac{\partial c^2}{\partial x^2} = -k_1 c - \frac{\partial \omega}{\partial t} \frac{\rho_b}{\theta} \tag{12-51}$$

where $\rho_b = \rho_s(1 - \theta)$, ρ_b is the bulk density of the soil, ρ_s the density of the soil grains, and ω_{Cs} the mass of chemical adsorbed per mass of soil. The solution and adsorbed chemical concentrations can be related using an adsorption isotherm such as the Freundlich adsorption isotherm

$$\omega_{Cs} = Bc^b \tag{12-52}$$

where B and b are empirical constants. For sparingly soluble hydrophobic chemicals partitioning into soil organic matter, $b = 1$ and B is usually written as a constant $B = K_d$, where K_d is called a soil distribution coefficient having units $[L^3 \, M^{-1}]$. Under these conditions

$$\frac{\partial \omega_{Cs}}{\partial t} = B \frac{\partial c}{\partial t} = K_d \frac{\partial c}{\partial t}, \quad b \equiv 1 \tag{12-53}$$

With this simplified form of the adsorption isotherm the transport equation can be written as

$$\left(1 + \frac{\rho_b}{\theta} K_d\right) \frac{\partial c}{\partial t} + u \frac{\partial c}{\partial x} - E_L \frac{\partial c^2}{\partial x^2} = -k_1 c \tag{12-54}$$

The term in large parentheses indicates the velocity of the chemical in relation to the velocity of water and is known as the retardation coefficient **R**, or

$$\mathbf{R} = \left(1 + \frac{\rho_b}{\theta} K_d\right) \tag{12-55}$$

The Retardation Coefficient In order to understand the meaning of **R** in chemical transport calculations, consider the following column experiment designed to measure **R**. A column of length L is packed with soil of porosity θ, as shown in

Figure 12.4. The column is pumped for a long period of time with water before the feed is switched to a reservoir containing a chemical dissolved in water at $t = 0$. One pore volume of this chemical solution is then pumped into the column. Assuming chemical dispersion is small, the water front will have traveled a distance $L = L_w$, while the chemical will have traveled a shorter distance L_C due to its adsorption to the soil. It is further assumed that the chemical front is a sharp line as in plug flow; in reality, chemical concentration will drop off over a longer distance due to dispersion and mass transfer limitations, with the thickness of this zone defined as the mass transfer zone.

The retardation coefficient can now be derived from a mass balance around this column. The mass added to the column (m_{add}) is equal to the mass of chemical in the water, or in the pores (m_{pore}), and the mass adsorbed to the soil (m_{ads}), or

$$m_{add} = m_{pore} + m_{ads} \qquad (12\text{-}56)$$

The mass added is just the product of the chemical concentration and volume of fluid added, equal to one column, or

$$m_{add} = (AL_w\theta)c \qquad (12\text{-}57)$$

Chemical adsorption occurs only up to a distance L_c; thereafter, there is no chemical on the soil surface or in solution. Since the chemical has only traveled a distance L_c, the mass of chemical in the pores is just

$$m_{pore} = (AL_c\theta)c \qquad (12\text{-}58)$$

The mass adsorbed is the product of the mass of media within the column up to length L_c and $\omega = K_d c$, or

$$m_{ads} = (AL_c\rho_b)(K_d c) \qquad (12\text{-}59)$$

Combining Eqs. 12-56 to 12-59

$$AL_w\theta c = AL_c\theta c + AL_c\rho_b K_d c \qquad (12\text{-}60)$$

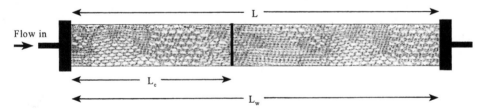

Figure 12.4 In a column test used to measure a retardation coefficient one pore volume of the water has passed through the column of length L, and therefore the water has traveled a distance L, while the chemical has only penetrated a shorter distance L_c.

and rearranging, we have the definition of the retardation coefficient

$$\frac{L_w}{L_c} = \left(1 + \frac{\rho_b}{\theta} K_d \right) \equiv \mathbf{R} \qquad (12\text{-}61)$$

Since we have assumed that dispersion was negligible in this calculation, we can see that $\mathbf{R}$ also relates the water velocity (u) to the chemical velocity (u_c), or $\mathbf{R} = u/u_c$.

The Distribution Coefficient, K_d   Chemical association with the porous medium can be a result of many factors, all incorporated into a single constant, K_d. For hydrophobic chemicals adsorbed to natural soils the single most important property of a soil affecting the extent of adsorption is the concentration of organic matter. It was shown in Section 2.5 that the mass of chemical adsorbed to a soil, ω_{Cs}, was only a linear function of water concentration when the mass of chemical adsorbed to the organic matter, ω_{Cm}, was written as a function of the fraction of organic matter (phase m) in the soil (phase s), ω_{ms}, or

$$\omega_{Cs} \left[\frac{\text{mass } C \text{ adsorbed}}{\text{mass soil}} \right] = \omega_{Cm} \left[\frac{\text{mass } C \text{ adsorbed}}{\text{mass organic matter}} \right] \omega_{ms} \left[\frac{\text{mass organic matter}}{\text{mass soil}} \right] \quad (12\text{-}62)$$

where ω_{ms} is the fraction of organic mattter (by weight) in the soil and is frequently given the symbol f_{om} (Swartzenbach et al., 1993). Combining Eq. 12-62 with the definition of K_d (Eq. 12-52) and $b = 1$, or $\omega_{Cs} = K_d c$, produces

$$K_d = \frac{\omega_{Cm}\omega_{ms}}{c} \qquad (12\text{-}63)$$

For a specific chemical the ratio of chemical in the soil and water phases, ω_{Cm}/c, is a constant, designated here as K_{Cms}, and therefore we can write more simply

$$K_d = K_{Cms}\omega_{ms} \qquad (12\text{-}64)$$

As a result of the large number of factors that affect chemical partitioning, K_{Cms} should be experimentally determined for each soil. However, in many cases once the fraction of organic matter is known for a soil, K_{Cms} is found to be constant, as shown in Figure 12.5. For many neutral, nonpolar chemicals the partitioning of a chemical between the organic matter and water phases is similar to the partitioning between octanol and water, and it has been possible to derive relationships between the organic content of soil and the octanol–water partitioning coefficient, K_{Cow}, such as

$$K_d[\text{ml g}^{-1}] = 0.63 \, K_{Cow}\omega_{Cs} \qquad (12\text{-}65)$$

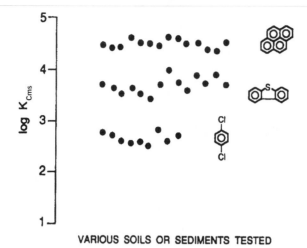

VARIOUS SOILS OR SEDIMENTS TESTED

Figure 12.5 The distribution coefficient, K_{Cms} for some nonpolar organic chemicals are constant for a variety of soils and sediments (From Swartzenbach et al., © 1993, Environmental Organic Chemistry. Reprinted by permission of John Wiley & Sons, Inc.).

where ω_{Cs} is the weight fraction of organic carbon of the soil, and 0.63 is an empirical constant developed for several chlorinated hydrocarbons and aromatic hydrocarbons. Many other such correlations have been developed for K_d between K_{Cms} and K_{Cow}. For example, for a series of neutral organic compounds, Swartzenbach et al. found

$$\log K_{Com} = 0.82 \log_{10} K_{Cow} + 0.14 \tag{12-66}$$

as shown in Figure 12.6. Examples of other correlations are shown in Table 12.2.

Intra-Aggregate and Surface Diffusion with Nonlinear Adsorption The heterogeneity of soil particle sizes can result in preferential flow paths through a porous medium. Some sections of the medium will have relatively little flow through them, producing areas of essentially immobile water, while other parts of the medium (macropores) will carry the bulk of the flow (Fig. 12.7). Chemicals in water transported through this soil can diffuse through bulk fluid, immobile water, on the particle surface and within particles, and can adsorb in a nonlinear manner as accounted for by an empirical Freundlich adsorption isotherm. A solution of such a case may involve 12 separate equations and must be done numerically (Crittenden et al., 1986) using techniques beyond the scope of this text. However, examination of the dimensionless groups developed to analyze chemical transport in soils under these conditions can provide some real insight into the factors controlling chemical transport in soils.

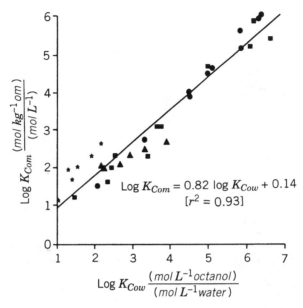

Figure 12.6 There is a linear relationship between the partitioning of chemicals between octanol–water (ow) and organic matter (om) for a variety of neutral organic compounds: ●, aromatic hydrocarbons; ■, chlorinated hydrocarbons; ▲, chloro-S-triazines; and *, phyenyl ureas (Adapted from Swartzenbach et al, © 1993, Environmental Organic Chemistry. Reprinted by permission of John Wiley & Son, Inc.).

TABLE 12.2 Different Correlations Developed to Relate K_d to K_{ow}

Equation for log K_d	$n(r^2)$	Chemicals
$0.544 \log K_{ow} + 1.377$	45 (0.74)	Wide variety, mostly pesticides
$0.937 \log K_{ow} - 0.006$	19 (0.95)	Aromatics, polynuclear aromatics, triazines, and dinitroaniline herbicides
$1.00 \log K_{ow} - 0.21$	10 (1.00)	Mostly aromatic or polynuclear aromatics; two chlorinated aromatics
$0.94 \log K_{ow} + 0.02$	9 (—)	S-Triazines and dinitroaniline
$1.029 \log K_{ow} - 0.18$	13 (0.91)	Variety of insecticides, herbicides, and fungicides
$0.524 \log K_{ow} - 0.855$	30 (0.84)	Substituted phenylureas, and alkyl-N-phenylcarbamates

Source: Swartzenbach et al. (1993).

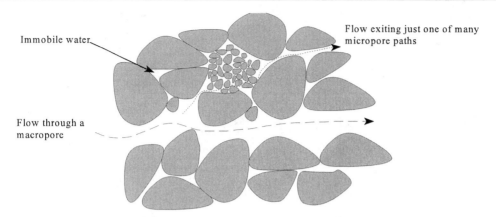

Figure 12.7 The heterogeneous nature of the material can produce larger macropores, which carry the bulk of the flow, and micropores, which carry smaller flows. There may also be stagnant or immobile zones through which no flow occurs.

Chemical transport in heterogeneous and adsorptive soils involves at least seven dimensionless groups: the Peclet number, Pe; Stanton number, St; pore diffusion modulus, Po; surface diffusion modulus, Su; immobile water chemical distribution, Im; the adsorbed chemical distribution ratio, Ad; and the Freundlich adsorption exponent, b (Table 12.3). Three of the dimensionless numbers are equilibrium relationships (Ad, Po, b), while the others are mass transport relationships.

When a chemical can both adsorb and diffuse into intra-aggregate pores, the retardation coefficient becomes

$$\mathbf{R} = 1 + Im + Ad \tag{12-67}$$

indicating that chemical breakthrough will be delayed as a result of either process. Im is a dimensionless ratio of the chemical mass in the immobile and mobile water phases, and Ad is a ratio of mass adsorbed to the mobile-phase concentration.

The effect of the power of the adsorption isotherm, b, on chemical transport depends on the magnitude of b relative to unity. When $b < 1$, the higher the chemical concentration, the faster the chemical will penetrate a distance L because lower concentrations will preferentially adsorb to a greater extent onto the soil grains. For this favorable adsorption ($b < 1$), a step increase in c will produce a self-sharpening front, while a step down in concentration will lead to a spreading wave; for $b > 1$, the reverse is true. There is no sharpening or spreading for a linear adsorption isotherm ($b = 1$); so a chemical moves in a plug-flow fashion.

In order to determine which mass transport resistance will limit the overall transport from the mobile to particle phases, the Biot numbers (Bi) for surface and pore

TABLE 12.3   Dimensionless Groups Used to Characterize Flow through Highly Heterogeneous Soil[a]

Group	Name	Meaning	Expression
Pe	Peclet number	Ratio of mass transport by advection to dispersion	$Pe = \dfrac{uL}{E_L}$
St	Stanton number	Ratio of mass transport by film to advection	$St = \dfrac{kL(1 - \theta)}{u\,Re}$
Po	Pore diffusion modulus	Ratio of intra-aggregate pore diffusion to advection	$Po = \dfrac{D_{Cw,p}\,Im\,L}{uR^2}$
Su	Surface diffusion modulus	Ratio of intra-aggregate surface diffusion to advection	$Su = \dfrac{D_{Cs,s}\,Ad\,L}{uR^2}$
Im	Immobile water chemical distribution ratio	Ratio of chemical masses in the immobile and mobile phases	$Im = \dfrac{(1 - \theta_{w,m})\theta_a}{\theta_{w,m}}$
Ad	Adsorbed chemical distribution ratio	Ratio of chemical mass adsorbed to mobile phase concentration c_0	$Ad = \dfrac{(1 - \theta_{w,m})\rho_a \omega_{Cs}}{\theta_{w,m} c_0}$
b	Freundlich adsorption isotherm exponent	If $b = 1$, adsorption is linear with chemical concentration in water	$\omega_{Cs} = Bc^b$

[a]Meaning of variables used in equations: R, radius of aggregated particle; $D_{Cw,p}$, diffusivity of chemical in water in pore; $D_{Cs,s}$, diffusivity of chemical on surface of soil grain; ρ_a, aggregate density which includes micropore volume; θ, total porosity; $\theta_{w,m}$, void fraction filled with mobile water; θ_a, void fraction of aggregate; c_0, inlet concentration.
Source: Adapted from Crittenden et al., 1986.

diffusion can be compared, where these numbers are defined as

$$Bi_s = \frac{St}{Su}$$

$$Bi_p = \frac{St}{Po} \tag{12-68}$$

If both Bi numbers are large ($Bi > 20$) then intra-aggregate diffusion resistance will control the mass transport rate into the aggregate. If either Bi_s or Bi_p is small ($Bi < 1$), film resistance controls, and intra-aggregate transport takes place by either surface or pore diffusion (since they act in parallel).

The controlling mass transport mechanism can be evaluated by comparing the relative size of the mass transport zone (MTZ) produced by each mechanism. The MTZ is defined as the distance over which a chemical concentration changes from c_0 to 0; for plug-flow conditions, the MTZ = 0. When the MTZ is defined for when all mass transport limitations are equal (Crittenden et al., 1986)

$$Pe = 3\,St\left(1 + \frac{1}{Ad}\right)^2 = 15\,Su\left(1 + \frac{1}{Ad}\right)^2 = 15\,Po\left(1 + \frac{1}{Ad}\right)^2 \tag{12-69}$$

This equation can be used to determine which transport mechanism is controlling in a system. For a strongly adsorbing chemical ($Ad > 20$) with $b = 1$, pore diffusion will control mass transport in the MTZ when $Po = 20$ ($Bi_p = 50$), $Su = 2$ ($Bi_s = 500$), $Pe = 3000$, and $St = 1000$. Pore diffusion is $\sim 10\times$ faster than surface diffusion, or $Po/Su = 10$, and $\sim 10\times$ slower than dispersion ($Pe/15\,Po = 10$) and film transport ($3\,St/15\,Po = 10$).

Analytical Solutions for Special Cases   Although an analytical solution does not exist for the above nonequilibrium conditions, the previously presented solutions can be adapted for column studies for cases where adsorption is important if local (and instantaneous) equilibrium is assumed. The transport equation for adsorption and reaction shown in Eq. 12-51 can be written as

$$\frac{\partial c}{\partial t^+} + u\,\frac{\partial c}{\partial x} - E_L\,\frac{\partial c^2}{\partial x^2} = -k_1 c \tag{12-70}$$

where the retardation coefficient has been included into the time variable, or $t^+ = t/\mathbf{R}$. By writing our governing equation in this form our previous solutions for chemical breakthrough concentrations are effectively normalized for the delay in chemical breakthrough due to the retardation coefficient. Since the only difference between this equation and the previous equations for nonsteady transport without (Eq. 12-14) and with (Eq. 12-31) reaction is a constant, our previous solutions (12-16–18, 12-33–40) apply.

Solutions to the advection–dispersion equation are often presented in dimensionless form. For the case of adsorption, time can be normalized by the column detention time, $\tau = L/u$, so that a dimensionless time variable, t^*, based on the breakthrough of an adsorbing chemical is

$$t^* = \frac{t}{\mathbf{R}\tau} = \frac{tu}{\mathbf{R}L} \tag{12-71}$$

For an adsorbing chemical the breakthrough will be delayed due to adsorption and will occur at a time $t = \mathbf{R}\tau$ compared to that of water. Therefore, t^* is the ratio of mass of chemical added to the column to the mass of chemical in a column at its saturated capacity (when c leaving the column is c_0). Defining dimensionless concentration and distance variables as $x^* = x/L$ and $c^* = c/c_0$, Eq. 12-16 for a nonadsorbing chemical (Lapidus and Amundson, 1952; Crittenden et al., 1986) becomes

$$c^*(x^*,\,t^*) = \frac{1}{2}\left[\operatorname{erfc}\left(\frac{Pe^{1/2}(x^* - t^*)}{2t^{*1/2}}\right) + \operatorname{erfc}\left(\frac{Pe^{1/2}(x^* + t^*)}{2t^{*1/2}}\right)\exp\,(Pe\,x^*)\right] \tag{12-72}$$

Chemical Desorption: A Rate-Limiting Process   Although adsorption can be assumed to occur rapidly relative to groundwater transport velocities, chemical desorption can be an extremely slow process. Slow desorption of hydrophobic chemicals in soils may ultimately limit the effectiveness of all pump-and-treat soil remediation efforts. The physical–chemical interactions between a hydrophobic chemical and soil particles is not well understood. Repeated measurement of adsorption isotherms over a 4 to 5 log range in concentration, even after allowing for equilibrium periods of 3 months, revealed hysteresis in chemical adsorption. The amount of chemical adsorption of trichloroethylene (TCE) on a soil (Santa Clara solids) was lower by a factor of two and increased at lower chemical concentrations (Farrell and Reinhard, 1994a), suggesting not all the chemical adsorbed to the surface was released over long time periods (as long as 11 months).

Chemical desorption rates measured for a completely inorganic porous medium over a four-week period reveal the presence of several factors controlling chemical adsorption at low concentrations of adsorbed chemical (Fig. 12.8). Farrell and Reinhard (1994b) equilibrated silica gels with TCE in air (100% relative humidity) at a concentration 79% of the vapor pressure of TCE. Desorption of the unsaturated media into a water-saturated nitrogen gas stream (2–3 pore volumes per min) revealed that 95% of the chemical desorbed within 5 to 10 min but that the remaining 5% would take years to desorb completely. Release of the fast-desorbing fraction could be well described by diffusion mechanisms that account for sorption, retarded aqueous diffusion, and pore tortuosity effects. Slow desorption of chemicals has been attributed to a combination of slow diffusion through organic matter and diffusion through small pores, but other factors must contribute to this phenomenon. There was no organic component to this medium, and pore diffusion could not account for the extremely slow desorption period that spanned $\sim$10 to 4×10^4 min (4 weeks).

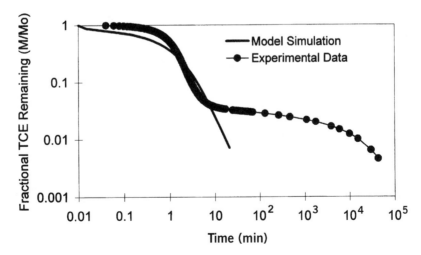

Figure 12.8 Comparison of measured and predicted rates of trichloroethylene (TCE) desorption from a silica gel. The prediction of model B, based on pore diffusion, works for the initial periods, but the third desorption period (occurring after $\sim$10 min) is not predicted by the model. (Data from Farrell and Reinhard, 1994).

It was hypothesized that adsorption in micropores of molecular dimensions were responsible for the desorption-resistant fraction. Adsorption in these micropores is associated with greatly increased adsorption energies and desorption activation energies. Whatever the mechanism of slow chemical release, these studies indicate that any inorganic medium contaminated with chemicals such as TCE will require extremely long periods of time to be cleaned up by pump-and-treat (air or water) technologies.

12.6   FORMATION OF GANGLIA OF NON-AQUEOUS-PHASE LIQUIDS

Chemicals leaked, spilled, or applied to soils will migrate down through the soil, leaving a trail of pure-phase chemical "ganglia" trapped in the porous medium (Fig. 12.9). All pure phases of chemical remaining in the soil are referred to as non-aqueous-phase liquids (NAPLs). If the volume of the applied chemical is large, the chemical will reach the water table and either float on the water, forming a LNAPL, or sink down through the water to an impermeable layer, forming a DNAPL, where L indicates the chemical is lighter than water, and D that the chemical is more dense than water. Ganglia are held in place between the soil grains by capillary forces in exactly the same way that water is held above the level of the water in a straw sitting in a glass of water.

   In order to understand how chemicals are trapped in soils, we need to understand a bit more about capillary forces and the energy at interfaces between different phases. When an air bubble is trapped in water (or alternatively when a water droplet

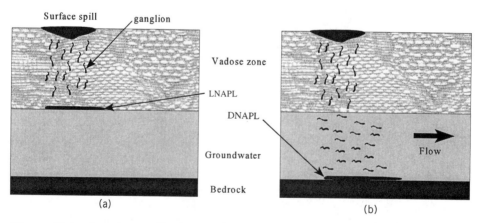

Figure 12.9   A surface spill of pure chemical will travel down through the vadose zone leaving behind pure phase ganglia and chemical adsorbed to the soil. (A) If the chemical is lighter than water, it will form a pool of pure phase chemical on the surface of the groundwater table. (B) If the chemical is denser than water, it will sink into the groundwater and fall until it reaches an impermeable surface. Ganglia and adsorbed chemical will also be present in the groundwater soil.

is in air), the water molecules at the air–water interface must be at a higher energy level than those in the bulk water. Water molecules in the bulk fluid are at a lower energy level because they can associate with other water molecules through hydrogen bonding. Molecules at the air–water interface bind to fewer water molecules and therefore are at a higher energy state. The total energy at the interface is a product of the total surface area (A) and the surface tension between the two phases, or

$$E = \sigma_{aw}A \qquad (12\text{-}73)$$

where $\sigma_{aw} = 72$ dyn cm^{-1} is the surface tension between air and clean water (1 dyn cm^{-1} = 1 g s^{-2}). Air bubbles in water are spheres because this shape is the lowest energy state (smallest area per volume) and therefore the most thermodynamically favorable shape for a small volume of one material in another.

As a result of the higher energy state of the interface, the pressure in the air bubble in water is slightly larger than in the surrounding fluid, much like the air pressure inside a balloon is higher than that outside the balloon. The pressures in the air bubble and water are related by

$$P_w = P_a - P_c \qquad (12\text{-}74)$$

where the capillary pressure, P_c, is a function of the surface tension between the two phases, P_w the pressure of the surrounding water, and P_a the air pressure in the bubble. If there is no surface tension between two phases, they are at equal pressures and will mix.

The relationship between the capillary pressure and surface tension can be derived from an energy balance. For an air bubble in water, any change in pressure in the bubble results in a change in the volume and interfacial energy, E, or

$$d(P_c V) = d(E) \qquad (12\text{-}75)$$

Using Eq. 12-73 and assuming the bubble of radius R_b maintains a spherical shape, $V = 4\pi R_b^3/3$, and $A = 4\pi R_b^2$

$$\frac{d}{dr}\left(P_c \frac{4}{3}\pi R_b^3\right) = \frac{d}{dr}(\sigma_{aw}4\pi R_b^2) \qquad (12\text{-}76)$$

Simplifying, we obtain a direct relationship between surface tension and capillary pressure as

$$P_c = \frac{2\sigma_{aw}}{R_b} \qquad (12\text{-}77)$$

This approach can be extended in order to analyze the case of water in a straw or water held in a porous medium by capillary forces. In both these situations we have a three-phase system: water, air, and solid. In the absence of the solid phase the air and water take on the lowest energy state (a flat plane). When the solid phase creates a small interfacial area between the air and water phases, the water phase in the constricted area will move to take on the shape of lowest energy (a sphere), and the water will be pulled into the constriction by capillary forces (Fig. 12.10). The height of water in a straw in water is dictated by hydrodynamic and capillary forces. The hydraulic head of water in a tube can be calculated in the normal way as

$$H_v = z + \frac{P_w}{\rho_w g} + \frac{u^2}{2g} \tag{12-78}$$

where z is the distance above some reference point, P_w the pressure of water, and u the water velocity. Water will be pulled by capillary forces into the tube and come to rest ($u = 0$) after reaching a height H_v in the vertical direction in the capillary tube of radius R_h (the throat radius). Defining the reference air pressure to be 0 atm with respect to a reference air pressure (1 atm), $P_w = -P_c$ from Eq. 12-74. Using this result in Eq. 12-78 with Eq. 12-77

$$H_v = 0 - \frac{P_c}{\rho_w g} + 0 = -\frac{2\sigma_{aw}}{\rho_w g R_h} \tag{12-79}$$

where g acts in the downward direction. This negative head is of course a consequence of the capillary forces that pull the water to a height H_v in a tube.

For porous media the analysis of the rise of water in a dry soil above a saturated soil is assumed to occur in the same manner. The dry soil is viewed as consisting of a series of capillary tubes that draw water up into the soil. The height of water rise in the soil is a function of the "pore" size. Based on the analysis of several soils, Berg (1975) found that an average tube radius, or the throat radius R_h, was

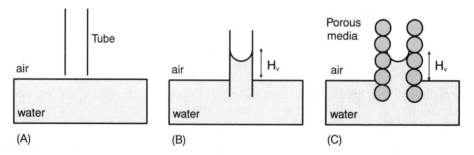

Figure 12.10 Capillary rise. (A) When an air-filled tube is lowered into contact with water, (B) the water is pulled up into the tube by capillary forces to a height H_v. (C) In the same manner, when an air-filled porous medium is brought into contact with water (or water-saturated porous media), there will be a capillary rise of the water in the void spaces in the porous medium.

related to the soil grain size by

$$R_h = 0.15R_g \qquad (12\text{-}80)$$

A soil grain radius of 100 μm would imply an average throat radius of 15 μm, and from Eq. 12-79, a capillary rise of water in the unsaturated soil (σ = 72 dyn cm^{-1} = 72 g s^{-2}) of H_v = 98 μm. The relationship between soil grain and throat size is subject to some uncertainty. For example, Ng et al. (1978) have found that R_h = 0.5 R_g, indicating that a characteristic throat radius is a function of the soil type.

Estimating the Sizes of Ganglia Formed under Stagnant Flow Conditions

The introduction of a separate, immiscible phase into a porous medium can result in the formation of ganglia held in place by capillary forces. The sizes of these ganglia can be quite variable, and there is no consensus on the best method for defining a characteristic ganglion size, but this parameter is necessary for mass transport calculations.

The maximum sizes of ganglia can be reasoned based on hydrostatic calculations and the properties of the chemical in water or air. The first consideration is whether the water or the chemical will preferentially associate with the mineral surface. Most mineral surfaces, unless they are completely dry, will have water-wet surfaces and will contain water in pores in the soil grains (intraparticle water). In an unsaturated soil, water can completely wet the mineral surface (Fig. 12.11a). When a hydrophobic chemical phase is added to the system the water, and not the chemical, will wet the surface (Fig. 12.11b). The chemical phase will be designed N for NAPL in the discussions that follow.

If there were no gravity forces acting on the three-phase system (NAPL, water, soil), there would be no pressure differences anywhere in the phase, and the curvature of the chemical phase would be the same at each water–chemical interface. However, in the more realistic case of gravity acting in the downward direction, the ganglion is pulled to form a ganglion of maximum length $L_{n,v}$. If the ganglion gets longer than $L_{n,v}$, its own weight will cause it to separate, forming a shorter ganglion. The curvature of the surface of the NAPL phase at the top of the ganglion is decreased due to the weight of the NAPL phase increasing the radius of curvature (Fig. 12.12).

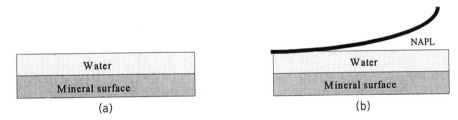

Figure 12.11 (A) Water will wet a mineral surface, and (B) when a pure chemical phase (NAPL) is added the water, and not the NAPL, will still wet the mineral surface.

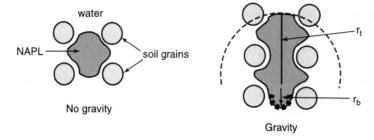

Figure 12.12 NAPL configurations in porous media in the presence and absence of gravity. Notice that gravity increases the radius of curvature of the NAPL at the top of the ganglion (r_t) relative to that at the bottom of the ganglion (r_b).

At the bottom, the radius of curvature of the ganglion, at the instant before it snaps off, is equal to that of the throat radius. In order to deduce the maximum length of this ganglion, we need to conduct a hydrostatic calculation, and we will find these assumptions concerning the curvature of the NAPL phase useful in simplifying our calculations.

Maximum Vertical Ganglia Lengths in Saturated Porous Media Let us assume that we have a ganglion of its maximum height in a porous medium composed of uniform grains of radius R_g. In the absence of motion the water pressure at the top of the ganglion can be assumed to be zero relative to pressure at the bottom of the ganglion. Using Eq. 12-74, the pressure of the NAPL at the ganglion top is the sum of water and capillary pressures, or

$$\text{at top: } P_{w,t} = 0$$

$$P_{n,t} = P_{w,t} + P_c = \frac{2\sigma_{nw}}{R_t} \tag{12-81}$$

where t indicates the position and σ_{nw} is the surface tension between the NAPL and water. At the bottom of the ganglion, the water pressure is larger due to the elevation change of $L_{nw,v}$. Similarly, the pressure in the NAPL has increased by $\rho_n g L_{nw,v}$. At the bottom of the ganglion, we therefore have

$$\text{at bottom: } P_{w,b} = P_{w,t} + \rho_w g L_{nw,v} = \rho_w g L_{nw,v} \tag{12-82}$$

$$P_{n,b} = P_{n,t} + \rho_n g L_{nw,v} = \rho_n g L_{nw,v}$$

We also know from our calculation of the capillary forces that at the bottom

$$P_{n,b} = P_{w,b} + P_c = P_{w,b} + \frac{2\sigma_{nw}}{R_b} \tag{12-83}$$

Combining Eqs. 12-81 to 12-83, we obtain after a bit of substitution (Hunt et al., 1988a)

$$L_{nw,v} = \frac{-2\sigma_{nw}}{g(\rho_n - \rho_w)} \left(\frac{1}{R_t} - \frac{1}{R_b} \right) \qquad (12\text{-}84)$$

We can now make use of our earlier assumptions about the radius of curvature of the ganglion at the top and bottom of the ganglion. For the largest ganglion possible, at the ganglion bottom the radius of curvature of the ganglion is approximately the same as the throat radius, or $R_b \approx R_h$: any smaller and the ganglion would separate from itself. At the ganglion top, the radius of curvature is much larger than the throat radius due to the weight of the ganglion, so that $R_t \gg R_h$. With these assumptions, it follows that $(1/R_t) \ll (1/R_b)$, and Eq. 12-84 can be simplified to

$$L_{nw,v} = \frac{2\sigma_{nw}}{g(\rho_n - \rho_w)R_h}, \quad \text{saturated zone} \qquad (12\text{-}85)$$

Assuming a soil grain radius of 100 μm, $\sigma_{nw} = 35$ g s^{-2} (for TCE), and Eq. 12-80, a NAPL ganglion could be as large as 104 cm long, or a length 10^4 times as long as a single soil grain.

The maximum ganglion length can also be expressed in terms of a dimensionless Bond number, Bo, defined as the ratio of gravity to capillary forces, and calculated as

$$Bo = \frac{|\rho_w - \rho_n|gR_h^2}{\sigma_{nw}} \qquad (12\text{-}86)$$

Inserting the Bond number into Eq. 12-85 produces

$$L_{nw,v} = \frac{2R_h}{Bo} \qquad (12\text{-}87)$$

Measuring Ganglia Size Distributions   A common technique for measuring the sizes of ganglia formed by immiscible liquids is to infuse a porous medium with a liquid chemical, such as styrene, that is then solidified with a polymerization initiator such as benzoyl peroxide. The porous medium is then dissolved using a strong acid such as hydrofluoric acid. The removal of the porous medium produces ganglia blobs that can be viewed with a microscope, sieved, or analyzed for the glob size distribution. Ganglia isolated in this manner by Conrad et al. (1992) in a Sevilleta sand (mean grain diameter of 200 μm) range in size from small, nearly spherical blobs, to longer wormlike shapes with many protrusions that reach lengths of 800 μm (Fig. 12.13). While this casting technique produces ganglia for direct analysis, it is not known to what extent the blobs break during handling or change size during a polymerization step that requires hardening in a pressurized vessel at elevated temperatures (for example, 85°C for 40 h).

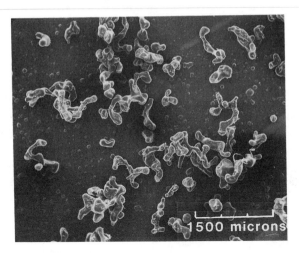

Figure 12.13   Photograph of styrene blobs generated in a column filled with Sevilleta sand (grain diameter of 200 μm) taken using a scanning electron microscope (from Conrad et al., 1992, copyright by the American Geophysical Union).

An alternative approach used to size ganglia is to infuse a transparent porous medium with a dyed organic liquid, to photograph the ganglia in the transparent medium, and to calculate ganglia sizes using an image analysis system. Mayer and Miller (1993) used this approach to measure ganglia size distributions in a thin box filled with glass beads 0.95 mm in diameter. Ganglia lengths in the vertical direction averaged 2 mm (50% cumulative frequency) and reached lengths of ~20 cm (95% cumulative frequency; Fig. 12.14). The maximum sizes of the perchloroethylene ganglia is predicted to be 91 cm using Eq. 12-87 for $Bo = 0.031$, and assuming $R_h = 0.5\, R_g$. Since the box was 30.48 cm in height, the box dimensions limited the size of the ganglion.

Maximum Vertical Ganglia Lengths in Unsaturated Porous Media   The addition of a pure chemical phase to unsaturated porous media results in four different phases: air, water, soil, and chemical (NAPL). Water will still wet the mineral surface, but we need to calculate whether the air or NAPL phase will "wet" the water phase covering the mineral phase. A force balance made for the three-phase system (air and NAPL on a water film) produces

$$\sigma_{nw}w = \sigma_{aw}w + \sigma_{an}w \cos \vartheta_{aw} \qquad (12\text{-}88)$$

where w is the unit width of the surface, ϑ_{aw} the angle between the air and water, and σ_{an} is the surface tension between air and the NAPL phase (Fig. 12.15A). Dividing by w, we obtain a result known as Young's equation

$$\sigma_{nw} = \sigma_{aw} + \sigma_{an} \cos \varphi_{aw} \qquad (12\text{-}89)$$

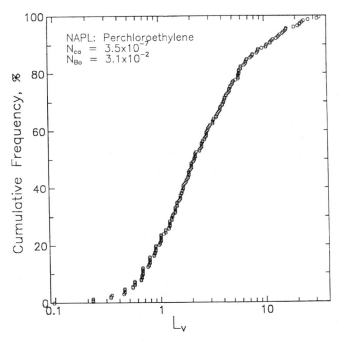

Figure 12.14 Ganglia lengths of perchloroethylene in the vertical direction in a box filled with glass beads (0.95 mm); lengths measured using photographs and an image analysis system (Reprinted from Mayer and Miller, 1993, with kind permission from Kluwer Academic Publishers.).

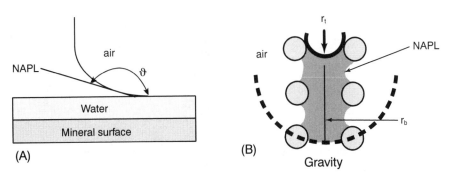

Figure 12.15 (A) There will usually be some water present in unsaturated media and water, not air, will wet the mineral surface. Using Young's equation (see text) the angle ϑ between the NAPL and air will be $\sim 180°$ so that the NAPL phase will appear to wet the media, not the air. (B) In the presence of gravity in unsaturated media, the radius of curvature of the at the ganglion top (r_t) will be smaller than that at the ganglion bottom (r_b).

By applying this equation to a water–air–TCE system using values of surface tension found in Table 12.4, we find that the term $\sigma_{an} \cos \vartheta_{aw} = 29 \cos \vartheta_{aw}$ must be negative, since $\sigma_{nw} = 35$ g s^{-2} $< \sigma_{aw} = 72$ g s^{-2}. For TCE, $\cos \vartheta_{aw}$ must be as negative as possible, or $\vartheta_{aw} = \pi$, since $\cos \pi = -1$.

   These calculations indicate that NAPL is the "wetting fluid" or that the system takes on a configuration that minimizes the air–water interface (Fig. 12.15B). In contrast to the case of fully saturated media, in unsaturated media the NAPL phase has a negative curvature where the ganglion meets the water phase. The pressure in the air is larger than the pressure in the NAPL phase because the NAPL phase is now a wetting phase relative to air (recall that the NAPL was a nonwetting phase relative to water). The air pressure is therefore

$$P_a = P_n + \frac{2\sigma_{an}}{R_h} \qquad (12\text{-}90)$$

Conducting the same analysis as above and making similar assumptions about the radius of curvature of the NAPL phase in the capillary tube results in an estimate of

TABLE 12.4   Chemical Properties of Groundwater Contaminants in California

Chemical	$c_{Nw,eq}$ (mg L^{-1})	ρ_n (g cm^{-3})	K_{ow}	v_n (cp)	σ_{na} (dyn cm^{-1})	σ_{nw} (dyn cm^{-1})
1,1,1-Trichloroethane (TCA)	700–1400	1.31	300	0.83	25	45
Trichloroethylene (TCE)	1100	1.46	200	0.59	29	35
Trichlorfluoromethane (Freon 11)	1100	1.49	340	—	—	—
Tetrachloroethylene (PCE)	150	1.63	760	0.89	31	44
o-Xylene	175	0.88	590	0.85	31	36
cis-1,2-Dichloroethylene	800	1.28	30	0.42	24	30
trans-1,2-Dichloroethylene	600	1.26	30	0.42	24	30
Toluene	500	0.87	490	0.59	29	36
Chloroform	8000	1.49	90	0.56	27	33
1,1-Dichloroethane	550	1.17	60	0.50	25	—
1,1-Dichloroethylene	400–2500	1.22	30	0.33	24	37
1,2-dibromo-3-chloropropane (DBCP)	1000	2.08	—	—	—	—
1,2-Dibromoethane ethylene dibromide (EDB)	2700–4300	2.18	—	1.71	39	36
cis-1,3-Dichloropropylene	2700	1.22	—	—	—	—
trans-1,3-Dichloropropylene	2800	1.22	—	—	—	—
Gasoline	100	0.73	—	0.45	20	50

Source: Hunt et al., 1988a.

the largest stable ganglion in the unsaturated zone of

$$L_{na,v} = \frac{2\sigma_{an}}{g\rho_n R_h}, \quad \text{unsaturated zone} \tag{12-91}$$

where we have assumed that $\rho_n \gg \rho_a$. For TCE and the same media conditions used above, a ganglion of 28 cm long is calculated to be stable. Since this ganglion size is smaller than that possible in the water-saturated medium, movement of the groundwater table into the unsaturated zone will not diminish the ganglion size.

Ganglia Sizes Produced by Groundwater Flow

The motion of water (or air) relative to a ganglion held in a porous medium presents an interesting situation relative to the size of a stable ganglion. The flow of water will stretch out immiscible liquids in the direction of the groundwater flow, potentially producing extremely long ganglia in the horizontal direction. The maximum sizes of these ganglia can also be estimated using hydrostatic calculations, and the size distributions measured using the previously described techniques.

Maximum Horizontal Ganglia Lengths   The flow of groundwater is driven by a hydraulic gradient, resulting in a pressure drop in the direction of flow. It is this pressure change in the direction of motion that now controls the length of the ganglion. Rearranging Darcy's Law given in Eq. 12-4, the pressure drop over a length of flow is

$$\Delta p = -\frac{U\mu_w \, \Delta l}{\kappa} \tag{12-92}$$

where κ is the permeability of the soil for water, μ_w the dynamic viscosity of water, $U = u\theta$ the Darcy velocity, and u the pore velocity. Defining the length of flow as the size of the longest stable ganglion, $L_{nw,h}$, in this horizontal direction, we can write the above equation as

$$\Delta P_w = -\frac{u\theta\mu_w L_{nw,h}}{\kappa} \tag{12-93}$$

To calculate the maximum length of a NAPL in an aquifer water flowing either under a natural gradient or due to pumping, we again examine the pressure drop along the length of the ganglion in the NAPL and water phase. Defining the "upstream" water pressure as zero, or $P_{w,up} = 0$, we can relate the upstream and downstream pressures for water using Eq. 12-93 as

$$\text{water: } P_{w,up} = 0 \tag{12-94}$$

$$P_{w,down} = P_{w,up} + \Delta P = -\frac{\mu_w u\theta L_{nw,h}}{\kappa}$$

For the NAPL, upstream and downstream pressures on the NAPL are

$$\text{NAPL: } P_{n,up} = P_{w,up} + P_c = \frac{2\sigma_{nw}}{R_{up}} \tag{12-95}$$

$$P_{n,down} = P_{w,down} + P_c = P_{w,down} + \frac{2\sigma_{nw}}{R_{down}}$$

Gravity is not acting in the horizontal direction, so that the pressure is equal in the NAPL phase in the direction of groundwater flow, or

$$P_{n,up} = P_{n,down} \tag{12-96}$$

Using this equality and Eq. 12-95 and 12-96, we can now solve for the length of the ganglion in the direction of flow (Hunt et al., 1988a) as

$$L_{nw,h} = \frac{2\sigma_{nw}\kappa}{\mu_w \theta u} \left(\frac{1}{R_{down}} - \frac{1}{R_{up}} \right) \tag{12-97}$$

Assuming that $(1/R_{up}) \ll (1/R_{down})$ and that $R_{down} = R_h$, this solution becomes

$$L_{nw,h} = \frac{2\sigma_{nw}\kappa}{\mu_w \theta u R_h} \tag{12-98}$$

Example 12.2

Starting with Eq. 12-96 derive an expression for the length of a ganglion as a function of the hydraulic gradient, $i = -\Delta h/\Delta l$. Using this result, calculate the maximum length of TCE ganglia for hydraulic gradients of 10^{-3} to 10^{-1} in soil with a grain radius of 100 μm. What do such calculations suggest about the effect of pumping on ganglia of NAPLs?

From Darcy's law (Eq. 12-4)

$$U = u\theta = -\frac{\kappa}{\mu} \frac{\Delta p}{\Delta l} \tag{12-99}$$

Since $\Delta p = \rho_w g \Delta h$, and $i = -\Delta h/\Delta l$, we can combine Eqs. 12-96 and 12-97 to obtain

$$L_{nw,h} = \frac{2\sigma_{nw}}{\rho_w g R_h i} \tag{12-100}$$

For TCE $\sigma_{nw} = 35$ g s^{-2}, resulting in

$$L_{nw,h} = \frac{2(35 \text{ g s}^{-2})}{(1 \text{ g cm}^{-3})(980 \text{ cm s}^{-2})[0.15 \times (100 \times 10^{-4} \text{ cm})]i} = \frac{48}{i} \text{ [cm]} \quad (12\text{-}101)$$

For the indicated hydraulic gradients, this produces

$$L_{nw,h}[i = 10^{-3}] = 480 \text{ m}$$

$$L_{nw,h}[i = 10^{-1}] = 4.8 \text{ m} \quad (12\text{-}102)$$

These ganglia are quite long, indicating that groundwater flow will produce much longer ganglia in the direction of flow than in the vertical direction under stagnant flow conditions. Increasing the hydraulic gradient, through pumping, will not shrink the sizes of ganglia. If fact, if there are any pools of NAPL in the aquifer, pumping will spread it out over distances in proportion to the hydraulic gradient, and therefore in proportion to the pumping rate. Thus pump-and-treat technologies offer little hope of shrinking the size of any ganglia already present in the system.

The maximum length of a ganglion in the direction of flow in a uniform flow field can also be expressed as a function of the dimensionless capillary number, Ca, defined as the ratio of viscous to capillary forces, and calculated as

$$Ca = \frac{u_w \mu_w}{\sigma_{nw}} \quad (12\text{-}103)$$

Substituting this result into Eq. 12-98, we have

$$L_{nw,h} = \frac{2\kappa}{Ca \ \theta R_h} \quad (12\text{-}104)$$

Substituting in the Kozeny–Carmen equation (Eq. 12-7) for the porous medium permeability and $R_h = 0.15 R_g$ produces

$$L_{nw,h} = \frac{1.9 R_g \theta^3}{Ca(1 - \theta)^2} \quad (12\text{-}105)$$

Measured Ganglia Size Distributions   The size distribution of ganglia produced by pumping NAPL and water through columns filled with porous media is a function of the soil type, the grain size distribution of the soil, and the pumping conditions. Polystyrene ganglia obtained by polymerizing polystyrene in four different sandy soils by Powers et al. (1992) varied widely in size and shape but were not the long, wormlike ganglia expected from the calculations of maximum ganglia developed above (Fig. 12.16).

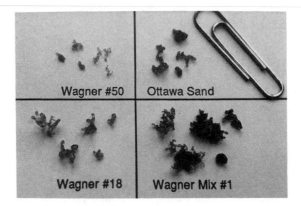

Figure 12.16 Photographs of the larger polystyrene ganglia solidified in four different sand samples. The paper clip is 3.5 cm long (from Powers et al., 1992, copyright by the American Geophysical Union).

The ganglia sizes measured by Mayer and Miller in the thin box filled with glass beads (as described above) using a photographic technique varied over a wide range. Ganglia averaged ~3 mm in length (50% cumulative frequency) with maximum sizes of ~20 mm (95% cumulative frequency) as shown in Figure 12.17. These ganglia are much smaller than those predicted using Eq. 12-105 of ~200 m assuming $Ca = 3.5 \times 10^{-7}$, $\theta = 0.38$, and $R_g = 0.475$ mm. However, ganglia of this length could not have been measured in the 15-cm-long sand box used by Mayer and Miller (1993). In a separate set of experiments the average sizes of ganglia were shown by Mayer and Miller (1992) to be a function of the porous media grain sizes, as shown in Figure 12.18. The geometric means of blob sizes were 206 μm ($d_g = 115$ μm), 666 μm ($d_g = 385$ μm), and 1380 μm ($d_g = 777$ μm) in glass bead porous media.

12.7 MASS TRANSPORT CALCULATIONS OF CHEMICAL FLUXES FROM NAPL GANGLIA

The chemical flux from a NAPL ganglion can be calculated in the usual manner in terms of a mass transport coefficient, $\mathbf{k}_w$, as

$$J_{n,x} = \mathbf{k}_w(c_{eq} - c) \tag{12-106}$$

where $c_{eq} = c_{nw,eq}$, the effective solubility of the NAPL chemical in water, and $c = c_{nw,\infty}$ is the concentration of the NAPL in the bulk water. This flux can be included into a governing transport equation that describes the rate of chemical transport from a NAPL phase into water moving at velocity $U = u\theta_w$, with dispersion coefficient

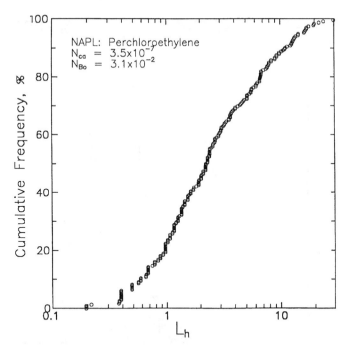

Figure 12.17 Ganglia lengths of perchloroethylene in the horizontal direction in a box filled with glass beads (0.95 mm); lengths measured using photographs and an image analysis system (Reprinted from Mayer and Miller, 1993, with kind permission from Kluwer Academic Publishers).

E_L, as

$$\theta_w \frac{\partial c}{\partial t} = \theta_w E_L \frac{\partial c^2}{\partial x^2} - U \frac{dc}{dx} + \mathbf{k}_w a_{nw}(c_{eq} - c) \qquad (12\text{-}107)$$

where a_{nw} is the specific surface area, or total interfacial area per volume between the NAPL and water. A subscript is added to the porosity term to indicate water flow is only through part of the soil porosity because $\theta = \theta_n + \theta_w$, where θ_w is the water-filled porosity and θ_n is the NAPL-filled porosity. In dimensionless form, this can be written as

$$\theta_w \frac{\partial c^*}{\partial t^*} = \left(\frac{\theta_w}{Pe} \frac{\partial^2 c^*}{\partial x^{*2}} \right) \frac{\partial c^*}{\partial x^*} + \frac{Sh_L}{Pe_L} a^*(1 - c^*) \qquad (12\text{-}108)$$

where $c^* = c_{nw}/c_{nw,eq}$, $x^* = x/L$, $t^* = tu/L$, $Pe_L = uL/E_L$, $Sh_L = \mathbf{k}_w L/E_L$, and L is the length of flow (for example, the length of a column).

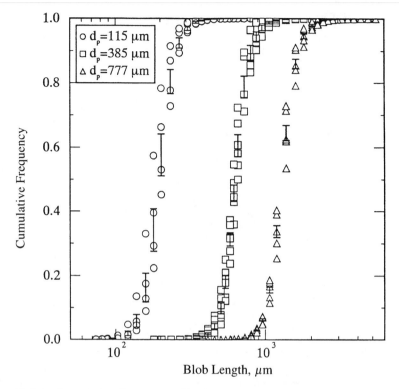

Figure 12.18 Cumulative frequency distributions of ganglia lengths parallel to the direction of water flow in a box filled with three different-sized glass beads. Ganglia lengths were measured using photographs and an image analysis system (Reprinted from Mayer and Miller, 1992, with kind permission of Elsevier Science-NL, Sara Burgerhartstraat 25, 1055 KV Amsterdam, The Netherlands).

For transport over short distances, such as laboratory columns, dispersion is often neglected because the magnitude of the dispersion term is negligible in comparison to the magnitude of the other terms. Retardation is neglected since the soil surfaces are already in equilibrium with the solute from the dissolving NAPL. This results in a simple expression for NAPL dissolution under steady conditions of

$$U \frac{dc}{dx} = \mathbf{k}_w a_{nw}(c_{eq} - c) \qquad (12\text{-}109)$$

The main difficulty of applying the above mass transport equations is that a_{nw} is not known. Since the porous medium blocks regions of the NAPL–water interface, and the geometry of the NAPL is poorly known, there is no exact method for specifying the interfacial area in a soil containing a distribution of ganglia sizes. Since we must somehow specify an interfacial area, a typical approach is to define a_{nw} as

$$a_{nw} = \frac{A}{V} = \frac{A_{nw}N_n}{V} \qquad (12\text{-}110)$$

where A is the total interfacial area of all NAPL ganglia per volume V of porous media. The total area of all NAPL ganglia is equal to the product of the interfacial area of a single ganglion, A_{nw}, and the number of ganglia in that volume, N_n. Since we must assume something about the nature of the interfacial area, it is usually assumed for convenience that the NAPL is a spherical ganglion (which is probably not true!). It is also sometimes assumed that the ganglia diameter will be much larger than that of a single soil grain, and therefore, that a single ganglion will fill the soil porosity around many soil particles and produce an interfacial area $A_{nw} = \theta_w \pi d_n^2$, where d_n is the diameter of a NAPL ganglion. The interfacial area is lower than that of a single, unobstructed sphere, because much of the ganglion is in contact with the soil grain and not the water.

If ganglia did not include the soil particles, the number of ganglia would be calculated as the total volume of NAPL divided by the volume of a single ganglion, V_n, or $N_n = V_n/(\pi d_n^3/6)$. However, since the ganglia include some soil grains, the volume of the NAPL phase that forms a ganglion of diameter d_n must be reduced in proportion to the volume of the soil inside the ganglion. Therefore, when the ganglia are assumed to include the soil particles, the number of ganglia is calculated as $N_n = V_n/(\pi d_n^3/6)$. For a total NAPL volumetric solid fraction of $\theta_n = V_n/V$, the interfacial area is therefore

$$a_{nw} = \frac{(\theta_w \pi d_n^2)[V_n/(\theta_w \pi d_n^3/6)]}{V} = \frac{6\theta_n}{d_n} \qquad (12\text{-}111)$$

It can be shown that this same result is obtained if it is assumed that the ganglion is much smaller than the soil grains and that the interfacial area between the ganglion and water is unobstructed by the soil grains.

This analysis also does not allow for the fact that the NAPL is not uniformly distributed in the soil. Using x-ray attenuation, Imhoff et al. (1996) demonstrated that trichloroethylene formed "fingers" of NAPL in sandy media (Fig. 12.19). Because the exact ganglia distribution is not predictable, the assumption of uniform distribution is at present unavoidable.

Mass Transport Correlations for Calculating Ganglia Dissolution Rates

Many mass transport correlations have been developed for predicting the mass transport coefficient, $\mathbf{k}_w$, for dissolving materials in packed beds. Most have the form

$$Sh = \beta_0 \, Re^{\beta_1} \, Sc^{\beta_2} \qquad (12\text{-}112)$$

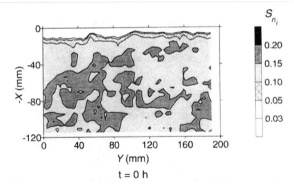

Figure 12.19   Contours of trichloroethylene in a box filled with sand (grain diameter of 0.36 mm) determined by x-ray attenuation (from Imhoff et al., 1996, copyright by the American Geophysical Union).

where β_n are constants developed from laboratory experiments. A frequently used correlation, developed by Wilson and Geankoplis (1966), is

$$\mathbf{k}_w = \frac{1.09 U^{1/3}}{\theta_w} \left(\frac{D_{Nw}}{d_n} \right)^{2/3} \tag{12-113}$$

or in dimensionless form

$$Sh_d = \frac{1.09}{\theta_w} Pe_d^{1/3} \tag{12-114}$$

This equation was developed to describe dissolving spheres in packed beds. The subscript d on the $Sh_d = \mathbf{k}_w d_n / D_{Nw}$ and $Pe_d = U d_n / D_{Nw}$ numbers is the diameter of the dissolving chemical, defined here as a NAPL ganglion of diameter d_n. Other correlations that predate those recently developed to predict NAPL dissolution in soils are summarized in Table 12.5.

Since the interfacial area is not known, and a mass transport coefficient must be experimentally measured, most researchers have found it easier to include the interfacial area term directly into a mass transport coefficient. Including the specific area into a modified mass transport $\mathbf{k}^+ = \mathbf{k}_w a_{nw}$ results in a mass transport coefficient with units of $[1/t]$. Therefore, we must define a modified Sherwood number, Sh^+, as

$$Sh^+ = \frac{\mathbf{k}^+ d_{50}^2}{D_{Nw}} \tag{12-115}$$

where a subscript 50 has been added the soil grain diameter to indicate it is defined in these correlations as a median particle diameter of the porous medium. For spherical ganglia it can be shown that $Sh^+ = 6\theta_n Sh$.

TABLE 12.5 Mass Transport Correlations for Spherical Particles in Packed Beds

Correlation[a]	Conditions	Limitations	Reference
$Sh = \dfrac{0.89}{\theta_w} Pe^{0.33}$	Isolated spheres	$Re < 5,\ (Pe/\theta_w) > 10^3$	Freidlander (1957)
$Sh = 2$	Isolated spheres	$Pe < 1$	Bowman et al. (1961)
$Sh = \dfrac{0.978}{\theta_w} Pe^{0.33}$		$Pe > 10$	
$Sh = 2.50\, Pe^{0.33}$	Solid sphere in packed bed	$0.035 < Re < 55$	Williamson et al. (1963)
$Sh = 2.40\theta_w^{0.67}\, Re^{0.33}\, Sc^{0.42}$		$0.08 < (Re/\theta_w) < 125$	
$Sh = \dfrac{1.09}{\theta_w} Pe^{0.33}$	Solid sphere in packed bed	$0.0016 < Re < 55,$ $0.35 < \theta_w < 0.75$	Wilson and Geankoplis (1966)
$Sh = \dfrac{1.1068}{\theta_w} Re^{0.28}\, Sc^{0.33}$	Solid spheres, cylinders, and flakes in packed and fluidized beds	$Re < 10$ All Re	Dwivedi and Updbay (1977)
$Sh = \dfrac{Sc^{0.33}}{\theta_w}(0.765\, Re^{0.18} + 0.365\, Re^{0.614})$			
$Sh = \dfrac{1.110}{\theta_w} Re^{0.2814}\, Sc^{0.33}$	Solid cylinders in packed bed	$0.016 < Re < 10,$ $0.26 < \theta_w < 0.632,$ $123 < Sc < 70600$	Kumar et al. (1977)
$Sh = 0.55 + 0.25\, Pe^{1.5}$	Oil dissolution in packed bed	$0.5 < Pe < 100$	Pfannkuch (1984)

[a] $Sh = k_w d/D_{Nw}$, where $d = d_n =$ dissolving particle diameter except for the correlation of Pfannkuch, where $d = d_g =$ soil grain diameter; $Pe = Ud_n/D_{mw}$, $Re = Ud_n/\nu_w$, $Sc = \nu_w/D_{mw}$.
Source: Powers et al., 1991.

For ganglia of toluene in a column packed with glass beads, Miller et al. (1990) obtained

$$Sh^+ = 12(\pm 2)\theta_w \, Re^{0.75(\pm 0.08)} \, \theta_n^{0.60(\pm 0.21)} \, Sc^{1/2} \tag{12-116}$$

This result can be compared to the correlation of Wilson and Geankoplis (1966) by writing their correlation on the basis of Sh^+ as

$$Sh^+ = \frac{6.54}{\theta_w} \, Re_d^{1/3} \, \theta_n^1 \, Sc^{1/3} \tag{12-117}$$

An important difference between these two correlations is the different dependence of the mass transport correlation on the NAPL volume fraction, θ_n.

Other researchers have similarly developed complex correlations to predict NAPL dissolution. Powers et al. (1994) investigated NAPL (trichloroethylene) dissolution in columns packed with irregularly shaped sands and therefore included into their correlations a uniformity index, ι, defined as the ratio of the diameter of 60% of the soil grains to 10% of the soil grains, or $\iota = d_{60g}/d_{10g}$. Their correlation was

$$Sh^+ = \frac{4.13}{\theta_w} \, Re^{0.60} \, \iota^{0.37} \left(\frac{\theta_n}{\theta_{n,i}}\right)^{0.75} \left(\frac{d_p}{d_m}\right)^{0.67} \tag{12-118}$$

Other correlations based on a modified Sherwood number for the mass transport coefficient are summarized in Table 12.6.

Example 12.3

In order to see how widely these mass transport correlations vary in their predictions of a mass transport coefficient, compare the following Sh^+ correlations as a function of Re: Wilson and Geankoplis (1966), Miller et al. (1990), Geller and Hunt (1993), and Powers et al. (1994). Assume the following parameters: $\theta_w = 0.38$ and $\theta_n = 0.045$, $D_{Nw} = 8 \times 10^{-6}$ cm^2 s^{-1}, $d_g = 0.05$ cm, and uniform media ($\iota = 1$).

As shown in Figure 12.20, these correlations differ by an order of magnitude at low Re numbers, but at Re = 0.1 all the correlations except the one developed by Powers et al. (1994) show good agreement. At the highest Re, where only two of the correlations are valid, there is also good agreement.

Estimating the Lifetimes of Ganglia in Porous Media

NAPLs dissolve in ground water very slowly due to their low solubility and limited interfacial surface area for mass transport into the liquid. The time for the ganglion to dissolve, defined as the ganglion lifetime, can be estimated by calculating the

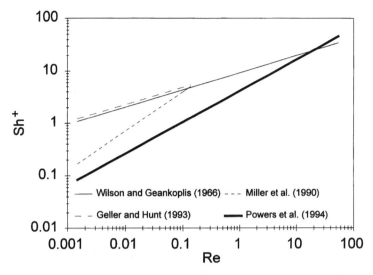

Figure 12.20 Comparison of mass transport correlations developed to describe NAPL ganglia dissolution rates. The Re range for the correlation developed by Powers et al. (1994) is not known.

chemical dissolution rate over time (Hunt et al., 1988a). The dissolution rate is simply the change in mass of a ganglion, dm_n/dt, written in terms of a mass transport coefficient as

$$\frac{d(m_n)}{dt} = \frac{d[\theta_w \rho_n (\pi/6) d_n^3]}{dt} = -\mathbf{k}_w (c - c_{eq}) \pi d_n^2 \theta_w \qquad (12\text{-}119)$$

The ganglion diameter d_n is assumed to be much larger than the soil grain diameter so that the ganglion volume includes soil grains. Therefore, the volume of NAPL present in a ganglion of size $(\pi/6) d_n^3$ is the ganglion volume multiplied by the porosity or $\theta_w (\pi/6) d_n^3$, as explained above. Similarly, the area for mass transport must be reduced by θ_w. This expression can be easily solved if some assumptions are made. First, the change in θ_w must be assumed to be small over time, so that $d\theta_w/dt = 0$. Second, we assume that the concentration in the dissolving water is much less than that of equilibrium or $c \ll c_{eq}$. Third, since the ganglion size is changing, we must be able to specify an expression for the mass transport coefficient as a function of diameter. Making these assumptions, and using the mass transport correlation proposed by Wilson and Geankoplis (1966) for the mass transport coefficient, Eq. 12-119 becomes

$$\frac{d(d_n)}{dt} = -\frac{2.18 U_w^{1/3} D_{Nw}^{2/3} c_{nw,eq}}{\rho_n \theta_w} d_n^{-2/3} \qquad (12\text{-}120)$$

TABLE 12.6 Mass Transport Correlations Developed for NAPLs in Packed Beds

Correlation	Conditions	Reference
$Sh^+ = 12\theta_w\, Re^{0.75}\, \theta_n^{0.60}\, Sc^{1/2}$	Toluene dissolution in glass bead columns	Miller et al. (1990)
$Sh^+ = 1240\theta_w\, Re^{0.75}\, \theta_n^{0.60}$	—[b]	Parker et al. (1990) (in Mayer and Miller, 1996)
$Sh^+ = 57.7\theta_w^{0.61}\, Re^{0.61}\, \iota^{0.41}\, d_{50}^{0.64}$	Styrene and trichloroethylene in sand columns	Powers et al. (1992)
$Sh^+ = 70.5\, Re^{1/3}\, \theta_n^{4/9}\, \theta_{n,i}^{5/9}\, \theta^{-11/9} \left(\dfrac{d_g}{d_{n,i}}\right)^{5/3}$	Toluene dissolution in columns packed with glass beads	Correlation by Imhoff et al. (1994) using data from Geller and Hunt (1993)
$Sh^+ = 340\, Re^{0.71}\, \theta_n^{0.87} \left(\dfrac{d_g}{x_n}\right)^{0.31}$	Trichloroethylene dissolution in sand columns	Imhoff et al. (1994)
$Sh^+ = 4.13\, Re^{0.60}\, \iota^{0.37} \left(\dfrac{\theta_n}{\theta_{n,i}}\right)^{0.75} \left(\dfrac{d_p}{d_m}\right)^{0.67}$	Styrene and trichloroethylene in sand columns	Powers et al. (1994)

[a] $Sh^+ = k_w^+ d_n^2/D_{mw}$, $Pe = Ud_n^2/D_{mw}$, $Re = Ud_n/\nu_w$, $Sc = \nu_w/D_{mw}$, θ_n = NAPL void fraction; θ_w = water void fraction; $\theta_{n,i}$ = initial void fraction filled with NAPL; θ = total void fraction; x_n = distance into region containing NAPL; d_n = diameter of spherical NAPL ganglion; d_g = porous medium grain diameter; d_m = porous medium reference diameter of 0.05 cm; uniformity index, ι, defined as the ratio of the diameter of 60% of the soil grains to 10% of the soil grains, or $\iota = d_{60g}/d_{10g}$.

[b] Conditions of correlation not given by Mayer and Miller (1996).

The lifetime of the ganglion, τ, can be found by integrating from $d_n(t = 0) = d_{n0}$ to $d_n(\tau) = 0$, to produce

$$\tau = \frac{0.275\rho_n\theta_w d_{n0}^{5/3}}{U_w^{1/3}D_{Nw}^{2/3}c_{eq}} \tag{12-121}$$

As shown in Figure 12.21, the lifetimes of NAPL ganglia such as trichloroethylene (TCE) can be quite long. For example, at groundwater velocities of ~ 1 m d^{-1} ganglia may persist for 1 to 100 years for ganglia of sizes 1 to 100 cm in diameter.

The concentration of chemical in the liquid leaving an area of soil contaminated with the NAPL can be calculated from the NAPL volume fraction in the soil and the NAPL dissolution rate. Using the transport equation for NAPL dissolution (Eq. 12-109) for a spherical ganglion, we have

$$\frac{dc}{dx} = \frac{\mathbf{k}_w 6\theta_n}{d_n U_w}(c_{eq} - c) \tag{12-122}$$

Hunt et al. (1988a) demonstrated that under typical groundwater conditions the concentration of the chemical in solution is much less than the equilibrium concentration. This can be seen by examining the initial concentration of the chemical in water, or when the ganglion size has not changed very much. Under these conditions that d_n is a constant, $\mathbf{k}_w$ is also constant, and Eq. 12-122 can be solved by separating terms

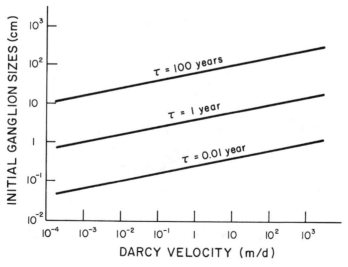

Figure 12.21 Predicted lifetimes (τ) of TCE ganglia assuming that the concentration in the bulk water phase is much less than solubility (from Hunt et al., 1988a, copyright by the American Geophysical Union).

and integrating from $c(t = 0) = 0$ over some distance L

$$\frac{c}{c_{eq}} = 1 - \exp\left(-\frac{\mathbf{k}_w 6\theta_n L}{d_{n0} U_w}\right) \tag{12-123}$$

For large ganglion sizes, and for low groundwater velocities (1 m d^{-1}), the concentration of the chemical in water is much less than saturation even during the time of the initial ganglion dissolution when the volume fraction of NAPL is greatest (Fig. 12.22). Concentrations of NAPL chemical in the liquid phase much less than saturation are necessary for calculations of NAPL lifetimes.

Assuming that the concentration in the dissolving water is much less than that of equilibrium or $c \ll c_{eq}$, we can write the transport equation for ganglia dissolution as a zero-order rate expression

$$\frac{dc}{dx} = \frac{6\mathbf{k}_w \theta_n c_{eq}}{d_n U_w} \tag{12-124}$$

Solving for a downstream concentration at a distance L, we have

$$c(L,t) = \frac{6\mathbf{k}_w(t)\theta_n(t)c_{eq}L}{d_n(t)U_w} \tag{12-125}$$

where the terms $\mathbf{k}_w$, θ_n, and d_n are indicated to be a function of time.

The downstream NAPL concentration in the water can be obtained as a function of time by specifying the terms in the above equation as a function of time. Using

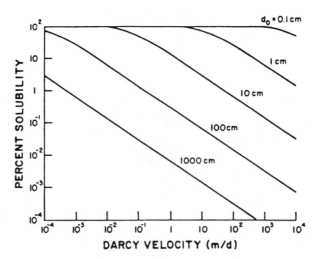

Figure 12.22 Initial liquid-phase solubilities (percent saturation) for NAPL ganglia dissolution into water for ganglia of initial size $d_{no} = V_{no}L = 0.1$ m (from Hunt et al., 1988, copyright by the American Geophysical Union).

the definition of the ganglion lifetime in Eq. 12-121, Eq. 12-120 can be solved for the diameter of a ganglion at any time t, as

$$d_n(t) = d_{n0} \left(1 - \frac{t}{\tau} \right)^{3/5} \qquad (12\text{-}126)$$

As discussed above, $V_n = N_n(\pi d_n^3/6)$. It therefore follows that the ganglia volume fraction is

$$\theta_n(t) = \theta_{n0} \left(1 - \frac{t}{\tau} \right)^{9/5} \qquad (12\text{-}127)$$

With a little effort it can also be shown that

$$\mathbf{k}_w(t) = \mathbf{k}_{w,0} \left(1 - \frac{t}{\tau} \right)^{-2/5} \qquad (12\text{-}128)$$

where the mass transport coefficient $\mathbf{k}_{w,0}$ is calculated based on the initial ganglion diameter d_{n0}. Combining the above equations, we have the final expression for the downstream NAPL concentration (Hunt et al., 1988a) of

$$c(t) = \frac{9\rho_n \theta_{n0} L}{5 U_w \tau} \left(1 - \frac{t}{\tau} \right)^{4/5} \qquad (12\text{-}129)$$

For a chemical such as TCE we can see in Figure 12.23 that the decrease in the water concentration is nearly linear with time.

Methods for Enhancing NAPL Dissolution Rates

A variety of methods have been proposed to increase the dissolution rates of NAPL ganglia, including: hot water, steam, methanol, and surfactants. Injecting hot water will decrease water viscosity and may increase chemical solubility, but Imhoff et al. (1995a) found that flushing NAPL-contaminated soils with hot water had little effect on the dissolution rate. The mass transport coefficient was increased by only a factor of two by increasing water temperatures from 5°C to 40°C. A best-fit mass transport correlation for hot water dissolution (Imhoff et al., 1995a) was

$$Sh^+ = 5.4(\pm 3.6)\theta_w \, Re^{0.75} \, \theta_n^{0.90} \, Sc^{0.48 \pm 0.09} \qquad (12\text{-}130)$$

where the correlation constants are $\pm 95\%$CI.

Steam injection into columns packed with a porous medium (sand) was demonstrated by Hunt et al. (1988b) to be a very effective method of removing immiscible liquids. Pure-phase liquids of trichloroethyelene, toluene, and gasoline were displaced as separate phases just ahead of a steam front in laboratory column experi-

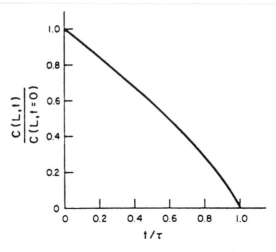

Figure 12.23   Concentration of chemical leaving an area of porous medium contaminated with NAPL ganglia. The variables are made dimensionless by dividing the time by a ganglion of half-life, and the water concentration by the initial concentration. It is assumed that all water concentrations are much less than solubility (from Hunt et al., 1988, copyright by the American Geophysical Union).

ments. Only one pore volume of steam was sufficient to remove most of the NAPL phase in column experiments. This treatment process is effective, but expensive. Based on these experiments, it was calculated that the energy from 6.8 L of fuel oil would be necessary to clean 1 m^3 of a contaminated aquifer.

A variety of solvents have been used to increase the solubility of a NAPL chemical in water. Imhoff et al. (1995b) examined flushing with methanol–water solutions (0, 20, 40, and 60% methanol by volume). Methanol increases the removal of NAPL by mobilizing the NAPL phase through decreases in the interfacial tension and by increasing chemical solubility in the water–methanol phase versus that of water. Surface tension decreased from 44.5 dyn cm^{-1} in pure water to 11.9 dyn cm^{-1} in a 60% methanol solution. In experiments using a 60:40 methanol:water solution Imhoff et al. (1995b) determined the mass transport correlation

$$Sh^+ = 4.8(\pm 3.4)\theta_w\, Re^{0.75}\, \theta_n^{1.0 \pm 0.2}\, Sc^{1/2} \tag{12-131}$$

Surfactants can increase NAPL dissolution rates by lowering the interfacial tension of the solution and increasing the chemical solubility through partitioning of NAPL into surfactant micelles (Fig. 12.24). Micelles form above a chemical's critical micelle concentration (CMC). The size of these micelles vary, but reported micelle diffusivities range from 2.2 to 2.7 $\times$ 10^{-7} m^2 s^{-1} for Triton X-100 (TX-100) and from 1.4 to 6.3 $\times$ 10^{-7} m^2 s^{-1} for sodium dodecyl sulfate (SDS) (Grimberg et al., 1996). Disadvantages of surfactants are: The NAPL phase may mobilize downward

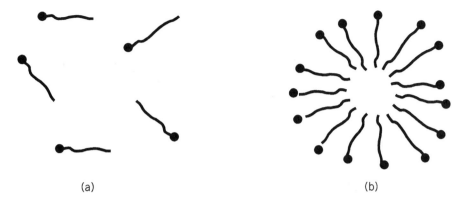

(a)   (b)

Figure 12.24 Surfactant molecules are composed of a hydrophobic tail and a hydro-philic head (which can be nonionic, positively, or negatively charged, depending on the surfactant). (A) At low concentrations the surfactants remain unassociated in water, but (B) at high concentrations they spontaneously form micelles. As surfactants are added to water the surface tension of the solution decreases steadily with surfactant concentration until the critical micelle concentration (CMC) is reached and micelles form; the surface tension then remains constant with further addition of the surfactant.

during pumping due to the decreased surface tension; large quantities of surfactants are necessary to produce rapid dissolution rates. Okuda et al. (1996) reported that 95% of tetrachloroethylene present as a NAPL in glass bead columns could be re-moved in 24 pore volumes at a surfactant (TX-100) concentration of 1% and that a 10% surfactant concentration would be necessary to achieve this removal in 2.5 pore volumes. Obviously the surfactant would need to be recovered and recycled in order for such surfactant-based systems to be cost effective in the field.

Example 12.4

Temperature and the addition of methanol to water can change the rate of NAPL dissolution in two ways: by changing the magnitude of the mass transport coef-ficient and by changing the chemical solubility. In order to see the magnitudes of these different factors on NAPL dissolution, compare the effects of hot water washing and methanol on chemical flux for PCE. Assume the conditions given in the previous example and the additional parameters: $v_w(40°C) = 0.00658$ cm^2 s^{-1}, $c_{Pw,eq}(40°C) = 240$ mg L^{-1}, $c_{Pw,eq}(20°C) = 235$ mg L^{-1}, $c_{Pw\text{-}m,eq}(20°C) = 10,000$ mg L^{-1}, and $u = 1.14$ m d^{-1}.

The rate of a dissolution of a PCE (chemical P) ganglion is

$$W_{Pw} = \mathbf{k}^+(c_{eq} - c) \tag{12-132}$$

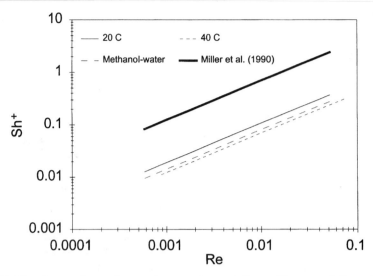

Figure 12.25   Comparison of mass transport correlations. The correlation of Miller et al. (1990) is based on water at 20°C.

Assuming $c = 0$, and from the definition of the mass transport coefficient

$$W_{Pw} = \frac{Sh^+ D_{Pw}}{d_g^2} c_{eq} \qquad (12\text{-}133)$$

The diffusion coefficient of PCE in water can be scaled with temperature using $(D_{Pw}T/\mu_w) = $ constant, according to the Stokes–Einstein equation. The mass transport correlations for the case of water (20°C), hot water (40°C), and a 60:40 mixture of methanol–water, obtained from the mass transport correlations of Miller et al. (1990), Imhoff et al. (1995a,b) are shown in Figure 12.25.

   To compare the chemical dissolution rates, we tabulate the data on solubility and mass transport coefficients via the Sh^+ number (Table 12.7). The correlation developed by Miller et al. predicts a much larger PCE dissolution rate than the correlations of Imhoff et al. (1995a,b). Based on a comparison of only the Imhoff

TABLE 12.7   Mass Transport Correlation Data

Condition	c_{eq} (mg L^{-1})	Re	Sh^+	k^+ (s$^{-1} \times 10^3$)	Dissolution rate W (µg s^{-1})
Miller et al. (1990), 20°C	220	0.00058	0.082	0.264	0.058
Imhoff et al. (1995a) Water, 20°C	220	0.00058	0.013	0.041	0.0089
Hot water, 40°C	235	0.00088	0.11	0.059	0.014
Methanol–water, 20°C	10,000	0.00058	0.0095	0.031	0.31

et al. correlations, we can see that heating the water increases the dissolution rate by a factor of 1.5, primarily as a result of the increase in the mass transport coefficient, but that the increase in PCE dissolution rate in the methanol–water solution is due almost entirely to the increased solubility of PCE in the methanol–water solution.

Ganglia Volatilization Rates in Unsaturated Porous Media

If a pure-phase chemical is applied to the ground, it will initially fall through the unsaturated zone, leaving NAPL ganglia in the soil. Szatkowski et al. (1995) examined volatilization rates of a NAPL (toluene) in glass bead (d_g = 0.40 mm) columns at water velocities of 0.2 to 20 m d^{-1} with a longitudinal dispersivity of 0.199 ($E_L/u \approx 5d_g$). Based on their experiments, they developed the mass transport correlation

$$Sh^+ = a_{aw}d_g[0.023(\pm0.014)\,Sc^{1/2} + 0.85(\pm0.65)\,Re^{0.86(\pm0.34)}\,Sc^{1/2}] \quad (12\text{-}134)$$

where $Sh^+ = \mathbf{k}_{aw}^+ d_g^2/D_{aw} = \mathbf{k}_{aw}a_{aw}d_g^2/D_{Nw}$, $Sc = v_w/D_{Nw}$, $Re = ud_g/v_w$, and $0.001 \leq Re \leq 0.1$. The specific interfacial area, a_{aw}, was calculated from

$$a_{aw} = \frac{P_c\theta_w}{\sigma_{aw}} \quad (12\text{-}135)$$

where the capillary pressure P_c is calculated as the difference between the air and water pressures, or $P_c = P_a - P_w$. This mass transfer correlation was compared to other correlations in the literature (see Table 12.8.). Szatkowski et al. (1995) found their equation predicted mass transfer rates of trichloroethylene about 10% higher than the correlation developed for low-velocity trickle-bed reactors by Turek and Lange (1981) and rates about an order of magnitude larger than a correlation developed from infiltration experiments Cho et al. (1994).

Wilkens et al. (1995) investigated the volatilization of a NAPL (styrene) at residual NAPL saturations of 8 to 16% in sand columns at 4 to 10% water saturation at pore velocities of 0.25 to 1.5 cm s^{-1}. Their mass transport correlation for volatilization was

$$Sh_a^+ = 10^{-2.8}\,Pe^{0.62}\left(\frac{d_g}{d_m}\right)^{1.82} \quad (12\text{-}136)$$

where the mass transport properties are now based on the air phase, so that $Pe = u_ad_g/D_{Na}$, $Sh_a^+ = \mathbf{k}_{na}d_g^2/D_{Na}$, and $d_m = 0.05$ cm. The soil grain sizes are based on a median size (d_{50}), and the NAPL sizes are assumed to be a function of the grain sizes so that only the grain size is used in the mass transport correlation. These gas-phase mass transport correlations can be used to calculate NAPL volatilization using

TABLE 12.8 **Mass Transport Correlations for NAPL Dissolution and Volatilization Rates**

Correlation[a]	Conditions	Reference
$Sh^+ = a_{aw} d_g (0.023\, Sc^{1/2} + 0.85\, Re^{0.86}\, Sc^{1/2})$	$0.001 \leq Re \leq 0.1$	Szatkowski et al. (1995)
$Sh^+ = 16.8 d_g^2 (\theta_w f_{w,m})^{0.25}\, Ga^{-0.22}\, Re^{0.25}\, Sc^{1/2}$	$0.1 \leq Re\ \theta_w f_{w,m} \leq 5$	Turek and Lange (1981)
$Sh^+ = \dfrac{0.94 d_g^2}{D_{Cw}} \left(\dfrac{v_w}{d_g}\right)^{0.151} \theta(1 - \theta_w)(\theta_w f_{w,m})^{0.151}$ $\times Re^{0.151}\, (1 - f_{aw,im})^{22.9}(1 - f_{w,m})^{1.22}$	$0.48 \leq f_{w,m} \leq 0.75,$ $0.9 \leq f_{aw,im} \leq 0.24,$ $0.0048 \leq Re \leq 0.026$	Cho et al. (1994)
$Sh_a^+ = 10^{-2.8}\, Pe^{0.62} \left(\dfrac{d_g}{d_m}\right)^{1.82}$	$0.04 \leq \theta_w \leq 0.10$ $0.08 \leq \theta_w \leq 0.16$ $0.06 \leq Pe \leq 2$	Wilkins et al. (1995)

[a]$Ga = d_g^3 g / v_w^2$ (Galileo number), $f_{w,m}$ = fraction of water which is immobile, $f_{aw,im}$ = fraction of immobile water in contact with both the air and mobile water phases; $d_m = 0.05$ cm.
Source: Szatkowski et al., 1995; Wilkens et al., 1995.

the transport equation written about the air phase

$$u_a \frac{dc_{na}}{dx} = k_{na}^+ (c_{na,eq} - c_{na}) \qquad (12\text{-}137)$$

where the interfacial area between the NAPL and air phases is included in the mass transport coefficient. Note that this form of the equation does not include a term for a variable amount of water in the column.

PROBLEMS

12.1 A lighter-than-water solvent is floating on a water table under stagnant (air and water) conditions. Using the typical diffusion coefficients (e.g., $D_{Na} = 0.2 \times 10^{-4}$ m^2 s^{-1} and $D_{Nw} = 10^{-9}$ m^2 s^{-1}) and penetration theory, plot the relative water and air concentrations away from the stagnant interface after one year. Use semilog axes. What does this suggest about the usefulness of soil gas analysis for determining groundwater contamination?

12.2 A heavier-than-water liquid initially occupies all the pore space in a porous medium following a spill. The compound has a fluid solubility in water of c_{eq}, and a dissolution flux controlled by diffusion through the fluid-filled medium pores as the liquid contaminant partitions into the fluid and diffuses away. In this situation the flux is $D\theta c_{eq}/L$, where L is the depth of penetration into the medium (see Fig. P-12.2). (a) What does this assumption imply about

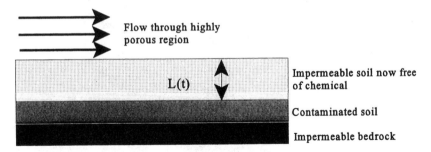

Figure P-12.2

the mass transfer coefficient in groundwater systems? (b) Set up the differential equation describing the depth to the pure liquid contaminant and solve it. Note that the mass lost ($A\rho_A\theta L d$) must be equal to the flux multiplied by ($A\theta\ dt$), where ρ_A as the liquid-phase density of the contaminant. (c) For a liquid TCE spill that completely saturates 2 m of a porous medium ($\theta = 0.4$), calculate the time required to remove the liquid TCE if the TCE vaporizes into air ($p_T^\circ = 60$ mm Hg). (d) Repeat part (c) for TCE dissolving into water.

12.3 A 55-gal drum of chemical C is accidentally spilled into the ground and seeps down into an aquifer. The chemical has the following characteristics: $\sigma_{Cw} =$ 25 dyn cm^{-1} (10^5 dyn = 1 kg m s^{-2}), $\rho_C = 1.05$ g cm^{-3}, $c_{Cw,eq} = 10^{-4}$ mol L^{-1}, $M_C = 150$ g mol^{-1}, $D_{Cw} = 0.75 \times 10^{-5}$ cm^2 s^{-1}. Assume for water $\rho_w = 1.0$ g cm^{-3} and $\mu_w = 0.01$ g cm^{-1} s^{-1}, and assume for soil $\theta = 0.4$, $\rho_s = 2.1$ g cm^{-3}, $d_g = 2$ mm, and $R_h = 0.077\ d_g$. (a) What is the largest ganglion that is stable with respect to buoyant forces in the aquifer? (b) Calculate if there will be NAPL migration for a 20-cm-long ganglion if the groundwater pore velocity is 100 m yr^{-1}. (c) What would be the lifetime (in years) of a 20-cm-diameter ganglion? Assume the mass transport correlation of Wilson and Geankoplis (1966) applies.

12.4 Porous ceramic materials are used in subsurface sampling of fluids (Fig. P-12.4). The ceramic material can be modeled as a collection of parallel tubes, each with a radius of 1 μm. The ceramic surface is preferentially wetted by water. Calculate the pressure (based on your knowledge of capillary forces) that you would need to collect a sample of xylene that is at atmospheric pressure (101 kPa = 1.01×10^6 dyn cm^{-2}). Assume that the sampler is initially saturated with water. For xylene, $\sigma_{Ca} = 31$ dyn cm^{-1}, $\sigma_{Cw} = 36$ dyn cm^{-1}. (From Hunt, 1998.)

12.5 Pollutant transport in landfills is restricted through the use of clay liners with low hydraulic conductivities. You are trying to figure out when a landfill will leak if 10 cm of water accumulates on top of a 1-m-thick clay liner ($K = 5.8 \times 10^{-8}$ m s^{-1}, $\theta = 0.35$, $d_g = 1$ μm). (a) How long until water reaches

Water

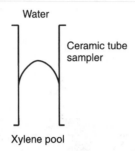

Ceramic tube
sampler

Xylene pool

Figure P-12.4

the end of the liner if only advective transport moves a chemical through the liner? If the retardation coefficient for this chemical and clay is 2, how long would the chemical take to penetrate the liner? (b) Using the data from Sweerts et al. in Figure 3.5 for fine sediments, calculate the effective diffusivity of a chemical in the clay if the molecular diffusivity in water was 2×10^{-5} cm^2 s^{-1}? (c) How long before a plume, moving by diffusion only, would reach the bottom of the liner? Comment on the time calculated by diffusion relative to that for advection calculated in part (a). (d) Assuming $\sigma_{nw} = 30$ dyn cm^{-1}, could a ganglion of pure phase of a chemical reach a length equal to or greater than the liner thickness?

12.6 You are designing a column experiment ($L = 10$ cm, $\theta = 0.4$) to evaluate a proposed treatment method for cleaning up NAPLs in groundwater. (a) What size sand grains would you use to produce maximum ganglia lengths of 5 cm in an unsaturated vertical column? (b) You now decide to place the column in the horizontal direction, and to flood the column with water. What pressure (psia) will develop to produce a saturated water flow velocity of 0.1 cm s^{-1}? [14.5 psia = 10^5 kg m^{-1} s^{-2}]. (c) How large is a stable ganglion under these saturated flow conditions? Will the ganglion produced in unsaturated media be stable? Assume $\sigma_{aw} = 20$ g s^{-2}, $\sigma_{nw} = 50$ g s^{-2}, and $\rho_n = 0.73$ g cm^{-3}.

12.7 Venting, or vacuum extraction, is a method used to clean up contaminated soils in the unsaturated zone. In this question we explore remediation times for NAPLs by venting. Let us assume that we have a contaminated zone with trichloroethylene (TCE) with the following properties: $M_T = 130$ g mol^{-1}, $\sigma_{Tw} = 35$ dyn cm^{-1}, $\sigma_{Ta} = 29$ dyn cm^{-1}, vapor pressure 57.9 mm Hg, and $\rho_T = 1.46$ g cm^{-3}. For the soil, $\rho_b = 1.5$ g cm^3, $\theta = 0.35$, and $d_g = 0.1$ cm. Assume for air $\nu_a = 1.8 \times 10^{-4}$ g cm^{-1} s^{-1}, $\rho_a = 1.5 \times 10^{-5}$ kg m^{-3}, all properties at 20°C. Use the correlation of Wilson and Geankoplis (1966) to calculate a mass transfer coefficient for dissolution or evaporation. (a) What is the size of the largest ganglion in the unsaturated media prior to any advective motion? (b) When air is pumped at a rate of 0.57 m^3 min^{-1} through a cylindrical area of radius 12 m, what is the size of the largest ganglion in the unsaturated media? (c) What is the lifetime of a ganglion in the unsaturated media? Use the throat radius as the characteristic radius of the ganglion.

12.8 The chemical methyl chloride (log K_{ow} = 2.89) has contaminated a groundwater flowing toward a town. The hydraulic conductivity of the aquifer is 1.5 $\times$ 10^{-5} m s^{-1}, θ = 0.37, ρ_b = 1.7 g cm^{-3}, and organic carbon content of 1% by weight. (a) What is the soil permeability in darcy? (b) How far will the chemical have traveled after 1 year if the hydraulic gradient is 1.19?

12.9 Ball and Roberts (1991) studied the long-term adsorption of tetrachloroethylene, or perchloroethylene (PCE), and 1,2,4,5-tetrachlorobenzene (TCB) in water-saturated soils from Borden, Ontario, and found that contact times of tens to hundreds of days were required to reach equilibrium conditions. They attributed the long adsorption times to low effective pore diffusion coefficients (~2–3 orders of magnitude less than those in water), and to constrictive effects of pore sizes (an additional 3- to 20-fold reduction). The apparent diffusion coefficients in the soil grain matrix, $D_{Cw,s}$ (the intraparticle diffusivity) were related to the pore diffusivity, $D_{Cw,p}$, by

$$D_{Cw,s} = \frac{D_{Cw,p}}{[1 + (\rho_b K_{Cs,i}/\theta_i)]} = \frac{D_{Cw,p}}{\mathbf{R}_{Cs,i}} \qquad (12\text{-}138)$$

where ρ_b is the bulk soil density, $K_{Cs,i}$ a soil–water partition coefficient for the interior of the soil grain, θ_i the internal porosity of the soil grain, and $\mathbf{R}_{Cs,i}$ the retardation factor for intraparticle diffusion. (a) Calculate the retardation factors, $\mathbf{R}_{Ps,i}$ and $\mathbf{R}_{Ts,i}$, for PCE and TCB in soil grains 0.6 mm in diameter using the data for a soil size fraction of 0.42 to 0.85 mm: ρ_b = 2.66 g cm^{-3}, θ_i = 0.0191, $K_{p,si}$ = 3.1 cm^3 g^{-1}, and $K_{Ts,i}$ = 111 cm^3 g^{-1}. (b) Calculate the pore diffusivities of PCE ($D_{Pw,p}$) and TCB ($D_{Tw,p}$) based on the measured effective diffusivities of ($D_{Pw,s}/a_s^2$ = 8.8 $\times$ 10^{-8} s^{-1}. Compare these with their diffusivities in water at 20°C (see Problem 3.13) of: 800 $\times$ 10^{-8} cm^2 s^{-1} and 640 $\times$ 10^{-8} cm^2 s^{-1}. (c) The pore diffusivity is often related to the bulk diffusivity by $D_{Cw,p} = D_{Cw}K_{con}/\tau$, where K_{con} is a factor that accounts for the effect of constrictions of diffusivity and τ is the soil tortuosity. Chantong and Massoth (1983) reported for polyaromatic compounds in alumina that K_{con} = 1.03 exp($-4.5R^*$) where R^* is the ratio of the molecule and pore diameter. Using this result, calculate the size of the pores that would be necessary to account for the observed pore diffusivities assuming τ = 1/θ_i.

12.10 An experiment on soil venting is being planned in a large lysimeter (an unsaturated soil column) 4 m in height and 2.5 m in diameter. The soil has an average grain diameter d_g = 127 μm, and gas flow rates of Q = 0.6 m^3 min^{-1} are planned. The soil has 10% water and 10% TCE saturation with a total soil porosity of 0.40. (a) For these conditions, what distance will be necessary to reach an air concentration of TCE of 95% saturation? (b) Assuming the chemical volatilization rate calculated in part (a) is constant, how long will it take to volatilize all the TCE over that length of air flow? (c) Calculate the number of pore volumes of air for part (b). (d) If TCE is present over a much smaller distance of 30 cm, calculate the concentration of TCE in the air. (e) How long will it take to volatilize the TCE in part (d)?

12.11   Surfactant flushing has been proposed as a means of cleaning up NAPLs in groundwater aquifers. Mason and Kueper (1996) reported that perchloroethylene (PCE) solubility increased from 150 to ˜15,000 mg L^{-1} and surface tension decreased from 44 to ˜3 dyn cm^- (both the solubility and surface tension continued to change in their experiments with time) with the addition of a 2% (by weight) surfactant solution (1:1 mixture of nonylenyl ethoxylate and a phosphate ester of a nonylphenyl ethoxylate). Assume the coarse grain sand has a diameter of 500 μm and a porosity of 0.32, and the column (47 cm long, 5.4 cm diameter) is pumped at a pressure of 98.8 Pa, a hydraulic gradient of 0.3, and a permeability of 1.58×10^{-10} m^2. PCE has a density of 1.63 g cm^{-3} and a diffusivity in water of $D = 850 \times 10^{-8}$ cm^2 s^{-1}. (a) What is the horizontal length and diameter of a ganglion in this experiment in the absence of a surfactant flush? (b) What would be the lifetime of the ganglion, assuming the diameter of the ganglion is the diameter of a sphere based on the volume of the ganglion calculated in part (a)? (c) How many pore volumes of pure water would be necessary to remove this single ganglion? (d) If the surfactant is added to the feed water, calculate the lifetime and the number of pore volumes of fluid necessary after the addition of the surfactant solution.

REFERENCES

Ball, W. P. and P. V. Roberts. 1991. *Environ. Sci. Technol.* **25**(7):1237–49.

Berg, R. R. 1975. *Am. Assoc. Pet. Geol. Bull.* **59**:939–56.

Bowman, C. W., D. M. Ward, A. I. Johnson, and O. Trass. 1961. *Can. J. Chem. Eng.* **39**(1): 9–13.

Brenner, H. 1990. *Langmuir* **6**(12):1715–24.

Chantong, A., and F. E. Massoth. 1983. *AIChE J.* **29**(7):725–31.

Cho, H. J., P. R. Jaffé, and J. A. Smith. 1994. *Water Resour. Res.* **29**(10):3329–42.

Conrad, S. H., J. L. Wilson, W. R. Mason, and W. J. Peplinski. 1992. *Water Resour. Res.* **28**(2):467–78.

Crittenden, J. C., N. J. Hutzler, and D. G. Geyer. 1986. *Water Resour. Res.* **22**(3):271–84.

Danckwurtz, P. V. 1953. *Chem. Eng. Sci.* **2**(1):1–13.

Dwivedi, P. N., and S. N. Updhay. 1977. *Ind. Eng. Chem. Process Design Dev.* **16**(2):157–65.

Farrell, J., and M. Reinhard. 1994a. *Environ. Sci. Technol.* **28**:53–62.

Farrell, J., and M. Reinhard. 1994b. *Environ. Sci. Technol.* **28**:63–72.

Freeze, R. A., and J. A. Cherry. 1979. *Groundwater*. Prentice-Hall, Englewood Cliffs, N.J.

Friedlander, S. K. 1957. *AIChE. J.* **3**(1):43–48.

Geller, J. T., and J. R. Hunt. 1993. *Wat. Resour. Res.* **29**(4):833–45.

Grimberg, S. J., C. T. Miller and M. D. Aitken. 1996. *Environ. Sci. Technol.* **30**(10):2967–74.

Hunt, J. R. 1998. Lecture notes for contaminant transport processes, Civil and Environmental Engineering 219, University of California, Berkeley.

Hunt, J. R., N. Sitar, and K. S. Udell. 1988a. *Water Resour. Res.* **24**(8):1247–58.

Hunt, J. R., N. Sitar, and K. S. Udell. 1988b. *Water Resour. Res.* **24**(8):1259–69.

Imhoff, P. T., A. Frizzell, and C. T. Miller. 1995a. CMR News, School of Public Health, University of North Carolina at Chapel Hill, **2**(1):1–4.

Imhoff, P. T., S. N. Gleyzer, J. F. McBride, L. A. Bancho, I. Okuda, and C. T. Miller. 1995b. *Environ. Sci. Technol.* **29**(8):1966–76.

Imhoff, P. T., P. R. Jaffé, and G. F. Pinder. 1994. *Water Resour. Res.* **30**(2):307–20.

Imhoff, P. T., G. P. Thyrum, and C. T. Miller. 1996. *Wat. Resour. Res.* **32**(7):1929–42.

Kumar, S., S. N. Updhay, and V. K. Mathur. 1977. *Ind. Eng. Chem. Process Design Dev.* **16**(1):1–8.

Lapidus, L., and N. R. Amundson 1952. *J. Phys Chem.* **56**:984–88.

Mason, A. R., and B. H. Kueper. 1996. *Environ. Sci. Technol.* **30**(11):3205–15.

Mayer, A. S., and C. T. Miller. 1992. *J. Contamin. Hydrol.* **11**:189–213.

Mayer, A. S., and C. T. Miller. 1993. *Transport in Porous Media* **10**:57–80.

Mayer, A. S., and C. T. Miller. 1996. *Water Resour. Res.* **32**(6):1551–67.

Miller, C. T., M. M. Poirier-McNeill, and A. S. Mayer. 1990. *Water Resour. Res.* **26**(11):2783–96.

Ng, K. M., H. T. Davis, and L. E. Scriven. 1978. *Chem. Eng. Sci.* **33**:1009–17.

Ogata, A. 1970. Prof. Paper No. 411-I, U.S. Geological Survey, 1–34.

Ogata, A., and R. B. Banks. 1961. Profl. Paper No. 411-A, U.S. Geological Survey, 27–79.

Okuda, I., J. F. McBride, S. N. Gleyzer, and C. T. Miller. 1996. *Environ. Sci. Technol.* **30**(6): 1852–60.

Pfannkuch, H. O. 1984. *Proc. NWWA Conf. on Petrol. Hydrocarbons and Organic Chem. in Groundwater,* pp. 111–29, National Well Water Assoc., Dublin, Ohio.

Powers, S. E., L. M. Abriola, and W. J. Weber. 1992. *Water Resour. Res.* **28**(10):2691–705.

Powers, S. E., L. M. Abriola, J. S. Dunkin, and W. U. Weber, Jr. 1994. *J. Contam. Hydrol.* **16**:1–33.

Powers, S. E., C. O. Loureiro, L. M. Abriola, and W. J. Weber, Jr. 1991. *Water Resour. Res.* **27**(4):463–77.

Rose, D. A. 1977. *J. Hydrol.* **32**:399–400.

Runkel, R. L. 1996. *J. Envir. Engng.* **122**(9):830–32.

Swartzenbach, R. P., P. M. Gschwend, and D. M. Imboden. 1993. *Environmental Organic Chemistry.* Wiley, NY.

Szatkowski, A., P. T. Imhoff, and C. T. Miller. 1995. *J. Contam. Hydrol.* **18**:85–106.

Tchoganoglous, G., and E. D. Schroeder. 1985. *Water Quality.* Addison Wesley, Reading, Mass.

Turek, F., and R. Lange. 1981. *Chem. Eng.* **36**:569–79.

Wilkins, M. D., L. M. Abriola, and K. D. Pennell. 1995. *Water Resour. Res.* **31**(9):2159–72.

Williamson, J. E., K. E. Bazazaire, and C. J. Geankoplis. 1963. *Ind. Eng. Chem. Fund.* **2**(2): 126–29.

Wilson, E. J., and C. J. Geankoplis. 1966. *Ind. Eng. Chem. Fund.* **5**(1):9–14.

CHAPTER 13

PARTICLES AND FRACTALS

13.1 INTRODUCTION

It is not always clear at what size molecules become colloids, colloids become particles, or even whether particulate material should be described as particles or aggregates. Marine chemists operationally (using ultrafiltration separations) define colloids as suspended organic matter with apparent molecular weights >1000 daltons (Guo and Santschi, 1996). Wastewater engineers define particulate BOD as the difference between the total BOD of a sample and the soluble BOD (sBOD), where the sBOD is measured on samples passed through a 0.45-μm filter. From a microbiological perspective the sizes of organic matter may be distinguished based on whether the material can pass a bacterial membrane. Molecules larger than ~1000 daltons must be cleaved to smaller sizes before being transported across the outer membrane (Law, 1980). Organic matter larger than 1000 amu, but able to pass a 0.2-μm membrane filter, can therefore often considered to be a macromolecular size fraction.

Most of the discussion of chemical transport in the preceding chapters has dealt with defined chemicals with small molecular weights. On a mass basis, however, a large fraction of organic matter in water is composed of particles. In the ocean, for example, ~50% of the organic carbon is >0.2 μm. The biodegradable organic matter in wastewaters entering a treatment plant are approximately two-thirds particulate (>0.45 μm). In early studies of wastewater size distributions a "colloidal" size fraction was defined as all OM that would pass a filter (0.2 μm pore diameter) but could be removed by centrifugation (Rickert and Hunter, 1967). In the remaining chapters we will, for the purposes of discussion, define OM > 100,000 dalton (100K) and <0.2 μm as colloids, with particles being defined as OM > 0.2 μm.

A rigorous treatment of particulate matter in water is important for environmental engineers designing and operating water treatment and wastewater treatment systems because these treatment systems are designed to form and remove particles. In a conventional water treatment plant small, nonsettleable particles are made larger through their coagulation into particles that will settle out, over a period of hours, in large basins (clarifiers). In activated sludge wastewater treatment systems particulate OM is converted by bacteria into smaller molecules able to be taken across their membrane. This material, and other dissolved organic matter (DOM), is then converted either into bacterial biomass or CO_2. Although many different bacteria participate in the degradation of OM, only bacteria that form fast-settling flocs that settle in the clarifier will be recycled and can remain in the system. The coagulation and filtration of bacteria and other particles are dependent on both the chemical and physical properties of the particles, such as their size, density, and surface charge. In this chapter we explore the physical properties of particles and develop methods to describe the bulk properties of the particles in terms of total mass, size distribution, and fractal nature of the particles.

13.2 SOLID PARTICLES AND FRACTAL AGGREGATE GEOMETRIES

A water sample consists of a mixture of particle types. Particles of the same size may be quite dissimilar in properties. Nonviable particles are assumed to form from debris in the water. Even submicrometer colloids are envisioned as loose assemblages of macromolecules, inorganic clays, bits of animals such as single-celled crustaceans, and many other smaller particles. Thus inorganic particles are formed by coagulation processes into larger and larger particles. Viable particles, such as bacteria and single-celled algae (phytoplankton), can be formed de novo through cell division. Of course viable cells such as bacteria also change size during growth, but the sudden creation of a new particle during cell division influences the particle concentration, and therefore can affect the coagulation and settling rates of natural particles.

A description of the particles in a water sample should probably distinguish between aggregated water column debris and solid viable particles such as bacteria, but such distinction is usually not made except in instances where there is a need to identify the properties of individual components or species in the water sample. For example, bacteria or phytoplankton can be distinguished from other particles using a microscope, and therefore, their biomass can be compared to the total mass of particles in the water sample. Except in unusual cases, individual components of the particle suspension, such as bacteria, will comprise less than 50% of the mass of similar-sized particles in the water sample. In practice, both detrital particles and viable cells are classified into the same size category in order to describe the average properties of particles in terms of their mass, density, and settling velocity.

Until recently, the properties of particles in water (often referred to as aquasols) have been described as if all the particles were spheres. The size of a particle is therefore given as a diameter, or equivalent diameter, d_p, based on some property of the particle such as cross-sectional area, A_p, or mass, m_p. From Euclidean geometrical relationships, we have for objects assumed to be spheres that

$$A_p = \frac{\pi}{4} d_p^2 \tag{13-1}$$

$$m_p = \rho_p \frac{\pi}{6} d_p^3 \tag{13-2}$$

where ρ_p is the particle density. Spheres are Euclidean objects because their properties always scale with size raised to integer values. The main problem with using Euclidean geometric equations based on spheres is that these equations usually fail accurately to describe properties of real particles as a function of size. When cross-sectional areas and sizes of aggregated particles are related using a power-law relationship, it is found that $A_p \sim d_p^b$ but that b is less than 3 (Meakin, 1988). Similarly, $m_p \sim d_p^3$ only if particle density is constant, and for particles found in natural systems $b < 3$. This indicates that a different approach is needed to describe the shapes and properties of particles—particularly when these particles are aggregates of many smaller particles.

In 1982 Mandelbrot used the word *fractal* to describe rugose (wrinkled or corrugated) objects such as mountain landscapes, shorelines, clouds, and even the distribution of eddies due to fluid turbulence. What Madelbrot realized was that not all objects and patterns in nature were constructed from, or could be defined as, orderly objects that fit nicely into Euclidean geometrical patterns. The nonlinear and non-integer mathematics that have evolved out of Mandelbrot's observations have come to be known as *fractal geometry*. While the subject of fractal mathematics is too broad to be covered here fully, aspects of fractal geometry are critical to the description of the shapes, sizes, and properties of particles in water. In the section to follow, a mathematical description of fractal particles is built using the conventional approach of Euclidean relationships to describe aggregated particles. At each step this Euclidean description is modified to include a fractal geometrical equation.

The Self-Similar Structure of a Fractal Aggregate

When an object is described as being a sphere or cube, with a length of 1 cm, two different people will immediately know exactly what the shape of the object is and how to describe its properties. Because all spheres are (geometrically) exactly the same, we immediately know that the mass of the sphere will increase as the cube of the length (diameter), the surface area as the square of the diameter, etc. But if two people try to describe exactly the shape of an object with fractal properties, for example, a cloud, it is likely that the objects would look quite different. In addition, the mass and densities of the clouds would probably increase with size at slower rates than those for solid spheres. The same exercise can be repeated for any fractal object, including aggregates of particles. If an aggregate obeys Euclidean geometry, then the particles forming the aggregate must be arranged so that they are close-packed. Two people describing this close-packed aggregate would have exactly the same mathematical description of the aggregate and its properties. Thus only close-packed aggregates obey Euclidean scaling relationships. However, aggregates that are nonhomogeneous (nonuniformly packed) possess fractal attributes, and their properties (such as mass and density) scale with size raised to noninteger values.

Fractal geometry developed from the observation that although the *exact* shape of an object could not be described in a simple manner, the *general* shape of an object, or some property of the object, would change in a predictable manner as the viewing scale changed. Thus the irregularity of a shoreline viewed over a distance of miles at a resolution of hundred feet was similar in character to a smaller piece of that shoreline measured at a resolution of inches. This irregularity of the shoreline could not be precisely specified at every location, but using fractal geometry it was shown that the essential nature of the irregularity of the shoreline could be defined. By measuring the irregularity over different length scales, it was possible mathematically to quantify the extent that one shoreline was more or less irregular than the other. This self-similarity of irregular objects, when analyzed at different measurement scales, is a critical property of a fractal object.

Compare the different two-dimensional objects shown in Figure 13.1 at two different scales of resolution. A close-packed, two-dimensional object formed from small spheres obeys scaling relationships of Euclidean geometry: As the object gets larger the number of spheres in the object, N^*, increases as the length squared. For example, at two particle diameters $N^* = 4$ and at five particle diameters $N^* = 25$. Therefore, $\ln(4)/\ln(2) = \ln(25)/\ln(5) = 2$. Thus we see that $N^* \sim d_a^2$, where d_a is the diameter of the circle enclosing the particles. For a fractal object the number of particles does not scale according to an integer, but instead scales to a fracational or fractal power, according to

$$N^* = b_n d_a^{D_n} \tag{13-3}$$

where D_n is the fractal dimension and the subscript n indicates the number of dimensions. When $D_n = n$, the object obeys Euclidean geometry; when $D_n < n$, the object may be fractal depending on whether it is self-similar at different length scales. The number of primary particles in the aggregate shown in Figure 13.1B scales with length as $\ln(5)/\ln(3) = \ln(25)/\ln(9) = 1.465$, or $D_2 = 1.465$.

In order for an object to be fractal it must, on average, look identical even at different locations on the object. The overall shape of the whole object enclosed in the larger circle in Figure 13.1B looks similar to the shapes of the smaller circle. The smallest part of the object that has the same geometrical shape of the aggregate is defined as the fractal *generator*. The five-ball object shown in Figure 13.1B is the fractal generator; any smaller part of this object will not resemble the shape of increasingly larger objects. The aggregate shown in Figure 13.1C is not fractal; although it appears to obey a noninteger scaling law, the object is not self-similar at different length scales (shown by the two circles).

The fractal object shown in Figure 13.1 is a *deterministic* fractal since the location of all the particles that make up the aggregate have the same, precise locations relative to one another everywhere in the aggregate. Most of the fractal aggregates

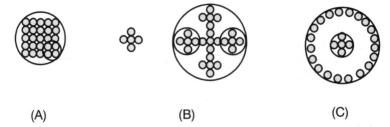

(A)   (B)   (C)

Figure 13.1   Three different aggregates composed of 25 particles. The circles compare the aggregate characteristics at small and large scales. (A) This aggregate has an integer scaling power or fractal dimension of 2 and therefore is not fractal. (B) This aggregate is a fractal since it is self-similar at the smaller and larger scales and has a fractal dimension less than two in two dimensions. The fractal generator, the smallest particle having fractal characteristics, is shown on the left. (C) This object is not self-similar. The arrangement of the spheres in the smaller circle is different from that in the larger circle.

that are produced through the coagulation of particles in nature are *statistical* fractals. This means that, on average, the concentration of particles in the aggregate changes in a manner that obeys fractal geometry over the scale of the aggregate. The aggregate shown in Figure 13.2 is a statistical fractal. The fractal nature of this aggregate was demonstrated by Meakin (1988) by counting the number of particles contained in the aggregate at different radii starting at a point on the aggregate defined as the radius of gyration.

The Properties of Fractal Aggregates

The fractal dimensions of particles analyzed in one, two, and three dimensions can be measured using a variety of techniques, but the starting point in the calculation is the definition of a length scale. If a single aggregate is analyzed, the properties of the aggregate are measured from its radius of gyration, or the root-mean-square radius, defined in terms of the number of particles in the aggregate as

$$R_g = \left(\frac{1}{N^*} \sum_{i=1}^{N^*} (x_i^2 + y_i^2 + z_i^2) \right)^{1/2} \tag{13-4}$$

where x_i, y_i, and z_i are the coordinates of particle i relative to the center of the aggregate containing N^* particles.

Instead of measuring the property of a single aggregate, it is also possible to measure the properties of a population of aggregates having different sizes. For this case other length scales can be used such as shortest, longest, geometric mean, and equivalent radius (based on area), although these length scales must always be larger than the size of the fractal generator. For a population of fractal particles, we can express the perimeter (P_p), cross-sectional area (A_p), and solid volume (v_p) as a

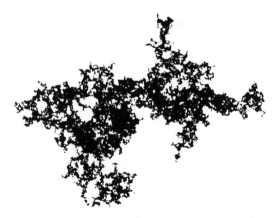

Figure 13.2 Aggregate formed from a colloidal suspension of 10-nm gold particles. (Reprinted with permission from Witten and Cates (1986). Copyright 1986 American Association for the Advancement of Science.)

function of total particle length, l, as

$$P \propto l^{D_1} \tag{13-5}$$

$$A \propto l^{D_2} \tag{13-6}$$

$$v \propto l^{D_3} \quad \text{or} \quad v \propto l^D \tag{13-7}$$

where the subscripts on D_n indicate the integer value of D_n for Euclidean particles. If the subscript is omitted, it is understood that its value is 3, or $D_3 = D$. The fractal dimension can be obtained from the slopes of log–log plots of the respective aggregate properties in Eqs. 13-5 to 13-7 and aggregate length.

The perimeter of irregular objects is larger, and the cross-sectional area is smaller, than the corresponding properties of spheres of the same length. Because the perimeter of fractal objects increases with particles size, $D_1 > 1$. For example, the one-dimensional fractal dimensions of particles in the ocean average $D_1 = 1.28$, indicating that perimeter increases slightly faster than aggregate size (Table 13.1). Fractal dimensions for different types of particles ranged from $D_1 = 1.82$ for agglomerations of nonidentifiable particles (amorphous) to $D_1 = 1.11$ for debris in the water column (miscellaneous). Two-dimensional fractal dimensions of these marine snow particles are lower than that of a sphere ($D_2 = 2$), indicating that these particle become less dense and more irregular as they increase in size. Aggregates formed primarily from single-celled phytoplankton (in this case, diatoms) had fractal dimensions of $D_2 = 1.86$, while those formed primarily from fecal pellets of microscopic and very small animals had fractal dimension of $D_2 = 1.34$.

For an aggregate made up of spherical monomers of diameter d_p, the number of particles in the aggregate is a function of the solid volume according to $N^* = v_{ag}/$

TABLE 13.1   One- and Two-Dimensional Fractal Dimensions of Aggregates Formed in the Ocean (Marine Snow) Measured from Photographs Taken in situ

Type	Size Range (mm)	D_1	D_2
Diatoms	2–60	1.30 ± 0.08	1.86 ± 0.13
Amorphous particles	2–7	1.82 ± 0.53	1.63 ± 0.72
Fecal pellets	1–17	1.19 ± 0.19	1.34 ± 0.16
Miscellaneous	1–23	1.11 ± 0.07	1.28 ± 0.11
All groups	1–60	1.28 ± 0.04	1.72 ± 0.07

Source: Kilps et al., 1994.

$(\pi d_p^3/6)$. Since $N^* \sim l$, Eq. 13-6 can also be written in terms of particle number as

$$N^* \propto l^D \qquad (13\text{-}8)$$

which is the same scaling relationship used in Eq. 13-3, except in Eq. 13-8 we have not specified the proportionality constant. From the primary particle density we can also develop the scaling relationship

$$m_{ag} \propto l^D \qquad (13\text{-}9)$$

Example 13.1

The number of particles in bacterial aggregates developed from pure cultures of a common floc-forming bacterium found in activated sludge reactors, *Zoogloea ramigera*, was examined by Logan and Wilkinson (1991). Individual aggregates were formed in test tubes rotated at an angle in a laboratory tube rotator. Individual aggregates were sized and dispersed using an enzyme (cellulase) to break the fibers holding the cells together. The number of cells in the suspension were then counted using standard acridine orange direct count (AODC) methods. Using the data in Table 13.1A, calculate the fractal dimension (D) of these aggregates.

To calculate the fractal dimension we will use the scaling relationship in Eq. 13-8, $N^* \sim l^D$. Taking the logs of the given data, we plot log N^* versus log l in

TABLE 13.1A

Size (mm)	$N_p \times 10^6$	Size (mm)	$N \times 10^6$
0.78	5.019	1.17	10.80
0.84	4.508	1.24	7.936
0.94	6.495	1.33	11.20
1.06	8.522	1.44	1.590

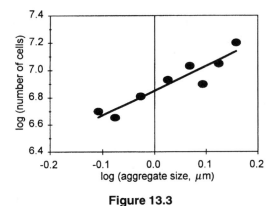

Figure 13.3

Fig. 13.3. From a linear regression, we obtain the fractal dimension from the slope, or $D = D_3 = 1.81$.

It is also possible to describe other properties of fractal aggregates, such as density, porosity, and settling velocity, but in order to do that we must explore the proportionality constants in the equations used above. These origin of these constants for fractal objects will be easier to follow if we first develop them for Euclidean objects.

The most basic property of an aggregate composed of monomers is the number of particles in the aggregate. This can be calculated for a Euclidean aggregate from the aggregate size and two factors: a shape factor ($\dot{\xi}$) and a packing factor ($\dot{\zeta}$). The overdot is used to denote variables based on Euclidean geometry. The shape factor is defined as the ratio of the encased volume of an aggregate to its characteristic length cubed, or $\dot{\xi} = v_{ag,e}/l^3$. For a cube $\dot{\xi} = 1$ and for a sphere $\dot{\xi} = \pi/6$. When the particles in an aggregate are uniformly distributed, the packing factor is calculated as

$$\dot{\zeta} = \frac{\dot{v}_{ag}}{\dot{v}_{ag,e}} = \frac{\dot{N}^* \dot{v}_p}{\dot{v}_{ag,e}} \tag{13-10}$$

where $\dot{v}_{ag}$ is the total solid volume of all particles in the aggregate, and $\dot{v}_p$ is the volume of a single particle. The number of particles in a close-packed aggregate containing uniformly distributed monomers is therefore (Jiang and Logan, 1991),

$$\dot{N}^* = \dot{\zeta}\, \frac{\dot{v}_{ag,e}}{\dot{v}_p} \tag{13-11}$$

This relationship can also be written in terms of characteristic lengths (diameters) using $\dot{v}_{ag,e} = \dot{\xi}_{ag} d_{ag}^3$ for the aggregate, and $\dot{v}_p = \dot{\xi}_p d_p^3$ for the primary particles, to obtain

$$\dot{N}_p^* = \frac{\dot{\zeta}\dot{\xi}_{ag}}{\dot{\xi}_p} \left(\frac{d_{ag}}{d_p} \right)^3 \tag{13-12}$$

where d_{ag} is the aggregate diameter, d_p the diameter of primary particles composing the aggregate, and $\dot{\xi}_{ag}$ and $\dot{\xi}_p$ are shape factors of the aggregate and primary particles. If the monomers are spherical, then $\dot{\xi}_{ag} = \dot{\xi}_p = \pi/6$. The magnitude of the packing factor depends on the arrangement of the particles relative to one another. For example, for a cubic packing of uniform spheres $\dot{s} = 0.5336$, and for a rhombohedral packing $\dot{\zeta} = 0.7404$. Equation 13-12 does not describe the number of particles in fractal aggregates because N_p^* is not proportional to d_a^3, and the coefficients are not constant, as shown below.

In order to derive an equation that includes scaling coefficients for a fractal aggregate, it is necessary to specify the nature of the fractal generator. A fractal aggregate is self-similar over many sizes, but at some very small size it must lose this self-similarity property because the shape of the fractal aggregate is not the same as that of the individual particles that form the aggregate. Let us assume that the aggregate is composed of spheres. The fractal generator is the smallest assemblage of these spheres that have the same fractal shape and properties as the large aggregate. The volume of the generator particle is larger than that of the primary particle by $v_g = v_p/\zeta_g$. Defining the generator length as l_g, the generator volume can be written as

$$v_{g,e} = \frac{v_p}{\zeta_g} = \xi_g l_g^3 \tag{13-13}$$

Using the definition of the particle volume as $v_p = \xi_p l_p^3$ and the above equation, we obtain

$$l_g = \left(\frac{\xi_p}{\xi_g \zeta_g}\right)^{1/3} l_p \tag{13-14}$$

Thus the basic relationships between the spherical monomers and the fractal generator is now known. From the scaling relationship in Eq. 13-8 and the assumption that l_g is a constant, we can write the scaling relationship in terms of the fractal generator as

$$N^* = b_g \left(\frac{l_{ag}}{l_g}\right)^D \tag{13-15}$$

Since $N^* = 1$ when $l_{ag} = l_g$ we conclude that $b_g = 1$. Combining Eqs. 13-14 and 13-15, we have the final result

$$N = \left(\frac{\xi_g \zeta_g}{\xi_p}\right)^{D/3} \left(\frac{l_{ag}}{l_g}\right)^D \tag{13-16}$$

Since the constants in the first term on the right-hand size of Eq. 13-15 are probably not known, we substitute a single constant defined as $b_D = (\zeta_g \xi_g/\xi_p)^{D/3}$, or

$$N^* = b_D \left(\frac{l_{ag}}{l_g}\right)^D \tag{13.17}$$

Notice that the scaling coefficient b_D is a function of the fractal dimension, and therefore this scaling factor will be different for aggregates with different fractal dimensions.

Solid Volume   The total solid volume of particles in an aggregate is just $N^* v_p$. Using Eq. 13-17 and the definition of the primary particle, the solid volume is

$$v_{ag} = \xi_p l_p^3 b_D \left(\frac{l_{ag}}{l_g} \right)^D \tag{13-18}$$

This form of the equation for volume reduces to the expected relationship for a nonfractal aggregate of $\dot{v}_{ag} = \dot{\zeta} \dot{\xi}_{ag} d_{ag}^3$ when $D = 3$, $l_{ag} = d_{ag}$, and the packing factors are the same for the generator and aggregate.

Aggregate Mass   If the primary particles forming the aggregate each have a mass m_p, where $m_p = \rho_p \xi_p d_p^3$, then the total mass of particles in an aggregate is just $N^* m_p = \rho_p v_{ag}$ or

$$m_{ag} = \rho_p \xi_p l_p^3 b_D \left(\frac{l_{ag}}{l_g} \right)^D \tag{13-19}$$

Aggregate Density   The calculated density of a fractal aggregate depends on the manner in which the aggregate encased volume is calculated. Since the border of a fractal aggregate is irregular, and the overall shape of the aggregate cannot be easily defined, the volume of the aggregate is sometimes calculated from an equivalent length based on the projected area, or $l = (4A/\pi)^{1/2}$. Other approaches are to use the maximum length, averages of maximum and minimum lengths, etc. However, once the length scale is defined for the aggregate, the density is calculated based on an encased spherical volume, $v_{ag,e} = (\pi/6) l_{ag}^3$. Notice that this volume must be larger than the solid volume calculated above for the aggregate. The density of the aggregate is $\rho_{ag} = m_{ag}/v_{ag,e}$, or using Eq. 13-18,

$$\rho_{ag} = \frac{\pi}{6} \rho_p \xi_p l_p^3 l_g^{-D} b_D l_{ag}^{D-3} \tag{13-20}$$

If $D = 3$ then aggregate density is constant, as expected for close-packed, nonfractal aggregates. When $D < 3$, aggregate density decreases in proportion to aggregate size, but the property of self-similarity must be maintained. This means that on any portion of a single aggregate, for example, near the center, the density of particles within some volume must be the same as the density of particles within that same volume at another location on the aggregate. This self-similarity property is not included in some models of fractal aggregates (Chellam and Wiesner, 1992).

Aggregate Porosity   The average porosity of an aggregate can be calculated in the usual manner from the solid and encased volumes as $\theta_{ag} = 1 - (v_{ag}/v_{ag,e})$. Using the definitions of these volumes

$$\theta_{ag} = 1 - \frac{\pi}{6} \xi_p l_p^3 l_g^{-D} b_D l_{ag}^{D-3} \tag{13-21}$$

Aggregate porosity, like density, is constant when $D = 3$. For fractal aggregates, however, the average porosity decreases as aggregate size increases.

Comparison of Aggregate Properties   The equations used to describe the relationship between particles size and properties, including mass, volume, encased volume, density and porosity, are compared in Table 13.2 for Euclidean and fractal aggregates. The most common method of calculating three-dimensional fractal dimensions for populations of particles is to use the size–particle number relationship shown in Table 13.2. If all particles forming the aggregate have the same density, linear regressions based on mass or porosity and aggregate size should also provide the same fractal dimension. A general relationship between size and settling velocity is given in Table 13.1, but as we shall see below, there is no single, simple scaling relationship that can be used between aggregate size and settling velocities.

Three-dimensional fractal dimensions of a variety of particles are summarized in Table 13.3. Small colloidal aggregates, formed through the diffusion of the particles (or Brownian motions; see Chapter 14), have fractal dimensions of 1.8 to 2.2. Marine snow aggregates may have even lower D values, although there is great uncertainty in these estimates because these aggregates are formed from a variety of particles and have different sizes and shapes. Inorganic and microbial aggregates appear to have no typical range of fractal dimensions. Ferric aggregates formed in sheared reactors had D values of 2.61–2.65, while microbial aggregates had fractal dimensions of 1.7 to 3 depending on the type of environment the aggregate was formed in (rotating tubes, shake flask, or aerated chemostat).

Settling Velocity: Drag Coefficient Approach   Two different approaches can be used to develop a terminal settling velocity equation. The first is based on modifying geometrical relationships in Stokes' Law used to predict the settling velocity of a sphere. The second is to assume the aggregate is permeable and to multiply the Stokes' Law settling velocity by a correction factor. In order to use either of these approaches, it is helpful to first derive Stokes' Law and then proceed with the modifications.

The three forces acting upon a settling aggregate, gravity (F_{grav}), buoyant (F_{buoy}), and drag (F_{drag}), are related by

$$F_{grav} - F_{buoy} = F_{drag} \tag{13-22}$$

Gravity acts upon the aggregate due to the density difference between the fluid (water) and the total solid mass that displaces the water. Therefore, $F_{grav} - F_{buoy} = (\rho_{ag} - \rho_v)v_{a,e}g$, where ρ_{ag} is the aggregate density, ρ_w the water density, and g the gravitational constant. We can write the density difference in terms of the aggregate

TABLE 13.2 Comparison of Relationships that Defined Fractal versus Euclidean Objects

Property	Euclidean Objects	Fractal Objects[a,b]
Number of primary particles in the aggregate	$\dot{N}^* = \dfrac{\zeta \xi_{Sag}}{\dot{\xi}_p} \left(\dfrac{d_{ag}}{d_p}\right)^3$	$N^* = b_D \left(\dfrac{l_{ag}}{l_g}\right)^D$
Solid volume	$\dot{v}_{ag} = \dot{\zeta} \xi_{Sag} d_{ag}^3$	$v_{ag} = \xi_p l_p^3 b_D \left(\dfrac{l_{ag}}{l_g}\right)^D$
Encased volume	$\dot{v}_{ag,e} = \xi_{Sag} d_{ag}^3$	$\dot{v}_{ag,e} = \dfrac{\pi}{6} l_{ag}^3$
Mass	$\dot{m}_{ag} = \rho_p \dot{\zeta} \xi_{Sag} d_{ag}^3$	$m_{ag} = \rho_p \xi_{Sp} l_p^3 b_D \left(\dfrac{l_{ag}}{l_g}\right)^D$
Density	$\dot{\rho}_{ag} = \rho_p \dot{\zeta}$	$\rho_{ag} = \dfrac{6}{\pi} \rho_p \xi_{Sp} l_p^3 l_g^{-D} b_D l_{ag}^{D-3}$
Porosity	$\theta_{ag} = 1 - \dot{\zeta}$	$\theta_{ag} = 1 - \left(\dfrac{\pi}{6} \xi_{Sp} l_p^3 l_g^{-D} b_D l_{ag}^{D-3}\right)$
Settling velocity	$U_s = \dfrac{g\,\Delta\rho (1-p)}{18 \nu \rho_w} d_{ag}^2$	$U_s = \left(\dfrac{\pi^2 g \xi_{Sp} l_p^3}{18 b_{d1} \rho_w \xi_2 v^{b_D}} (\rho_p - \rho_w) b_D l_g^{D_2 - D - 2} l^{D - D_2 + b_D} l_{ag}\right)^{1/2 - B_D}$

[a] For a sphere, $\xi_{Sag} = \pi/6$.
[b] $b_D = (\xi_{Sg} \zeta / \xi_{Sp})^{D/3}$.

TABLE 13.3 Fractal Dimensions of Different Types of Aggregates[a]

Type	Conditions	D	Reference
Inorganic colloids	Formed by diffusion of particles (Brownian motion)	1.8–2.1	Lin et al. (1989)
Ferric	Stirred reactors	2.61–2.65	Li and Ganczarczyk (1989)
Yeast flocs (*Saccharomyces cerevisae*)	Shake flasks	1.70–2.25	Davis and Hunt (1986)
Bacteria (*Zoogloea ramigera*)	Rotating tubes	2.66 ± 0.34	Logan and Wilkinson (1991)
	Rotating tubes	1.8 ± 0.3	Logan and Wilkinson (1991)
	Aerated chemostat	3.0 ± 0.4	Logan and Wilkinson (1991)
Marine snow—all types	Aggregates formed naturally in the ocean	1.39 ± 0.15	Logan and Wilkinson (1989)
Marine snow—diatoms	Aggregates formed naturally in the ocean	1.52 ± 0.19	Logan and Wilkinson (1989)

[a] All fractal dimensions determined from the numbers of particles in the aggregates, except for marine snow aggregates (based on a regression of size and porosity).

porosity as

$$(\rho_{ag} - \rho_w) = (1 - \theta_{ag})(\rho_p - \rho_w) = (1 - \theta_{ag})\,\Delta\rho \qquad (13\text{-}23)$$

where θ_{ag} is the aggregate porosity and $\Delta\rho$ is the difference between the particle and fluid densities. The drag force exerted on an object is expressed as a function of the fluid density and the object's velocity (U), projected area (A), and an empirical drag coefficient (b_{drag}), or $F_{drag} = \frac{1}{2}\rho U_{sett}^2 A_{ag} b_{drag}$. Using these relationships in Eq. 13-21, we obtain

$$v_{ag}(1 - \theta_{ag})\,\Delta\rho\ g = \tfrac{1}{2}\rho_w U_{sett}^2 A_{ag} b_{drag} \qquad (13\text{-}24)$$

In order to simplify this expression for spheres, we use the following geometrical relationships

$$A_{ag} = \frac{\pi}{4}\,d_{ag}^2 \qquad (13\text{-}25)$$

$$v_{ag} = \frac{\pi}{6}\,d_{ag}^3 \qquad (13\text{-}26)$$

$$b_{drag} = \frac{24}{Re}, \qquad Re \ll 1 \qquad (13\text{-}27)$$

where the Reynolds number is calculated as $Re = U_{sett}d_{ag}/v_w$, d_{ag} is the aggregate diameter, and v_w the kinematic viscosity of the fluid (water). Combining Eqs. 13-23–13-26 produces Stokes' Law for a sphere

$$U_s = \frac{g\,\Delta\rho(1 - \theta_{ag})}{18 v_w \rho_w}\,d_{ag}^2 \qquad (13\text{-}28)$$

From the definition of aggregate porosity, or

$$(1 - \theta_{ag}) = \frac{N^* v_p}{v_{ag}} \qquad (13\text{-}29)$$

we can also write Stokes' Law in terms of the number of particles in the aggregate as

$$U_s = \frac{g\,\Delta\rho\ N^* v_p}{3\pi v_w \rho_w d_{ag}} \qquad (13\text{-}30)$$

One of the reasons why fractal geometry was used to describe aggregates produced in water and wastewater treatment operations was that it was commonly observed that U_s was not proportional to d_{ag}^2 for aggregates. Instead, $U_s \propto d_{ag}^c$, where c is some value less than 2 (Tambo and Watanabe, 1979). The reason for this difference from Stokes' Law was attributed by Li and Ganczarczyk (1989) only to the assumption that the aggregate porosity was not constant. If the only factor that is included in Stokes' law is the fractal relationship between aggregate size and porosity, or $(1 - \theta_{ag}) \sim l_{ag}^{D-3}$, then Stokes' Law leads to the scaling relationship between settling velocity and aggregate size

$$U_s \sim l_{ag}^{D-1} \tag{13-31}$$

Thus it was first proposed that the fractal dimension D could be derived from the slope of a log–log plot of settling velocity and aggregate size.

An examination of the equations used to derive Stokes' Law will show that other geometrical assumptions can be important. Logan and Wilkinson (1991) used the area–length equation (Eq. 13-6) instead of Eq. 13-24, which produced the scaling relationship

$$U_s \sim l_{ag}^{D-D_2+1} \tag{13-32}$$

This scaling relationship has an unusual property due to the relationship between D and D_2. When $D < 2$, a colloidal aggregate viewed in two dimensions is transparent, since all particles are visible, and $D_2 = D$. However, when $D \geq 2$, then $D_2 = 2$ (Vicsek, 1992). Thus for $D < 2$ Eq. 13-32 would predict that $U_s \sim l_{ag}^1$; unfortunately, this linear relationship between U_s and l_{ag} is not consistently observed.

Over a wider range of Re (for $Re > 0.1$), a similar approach can be used to develop settling velocity relationships for fractal aggregates (Jiang and Logan, 1991). Assuming that a drag coefficient can be expressed as a function of Re, it has been shown for spheres that

$$b_{drag} = b_{d1} Re^{-b_d} \tag{13-33}$$

where for spheres $b_{d1} = 24$, $b_d = 1$ ($Re \ll 1$), and $b_{d1} = 29.03$, and $b_d = 0.871$ ($0.1 < Re < 10$) based on a correlation in White (1974). Using this definition of the drag coefficient, and assuming for the projected area of the aggregate that $A_{ag} = \xi_2 l^{D_2}$, Jiang and Logan (1991) produced the relationship

$$U_s = \left(\frac{\pi^2 g \xi_p l_p^3}{18 b_{d1} \rho_w \xi_2 \upsilon^{b_d}} (\rho_p - \rho_w) b_D l_{D2-D-2}^{D_2-D-2} l_{ag}^{D-D_2+b_d} \right)^{1/(2-b_d)} \tag{13-34}$$

The scaling relationship between U_s and l_{ag} is therefore

$$U_{set} \sim l_{ag}^{D_3-D_2+b_d/2-b_d} \tag{13-35}$$

These relationships are summarized in Table 13.4 for different ranges of Reynolds numbers and fractal dimensions.

TABLE 13.4 Slopes between Settling Velocity and Aggregate Length as a Function of the Three-Dimensional Fractal Dimension Predicted from Scaling Relationships and Measured in Settling Experiments by Johnson et al. (1996)

Expression	Conditions	Slope from Scaling Equation		Slopes Observed or Calculated from Experimental Results		
		if $D < 2$	if $D \geq 2$	Exp. 1, $D = 1.79 \pm 0.10$	Exp. 2, $D = 2.19 \pm 0.12$	Exp. 3, $D = 2.25 \pm 0.10$
$-\!-\, \sim l_{ag}^{D-1}$	observed	—	—	1.04 ± 0.10	1.20 ± 0.11	1.33 ± 0.10
$U_s \sim l_{ag}^{D-1}$	Re $\ll$ 1	$D - 1$	$D - 1$	0.79	1.19	1.25
$U_s \sim l_{ag}^{D-D_2+1}$	Re $\ll$ 1	1	1	1	1.19	1.25
$U_s \sim l_{ag}^{D-D_2+b_D(2-b_D)}$	Re $\ll$ 1 $b = 1$	1	$D - 1$	1	1.19	1.25
$U_s \sim l_{ag}^{D-D_2+b_D(2-b_D)}$	0.1 < Re < 10 $b_D = 0.871$	0.77	$0.89(D - 1)$	0.77	1.06	1.11

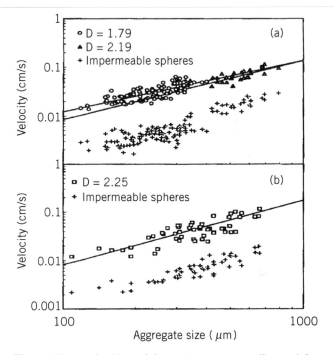

Figure 13.4 The settling velocities of fractal aggregates (formed from latex micro-spheres) with fractal dimensions of 1.79, 2.19, and 2.25, were found to be 4–8.3 times larger than predicted for either impermeable or spheres or permeable spherical aggregates. (Reprinted with permission from Johnson et al. (1996). Copyright 1996, American Chemical Society.)

Settling velocity experiments by Johnson et al. (1996) demonstrated that the drag coefficients for spheres overestimated the drag on settling fractal aggregates, although the fractal dimension could be estimated with some accuracy using Eq. 13-34. In their experiments, aggregates were grown from latex microspheres in paddle mixers and settled, one at a time, in a water column. Each individual aggregate was recovered and analyzed for size and cross-sectional area before the aggregate was broken up and the number of particles in the aggregate counted. By using a range of aggregate sizes, the fractal dimensions D_2 and D were calculated from the slopes of log–log plots using Eqs. 13-6 and 13-7. On average, fractal aggregates settled 4–8.3 times faster than predicted by Stokes' Law using drag coefficients for spheres (Fig. 13.4). From their data, they calculated the constants for a drag coefficient in Eq. 13-32 shown in Table 13.5. Based on their limited data set, it was not possible to construct a more general drag correlation for fractal aggregates.

Settling Velocity: Aggregate Permeability Approach If an aggregate is assumed to be permeable to advective flow, then the settling velocity of the aggregate

TABLE 13.5 Fractal Dimensions and Drag Coefficients Used in the Equation $b_{drag} = b_{d1} \, Re^{-b_d}$ Measured in Settling Velocity Experiments Using Aggregates Formed from Latex Microspheres in Paddle Mixers (Data from Johnson et al., 1996)

Constant	Impermeable Sphere	Data from Experiments		
		Exp. 1	Exp. 2	Exp. 3
D	3	1.79	2.19	2.25
b_{d1}	24	0.14	0.75	0.52
b_d	1	1.31	1.04	1.05

can also be calculated from the aggregate permeability. As described in Chapter 8, the settling velocity of a permeable and impermeable aggregate are related (Matsumoto and Suganuma, 1977) by

$$\frac{U_s}{\dot{U}_s} = \left(\frac{\kappa^*}{\kappa^* - \tanh(\kappa^*)} + \frac{3}{2\kappa^{*2}} \right) \tag{13-36}$$

where κ^* is the dimensionless permeability defined as $\kappa^* = \kappa/\sqrt{d_{ag}}$ and we have assumed that the settling velocity of the impermeable aggregate, $\dot{U}_{set}$, is calculated using Stokes' Law, and that the permeable aggregate is a fractal aggregate with a settling velocity U_s.

The permeability of a porous medium is a function of size of the particles that compose the medium and the porosity. In order to develop a permeability correlation for a fractal aggregate, Li and Logan (1997a) modified a correlation developed by Brinkman (Adler, 1981). For highly porous and homogeneous (nonfractal) media, Brinkman's model is

$$\kappa = \frac{d_p^2}{72} \left(3 + \frac{4}{(1 - \theta_{ag})} - 3 \sqrt{\frac{8}{(1 - \theta_{ag})} - 3} \right) \tag{13-37}$$

where for a settling aggregate d_p is the diameter of particles within the aggregate. This equation is only valid when all particles in the medium are uniformly packed or equally distant from each other. A fractal aggregate consists of clusters of smaller aggregates, which in turn are composed of increasingly smaller clusters of particles. This hierarchical structure, such as the self-similar structure shown in Figure 13.1, results in smaller gaps or pores formed by the clusters at each decreasing hierarchical level. If it is assumed that the size of the largest clusters, defined here as the principal clusters, that form the aggregate primarily dictate the overall permeability of the aggregate (i.e., the smaller clusters are impermeable), it is possible to modify the Brinkman model to describe fractal aggregates. A power-law relationship between the sizes of the principal clusters, d_{cl}, and the aggregate d_{ag} is defined (Li and Logan, 1997a) as

$$d_{cl} = b_{cl1} d_{ag}^{b_{cl}} \tag{13-38}$$

where b_{cl1} and b_{cl} are empirical coefficients. Replacing d_p in Eq. 13-36 with d_{cl}, the permeability of a fractal aggregate becomes

$$\kappa = \frac{b_{cl1}^2 d_{ag}^{2b_{cl}}}{72} \left(3 + \frac{4}{(1 - \theta_{ag})} - 3 \sqrt{\frac{8}{(1 - \theta_{ag})} - 3} \right) \qquad (13\text{-}39)$$

According to this model, the permeability of a fractal aggregate is solely a function of its size, porosity, and distribution of the largest (principal) clusters within the aggregate.

The permeability of fractal aggregates composed of latex microspheres was found to be about three orders of magnitude larger than that predicted by the Brinkman model (Fig. 13.5). Li and Logan (1997a) measured aggregate properties (size, area, number of particles) and settling velocities of microsphere aggregates in settling columns. Solid volume was calculated as a function of the aggregate size using Eq. 13-17 with all the constants expressed as a single constant, b_3, so that the solid volume is $v_{ag} = b_3 l_{ag}^D$. Using this relationship, the porosity can be calculated as $(1 - \theta_{ag}) = (6b_3/\pi) l_{ag}^{D-3}$. For two sets of aggregates, with fractal dimensions of $D = 1.81$ ($b_3 = 28.7$) and $D = 2.33$ ($b_3 = 1.68$), where $b_{cl1} = 0.056$ and $b_{cl} = 0.44$, l [μm], v_a [μm^3]. Settling velocities in their experiments averaged 2–3 times those for impermeable spheres.

Fractal dimensions reported in the literature that have been determined from settling velocities vary over a wide range, but almost all of the reported values for D are less than 2.0 (Table 13.6). Depending on the method used, the fractal dimension will be underestimated (see Table 13.4). Thus, while the fractal dimensions reported

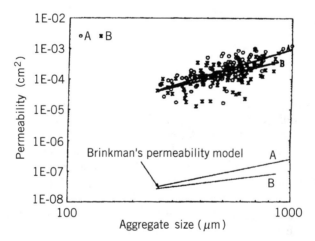

Figure 13.5 Fractal aggregates permeabilities were measured to be three orders of magnitude larger than those predicted based on the Brinkman model relating permeability to porosity for a medium with a homogeneous distribution of particles. (Reprinted with permission from Li and Logan (1997). Copyright 1977, American Chemical Society.)

TABLE 13.6 Fractal Dimensions (D) of Aggregates Derived from Settling Velocity Data Reported in the Literature

Conditions	Type	D
Activated sludge	normal	1.3
		1.45–2.0
		1.44–1.49
		1.70–2.07
	filamentous	1.0
Trickling filter	sloughed biofilm	1.73
Marine snow		1.26 ± 0.06
Aggregates from natural systems	estuarine	1.78
	lacustrine	1.39–1.69
	oceanic	1.94, 2.14
	recoagulated sediments	1.54
Inorganic aggregates (coagulated in stirred reactors)	alum	1.59–1.97
	clay–iron	1.92
	clay–magnesium	1.91

Source: Reported in Li and Ganczarczyk, 1989; Logan and Wilkinson, 1989, 1991.

in Table 13.6 may be slightly low, these experimental data do suggest that fractal dimensions of a wide range of many particles are low and less than 2.

13.3 PARTICLE SIZE SPECTRA

Particles in natural waters vary considerably in size and density, and therefore it is important to quantify the size distribution in order to predict the fate and transport of these particles. Characterizing particle size distributions requires the introduction of additional notation for particle concentrations. We have already defined N^* (a dimensionless number) as the number of particles in an aggregate. When N is used with a subscript, for example, p, it will refer to the number *concentration* of particles; therefore, N_p has units of [# mL^{-1}] or just [mL^{-1}]. The mass concentration of particles is still described using c or ρ [mg L^{-1}]. When there are particles with a range of sizes, N_p must refer to particles of a certain size; for spherical particles of diameter d_p we can write $N_p(d_p)$. The total number of particles in the size distribution will be defined as N_T. When the distribution is monodisperse (all particles have the same size), $N_p = N_T$; when the system is polydisperse, the total number of particles is written as N_T, but the range of particle sizes described by N_T must either be understood from the context of the calculation or it must be designated along with N_T. For example, the number of particles larger than 1 μm, but smaller than 100 μm, can be written as N_T, $N_T(>1 \text{ μm})$, $N_T(<100 \text{ μm})$, or $N_T(1 \le d_p \le 100 \text{ μm})$, depending on the extent to which the limits of the statement need to be defined.

In practice, it is unusual to be able to specify exactly the number concentration of particles of a specific size. More typically, when we give a particle size of, say

d_p, we are either referring to particles of size $d_p \pm \Delta d_p$, where the Δd_p is a standard deviation in size (of a monomer suspension), or to the size interval allowed for a particle to be classified as a size d_p. In order to be able to process data on a large number of particles, instruments used to measure size distributions (see below) are designed to categorize particles into different size classes. From measurements of particle sizes in different size intervals, we can define $n(d_p)$, the size density function, as

$$n(d_p) = \frac{\Delta N_p}{\Delta d_p} \qquad (13\text{-}40)$$

so that n has units of number concentration of particles in the size interval Δd_p, or, for example, [# mL^{-1} μm^{-1}]. As the size intervals become small, or taking the limit as $\Delta d_p \rightarrow 0$, the size density function becomes

$$n(d_p) = \frac{dN_p}{dd_p} \qquad (13\text{-}41)$$

The size density function can be used to describe particle concentrations in terms of either discrete or cumulative size distributions. In a discrete size distribution, $n(d_p)$ refers to those only of size d_p (Fig. 13.6a). It can be difficult to determine the slope of a size density function from discrete size distributions because particle concentrations in adjacent size intervals may be substantially different. In a cumulative size distribution, the concentration of particles in a size class includes the particles in the previous size class. Therefore, the number concentration of particles of size d_p in a cumulative distribution is the sum of all particles larger (or smaller) than those in the current size interval (Fig. 13.6b). If the type of cumulative size distribution is not clear, the size density function can be written as $N_p(>d_p)$ for particles summed

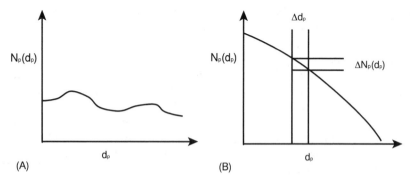

Figure 13.6 Different forms of particle number concentration distributions. Particle concentrations can be expressed in terms of (A) a discrete distribution, or (b) by a cumulative number distribution. In the cumulative number distribution shown here, N ($<d_p$) is summed from larger to smaller particles.

from smallest to largest sizes, or $N_p(<d_p)$ for particles summed from largest to smallest sizes. The cumulative size density function is often calculated in the form of a single-power function as $N_p(>d_p) = b_{nd}d_p^{-S_d}$, where the slope of the size distribution, S_d, is typically in the range of 2–5. Note that if the particle concentration is summed from largest to smallest, then the slope S_d will be negative, while if the summation proceeds from smallest to largest, S_d will be positive.

Total Number Concentration   If the size density function is known, a variety of distribution properties can be calculated. The total number concentration is obtained as the sum of the product of the density function and the size classes, or

$$N_T = \sum_{i=0}^{\infty} n(d_p)\,\Delta d_{p,i} \tag{13-42}$$

Taking the limit as the size interval becomes small

$$N_T = \int_0^{\infty} n(d_p)\,dd_p \tag{13-43}$$

where typical units of these variables are $n(d_p)$ [mL^{-1} μm^{-1}], d_p [μm], producing N_T [mL^{-1}]. The double d in the term dd_p is rather awkward, but its use is unavoidable due to the use of d for both particle diameter and the derivative.

Total Volume Concentration   If the relationship between particle size and volume is known, the size density function can be used to calculate the total solid volume of particles in a given size range. If the solid volume of particles (or aggregates) in a size interval is known, then the total volume *concentration* of particles is

$$V_T = \sum_0^{\infty} v_p n(d_p)\,\Delta d_p \tag{13-44}$$

If all particles are solid, then the geometric relationship $v_p = (\pi/6)d_p^3$ can be used to sum the total particle volume. Taking the limit as $\Delta d_p \rightarrow 0$, the total solid volume is

$$V_T = \int_0^{\infty} \frac{\pi}{6} d_p^3 n(d_p)\,dd_p \tag{13-45}$$

When volume concentrations are calculated over a smaller size range, then V_P is used instead of V_T. From the typical units used above, the volume concentration has the units [μm^3 mL^{-1}]. By noting that 1 μm^3 mL^{-1} = 10^{-12} cm^3 cm^{-3}, we can express V_p in units of *ppmv* using the relationships 1 *ppmv* = 10^6 μm^3 mL^{-1} = 10^{-6} cm^3

cm^{-3}. This calculation requires spherical solid particles, but as discussed below, particle size distributions obtained using resistance particle counters are made on this basis.

Total Mass Concentration   The total mass of particles can be obtained by multiplying the particle volume by the particle density and integrating over the size range of particles

$$m_T = \int_0^\infty \rho(d_p) \frac{\pi}{6} d_p^3 n(d_p) \, dd_p \tag{13-46}$$

This integration requires that all particles of the same size have the same density and that either the density is constant or the density is a known function of size.

Volume Distribution Function   Particle volume concentration can be obtained from summation (or integration) of particle volume over a range of particle sizes, but such an approach does little to convey important information about how volume is distributed in the size distribution. Many particle sizing instruments have software that presents particle volume as volume distributions (V_{dist}) in the form

$$V_{dist} = \frac{dV_p}{d(\log d_p)} \tag{13-47}$$

The advantage of this approach is that the particle volume over a log size interval of $d(\log d_p)$ can be seen to be the area under the curve.

Example 13.2

The discontinuous volume distribution shown in Figure 13.7 indicates that for this water sample there are three particle ranges over which the V_{dist} is constant. Using this distribution, calculate the concentration of particles in ppmv.

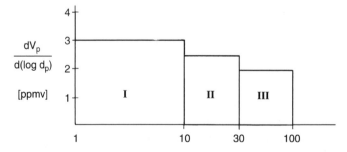

Figure 13.7   Volume distribution (V_{dist}) function in units of ppmv as a function of particle size (log units). The area under the three curves is the volume concentration in each of the sections.

Since the V_{dist} is constant, the area under the curve is the total volume, or

$$V_T = \left|\frac{dV_p}{d(\log d_p)}\right|_I \Delta(\log d_p)|_I + \left|\frac{dV_p}{d(\log d_p)}\right|_{II} \Delta(\log d_p)|_{II}$$

$$+ \left|\frac{dV_p}{d(\log d_p)}\right|_{III} \Delta(\log d_p)|_{III} \qquad (13\text{-}48)$$

where the subscripts I, II, and III indicate the three regions shown in Figure 13.7. Using the values shown, the total volume in ppmv is calculated as

$$V_T = (3 \times 1) + (2.5 \times 0.477) + (2 \times 0.523) = 5.24 \text{ ppmv} \qquad (13\text{-}49)$$

If the function $n(d_p)$ is continuous and known, and particles in the distribution are spherical, then it is possible to obtain an analytical expression for the volume distribution function. From Eq. 13-45, the volume concentration over the size range of 0 to some particle size d_p is

$$V_T = \int_0^{d_p} \frac{\pi}{6} d_p^3 n(d_p) \, dd_p \qquad (13\text{-}50)$$

In order to obtain the V_{dist} function using this expression for volume, we need to recall Leibnitz's rule for a function $F(x)$

$$F(x) = \int_0^{\psi(x)} g(y) \, dy \qquad (13\text{-}51)$$

that

$$\frac{dF(x)}{dx} = g(\psi) \frac{d\psi(x)}{dx} \qquad (13\text{-}52)$$

Using this approach, the V_{dist} function becomes

$$\frac{dV}{d(\log d_p)} = \frac{\pi}{6} d_p^3 n(d_p) \frac{dd_p}{d(\log d_p)} \qquad (13\text{-}53)$$

This can be simplified by manipulation of the last term so that

$$\frac{dd_p}{d(\log d_p)} = \frac{dd_p}{d\left(\dfrac{\ln d_p}{2.3}\right)} = \frac{2.3 \, dd_p}{d\left(\displaystyle\int \frac{1}{d_p} dd_p\right)} = 2.3 \, d_p \qquad (13\text{-}54)$$

Using this result, the volume distribution becomes more simply

$$\frac{dV_p}{d(\log d_p)} = \frac{2.3\pi}{6} d_p^4 n(d_p) \tag{13-55}$$

If the units of the variables are $n(d_p)$ [mL^{-1} μm^{-1}] and d_p [μm], then the V_{dist} function has the units [μm^3 mL^{-1}], which can be converted to ppmv using 1 ppmv = 10^6 μm^3 mL^{-1}.

13.4   MEASURING PARTICLE SIZE DISTRIBUTIONS

A variety of instruments can be used to measure particle sizes, but only a few of them are able to determine particle concentrations simultaneously (Table 13.7). The instruments that measure only particle sizes operate on many principles and employ different methods to track particles such as lasers, sound, and x-rays. Lasers are used in several instruments to illuminate particles and estimate their size from backscattering and diffraction. In one instrument (Galai CIS-100), the laser beam is rotated in a small circle so that the beam can cross the whole particle and detect the particle's edge. In some instruments particles are located and sized using sound waves, and some of these instruments are also capable of measuring a particle's electrophoretic mobility. If particles are all assumed to have the same density, particle size can be measured from settling velocities (usually using x-rays to detect the particle). The use of settling velocity requires the assumption that Stokes' Law is valid, which is a questionable assumption for porous particles. Particles can also be separated based on diffusivity in capillary tubes. Clearly, particle sizes can be measured in a number of different ways, but notice that particle size is not directly measured in most of these instruments. Each instrument measures some property of a particle, and often uses proprietary techniques and algorithms to produce estimates of particle size.

In order to measure particle concentrations, an instrument must have some means of tracking the volume sampled. In instruments capable of measuring particle concentrations, this is accomplished by pulling fluid through a relatively narrow orifice for a set time, at a known velocity. The most common orifice-type instruments, such as Coulter and Elzone particle counters, measure the size of the particle by the change in the fluid resistance as the particle moves through the orifice. A particle suspended in a conducting fluid displaces water and changes the conductance of the solution. As long as the particle suspension is relatively dilute, particle sizes can be measured from changes in conductance as fluid is pulled through the orifice. Particle resistance is correlated to particle volume or size by a calibration step with solid spheres (usually latex microspheres). The volume of the particle is calculated from the volume of displaced water and does not include in the size any water in the particle's interior. Therefore, when porous particles are measured using this instrument, the size of the porous particle that is recorded is the size if all the water was squeezed out of the particle. This size is referred to as the "solid equivalent diameter," d_s, and this size is always equal to or smaller than the particle's actual size.

TABLE 13.7 Different Instruments Used to Measure Particle Size Distributions

Instrument Type	Function	Size Range (μm)
Photon correlation spectroscopy; multiangle laser measurement (Coulter N4 Plus)	Size distribution	0.003–3
Laser diffraction (Coulter LS 100Q)	Size distribution	0.4–950
Laser diffraction (Malvern Mastersizer X)	Size distribution	0.1–2000
Laser time-domain analysis (Galai Cis 2)	Size distribution	0.7–1200
Aperture impedance systems with orifice detection (Coulter Multisizer, Elzone Particle Sizer)	Size distribution, particle concentration	0.4–1200
Acoustic techniques (Matec Applied Sciences)	Size distribution, zeta potential	0.08–10
Capillary hydrodynamic fractionation (Matec CDHF-2000)	Size distribution	0.015–1.1
Dual laser beam/photon correlation spectroscopy (Malvern Zetasizer 4)	Size distribution, zeta potential	0.010–30
Light transmission with orifice (HIAC/Royco Division; Pacific Scientific Co.)	Size distribution, particle concentration	2–400
Settling velocity with *x-ray* detection (SediGraph 5100, Micromeritics)	Size distribution	0.1–300
Image analysis of filtered particles or photographs	Size distribution, particle concentration	>0.2

Aperture impedance instruments are not always accurate. Problems can arise when two or more particles overlap during their passage through the orifice. Also, if the particles have different shapes, or contain immobile water (as in the case of aggregated materials), the accuracy of particle sizing is reduced. The extent to which particle volumes are inaccurately calculated is not well known. One review of aperture impedance studies concluded that particle volumes of porous particles should be accurate to within a factor of two compared to that for a solid sphere, indicating that d_p reported by resistance particle counters is accurate to within a factor of $2^{1/3}$ = 1.26 (Jackson et al., 1995). Aggregated and other delicate particles can be sheared apart as they are pulled through an orifice, possibly resulting in inaccurate reporting of the numbers and sizes of the particles. Most manufacturers recommend that particles have a size between 2% and 50% of the orifice diameter. Although particle sizes in the suspension may have sizes that fall outside this range, this typically is not a problem. Particles smaller than the orifice will fail to be recognized by the instrument, due to insufficient sensitivity of the detector to changes in conductance, and larger particles will simply not fit through the orifice. If high concentrations of particles larger than the 50% cutoff are recorded in the instrument's output, these data must be considered to be suspect and should be discarded from further analysis. Samples generally should not be filtered to remove particles larger than the orifice. Particles smaller than the orifice will also be removed during filtration, altering the final size distribution (see example in Chapter 15).

If coagulated particles are analyzed using an orifice-type instrument, it is usually necessary only to include data for particles with solid equivalent diameters even smaller than 50% of the orifice diameter. Highly porous fractal aggregates will have maximum sizes much larger than the solid equivalent diameter. Based on scaling equations presented above, the encased aggregate volume is related to the aggregate size by $v_{ag,e} = b_a l^D$, and the solid volume is a function of encased volume according to $v_{ag} = v_{ag,e}(1 - \theta_{ag})$. The aperture impedance instrument reports a solid volume in terms of the solid equivalent diameter as $v_{ag} = (\pi/6)d_s^3$. Using these relationships, we can see that the aggregate's actual size (based on the encased volume) is related to the solid volume according to

$$l = b_s d_s^{3/D} \tag{13-56}$$

where b_s is a new constant. Since $D < 3$ for fractal objects, the size (l) of an object can be much larger than the solid equivalent diameter. Let us assume, for example, that $b_f = 1.5$ and $D = 2$ and that particles are being measured using an orifice of 80 μm. When $d_s = 20$ μm is reported by an aperture impedance instrument, the actual size of the fractal particle would be 134 μm; if d_s were doubled to 40 μm, the length would increase by 184% to 380 μm (Li and Logan, 1995). These actual particle sizes are obviously much larger than the orifice, and therefore the output data from the instrument would likely be incorrect. For aggregated suspensions that are known to be fractal, a solid diameter of <20% of the orifice diameter is a more reasonable upper limit (Li and Logan, 1995).

A second common type of orifice-type system is a particle counter based on light transmission, such as the HIAC/Royco particle counter. A beam of light is directed perpendicular to the flow, and as the particles pass through the beam they are counted. As a result of this technique, particle sizes are measured in terms of an equivalent diameter based on the particle projected area, assuming $d_p = (4A/\pi)^{1/2}$.

Image analysis has been used in many different ways to size particles, and it is also possible using this technique to calculate particle concentrations. The technique is quite simple. If particles are large, then they can be viewed directly (or indirectly using photographs) with a camera linked to an image analysis system. This image is converted by computer software to a binary image (black and white or 0 and 1), and the number of pixels for each particle in the image calculated by the system. There are many different image analysis systems available, varying in: complexity of software, capability for color or only black and white, and maximum resolution (number of pixels). When an image analysis system is connected to a microscope, it is possible to image microscopic particles. Different stains can be used to increase particle contrast. Fluorescent stains provide the best contrast, although nonfluorescent surfaces (backgrounds) are needed (Li and Logan, 1995; Logan et al., 1994). At a magnification of 1000 × most systems can size particles as small as a bacterium ($\tilde{\ }$0.5 μm). The particle concentration in a sample is determined from the volume of sample applied and the viewing field. For example, if 1 mL of a sample is filtered onto a polycarbonate filter (a very flat surface) using a circular funnel 15 mm in diameter, the microscopic viewing field is 150 μm, and 300 bacteria-sized particles are counted over 10 viewing fields, the particle concentration of this size particle is 3×10^5 mL^{-1}. Since there are 10,000 possible viewing fields in this example on the filter surface, many fields must be viewed for accurate particle size distributions when particle concentrations are low. The main limitations of particle sizing using image analysis is that particles are only viewed in one direction. Also, when particles are filtered onto a surface, they may collapse during filtration, and this may alter the size and morphology of the particle. Different characteristic sizes of the particles can be used, such as maximum, minimum, and multiangle averaged lengths, or projected areas (A) can be converted to lengths assuming $l = (4A/\pi)^{1/2}$.

13.5   CALCULATING FRACTAL DIMENSIONS FROM PARTICLE SIZE DISTRIBUTIONS

There are a variety of methods for calculating fractal dimensions. For a single particle, a fractal dimension can be calculated by counting particles within successively larger concentric circles starting at the radius of gyration, or by measuring light scattering at different angles (Avnir, 1990). For a large number of particles having a distribution of sizes, we have already seen that fractal dimensions can be calculated as the scaling power in plots of aggregate size and some property, such as perimeter, area, solid volume, and settling velocity. Fractal dimensions can also be calculated from the slopes of size distributions for particles formed by coagulation. These methods are referred to as: the two-slope method (TSM), the particle concentration technique (PCT), and the multiple range method (MRM). If a coagulating system is at steady state, then it is also possible to calculate the fractal dimension from the slope of the steady-state size distribution (steady-state method, or SSM). The governing equations used for these approaches are presented below.

Fractal Dimensions Using the Two-Slope Method   In order to use the TSM, the size distribution of coagulated particles must be measured in terms of both length (maximum or average) and solid volume. These spectra must be obtained using two different instruments, such as an aperture impedance particle counter for solid volume and an image analyzer for length. The size distributions can be expressed in terms of two power-law equations as

$$N(l) = B_l l^{S_l} \tag{13-57}$$

$$n(v) = B_v v^{S_v} \tag{13-58}$$

where S_l and S_v are slopes of the cumulative length and solid volume size distributions, $N(l)$ and $N(v)$, obtained from log–log plots based on aggregate length and volume, and B_l and B_v are empirical constants. If the same population of particles are analyzed, then the number of particles in the size distributions must be equal, or

$$N(l) = N(v) \tag{13-59}$$

From the definition of solid volume (Eq. 13-17) and Eqs. 13-56–13-58, we have the equality

$$B_l l^{S_l} = B_v (\xi_p l_p^3 b_D l_g^{-D} l^D)^{S_v} \tag{13-60}$$

Since the exponent of l should be the same on both sides of the equation, we can equate the slopes of two size distributions as

$$S_l = D S_v \tag{13-61}$$

Rearranging, the fractal dimension can be obtained from the two slopes of the length and volume distributions (Jiang and Logan, 1996) as

$$D = \frac{S_l}{S_v}, \qquad \text{TSM} \tag{13-62}$$

or in terms of the length and solid equivalent diameter (Li and Logan, 1995) as

$$D = \frac{3S_l}{S_d}, \qquad \text{TSM} \tag{13-63}$$

If discrete size distributions are used, the size and volume relationships can be written (Logan and Kilps, 1995) as

$$n(l) = b_l l^{s_l} \tag{13-64}$$

$$n(v) = b_v v^{s_v} \tag{13-65}$$

where s_l and s_v are slopes of the discrete length and solid volume size distributions, $N(l)$ and $N(v)$, obtained from log–log plots based on aggregate length and volume, and a_l and a_v are empirical constants. The fractal dimension can be calculated from these distributions using an approach similar to the one above (Logan and Kilps,

1995) to yield

$$D = \frac{1 + s_l}{1 + s_v}, \qquad \text{TSM} \tag{13-66}$$

Fractal Dimensions Using the Particle Concentration Technique   In many instances the size distributions are not linear, and therefore a single slope cannot be fit to the size distribution over the whole size range of interest. Curved distributions may be a result of non-steady-state conditions or may be produced by particle breakup. When particles are broken into smaller pieces, these fragments are added into the distribution at smaller sizes, and disappear from the size distribution at larger sizes, producing changes in the size distributions dependent on the sizes of the fragmented particles. Even for nonlinear distributions, the fractal dimension can be calculated from size distributions using the PCT, as long as breakup does not alter the fractal nature of the particles.

In order to use the PCT, a population of particles must be characterized separately in terms of both solid equivalent diameter (or solid volume) and size. From Eq. 13-58 we know that $N(<d_s) = N(<l)$ or that when we have a certain number of particles of size d_s or smaller, and we have the same number concentration of particles in terms of a size l, that d_s corresponds to length l. Thus, using the concentrations, we can match a solid diameter with a length. The sizes l and d_s are not equal because d_s is a measure of the size of the particle if all the water in the particle were removed (i.e., if the particle had a porosity of 0), and l is the actual particle size; therefore, $d_s \leq l$. As an example of this procedure, consider the two size distributions shown in Figure 13.8a for a seawater sample taken from East Sound, off the Coast of Washington. At a particle concentration of $N = 30$ mL^{-1}, a particle with a solid diameter of $d_s \approx 54$ μm. In terms of particle length, the same number concentration is observed to have an equivalent length of $l \approx 76$ μm. Therefore, we have matched l and d_s at this particle concentration. If this procedure is repeated for other particle concentrations, we can obtain a data set of d_s versus l. We have already shown that $d_s = b_s l^{D/3}$ (Eq. 13-56). Taking the logarithms of both sides of this equation produces the linear equation

$$\log d_s = \log b_s + \frac{D}{3} \log l, \qquad \text{PCT} \tag{13-67}$$

D can therefore be calculated as the slope in Eq. 13-67 from a linear regression of the matched set of l versus d_s data produced over a wide range of l.

Example 13.3

Using the data in Figure 13.8, calculate the fractal dimension D using the PCT.

In order to calculate D, we must develop a data set of d_s versus l. To do this we compare the highest particle concentration for the length distribution, $N(l)$,

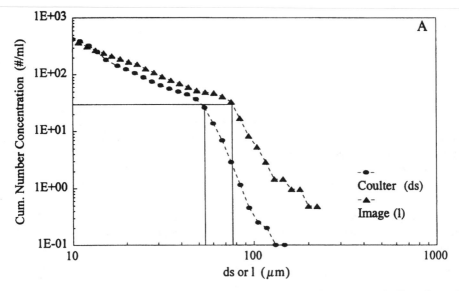

Figure 13.8 Cumulative size distribution of particles in surface waters in East Sound, WA (April 24, 1994) measured in terms of solid equivalent diameter, d_s (using a Coulter counter), and length, l (image analysis of acridine-orange stained particles on a poly-carbonate filter). (Reprinted from Li and Logan, 1997, with kind permission from El-sevier Science Ltd, The Boulevard, Landford Lane, Kidlington, OX5 1GB).

with those from the $N(d_s)$ distribution to identify the most similar concentration. Once we have this first pair of d_s and l, we then examine successive number concentrations, $N(l)$, separated into length scale intervals of $\Delta \log(l) = 0.047$, each time identifying a particle concentration $N(d_s)$ closest to $N(l)$. This produced a series of d_s versus l data such as shown in Figure 13.9. After excluding data from the ends of the size distribution (the least accurate portions), we obtain a particle size range of 15–200 μm in length. From the slope of the line, we calculate for this example that $D = 2.54$.

The preceding example demonstrates that it is possible to calculate fractal dimensions from curved size distributions using the PCT since the slopes of the size distributions are not used in the calculation. Although the size distributions were curved, the resulting plot of d_s versus l was linear. At particle lengths l of 50 to 80 μm, for example, the sudden change in slope in Figure 13.8 did not produce substantial deviations in the slope used to calculate D, as shown in Figure 13.9.

Fractal Dimensions Using the Multiple Range Method   Particle concentrations that occur in natural waters, such as the ocean, can range in sizes by orders of magnitude. Even if submicrometer particles are not included, particles can vary from ~1 μm for bacteria and unidentifiable debris, to tens of centimeters for fecal pellets

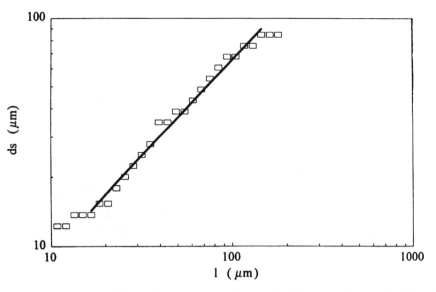

Figure 13.9 (Reprinted from Li and Logan, 1997, with kind permission from Elsevier Science Ltd, The Boulevard, Landford Lane, Kidlington, OX5 1GB).

from animals and marine snow (aggregated cells, fecal pellets, and debris). Because particles can range in size over such a wide range, it is currently impossible to use the same instrument to cover the whole particle size spectrum. Jackson et al. (1995) derived an equation to relate the slopes of size distributions of smaller particles (s_v) obtained from aperture impedance instruments, to slopes ($s_{l,lp}$) of larger particles obtained from *in situ* photographic systems. The derivation of the final equation is rather long, and will not be reproduced here. Assuming that the fractal dimension of all particles was constant over large ranges in particle size, these slopes were related by

$$D = D_2 \frac{3s_{l,lp} + 3}{2s_v + 2}, \qquad \text{MRM} \qquad (13\text{-}68)$$

When $D < 2$, then $D = D_2$ (Vicsek, 1992), and fractal dimensions cannot be determined from the slopes. However, when $D > 2$, then $D_2 = 2$, and D can be obtained from this scaling result.

Fractal Dimensions from Steady-State Size Distributions It will be shown in Chapter 14 that fractal dimensions can also be obtained from a single size distribution for coagulating particles. If the sizes of particle undergoing coagulation reach a steady condition, where the slope of the size distribution is constant, then the fractal dimension can be calculated from either a length distribution (Jiang and Logan, 1991) as

$$D = -2S_l - 3, \quad \text{SSM} \tag{13-69}$$

or from the solid equivalent diameter distribution as

$$D(\text{SS}) = \frac{9}{2S_d + 3}, \quad \text{SSM} \tag{13-70}$$

Comparison of Fractal Dimensions Using Different Size-Distribution Techniques   The use of fractal scaling techniques is a relatively new approach to characterizing particle sizes and shapes, and therefore, there are few studies available to compare these different techniques. Latex microspheres were coagulated in two different devices, a paddle mixer (jar test apparatus), and a horizontal rolling cylinder (a tube completely filled with water; Logan and Kilps, 1995). The application of the TSM indicated that on average the particles generated in the rolling cylinder ($D = 1.59$) were more fractal than those in the paddle mixer ($D = 1.92$; Table 13.8). For both devices, the fractal dimensions calculated assuming the size distributions at steady state were different, indicating that steady conditions were not achieved.

Both the MSM and PCT were used to analyze particles during a phytoplankton bloom in a mesocosm in a laboratory experiment. In most cases, D calculated with the MRM technique for particles 1 μm to 1 cm in length (Jackson et al., 1997) were larger than those calculated using the PCT for a smaller size range (15 to 200 μm) of particles (Fig. 13.10). This comparison suggests that the MRM produces higher fractal dimensions than the PCT. Field experiments conducted using seawater samples from Monterey Bay, CA, also demonstrated that the MRM provides higher

TABLE 13.8   Fractal Dimensions Calculated from Particle Size Distributions Using Different Techniques. The Lack of Agreement between the Two Methods Based upon the Assumption of a Steady-State Size Distribution SSM(l) and SSM(d_s) Indicates that the System Was Not at Steady State

Type	Method	D	Reference
Latex microspheres coagulated in a paddle mixer (cumulative size distributions)	TSM	1.92 ± 0.06	Logan and Kilps (1995)
	SSM(l)	1.31 ± 0.06	
	SSM(d_s)	2.37 ± 0.04	
Latex microspheres coagulated in a horizontal rolling cylinder (cumulative size distributions)	TSM	1.59 ± 0.16	Logan and Kilps (1995)
	SSM(l)	2.31 ± 0.65	
	SSM(d_s)	0.83 ± 0.07	
Particles from Monterey Bay, CA	PCT	1.77 ± 0.34	Li et al. (1997)
	TSM	1.67 ± 0.28	
	SSM(l)	0.66 ± 0.49	
	SSM(d_s)	2.61 ± 0.58	
Particles from Monterey Bay, CA	MRM	$2.24-2.3$	Jackson et al. (1997)

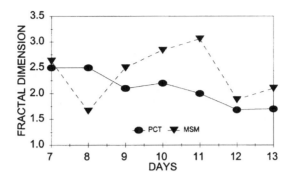

Figure 13.10 Comparison of fractal dimensions calculated in a laboratory mesocosm experiment using the particle concentration technique (PCT) and multiple size range (MSM) methods. (Data from Jackson et al. 1995; Li and Logan, 1995).

estimates of D than the PCT. Jackson et al. (1997) reported $D(\text{MRM}) = 2.24$ to 2.3 versus $D(\text{PCT}) = 1.77$ (Li et al., 1997).

At present, it is not known which of these methods provides a more accurate estimate of fractal dimensions of particles in seawater. Fractal dimensions of <2 have been reported for particles measured individually for D, as opposed to the size distribution approaches, and for fractal dimensions calculated from size–area relationships. Risović and Martinis (1996) analyzed suspended seawater particles larger than ˜0.5 μm using a light scattering technique. They found $D = 1.74 \pm 0.09$ ($n = 34$) for particles from eight sites at sampling depths of 10 to 800 m. At an unidentified coastal site off Southern California they measured $D = 1.8$, a result that compares favorably with the value of $D(\text{PCT}) = 1.77$ found in Monterey Bay. Also, the analysis of photographs of larger marine-snow aggregates, taken in situ by divers, indicated $D_2 = 1.72 \pm 0.07$ ($\pm$SD). Since $D_2 < 2$, it can be concluded that on average these larger particles have fractal dimensions in the same range as those indicated by the PCT for smaller particles. Therefore, it is likely that particles commonly found in seawater, on average, have fractal dimensions less than 2.

PROBLEMS*

13.1 In coastal and oceanic waters and in digested sewage sludge, a commonly observed particle size distribution is:

$$n(d_p) = ad_p^{-4}$$

for particle diameters in the range of 1 to 20 μm. For anaerobically digested sewage sludge, a is approximately 10^{10} μm³ cm⁻³ and $n(d_p)$ has units of # mL⁻¹ μm⁻¹ and d_p is in μm (Faisst, 1980). For the sludge particle size distribution given above: (a) Calculate N_T (# mL⁻¹), the total number concentration from 1 to 20 μm. (b) Calculate A_T (m² L⁻¹), the total surface area of all particles from 1 to 20 μm, assuming spherical particles. (c) Calculate V_T (in ppmv), the total volume of particles from 1 to 20 μm, assuming spherical particles. (d) Calculate and plot the volume distribution, $dV_p/d(\log d_p)$ versus $\log d_p$ for the above size distribution to obtain the total suspended particle volume in the size range 1 to 20 μm by graphically integrating the area under the volume distribution. Express $dV_p/d(\log d_p)$ in units of ppmv and $cm^3 \ cm^{-3}$. (From Hunt, 1997).

13.2   For ideal settling tanks (which do not exist in practice), the particle removal efficiency is given by

$$\eta = 1.0, \qquad U_s(d_p) > u_o$$

$$\eta = \frac{U_s(d_p)}{u_o}, \qquad U_s(d_p) < u_o$$

where $U_s(d_p)$ is the settling velocity of a particle of diameter d_p, $u_o = Q/A_b$ is the overflow rate for the settling basin, Q is the flow rate into the basin, and A_b is the total surface area of the basin. Assume that the settling velocity is given by Stokes' Law, or $v_p = g \ \Delta\rho \ d_p^2/(18 \ \upsilon\rho)$. For the size distribution of $n(d_p) = 10^7 d_p^{-4}$ in the size range of $1 \le d_p \le 60$ μm, $u_o = 28.2 \ m^3 \ d^{-1} \ m^{-2}$, and $\rho_p = 2.5 \ g \ cm^{-3}$ (typical of clay), calculate the following: (a) For the ideal settling basin calculate d_p for which $u_o = U_s$. (b) Calculate the effluent particle size distribution and plot the influent and effluent volume distributions, V_{dist}. Assume for water a kinematic viscosity of $0.01 \ cm^2 \ s^{-1}$ and density of 1.0 g cm^{-3}. (c) Calculate the overall removal efficiency of the tank based on total particle number and suspended mass in the range of 1 to 60 μm. (From Hunt, 1997.)

13.3   Oil-in-water emulsions are both pollutants from refinery operations and more frequently encountered as food products such as salad dressings. Oil (10 mL) was vigorously mixed with water (990 mL) producing a 1-L suspension with the size distribution:

$$n(d_p) = a d_p^{-4}, \qquad 1.0 \le d_p \le 100 \ \mu m$$

(a) Determine the constant a. (b) If the interfacial surface tension (energy due to droplet surface area) between the oil and water is 50 erg cm^{-2}, calculate the total interfacial energy in 1 L of solution. (From Hunt, 1997).

13.4   Sizes and porosities of particles from lake surface waters are shown in Table P-13.4. (a) Using these data, calculate the average fractal dimension of these particles. (b) These average particle sizes are much larger than the solid equivalent diameters reported by resistance-type particle counters. Calculate d_s for a particle with an average length of 1000 μm.

13.5   A cumulative size distribution obtained using a resistance-type particle counter can be expressed as a function of solid equivalent diameter as $N(d_s) =$

TABLE P-13.4

Length	Porosity
108	0.524
194	0.829
288	0.874
376	0.909
491	0.903
624	0.936
704	0.954
831	0.937

$B_d d_s^{S_d}$. Based on the relationships between porosity and average particle size, such as $(1 - \theta_{ag}) = b l_{ag}^{D-3}$, show that the cumulative volume size density function can be expressed in terms of particle length in a discrete size distribution as:

$$n(l) = \frac{B_d S_d D}{3} b^{S_d/3} l^{(S_d D - 3)/3}$$

13.6 Particles in seawater are fractal, and their porosity is the function of the particle size, or $\theta_{ag} = 1 - b l^{3-D}$. It is assumed that all particles are in the size range 0.5 to 10^4 μm, particle size density is $n(l) = 10^6 l^{-3}$, where $l[\mu m]$ and $n(l)[\mu m^{-1}\ mL^{-1})$, the suspended solids (SS) value of the water sample is 4 mg L^{-1}, and the fractal dimension is $D = 2$. If the average density of solid matter within particles is 1.2 kg L^{-1}, estimate the coefficient b.

13.7 Flocs produced in an activated sludge process may have different settling velocity characteristics than aggregates formed from latex microspheres due to the production of exopolymeric material that can clog internal pores. However, it is well established that the settling velocity does not follow the expected relationship of $U_s \propto d_{ag}^2$ expected based on Stokes' Law. (a) Using the settling velocity data shown in Table P-13.7, calculate the fractal dimension of these flocs assuming $U_s \propto l_{ag}^{D-1}$. (b) Compare the settling velocity of a fractal activated sludge floc 500 μm in size ($\theta_{ag} = 0.88$) with that of a solid floc ($\theta_{ag} = 0$) of the same size and density assuming Stokes' Law.

13.8 A coagulation test is conducted in a jar-test apparatus, resulting in the particle concentrations shown in Table P-13.8. Assume that mass is conserved during the experiment, and that the particle and water are densities $\rho_s = 1.6$ g cm^{-3} and $\rho_w = 1$ g cm^{-3}. (a) Estimate the fractal dimension of the flocs formed. (b) If the total mass concentration in the beaker is 10 mg L^{-1}, and all flocs have an average diameter of 100 μm, what is the average bulk density of the flocs?

13.9 The method used to characterize the size of a particle can affect the magnitude of the fractal dimension when properties of the aggregates scale at different

TABLE P-13.7

Length (μm)	Settling Velocity (cm min^{-1})
152	0.738
208	0.840
267	1.518
345	1.620
414	2.082
536	2.256
635	3.078
702	2.748
826	3.138
911	4.392

TABLE P-13.8

Time (min)	Length (μm)	N_T (mL^{-1})
1	2.1	299,157
2	4.2	25,667
3	8.9	15,633
4	12.5	32,170
5	23.9	25,492
10	51.4	241
15	94.6	104

rates. For example, if particles tend to become more elongated as they get larger, then the maximum length (l_m) of the particle may increase faster than the equivalent length (l_A), based on area, as aggregates become larger. If the maximum and equivalent lengths are related to each other according to $l_m = 1.25\ l_A^{1.10}$, and the fractal dimension based on the maximum particle length is 1.75, what is fractal dimension based on the equivalent length?

13.10 In some instances particle size distributions are very narrow, and we are more interested in an average size of a particle than the shape of the size distribution. Let us assume that a particle size distribution is measured using a resistance-type particle counter, with the results expressed in the form $n(d_s) = b_s d_s^{S_d}$. Using the expression derived for the porosity as a function of particle size ($\theta_{ag} = 1 - b_\theta l^{3-D}$), derive an expression for the mean size (length) of particles in the size range between l_I and l_{II}.

13.11 Particles in a water sample are measured using both an image analysis and a resistance-type particle counter, producing a two size distributions in terms of solid equivalent diameter, $N(d_p) = 4.0 \times 10^4 (\ln d_p)^{-5.5}$, and aggregate equivalent length (based on area) of $N(l) = 8 \times 10^4 (\ln l)^{-5.1}$. Using the PCT approach, calculate the fractal dimension of particles 3 to 500 μm in length.

*Problems 13.4–13.11 courtesy of X. Li.

REFERENCES

Adler, P. M. 1981. *J. Colloid Interface Sci.* **81**:531–35.

Avnir, D. 1989. *The Fractal Approach to Heterogeneous Chemistry*. Wiley, New York, 441 pp.

Chellam, S., and M. R. Wiesner. 1994. *Water Res.* **27**:1493–96.

Davis, R. H., and T. P. Hunt. 1986. *Biotechnol. Prog.* **2**:91.

Faisst, W. K. 1980. *Adv. Chem. Ser.* **189**:259–82.

Guo, L., and P. H. Santschi. 1996. *Mar. Chem.* **55**:113–27.

Hunt, J. R. 1997. Lecture notes for water treatment engineering, Civil and Environmental Engineering 211, University of California, Berkeley.

Jackson, G. A., B. E. Logan, A. L. Alldredge, and H. Dam. 1995. *Deep-Sea Res. II* **42**(1): 139–57.

Jackson, G. A., R. Maffione, D. K. Costello, A. L. Alldredge, B. E. Logan, and H. G. Dam. 1997. *Deep-Sea Res. I* **44**(11):1739–67.

Jiang, Q., and B. E. Logan. 1991. *Environ. Sci. Technol.* **25**(12):2031–38.

Jiang, Q., and B. E. Logan. 1996. *J. AWWA* **88**(2):100–13.

Johnson, C. P., X. Li, and B. E. Logan. 1996. *Environ. Sci. Technol.* **30**(6):1911–19.

Kilps, J. R., B. E. Logan, and A. L. Alldredge. 1994. *Deep-Sea Res.* **41**(8):1159–69.

Law, B. A. 1980. In *Microorganisms and Nitrogen Sources*. Payne, J. W., Ed. John Wiley and Sons. pp. 381–409.

Li, D.-H., and J. Ganczarczyk. 1989. *Environ. Sci. Technol.* **23**(11):1385–89.

Li, X., and B. E. Logan. 1995. *Deep-Sea Res. II* **42**(1):125–38.

Li, X., and B. E. Logan. 1997a. *Environ. Sci. Technol.* **31**(4):1229–36.

Li, X., and B. E. Logan. 1997b. *Environ. Sci. Technol.* **31**(4):1237–42.

Li, X., U. Passow, and B. E. Logan. 1998. *Deep-Sea Res. I* **45**(1):115–31.

Lin, M. Y., H. M. Lindsay, D. A. Weitz, R. C. Ball, R. Klein, and P. Meakin. 1989. *Nature* **339**:360–62.

Logan, B. E., and J. R. Kilps. 1995. *Water Res.* **29**(2):443–53.

Logan, B. E., and D. B. Wilkinson. 1989. *Limnol. Oceanogr.* **35**(1):130–36.

Logan, B. E., and D. B. Wilkinson. 1991. *Biotechnol. Bioengin.* **38**(4):389–96.

Logan, B. E., H.-P. Grossart, and M. Simon. 1994. *J. Plankton Res.* **16**(12):1811–15.

Mandelbrot, B. B. 1982. *The Fractal Geometry of Nature*. W. H. Freeman, New York.

Matsumoto, K., and A. Suganuma. 1977. *Chem. Eng. Sci.* **32**:445–47.

Meakin, P. 1988. *Adv. Colloid Int. Sci.* **28**:249–331.

Rickert, D. A., and J. V. Hunter. 1967. *J. Water Pollut. Control Fed.* **39**:1475–86.

Riscović, D., and M. Martinis. 1996. *J. Colloid Inter. Sci.* **182**:199–203.

Schaefer, D. W. 1989. *Science* **243**:1023–27.

Tambo, N., and Y. Watanabe. 1979. *Water Res.* **13**:409–19.

Vicsek, T. 1992. *Fractal Growth Phenomena*, 2nd ed. World Scientific. New Jersey, 488 pp.

White, F. M. 1974. *Viscous Fluid Flow*. McGraw-Hill, New York.

Witten, T. A., and M. E. Cates. 1986. *Science* **232**:1607–12.

CHAPTER 14

COAGULATION IN NATURAL AND ENGINEERED SYSTEMS

14.1 INTRODUCTION

The coagulation of small particles into larger, faster-settling aggregates is critical to the performance of most water and wastewater treatment systems. The goal in water treatment, for example, is to remove dissolved and particulate matter from the water source prior to final disinfection and distribution. Gravitational sedimentation is a very inexpensive method of removing particles from water, but very small, micrometer-sized particles do not settle fast enough to be removed in clarifiers. While larger particles can be formed through coagulation of smaller particles, many particles in natural waters persist because they are stable, and therefore infrequently stick to each other when they collide. In water treatment plants, particles are made sticky by the addition of chemicals, and then collision frequencies are increased by mixing the water with paddle mixers. In biological wastewater treatment plants such as activated sludge, only those bacteria that attach to form flocs are retained in the clarifer, and therefore the whole treatment system. Nonattaching cells are washed out of the system. Thus, while no chemicals are added to a biological wastewater treatment plant, the system selects for highly adhesive strains of bacteria.

The process of particle coagulation consists of two steps: particles moving by diffusion (Brownian motion), shear, or sedimentation strike other particles; particles either stick together, forming a larger particle, or the collision is unsuccessful and they separate (Fig. 14.1). The rate particles collide is roughly a second-order rate process and so collision rates increase in proportion to particle concentration squared. Particle collision rates can be modeled with some success, but the probability of attachment, or the particle stickiness, must be experimentally determined. Most coagulation models are based on the assumption of "hard spheres," or the assumption that all particles are nondeforming spheres. Upon collision, the volume of two col-

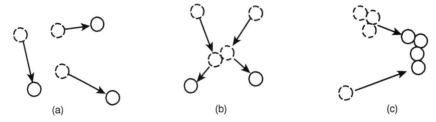

Figure 14.1   The motion of particles suspended in a fluid can (A) lead to no particle collisions, (B) produce ineffective particle collisions (particles collide, but do not stick), or (C) result in collision and attachment, producing fewer particles (two particles form one particle). Notice in all cases that only collisions between two particles is considered. The dashed circles indicate the former position of the particles drawn with solid lines.

liding particles is conserved, but a new particle is formed that is also a perfect, hard sphere. While this assumption seems rather unrealistic, it has led to models that describe the initial coagulation rates of particles under certain conditions. Fractal geometry has only recently been applied to calculations of particle coagulation rates. It is likely that future advances in coagulation models will increasingly rely upon the description of particle geometry using fractal dimensions.

14.2   THE GENERAL COAGULATION EQUATIONS: INTEGRAL AND SUMMATION FORMS

A general transport equation describing the change in particle concentration can be developed in the usual manner from a mass balance of particles in a control volume. Except under certain circumstances, we cannot simply designate the total concentration of particles with a single variable such as c because the suspension is likely undergoing coagulation, and therefore both particle sizes and concentrations will vary over time. In some instances it is possible to describe the particle number concentration of all the different-sized particles in terms of a single size distribution function, n. Because this function describes the concentrations of all types of particles in a measured size interval (typically #/mL μm), the subscripts ag (for aggregate) and p (for particle) are not necessary. As described in Chapter 13, the size distribution function can be expressed in terms of cumulative or discrete size distributions, with particle sizes measured in terms of volume or diameter. Assuming a particle concentration based on solid volume, the number concentration of particles in a size interval dv is related to $n(v)$ by $dN_p = n(v)\, dv$.

   The change in the concentration of suspended particles of volume v in the water is a function of the rate of particle production by coagulation of particles smaller than v, particle losses due to sinking and breakup, and coagulation into particles larger than v. Neglecting particle losses due to sinking and breakup, this can be expressed mathematically (Friedlander, 1977; Hunt, 1982) as

$$\frac{\partial n(v)}{\partial t} = \frac{1}{2} \int_0^v \alpha\beta(v', \, v - v')n(v')n(v - v') \, dv' - \int_0^\infty \alpha\beta(v, \, v')n(v)n(v') \, dv' \quad (14\text{-}1)$$

where β is the collision function describing the rate that particles are brought into contact by Brownian, shear, or differential sedimentation, α the sticking coefficient defined as fraction of collisions that result in particle attachment, and v' is the size of a particle smaller than v that upon collision with a particle of size $v - v'$ forms a particle of size v. Equation 14-1 cannot be solved analytically, and two approaches have been used to provide approximate solutions. These include using direct numerical solutions (which have not been experimentally verified) and asymptotic solutions for later times (Hunt, 1982). Partial solutions to Eq. 14-1 have been obtained using self-preserving transformations as described below (Hunt, 1982; Jiang and Logan, 1991), although these are only applicable to steady-state processes.

Advances in computer processing speeds have made it possible to follow particle concentrations during coagulation. Assuming an initial size distribution, or even a monodisperse suspension, a size distribution of particles will evolve over time due to particle coagulation. The general coagulation equation in summation form is

$$\frac{dn_h}{dt} = \frac{1}{2} \sum_{i+j=h} \alpha\beta(v_i, \, v_j)n_in_j - \sum_{i=1}^\infty \alpha\beta(v_i, \, v_h)n_in_h \quad (14\text{-}2)$$

where the subscripts are added to indicate different size particle classes. Notice that because sums are used instead of integrals, n does not have to be a continuous function; n must only describe the concentration of particles in a certain size interval. From Eq. 14-2 we can see that the change in the concentration of particles of size h is a function of the rate particles form smaller particles of size i and j, and the rate particles of size h disappear into other size classes through collisions with particles of all other sizes. The $\frac{1}{2}$ term is included because the formation of one particle of size h from two particles reduces the number of particles by half.

While either of the above equations can be used to simulate coagulation processes, the accuracy of these simulations is based primarily on defining the sticking coefficient, α, and the collision frequency function, β, for particles as a function of particle sizes and solution and particle chemistries. Because sticking coefficients cannot reliably be predicted, they are experimentally measured. Thus, any inaccuracies in β are hidden in the only adjustable parameter, α. The factors that contribute determine particle stickiness, and therefore α, and the equations used to calculate β are presented below.

14.3 FACTORS AFFECTING THE STABILITY OF AQUASOLS

The attraction or repulsion of two colloidal particles is known to be governed by electrical, van der Waals, Born, hydration, and hydrophobic forces. Whether particles attach to each other is dependent on the magnitude of these forces, the presence of surface polymers (steric effects), and the roughness of the particle's surface. While a detailed description of these factors contributing to particle attachment is beyond the scope of this book, the general nature and origin of these forces is reviewed below.

Electrically Charged Surfaces and Particles

The Gouy–Chapman Double-Layer Model   Almost all particles in water carry a net charge. Since the charge on the surface of a particle must remain separate from water, this separation of charge generates an electrical potential between the water and the particle. Most particles (or any surface for that matter) in water carry net negative charges, and this charge must be balanced by an equal number of counter-ions (in this case positive ions) in the water. Before we address the case of suspended particles in water, consider the case of a flat, completely uniform planar surface of infinite area. The ions in solution are point charges (have no physical size), and the solvent has properties that are independent of distance from the plate. The concentration of these negative charges on the surface produce a surface potential of Ψ_0, which can be calculated using the Nernst equation as

$$\Psi_0 = \left(\frac{\mathcal{R}T}{z_i\mathcal{F}}\right) \ln a_i + b_1 \tag{14-3}$$

where $\mathcal{R}$ is the gas constant, T the absolute temperature, $\mathcal{F}$ Faraday's constant, and z_i and a_i the valence and activity of the counterions. At 25°C, and for ideal (dilute) solutions where the activity can be replaced by the concentration of ions, changes in the surface potential are related to the solution concentration of ions by

$$\frac{d\Psi_0}{d(\log c_i)} = \frac{2.303\mathcal{R}T}{z_i\mathcal{F}} = \frac{59.2 \text{ mV}}{z_i} \tag{14-4}$$

The cloud of counterions surrounding a charged particle decreases in concentration exponentially with distance from the particle's surface (Fig. 14.2). The charge density in the solution, ρ_c[C m^{-3}] at any point is related to the potential, Ψ, by the Poisson equation

$$\nabla^2\Psi = \frac{\rho_c}{\varepsilon} \tag{14-5}$$

where ε is the permittivity of the solution. For water, $\varepsilon = 7.083 \times 10^{-12}$ C^2 cm^{-1} J^{-1}. The potential in an electrolyte solution, such as water, also depends on the local

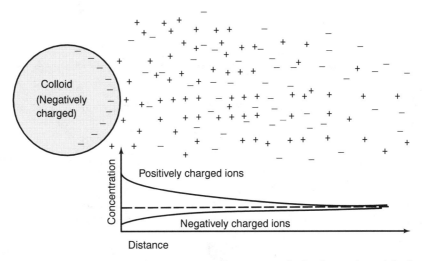

Figure 14.2 Concentration of ions surrounding a negatively charged particle (in one dimension). The positively charged ions are higher in concentration near the negatively charged particle surface, and decrease exponentially in concentration with distance from the particle surface. Very far from the charged surface the negative and positive ions balance in concentration.

concentration of anions and cations, as indicated by the Boltzmann equation

$$\frac{N_i}{N_{i\infty}} = \exp\left(-\frac{z_i e \Psi}{k_B T}\right) \tag{14-6}$$

where $N_{i\infty}$ is the number concentration of i ions far away from the particle surface (at infinity, or in the bulk solution), N_i the ion concentration at a location where the potential is Ψ, $e = 1.60219 \times 10^{-19}$ C, the electrical charge, and $k_B = 1.38066 \times 10^{-23}$ J K^{-1} (Boltzmann's constant). The exponential term is a ratio of the electrical energy of a particle, $z_i e$, to the thermal energy, $k_B T$. The ion concentration and charge density are related by

$$\rho_c = \sum_i z_i e N_i \tag{14-7}$$

Combining Eqs. 14-5–14-7, we obtain in one dimension

$$\frac{\nabla^2 \Psi}{dx^2} = \sum_i \frac{z_i e N_{i\infty}}{\varepsilon} \exp\left(-\frac{z_i e \Psi}{k_B T}\right) \tag{14-8}$$

From a Taylor series expansion of the exponential term, and for the case of low potentials where $|z_i e \Psi/(k_B T)| \ll 1$ (for a 1–1 electrolyte, this means Ψ_0 is less than

~25 mV), we can write Eq. 14-8 as

$$\frac{\nabla^2\Psi}{dx^2} = \sum_i \left(\frac{z_i e N_{i\infty}}{\varepsilon} - \frac{z_i^2 e^2 N_\infty \Psi}{\varepsilon k_B T}\right) \tag{14-9}$$

Because the solution must remain electrically neutral, $z_i e N_{i\infty}$ must be zero, and this result can be further simplified to

$$\frac{\nabla^2\Psi}{dx^2} = \frac{e^2 \Psi}{\varepsilon k_B T} \sum_i z_i^2 N_{i\infty} \tag{14-10}$$

This equation can be easily solved with the boundary conditions of $\Psi = \Psi_0$ at $x = 0$, and $d\Psi/dx = 0$ as $x \to \infty$, producing the expected exponential decay expression

$$\Psi = \Psi_0 e^{-hx} \tag{14-11}$$

where h is the Debye–Hückel parameter, defined as

$$h = \left(\frac{e^2 \sum_i z_i^2 N_{i\infty}}{\varepsilon k_B T}\right)^{1/2} \tag{14-12}$$

The surface charge density of the particle, σ_0, must balance the charge in the diffuse layer. Therefore, σ_0 can be obtained by integrating the charge density in the solution over the distance of the diffuse layer. For symmetrical electrolytes, and for small potentials, the result is

$$\sigma_0 = \varepsilon h \Psi_0 \tag{14-13}$$

The Debye–Hückel parameter has units of $[L^{-1}]$ and its inverse, h^{-1}, is called the double-layer thickness. This thickness is often used as a scaling parameter to indicate a characteristic size of the cloud of counterions surrounding a surface or particle in an electrolyte such as water. Because the potential decays exponentially from the surface, the double-layer thickness is the distance where the potential has decayed only by 37%. Other decay lengths can be calculated from Eq. 14-11 as

$$L_{0.5} = \frac{\ln 2}{h} \quad \text{or} \quad L_{0.1} = \frac{\ln 10}{h} \tag{14-14}$$

where $L_{0.5}$ and $L_{0.1}$ are the distances the potential has decreased by half, and by an order of magnitude.

The Stern Layer   Ions in solution have a finite size, and are not point charges as assumed above. Because the ion cannot reach a charged surface, they must remain separated from the planar surface at a distance δ_s, referred to as the Stern layer thickness. In water, the permittivity in the Stern layer may be lower than that of the bulk water (Fig. 14.3). According to the Stern–Grahame model, the Stern layer can be further subdivided into two layers, the inner Hemholtz plane occupied by adsorbed, unhydrated ions, and an outer Helmholtz lange where hydrated counterions are located. The diffuse layer begins only beyond the Stern plane.

The main effect of the Stern layer is to reduce the effective surface charge due to the capacitance of the Stern layer. Assuming a Stern layer thickness of $\delta_s = 0.3$ nm, a dielectric constant for water of ten times that of air, the capacitance of the Stern layer is $C_\delta = 0.5$ F m^{-2}. The potential drop across the Stern layer, from Ψ_0 to Ψ_δ, is estimated at 25°C (Elimelech et al., 1995) as

$$\Psi_0 - \Psi_\delta = \frac{0.0117c_{z-z}^{1/2}}{C_\delta} \sinh(19.4z\Psi_\delta) \tag{14-15}$$

where c_{z-z} [mol L^{-1}] is the concentration of the z–z electrolyte.

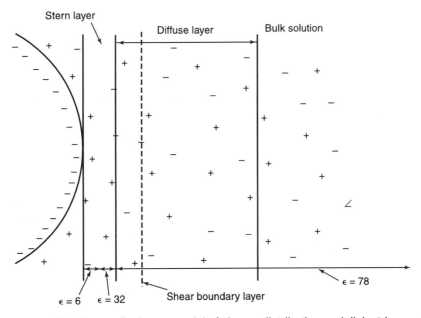

Figure 14.3 The Stern–Grahame model of charge distribution and dielectric constant of the water surrounding a particle. The Stern layer consists of inner and outer Helmholtz layers of reduced dielectric constants (values shown are 10^{-14} C^2 cm^{-1} J^{-1}) with relatively immobile hydrated and positively charged ions. At some point past the Stern layer, but inside the diffuse layer, charges become mobile (shear boundary). The decay of the diffuse layer is characterized by the inverse of the Debye parameter h^{-1}, where the length h^{-1} is referred to as the double layer thickness (adapted from Elimelech et al., 1995).

Charged Spherical Particles   The discussion so far has considered only planar surfaces, but we are more interested here in spherical particles. In spherical coordinates, the Poisson–Boltzmann equation becomes

$$\frac{1}{r^2}\frac{d}{dr}\left(r^2\frac{d\Psi}{dr}\right) = \sum_i \frac{z_i e N_{i\infty}}{\varepsilon}\exp\left(-\frac{z_i e\Psi}{k_B T}\right)$$ (14-16)

For low potentials, this reduces to

$$\frac{\Psi}{\Psi_0} = \frac{R}{r}e^{-h(r-R)}$$ (14-17)

where R is the radius of the sphere having a surface potential Ψ_0. Alternately, Ψ_0 can be replaced by the Stern layer potential Ψ_δ. The charge on the sphere is

$$Q = 4\pi\varepsilon\Psi_0 R(1 + hR)$$ (14-18)

and the surface charge density is

$$\sigma_0 = \Psi_0\varepsilon\frac{(1 + hR)}{R}$$ (14-19)

When the diffuse layer is thin compared to the particle radius, or $hR \gg 1$, then the equations for spherical particles reduce to those for flat plates.

Example 14.1

Calculate the characteristic sizes of the diffuse layer around particles 1 μm in diameter at 20°C in a synthetic groundwater of 120 mg/L solution of NaCl.

The solution concentration must first be converted to a molar basis. NaCl has a molecular weight of 58.35 mg/mmol, so that

$$c = \left(120\ \frac{mg}{L}\right)\frac{mmol}{58.35\ mg} = 2.06\ mM$$ (14-20)

For this simple 1–1 electrolyte solution, the ion concentration of Na^+ and Cl^- are both 2.06 mM. Using Eq. 14-12, the Debye-Hückel parameter h is

$$h^{-1} = \left(\frac{e^2\sum_i z_i^2 N_{i\infty}}{\varepsilon k_B T}\right)^{-1/2} = \left(\frac{e^2[(-1)^2(2.06\ mM) + (+1)^2(2.0\ mM)]}{\varepsilon k_B T}\right)^{-1/2} = \left(\frac{e^2[4.12\ mM]}{\varepsilon k_B T}\right)^{1/2}$$ (14-21)

and using Avagadro's number to convert from moles to ion concentrations.

$$= \left(\frac{(1.60 \times 10^{-19}\text{C})^2(4.12 \text{ mmol/L})(6.02 \times 10^{23}/\text{mol})(10^{-3} \text{ mol/mmol}) \, (10^{-3} \text{ L/cm}^3)}{(7.08 \times 10^{-12} \text{ C}^2/\text{J cm})(1.38 \times 10^{-23} \text{ J/K})(298 \text{ K})} \right)^{-1/2}$$

(14-22)

$$h^{-1} = 6.8 \times 10^{-7} \text{ cm} \times 10^7 \frac{\text{nm}}{\text{cm}} = 6.8 \text{ nm} \qquad (14\text{-}23)$$

The corresponding 50% and 1-log reduction thicknesses are:

$$L_{0.5} = 4.7 \text{ nm}$$
$$L_{0.1} = 15.7 \text{ nm}$$

(14-24)

Electrophoretic Mobility   A common method of evaluating the electrical charge of particles suspended in a fluid is to use electrokinetic techniques to measure the zeta potential of the particle. This is accomplished by measuring a particle's velocity per unit field strength, known as the electrophoretic mobility, relative to that of the fluid. A charged particle will respond to an electrical field based on the potential at the shear plane, a potential that is lower than the particle's surface potential. Thus the ions in the Stern layer appear fixed, while those in the diffuse layer are mobile. For particles that are large compared to the double-layer thickness ($hR \gg 1$), the relative velocity of a particle in an electric field can be calculated using the Smoluchowski equation

$$U_{p,e} = \frac{\varepsilon}{\mu_w} \zeta \qquad (14\text{-}25)$$

where $U_{p,e}$ is the electrophoretic mobility of the particle, ε is the solution permittivity, μ_w the fluid dynamic viscosity (water), and ζ the particle's zeta potential. For a particle in water at 25°C, this produces the relationship (in commonly used units) of $\zeta(\text{mV}) = 12.8 U_{p,e}$ [μm V s^{-1} cm^{-1}]. For very small particles relative to the double-layer thickness ($hR \ll 1$), typical of macromolecules such as proteins, the electrophoretic mobility is related to the zeta potential using the Hückel equation, or

$$U_{p,e} = \frac{2\varepsilon}{3\mu_w} \zeta \qquad (14\text{-}26)$$

which differs from the Smoluchowski result by only a factor of 2/3.

Particle Stability: Interactions between (Spherical) Charged Particles

Interactions between charged particles in water are modeled using DLVO theory, an approach named for the original investigations by Derjaguin and Landau (1941) and Verwey and Overbeek (1948). This approach originally included only electrical double-layer repulsive and van der Waals attractive forces, although Born repulsive forces are also frequently included. At short range, other interactions must be included to describe the motion of particles at short range. These so-called non-DLVO interactions include hydrophobic forces, the effect of hydrated ions on a particle's surface, and the presence of large macromolecules and polymers leading to particle attachment via polymer bridging at distances, relative to Debye layer thicknesses, that are large.

Double-Layer Interactions: DLVO Theory   Double-layer interactions can be quite complicated, and there are many models for calculating interaction forces. For a more detailed description of these calculations the reader should consult texts such as Elimelech et al. (1995). In calculations here, it will be assumed that particle interactions can be modeled by linear superposition of forces (Elimelech et al., 1995) as

$$E_T = E_R + E_A \tag{14-27}$$

where E_T is the total potential energy, E_R electrostatic double-layer repulsive forces, and E_A London–van der Waals attractive forces.

The double-layer potential can be calculated for interactions between spheres by one-dimensional integration of the Poisson–Boltzmann equations for plate–plate interactions for particles with constant surface potentials (Hogg et al., 1966) using

$$E_R = \frac{\pi \varepsilon R_1 R_2}{R_1 + R_2} \left[2\Psi_{01}\Psi_{02} \ln \left(\frac{1 + e^{-hs}}{1 - e^{-hs}} \right) + (\Psi_{01}^2 + \Psi_{02}^2) \ln(1 - e^{-2hs}) \right] \tag{14-28}$$

where R_1 and R_2 are the radii of the two particles having surface potentials Ψ_{01} and Ψ_{02}, where the two spheres are separated by a distance s. For interactions between identical spheres with small zeta potentials, the total electrical force between the particles of radius R is

$$E_R = 2\pi \varepsilon R \zeta^2 e^{-hs} \tag{14-29}$$

Equation 14-29 shows the effect of the particle zeta potential on electrostatic interactions, and demonstrates that electrical repulsion is a linear function of particle size R. For a 1-μm-diameter particle ($R = 0.5$ μm), $\zeta = 25$ mV, in a 0.1 M 1–1 electrolyte solution at 25°C, the electrical potential interaction energy is ˜250 $k_B T$ at a separation distance of 1 μm, and decays to <20$k_B T$ at 10 μm (Elimelech et al., 1995).

The attractive force between two colloidal particles at short distances is called the London–van der Waals force (Hamaker, 1937). This force is a function of the system geometry and the properties of the particle and suspending medium. The van der Waals attractive force, derived from the Lennard–Jones m–n interaction potential, has an s^{-6} dependence on the separation distance and for two spheres (Hamaker,

1937; Ryan and Gschwend, 1994) is

$$E_A = -\frac{\mathcal{A}}{12}\left[\frac{r^*}{s^{*2} + s^*r^* + s^*} + \frac{r^*}{s^{*2} + s^*r^* + s^* + r^*}\right.$$
$$\left. + 2\ln\left(\frac{s^{*2} + s^*r^* + s^*}{s^{*2} + s^*r^* + s^* + r^*}\right)\right] \tag{14-30}$$

where $r^* = R_1/R_2$, $s^* = s/2R_1$, and $\mathcal{A}$ is the Hamaker constant. For two identical spheres, this attractive force can be more simply calculated as

$$E_A = -\frac{\mathcal{A}R}{12s} \tag{14-31}$$

Equation 14-31 is inaccurate for $s > (0.1R)$. For particles of similar composition with a Hamaker constant $\mathcal{A}_p$ in water (Hamaker constant $\mathcal{A}_w$) the overall Hamaker constant can be approximated as

$$\mathcal{A} = (\mathcal{A}_p^{1/2} + \mathcal{A}_w^{1/2})^2 \tag{14-32}$$

Typical values of the Hamaker constant for aqueous dispersions are in the range of $0.3-10 \times 10^{-20}$ J. In general, mineral surfaces have large Hamaker constants, while low-density materials, such as biological surfaces, have values in the lower part of this range (Elimelech et al., 1995). Mica surfaces, for example, have an experimentally measured $\mathcal{A} = 2.2 \times 10^{-20}$ J (Israelachvili and Adams, 1978), while for quartz $\mathcal{A} = 0.8 \times 10^{-20}$ J (Ryan and Gschwend, 1994).

Dispersion forces are electromagnetic in character, and the finite time of propagation of this energy reduces interaction forces between particles. This reduction, or retardation, of the attractive force is important when particles are separated at short distances. The retardation effect can be included in the expression for the attractive force by introducing the factor $(1 + 14s/\lambda)^{-1}$ into Eq. 14-31, to produce

$$E_A = -\frac{\mathcal{A}R}{12s(1 + 14s/\lambda)} \tag{14-33}$$

where $\lambda = 2\pi v_L/\omega_d$ is the characteristic wavelength of the interaction, v_L is the velocity of light, and ω_d is the dispersion frequency, which typically has a value of 100 nm for most materials (Gregory, 1981). For particles 1 μm in radius, the attractive force is reduced by half at a distance of 10 nm (Elimelech et al., 1995).

DLVO theory can be used to describe colloid interactions and serves as a basis for understanding particle coagulation. Particles that readily stick to one another are referred to as *destabilized*, while a suspension that does not coagulate is defined as *stable*. These two cases only describe the limits of particle–particle interactions. Each time two particles approach each other there is some probability of attachment, quantified in coagulation theory by the sticking coefficient, α. For perfectly stable solutions, $\alpha = 0$, and for completely destabilized particles, $\alpha = 1$. Neglecting retardation forces, the total potential energy between two particles separated by a distance s is assumed to be additive, and can therefore be calculated as the sum of electrostatic and repulsive forces using Eqs. 14-29 and 14-31, or 14-28 and 14-30 in Eq. 14-27. For the former case, this produces

$$E_T = 2\pi\varepsilon R\zeta^2 e^{-hs} + \frac{\mathcal{A}R}{12s} \tag{14-34}$$

The net force between two particles of radius R is therefore either repulsive or attractive depending on the solution ionic strength, the particle's charge (or electrophoretic mobility), the separation distance s, and the Hamaker constant.

Interaction energies between two particles, each of radius 1 μm, is shown in Figure 14.4 for three different solutions strengths assuming the given chemical parameters. In all cases there is a barrier to attachment at separation distances of ˜1 to 2 nm. The net attractive force at smaller separation distances is termed the *primary minimum*. Particles that overcome the repulsive force and approach sufficiently close to experience this net attraction are assumed to be irreversibly trapped in the primary minimum. At low ionic strengths there is always a large repulsive force between two particles that must be overcome for irreversible attachment in the primary minimum, but there is also a smaller, but attractive, force at larger distances that may lead to attachment. For the conditions used to construct Figure 14.4, this secondary region extends from ˜2 to 10 nm. The point of highest net attractive

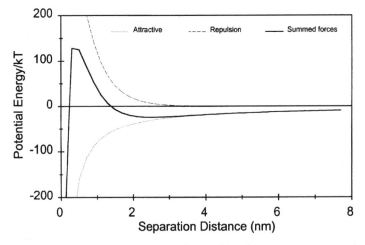

Figure 14.4 Typical interaction forces between two spherical particles each 1 μm in diameter (h^{-1} = 1.94 nm). The primary minimum is at <0.2 nm, but a secondary region of attraction exists at distances >1.4 nm with a secondary minimum at 2.5 nm. (Ionic strength, c = 0.1 M; water permittivity, ε = 7.1 × 10^{-12} C^2 J^{-1} cm^{-1}; 298 K; surface potential, Ψ_0 = −40 mV; Hamaker constant, $\mathcal{A}$ = 8 × 10^{-21} J).

energy (most negative potential energy), referred to as the secondary minimum, occurs in this case at 2.5 nm. Particles trapped within the secondary minimum are only weakly held by van der Waals forces and are therefore considered to be reversibly attached.

Born Repulsive Forces   Very short-range repulsion arises between atoms as they approach each other. This is thought to result from the interaction of their electron shells, but the origins of short-range repulsive forces are not well understood. In water systems most particles will have hydrated ions on their surface and may prevent closer separation distances than 0.3 nm. Born repulsive forces are very important in understanding particle detachment, particularly from surfaces, because inclusion of Born repulsion limits the attractive force due to London–van der Waals interactions.

Born repulsion can be included in the total potential energy between two particles as $E_T = E_E + E_A + E_B$. The Born potential for interactions between spheres has an s^{-12} dependence on distance (Feke et al., 1984; Ryan and Gschwend, 1994) according to

$$
\begin{aligned}
E_B = \frac{\mathscr{A}}{75600s^*} &\left(\frac{\sigma}{R_2}\right)^6 \left(\frac{-4s^{*2} - 14s^*(r^* - 1) - 6(r^{*2} - 7r^* + 1)}{(2s^* - 1 + r^*)^7}\right. \\
&+ \frac{-4s^{*2} + 14s^*(r^* - 1) - 6(r^{*2} - 7r^* + 1)}{(2s^* + 1 - r^*)^7} \\
&+ \frac{4s^{*2} + 14s^*(r^* - 1) + 6(r^{*2} + 7r^* + 1)}{(2s^* + 1 + r^*)^7} \\
&+ \left.\frac{4s^{*2} - 14s^*(r^* - 1) + 6(r^{*2} + 7r^* + 1)}{(2s^* - 1 - r^*)^7}\right)
\end{aligned}
\tag{14-35}
$$

A collision parameter of $\sigma = 0.5$ nm is usually assumed, although Ryan and Gschwend (1994) found that increasing this parameter to $\sigma = 2$ nm was necessary to explain the detachment of colloids (hematite) in a column packed with quartz media. Increasing σ to 2 nm would increase the calculated distance of "closest approach" of two colloids. In their study at a pH = 7, for example, the closest approach distance increased from 0.2 to 0.7 nm, a distance that is within the range of 0.4 to 1 nm calculated in other colloid release experiments.

For particles at surfaces, Born repulsion between a sphere and a plate can also be calculated using the expression derived by Ruckenstein and Prieve (1976)

$$
E_B = \frac{\mathscr{A}\sigma_c^6}{7560} \left(\frac{8R + s}{(2R + s)^7} + \frac{6R - s}{s^7}\right)
\tag{14-36}
$$

Hydrophobic Forces   Hydrogen bonding allows water molecules to achieve an overall lower energy state. When a surface is present in water that has no polar or ionic groups, the surface is said to be hydrophobic because the arrangement of water molecules is disrupted and the overall energy state of the solution is increased. Water molecules continually associate and realign with other water molecules, but somehow the presence of a hydrophobic surface disrupts the level of water molecule association relatively far from the hydrophobic surface. While it is currently believed that a disoriented water layer is only a few molecules thick, hydrophobic forces have been measured over a range of 80 nm between mica sheets coated with adsorbed hydrocarbon and fluorocarbon surfactants to make the mica surface hydrophobic (Israelachvili and Pashley, 1978; Claesson and Christenson, 1988). In this case, the hydrophobic interactions were much stronger, and operated over much larger distance than van der Waals forces. The importance and magnitude of hydrophobic forces in particle–particle adhesion is not well understood and is a subject of ongoing research.

Steric Interactions and Polymer Bridging   The addition of polymeric coagulants to water is a common method of increasing the efficiency of particle collisions. Electrostatic repulsion between particles can prohibit sufficiently close approach to one another to allow van der Waals attractive forces to pull the particles together. Positively charged polymeric coagulants added to water will readily attach to negatively charged particles and will have parts of the molecule that extend out into solution past the repulsive layer. Particles can attach via these polymers, by a mechanism referred to as *bridging*, since the particles are not repelled by electrostatic forces over the distances the polymers extend (Fig. 14.5). However, the addition of too much polymer will fill the surface of the particle and stabilize the solution (Fig. 14.6).

Not all polymers will destabilize particle suspensions. The presence of even small concentrations of macromolecules on the surface of particles can stabilize the particles if the exposed parts of the molecules do not adsorb to other particles. Most particles in natural waters are coated with organic matter. The attractive forces between the molecule and the surface, the repulsive forces between the water and the molecules, the types and concentrations of molecules in the water will all contribute to different orientations and interactions between these adsorbed molecules. The overall nature of these interactions makes it possible to have stable colloidal sus-

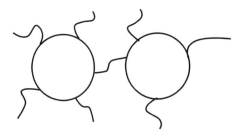

Figure 14.5   Polymer bridging allows particle attachment at distances that may be larger than the thickness of the electrostatic repulsive layer around the particles.

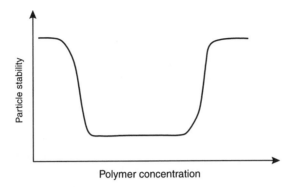

Figure 14.6 Particle stability as a function of polymer concentration. As the polymer dose is increased, particles become less stable and stick with high probabilities. However, if the polymer dose becomes too high, the particles become completely covered with the polymer and the particles are restabilized.

pensions even in high-ionic-strength solutions, such as seawater, despite high particle concentrations (see Example 14.8).

Implications for Coagulation   The above analysis can be used to understand whether two particles that approach each other will have a net attractive or repulsive force, but this does not directly translate to whether two particles will strike and stick to each other. If the particles are completely destabilized, and have no net repulsive force, then they should stick each time they collide ($\alpha = 1$). When particles have some net repulsive force between them, then the efficiency of collisions must be explained in terms of statistical probability ($\alpha < 1$). Particles with high energy barriers of $>100k_BT$ are unlikely to collide successfully, and will be perfectly stable ($\alpha = 0$). For the cases of coagulation by Brownian motion (orthokinetic coagulation) and fluid shear (perikinetic coagulation), equations have been derived that predict reduced coagulation efficiencies ($\alpha < 1$) as a function of electrostatics. In practice, however, unless the value of α is zero or unity, α must be determined empirically using models that predict the frequency of particle collisions.

14.4   COAGULATION KINETICS: COLLISION KERNELS FOR SPHERES

As we have discussed, particle coagulation is a function of three factors: particle concentration, collision efficiency (α), and collision frequency (β). The frequency of particle collisions can be solved analytically for special cases, such as a monodisperse suspension of perfectly spherical particles. Particles that coagulate quickly produce nonspherical (and fractal) aggregates, and there is no exact solution for describing the motion of particles in a mixed fluid. Despite these difficulties, it is possible to develop coagulation models that can be applied to coagulating suspensions. In this section, collision kernels are presented for spherical particles.

Rectilinear Collision Kernels

Particles suspended in an infinite fluid can collide by three mechanisms: Brownian motion, fluid shear, and differential sedimentation (Fig. 14.7). The overall collision function is calculated from the superposition of collision frequencies as

$$\beta = \beta_{Br} + \beta_{sh} + \beta_{ds} \tag{14-37}$$

where the subscripts refer to collisions generated by Brownian motion, Br, shear, sh, and differential sedimentation, ds. Although all three mechanisms contribute to collisions between particles, only one collision is assumed to dominate primarily as a function of particle size. The smallest particles (<1 μm) are assumed to collide by Brownian motion, while larger particles collide by fluid shear, when shear rates are relatively large ($G > 1-10 \text{ s}^{-1}$), or by differential sedimentation.

The kinetics of flocculation by Brownian motion and fluid shear were first described by Smoluchowski in 1917, while flocculation by differential sedimentation was added much later (Camp and Stein, 1943). Collisions by Brownian motion and fluid shear were assumed to occur as a result of the motion of the particles in straight lines, resulting in the collision functions being referred to as rectilinear collision functions or kernels. The motion of particles by Brownian motion was not assumed to be linear, but the same term was used for this collision mechanism when short-range forces were ignored. Later, all three collision kernels were modified by others to include changes in fluid motion (hydrodynamics) and short-range forces that were important as particles approached each other, resulting in a series of different models, all referred to as curvilinear models. In this section, the rectilinear models are presented. In subsequent sections, it is shown how these equations can be modified in order to account for short-range forces and the fractal geometry of aggregated (nonspherical) particles.

Brownian Motion   Water molecules and particles all move due to their thermal energy in random paths defined as Brownian motion. The net flux of particles across a given plane is described by Fick's law, in terms of a diffusion coefficient. For a

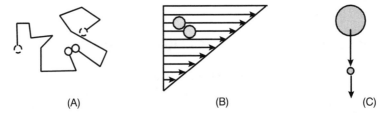

(A)   (B)   (C)

Figure 14.7   The three different coagulation mechanisms: (A) Brownian motion; two particles initially at the position shown with dashed lines move by Brownian motion and collide. (B) Shear; the particle moving on the faster streamline overtakes the particle on the slower-moving streamline. (C) Differential sedimentation. The larger, faster-settling particle overtakes the slower-settling particle and collides with it.

single particle, the random motion of the particle is therefore characterized by its diffusivity, given by the Stokes–Einstein equation, as

$$D_p = \frac{k_B T}{3\pi\mu_w d_p} \tag{14-38}$$

where D_P is the particle diffusivity, k_B Boltzmann's constant (1.38×10^{-16} erg K^{-1} = g cm^2 s^{-2} T^{-1}), T the absolute temperature, μ_w the fluid viscosity, and d_p the particle diameter. The fluid is assumed here to be water, which has a viscosity of 0.89 cp = 0.0089 g s^{-1} cm^{-1} at 25°C (298 K), and all subscripts referring to the fluid in this chapter will consider water to be the fluid for simplicity of notation. At a fixed temperature, particle diffusivity is a simple function of size. For example, at 25°C, D_p [cm^2 s^{-1}] = $4.9 \times 10^{-9} d_p$ [μm].

The Brownian collision function can be derived by solving for the flux of particles of size d_j towards one particle of size d_i in spherical coordinates, where the diffusion coefficient of the particle is given by the Stokes–Einstein equation. Multiplying the flux to a particle by the concentration of particles in the system produces the collision frequency

$$\beta_{Br} = \frac{2k_B T}{3\mu_w} \frac{(d_i + d_j)^2}{d_i d_j} \tag{14-39}$$

The term $(d_i + d_j)$ results from using a boundary condition that states that the particles cannot approach each other more closely than the sum of their individual radii $(R_i + R_j)$.

If electrostatics are included in the derivation of the collision function, then the rate of coagulation will be reduced from the case where $\alpha = 1$. The force experienced by a particle is a function of the radial distance from the surface of the other particle according to $d\Psi/dr$, where Ψ is the electrostatic potential energy between the two particles. Integrating this force over the separation distance from $2R$ to infinity results in the correction factor

$$W = 2R \int_{2R}^{\infty} \frac{e^{\Psi/k_B T}}{r^2} \, dr \tag{14-40}$$

where W is known as the Fuchs stability factor for slow coagulation. If the theoretical prediction of collision frequency is accurate, then $\alpha = 1/W$. The effects of electrostatic repulsion and van der Waals attraction on can be determined using DLVO theory. Assuming Ψ_{max} is the energy barrier in DLVO theory, then α can be approximated by

$$\alpha \approx 2Rh e^{-\Psi_{max}/k_B T} \tag{14-41}$$

For example, if $\Psi_{max} = 5k_B T$ and $R = 0.05$ μm for a 0.001 M NaCl solution at 25°C, then $h^{-1} = 9$ nm and $\alpha = 0.075$ (Amirthrajah and O'Melia, 1990). This demonstrates that even with very low energy barriers the sticking coefficient can be much less than unity.

Fluid Shear Two expressions are used to describe the coagulation rate of particles. These equations arise from whether the particle is suspended in a laminarly sheared fluid, or whether the fluid is assumed to be isotropically turbulent. The distinction is relatively unimportant in practice, since these two results differ only by a small factor that is a constant.

When two particles are suspended in a fluid having a velocity gradient du_x/dx, where u_x is a function of y, a particle moving with a faster streamline will overtake and collide with a particle on the smaller streamline (Fig. 14.7). The frequency of collisions is proportional to the shear rate, and the individual particle volumes, or

$$\beta_{sh} = \tfrac{1}{6}(d_i + d_j)^3 \frac{du_x}{dx}, \quad \text{laminar shear} \tag{14-42}$$

If the shear rate $G = du_x/dx$, then this can be written more simply as

$$\beta_{sh} = \frac{G}{6}(d_i + d_j)^3, \quad \text{laminar shear} \tag{14-43}$$

When particles are much smaller than the Kolmogorov microscale of turbulence, the fluid environment of a particle is that of laminar shear, but the turbulent eddies are in a constant state of generation and decay. For isotropic turbulence, the collision frequency between two particles can be expressed as

$$\beta_{sh} = \frac{1}{6.18}(d_i + d_j)^3 \left(\frac{\varepsilon_{diss}}{v}\right)^{1/2}, \quad \text{turbulent shear} \tag{14-44}$$

Since we have isotropic turbulence, and most of the energy is dissipated in small eddies, we assume that the mean shear rate can be related to the energy dissipation rate by $G = (\varepsilon_{diss}/v)^{1/2}$, so that we can write Eq. 14-42 as

$$\beta_{sh} = \frac{G}{6.18}(d_i + d_j)^3, \quad \text{turbulent shear} \tag{14-45}$$

Thus the expression for turbulent shear differs by only 3% from that derived for the simpler case of laminar shear.

Both of the above equations assume that the fluid does not interfere with the motion of the particles. However, when the hydrodynamics of two approaching particles is included in collision models, it becomes unlikely that particles with dissimilar sizes would strike each other. Such effects are included in the curvilinear models below. The uncertainties associated with modeling particle trajectories are much more important to coagulation kinetics than the 3% difference between turbulent and shear coagulation rates.

When particles are completely destabilized, and double-layer repulsion is negligible, van der Waals attraction and hydrodynamic retardation produce an $\alpha < 1$. van de Ven and Mason (1977) derived the empirical expression from numerical analysis of interacting spheres as

$$\alpha = 0.9 \left(\frac{\mathscr{A}}{36\pi\mu GR^3} \right)^{0.18} \tag{14-46}$$

For particles 0.5 μm in radius, $\mathscr{A} = 0.8 \times 10^{-13}$ J (quartz), $G = 10$ s^{-1}, and at 20°C, we obtain $\alpha = 0.54$. Values of α in this range are typically obtained for completely destabilized monodisperse particles in coagulation tests.

Differential Sedimentation A particle with a density different from that of water will rise or settle, producing collisions with particles in its path (Fig. 14.7). Even if both particles are settling, the faster-settling particle will overtake the other particle and collide with it. The collision diameter between the two particles is just the sum of their areas, or $(\pi/4)(d_i + d_j)^2$. When two both particles have the same density, collisions occur in proportion to their settling velocities (calculated using Stokes' Law), and therefore the square of their diameters, according to

$$\beta_{ds} = \left(\frac{g\pi \, \Delta\rho}{18v\rho} |d_i^2 - d_j^2| \right) \left(\frac{\pi}{4} (d_i + d_j)^2 \right) \tag{14-47}$$

where $\Delta\rho = \rho_p - \rho$ is the difference in density between the particle, ρ_p, and the fluid, ρ. Simplifying, we have

$$\beta_{ds} = \frac{g\pi \, \Delta\rho}{72v\rho} |d_i^2 - d_j^2|(d_i + d_j)^2 \tag{14-48}$$

For impermeable particles with large differences in diameters, this result overpredicts collision frequencies since small particles will move on streamlines around the larger, faster-settling particle (the so-called "Queen Mary" effect). The collision kernels for the three collision mechanisms are summarized in Table 14.1.

Example 14.2

Compare the Brownian, laminar shear, and differential sedimentation collision functions for a bacteria-sized particle (0.8 μm) and other particles in a flocculation device for particles 0.01 to 100 μm in diameter. Assume: 20°C, $\rho_p = 1.03$ g cm^{-3}, and shear rates of $G = 10$ and 50 s^{-1}.

TABLE 14.1   Collision Functions for Spherical Particles

	Collision Function	
Mechanism	$\beta(d_i, d_j)$	$\beta(v_i, v_j)$
Brownian motion	$\beta_{Br} = \dfrac{2k_BT}{3\mu_w} \dfrac{(d_i + d_j)^2}{d_i d_j}$	$\beta_{Br} = \dfrac{2k_BT}{3\mu_w} \dfrac{(v_i + v_j)^2}{v_i v_j}$
Fluid shear Laminar	$\beta_{sh} = \dfrac{G}{6}(d_i + d_j)^3$	$\beta_{sh} = \dfrac{G}{\pi}(v_i + v_j)^3$
Turbulent	$\beta_{sh} = \dfrac{G}{6.18}(d_i + d_j)^3$	$\beta_{sh} = \dfrac{G}{1.03\pi}(v_i + v_j)^3$
Differential sedimentation	$\beta_{ds} = \dfrac{g\pi \Delta\rho}{72v\rho}|(d_i^2 - d_j^2)^2|(d_i + d_j)^2$	$\beta_{ds} = \dfrac{6^{4/3}g \Delta\rho}{72\pi^{1/3}v\rho}|v_i^{2/3} - v_j^{2/3}|(v_i^{1/3} + v_j^{1/3})^2$

The collision functions, calculated using the equations given in Table 14.1, are shown in Figure 14.8. This comparison of collision kernels suggests that particle coagulation is slowest between similar sized particles, and greatest between either very large ($d_p > 10$ μm) or very small ($d_p < 0.1$ μm) particles. At the given shear rates, collisions by differential sedimentation are insignificant except for very large particles. As we will see, collision frequencies between particles that vary widely in size are overestimated.

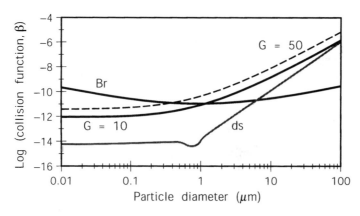

Figure 14.8

Example 14.3

In the previous example, only the magnitudes of the collision functions were compared, but the collision rate of particles is dependent on particle concentration as well as particle size. Compare the collision rates of 0.8-μm-sized particles with other particles assuming the particle size distribution is $n(d) = 10^7 d_p^{-3}$, where d_p [μm] and n [#/mL μm]. Assume the same conditions given in the previous example, but consider only the case of $G = 10$ s^{-1}.

The collision frequency of particles is a second-order function where $dn_i/dt \sim \beta n_i n_j d(d_j)$. The change in particle concentration of one-size particle of concentration n_i, normalized to the concentration n_i, is $(1/n_i)\, dn_i/dt \sim \beta n_j d(d_j)$. Therefore, we can approximate the collision rate for the given size distribution as $\beta n_j d(d_j)$. For a size range of Δd_j, this collision rate becomes $\bar{\beta}(0.8, d_j)\bar{n}_j(\Delta d_j)\Delta(d_j)$, or more simply $\bar{\beta}(0.8, d_j)\bar{N}_j(\Delta d_j)$, where $\bar{N}_j(\Delta d_j)$ is the number concentration of particles in the size interval Δd_j. Using the collision kernels from Example 14.2, and the given particle number distribution, we obtain the results shown in Figure 14.9. Because there are large concentrations of smaller particles, there will be many more collisions of 0.8-μm-sized particles with small particles than there will be with larger particles.

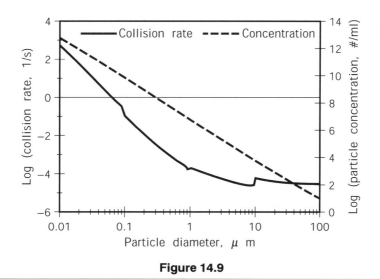

Figure 14.9

Monodisperse Coagulation Kinetics

A special case of particle coagulation arises when the particle suspension is initially monodisperse, or when all particles are initially the same size. For this case, there is only one size interval and so the size distribution function, $n(d_p)$, simply becomes

N_T, the number concentration of particles, or

$$\sum_{h=1}^{\infty} n_h = N_T \tag{14-49}$$

As particles coagulate, particles of added mass and diameter will of course be produced, but on *average* the size of particles will not substantially change. By neglecting the changes in particle size, which can only be done during the initial period of coagulation, the change in particle concentration can be described using the general coagulation equation (Eq. 14-2) for a monodisperse suspension as

$$\frac{dN_T}{dt} = -\frac{1}{2}\alpha\beta N_T^2 \tag{14-50}$$

This result can be seen more clearly by examining each term in Eq. 14-2 separately. If all particles in the distribution are summed over the population of n_h, then Eq. 14-2 becomes

$$\sum_{h=1}^{\infty}\left(\frac{dn_h}{dt}\right) = \sum_{h=1}^{\infty}\left(\frac{1}{2}\sum_{i+j=h}\alpha\beta n_i n_j\right) - \sum_{h=1}^{\infty}\left(\sum_{i=1}^{\infty}\alpha\beta n_i n_h\right) \tag{14-51}$$

where the notation in parentheses for β is no longer needed. The first term in Eq. 14-51 can be simplified by moving the summation inside the derivative, or

$$\sum_{h=1}^{\infty}\frac{dn_h}{dt} = \frac{d}{dt}\sum_{h=1}^{\infty}n_h = \frac{dN_T}{dt} \tag{14-52}$$

The two terms on the right-hand side of Eq. 14-51 are similarly simplified, resulting in

$$\sum_{h=1}^{\infty}\left(\frac{1}{2}\sum_{i+j=h}\alpha\beta n_i n_j\right) = \frac{1}{2}\alpha\beta\sum_{i=1}^{\infty}n_i\sum_{j=1}^{\infty}n_j = \frac{1}{2}\alpha\beta N_T^2 \tag{14-53}$$

$$\sum_{h=1}^{\infty}\left(\sum_{i=1}^{\infty}\alpha\beta n_i n_h\right) = \alpha\beta N_T\sum_{h=1}^{\infty}n_h = \alpha\beta N_T^2 \tag{14-54}$$

Combining the three terms produces

$$\frac{dN_T}{dt} = \frac{1}{2}\alpha\beta N_T^2 - \alpha\beta N_T^2 = -\frac{1}{2}\alpha\beta N_T^2 \tag{14-55}$$

which is the same as Eq. 14-50. The second-order dependence of the coagulation rate on particle concentration is now clearly seen in Eq. 14-50 or 14-55. The assumption of monodisperse coagulation appears to be valid for changes in particle concentration of less than ⁻30%.

For a monodisperse suspension, the individual collision functions become

$$\beta_{Br} = \frac{8k_B T}{3\mu} \tag{14-56}$$

$$\beta_{sh} = \frac{4}{3} G d_p^3 \tag{14-57}$$

$$\beta_{ds} = 0 \tag{14-58}$$

Collisions do not occur by differential sedimentation for a monodisperse suspension because all particles are the same size and density, and therefore they all have the same settling velocity.

Monodisperse Coagulation Due to Brownian Motion When particles are very small, $\beta_{Br} \gg \beta_{sh}$. Equations 14-50 and 14-56 can be combined to produce

$$\frac{dN_T}{dt} = -\frac{4\alpha k_B T}{3\mu_w} N_T^2 \tag{14-59}$$

Notice that Eq. 14-59 demonstrates that the coagulation rate is independent of the particle size. This second-order rate expression can be separated and integrated

$$\int_{N_{T0}}^{N_T} \frac{dN_T}{N_T^2} = -\frac{4\alpha k_B T}{3\mu_w} \int_0^t dt \tag{14-60}$$

to produce

$$N_T = \frac{N_{T0}}{1 + \dfrac{4\alpha k_B T}{3\mu_w} N_{T0} t} \tag{14-61}$$

Experiments conducted with particles completely destabilized using high-ionic-strength solutions have shown that coagulation rates are slightly less than the maximum predicted by Eq. 14-59. Because α is the only adjustable parameter in Eq. 14-59, maximum values of α found experimentally are is ~0.5 (Overbeek, 1977).

Monodisperse Shear Coagulation When coagulation occurs primarily by collisions generated by fluid shear ($\beta_{sh} \gg \beta_{Br}$), Birkner and Morgan (1968) have shown that the initial rate of coagulation can be approximated as a first-order reaction. Combining Eqs. 14-50 and 14-58, the change in particle concentration is

$$\frac{dN_T}{dt} = -\tfrac{2}{3}\alpha G d_p^3 N_T^2 \tag{14-62}$$

Solid volume is conserved during particle coagulation, and if the diameter of a particle is constant over short periods, then the initial solid volume, $V_T = N_0\pi d^3/6$, is constant. Incorporating the solid volume into Eq. 14-62 produces the first-order coagulation equation

$$\frac{dN_T}{dt} = -b_{sh}N_T \qquad (14\text{-}63)$$

where b_{sh} is a constant defined as $b_{sh} = 4\alpha GV_T/\pi$. Integrating Eq. 14-63 from an initial concentration N_{T0} at $t = 0$ produces

$$\frac{N_T}{N_{T0}} = e^{-b_{sh}t} \qquad (14\text{-}64)$$

This result indicates that particle concentrations of a monodisperse suspension due to shear motion initially obey a simple exponential decay.

The half-life, defined as the time for the particle's concentration to decrease by 50%, is obtained from this result as

$$t_{1/2} = -\frac{\ln 2}{b_{sh}} \qquad (14\text{-}65)$$

The half-life is a useful scaling parameter for evaluating the particle coagulation rate in a system. The collision functions, coagulation equations, and half-lives of monodisperse particles are summarized in Table 14.2.

The maximum coagulation rates predicted for shear coagulation by Eq. 14-63 have not been obtained in experiments using monodisperse suspensions of destabilized particles mixed in a jar-test (paddle mixer) apparatus. In practice, α must be less

TABLE 14.2   Collision Functions, Integrated Coagulation Equations and Half-Lives of Particles for Monodisperse Particles, Applicable to the Initial Coagulation Period

	Coagulation Mechanism	
Condition	Brownian Motion	Laminar Shear
Collision function	$\beta_{Br} = \dfrac{8k_BT}{3\mu_w}$	$\beta_{sh} = \dfrac{4}{3}Gd_p^3$
Coagulation equation	$N_T = \dfrac{N_{T0}}{1 + \dfrac{4\alpha k_BT}{3\mu_w}N_{T0}t}$	$N_T = N_{T0}e^{-b_{sh}t}$
Half-life	$t_{1/2} = \dfrac{3\mu}{4\alpha k_BTN_{T0}}$	$t_{1/2} = -\dfrac{\ln 2}{b_{sh}}$

than the theoretical maximum of unity. When latex microspheres were coagulated with a cationic polymer (polyethylenimine) at an optimum chemical dosage, Birkner and Morgan (1968) found a maximum value of $\alpha \approx 0.2$ that decreased to ~ 0.02 as mean shear rates increased from 11 to 120 s^{-1}. In suspensions destabilized with salt (1 M NaCl), α decreased by a much smaller factor from 0.45 to 0.29 over the same shear rate.

An alternative scaling parameter for shear coagulation is the Camp number, Cp, defined as

$$Cp = V_T G t \qquad (14\text{-}66)$$

The Camp number is often used to scale the efficiency of particle removal during coagulation with flocculator detention time and shear rate in water treatment operations. If particle removal due to coagulation is assumed to be a second-order process, or $dN_T/dt = -bN_T^2$, then the effluent particle concentration from a tank with an influent particle concentration of N_{T0} (Amirthrajah and O'Melia, 1990) is

$$N_T = \frac{N_{T0}}{1 + bN_{T0}t} \qquad (14\text{-}67)$$

For completely destabilized particles, this second-order result predicts an unreasonably rapid coagulation rate. A similar form of an equation can be developed from a steady-state analysis (see below) that does not require a monodisperse suspension. This more general result, which is based on a second-order coagulation rate, is

$$N_T = \frac{N_{T0}}{1 + b_{Cp}V_T G t} \qquad (14\text{-}68)$$

which can be written more simply in terms of the Cp number as $N_T = N_{T0}/(1 + b_{Cp}\,Cp)$, where b_{Cp} is an empirical constant. As shown in the example below, this equation predicts similar changes in particle concentrations with time (for short times) as the monodisperse equation above.

Example 14.4

Compare the particle concentrations of a monodisperse suspension predicted for short time periods using a first-order model with that calculated using a second-order model (Eq. 14-67). Assume an initial particle concentration of 1×10^6 mL^{-1}, $G = 20$ s^{-1}, $\alpha = 0.5$, $b_{Cp} = 1$ mL, and $d_p = 2$ μm.

For the given conditions, $V_T = N_0 \pi d_p^3/6 = 4.19 \times 10^{-6}$ mL^{-1}, and $b_{sh} = 4\alpha G V_T/\pi = 5.33 \times 10^{-5}$ s^{-1}. The half-life of a monodisperse suspension is $t_{1/2} = \ln(2)/b_{sh} = 217$ min. Using the half-life as an estimate of a "short time period," we have the result shown in Figure 14.10. The first-order result, which shows particles undergo exponential decay in concentration, produces similar results to

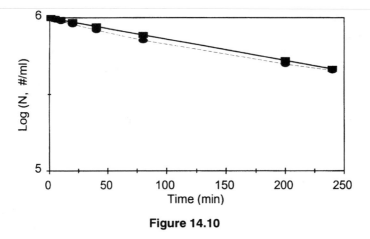

Figure 14.10

the more general coagulation result given in Eq. 14-65 over the indicated time period.

Curvilinear Collision Kernels

The above rectilinear models are based on the assumption that a particle sweeps out all other particles in the water in its path. However, the motion of two particles as they approach each other will not be linear and result in a collision area that is smaller than that calculated from the particle's size (Fig. 14-11). In addition, when two particles approach each other, the water between the two particles must be

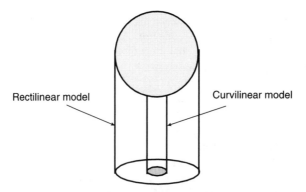

Figure 14.11   Different areas swept out by a settling particle. The assumption used for the rectilinear model is that a settling particle can contact all particles that cross its projected area. However, many particles will follow streamlines that go around the falling particle and therefore will not contact the falling particle. This results in a smaller cross-sectional area used to calculate collision frequencies in curvilinear model calculations.

squeezed out in order to allow the particles to contact each other. There are also short-range attractive forces that pull the particles together, and electrostatic repulsive forces that push the particles apart. Models that include these other factors are referred to as curvilinear collision models, while the mathematical description of the path is termed trajectory analysis.

Numerous curvilinear coagulation models have been presented in the literature, far more than can be presented and critically evaluated here. Han and Lawler (1992) examined these models and developed a series of equations that would be applicable to describe particle coagulation in water treatment plants. Their analysis is summarized here. Because of the assumed context of a water treatment process, all particles are assumed to be completely destabilized, and therefore, electrostatic repulsion is not included in their semiempirical models.

The sticking coefficient, α, is really a product of two terms: the sticking efficiency α_s and the collision efficiency α_m, where m refers to one of the three collision mechanisms (Brownian, Br; shear, sh; or differential sedimentation, ds). If all particles are destabilized, the sticking efficiency is unity. The collision efficiency, or the probability of a collision between two particles that are moving toward each other, can be modeled by considering the attractive forces between the particle and the fluid motion. Therefore, the approach of Han and Lawler (1992) was to develop solutions for the collision efficiency for destabilized particles ($\alpha_s = 1$). By superposition of collision mechanisms, the product $\alpha\beta$ for the three collision mechanisms becomes

$$\alpha\beta = \sum \alpha_s \alpha_m \beta_m = \alpha_{Br}\beta_{Br} + \alpha_{sh}\beta_{sh} + \alpha_{ds}\beta_{ds} \tag{14-69}$$

The equations for α_m for each collision mechanism are presented below.

Brownian Motion   The collision efficiency for Brownian motion is

$$\alpha_{Br} = b_1 + b_2 r^* + b_3 r^{*2} + b_4 r^{*3} \tag{14-70}$$

where r^* ($0 \leq r^* \leq 1$) is the size ratio of the two particles of radii R_1 and R_2, and the constants b_1 through b_4 are given in Table 14.3 for the case of $\mathcal{A} = 4 \times 10^{-13}$ g cm^2 s^{-2}. The results are relatively insensitive to Hamaker constants over the range typically encountered for aquasols. Collision efficiencies for Brownian coagulation using Eq. 14-70 agree with those calculated by Valioulis and List (1984) and Spielman (1970) when van der Waals attractive forces are included (Han and Lawler, 1992).

Shear   Empirical equations were developed by Han and Lawler (1992), based on work by Adler (1981), to calculate the collision efficiency for shear flow as

$$\alpha_{sh} = \frac{8}{(1 + r^*)^3} \, 10^{(b_1 + b_2 r^* + b_3 r^{*2} + b_4 r^{*3})} \tag{14-71}$$

where the four constants b_1 through b_4 for shear flow are given in Table 14.3 as a

TABLE 14.3 Regression Constants Used to Predict the Collision Efficiency (α_m), for Completely Destabilized Particles

α_m	Parameter	b_1	b_2	b_3	b_4
α_{Br}	Diameter (μm)				
	0.1	1.025	−0.626	0.516	−0.152
	0.2	1.007	−0.860	0.870	−0.322
	0.6	0.976	−1.155	1.342	−0.554
	1.0	0.962	−1.263	1.522	−0.645
	2.0	0.943	−1.383	1.725	−0.748
	6.0	0.916	−1.533	1.991	−0.887
	10.0	0.905	−1.587	2.087	−0.936
	20.0	0.891	−1.658	2.221	−1.009
	60.0	0.871	−1.739	2.371	−1.090
	200.0	0.863	−1.775	2.439	−1.125
α_{sh}	$\log b_{c,sh}$				
	1	−1.128	2.498	−2.042	0.671
	0	−1.228	2.498	−2.042	0.671
	−1	−1.482	3.189	−3.468	1.581
	−2	−1.704	3.116	−2.881	1.121
	−3	−2.523	5.550	−6.098	2.553
	−4	−3.723	10.039	−12.569	5.557
	−5	−5.775	18.267	−24.344	10.992
	−6	−7.037	20.829	−25.589	10.755
	−7	−8.733	25.663	−30.703	12.555
	−8	−9.733	30.663	−35.703	14.555
α_{ds}	$\log b_{c,ds}$				
	0	−0.840	0.445	−1.069	0.930
	−1	−1.320	1.318	−2.17	1.361
	−2	−1.757	2.137	−3.229	1.794
	−3	−2.152	2.880	−4.232	2.230
	−4	−2.505	3.547	−5.186	2.668
	−5	−2.815	4.137	−6.088	3.108
	−6	−3.084	4.652	−6.940	3.551
	−7	−3.310	5.090	−7.742	3.996
	$-\infty$	−3.928	6.423	−9.449	4.614

Source: Han and Lawler, 1992.

function of the Hamaker constant $\mathscr{A}$ and shear rate G included in the dimensionless attraction number, $b_{c,sh}$

$$b_{c,sh} = \frac{\mathscr{A}}{18\pi\mu G d_{p,l}} \tag{14-72}$$

where $d_{p,l}$ is the diameter of the larger particle and μ the fluid viscosity.

Differential Sedimentation   For differential sedimentation, the collision efficiency factor (Han and Lawler, 1992) is

$$\alpha_{sh} = 10^{(b_1 + b_2 r^* + b_3 r^{*2} + b_4 r^{*3})} \tag{14-73}$$

where the constants b_1 through b_4 for are given in Table 14.3 as a function of the Hamaker constant $\mathscr{A}$ and the dimensionless number, $b_{c,ds}$

$$b_{c,ds} = \frac{48\mathscr{A}}{\pi g (\rho_p - \rho) d_{p,l}^4} \tag{14-74}$$

These results for differential sedimentation agree with results of Davis (1984).

Example 14.5

Compare the rectilinear collision functions calculated in Example 14-3 with those predicted by the curvilinear model for $\mathscr{A} = 4 \times 10^{-13}$ g cm^2 s^{-2} for a 1-μm-diameter particle colliding with particles 0.01 to 100 μm in diameter, at a shear rate of 10 s^{-1}.

The solutions given in Eqs. 14-70 to 14-73 can be used to solve this problem. Although interpolation between tabulated values is necessary, we will round values of diameter ratios and the Hamaker constant (to the nearest integer) for simplicity here. The results are shown in Figure 14.12. We see that collisions between small particles are always dominated by Brownian motion and that the curvilinear equations predict slightly lower collision frequencies than the rectilinear equations. From 1 to 20 μm, all three collision mechanisms contribute to collisions, while collisions between particles very dissimilar in size are predicted to occur via differential sedimentation. A very interesting implication of these calculations is

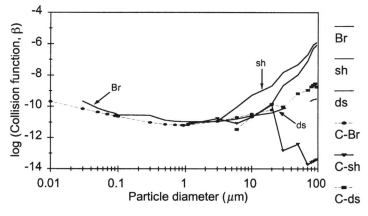

Figure 14.12 Calculated rectilinear (solid lines) and curvilinear (symbols) collision functions.

that if particles are neutrally buoyant ($\beta_{ds} = 0$), then Brownian motion would again be predicted by the rectilinear equations to become the dominant collision mechanism at large particle sizes.

Steady-State Solutions from Dimensional Analysis: Hard-Sphere Model

For systems with a continuous input of small particles coagulating into very large particles that leave the system through sedimentation, it is possible to develop steady-state conditions resulting in a size distribution that is constant. The shape of the steady size distribution can be obtained by dimensional analysis (Hunt, 1980), assuming that several conditions are met: (1) There is a constant source of very small particles; (2) particles enter the size distribution only at small sizes; (3) particles leave the size distribution by sedimentation; (4) sticking coefficients are constant and independent of particle size; (5) only one coagulation mechanism (Brownian, shear, differential sedimentation) dominates at a given particle size; (6) a single parameter can characterize each mechanism of coagulation.

The existence of a steady state indicates that there is a constant (solid) volume flux through the size distribution, defined here as J_v with dimensions $[L^3 \ l^{-3} \ t^{-1}]$ where $[L]$ is the fluid length and $[l]$ is the particle length. From the coagulation mechanisms derived above, the parameters that characterize each coagulation are chosen (Hunt, 1980) to be

$$\Lambda_{Br}[L^3 \ t^{-1}] = \frac{k_B T}{\mu_w} \tag{14-75}$$

$$\Lambda_{sh}[L^3 \ l^{-3} \ t^{-1}] = G \tag{14-76}$$

$$\Lambda_{ds}[L^3 \ l^{-3} \ t^{-1}] = \frac{g(\rho_p - \rho)}{\mu_w} \tag{14-77}$$

where Λ_m is a coagulation parameter having the indicated units for Brownian (Λ_{Br}), shear (Λ_{sh}), and differential sedimentation (Λ_{ds}). With these three chosen parameters, the size distribution becomes a function of four parameters, or $n(d_p, \Lambda_{Br}, \Lambda_{sh}, \Lambda_{ds})$ and three units (L, l, and t). With assumption (5) above, these parameters are used to obtain steady-state size distributions from dimensional analysis for each coagulation mechanism (Hunt, 1980) as

$$n(d_p)|_{Br} = B_{Br} \left(\frac{J_v}{\Lambda_{Br}} \right)^{1/2} d_p^{-2.5} \tag{14-78}$$

$$n(d_p)|_{sh} = B_{sh} \left(\frac{J_V}{\Lambda_{sh}} \right)^{1/2} d_p^{-4} \tag{14-79}$$

$$n(d_p)|_{ds} = B_{ds} \left(\frac{J_V}{\Lambda_{ds}} \right)^{1/2} d_p^{-4.5} \tag{14-80}$$

where B_{br}, B_{sh} and B_{ds} are constants. This analysis produces a different slope of the

TABLE 14.4 Slopes of Steady-State Size Distributions Derived from Dimensional Analysis, in Terms of Solid Volume, v, or Diameter, d_p

Coagulation Mechanism	Discrete Size Distribution		Cumulative Size Distribution	
	s_v	s_d	S_v	S_d
Brownian motion	$-3/2$	$-5/2$	$-1/2$	$-3/2$
Shear	-2	-4	-1	-3
Differential sedimentation	$-13/6$	$-9/2$	$-7/6$	$-7/2$

size distribution (-2.5, -4 and -4.5) for each coagulation mechanism when n is plotted as a function of particle diameter.

The same dimensional analysis can be conducted based on particle diameter, d_p, in terms of size distribution function $n(d_p)$. The slopes obtained in terms of solid volume, s_v and particle diameter, s_d, are related by

$$s_v = \frac{s_d - 2}{3} \tag{14-81}$$

The relationships for the size distribution function derived in terms of particle diameters, $n(d_p)$ and solid volumes, $n(v)$, are summarized in Table 14.4. The shape of a steady-state size distribution is a straight line on log–log axes with slopes that vary according to the coagulation mechanism (*Br, sh, ds*) and type of distribution (discrete or cumulative). The slopes predicted by dimensional analysis are summarized in Table 14.5. Hunt (1982) calculated different slopes from particle size distributions for clay particles coagulated in artificial seawater by Brownian motion and

TABLE 14.5 Steady-State Discrete Size Distributions in Terms of Diameter and Solid Volume

Coagulation Mechanism	Steady-State Size Distribution in Terms of	
	Particle diameter, d_p	Solid volume, v
Brownian	$n(d_p)\big|_{Br} = B_{Br}\left(\dfrac{J_V}{\Lambda_{Br}}\right)^{1/2} d_p^{-2.5}$	$n(v)\big|_{Br} = B'_{Br}\left(\dfrac{J_V}{\Lambda_{Br}}\right)^{1/2} v^{-3/2}$
Shear	$n(d_p)\big|_{sh} = B_{sh}\left(\dfrac{J_V}{\Lambda_{sh}}\right)^{1/2} d_p^{-4}$	$n(v)\big|_{sh} = B'_{sh}\left(\dfrac{J_V}{\Lambda_{sh}}\right)^{1/2} v^{-2}$
Differential sedimentation	$n(d_p)\big|_{ds} = B_{ds}\left(\dfrac{J_V}{\Lambda_{ds}}\right)^{1/2} d_p^{-4.5}$	$n(v)\big|_{ds} = B'_{ds}\left(\dfrac{J_V}{\Lambda_{ds}}\right)^{1/2} v^{-13/6}$

Source: Hunt, 1980, 1982.

fluid shear. He found good agreement between the measured slopes and those predicted by dimensional analysis based on solid volume.

Example 14.6

The steady-state dimensional analysis for shear coagulation can be used to develop
the previously presented expression (Eq. 14-67) for modeling particle removal in
coagulation–sedimentation processes (Hunt, 1997). Starting with the steady size
distribution function, Eq. 14-76, derive an expression for total particle volume,
V_T, for particles larger than 1 μm. Next, show that the particle volume removal
is a second-order process, and integrate this expression to obtain an equation of
the same form as Eq. 14-67. You will need to recognize that the volume flux
through the size distribution, J_V, the rate that particles are removed by settling, is
related to volume by $J_V = dV_p/dt$.

The total volume of particles in a size distribution was shown in Chapter 13
to be $V_T = \int n(d_p)v_p \, d(d_p)$. Using Eq. 14-76 for the size distribution function, and
substituting in $G = \Lambda_{sh}$, we have

$$V_T = \frac{B_{sh}J_V^{1/2}}{G^{1/2}} \int_1^{d_p} \frac{\pi}{6} d_p^3 d_p^{-4} \, d(d_p) \tag{14-82}$$

Solving the integral, and rearranging, the volume flux is

$$J_V = \left(\frac{6}{A_{sh}\pi \ln(d_p)}\right)^2 GV_T^2 \tag{14-83}$$

Substituting $J_V = dV_T/dt$ and $b_1 = [6/A_{sh}\pi \ln(d_p)]^2$, we obtain the second-order rate
equation

$$\frac{dV_T}{dt} = b_1 GV_T^2 \tag{14-84}$$

Separating, and integrating from $V_T = V_{T0}$ at $t = 0$, we obtain

$$V_T = \frac{V_{T0}}{1 + b_1 GV_{T0}t} \tag{14-85}$$

This equation contains the Camp number and has the same form as Eq. 14-67 if
V_T is a known function of N_T.

14.5 FRACTAL COAGULATION MODELS

Even if a particle suspension is initially monodisperse, as the suspension coagulates
aggregates will be formed that are not spheres, but fractals. The formation of fractal

aggregates by Brownian coagulation has been shown in both experiments and computer simulations to produce aggregates that have characteristic fractal dimensions that vary from 1.8 to 2.2, a range that depends on the particle stability, or particle stickiness. The magnitude of the fractal dimension of an aggregate also indicates the general shape of the aggregate. As $D \rightarrow 3$ the aggregate becomes more spherical in shape and more compact, while as $D \rightarrow 1$ the aggregate becomes highly amorphous and stretched out in length (Fig. 14.13).

The relationship between aggregate appearance and D can be understood by considering how different conditions affect how aggregates are formed. When particles are completely destabilized, they have a sticking coefficient of unity. During coagulation, as a particle (or an aggregate of particles) move toward an aggregate, it cannot penetrate into the aggregate without striking some particle on the aggregate exterior and sticking to it. This results in *diffusion-limited* aggregate formation with characteristic fractal dimensions of $D = 1.8$. As particles become more stable, many collisions may occur before the particle will stick to another particle in the aggregate. In this *reaction-limited* aggregation case, the particle can penetrate more fully into the aggregate because, although the particle will collide many times with particles on the exterior, it will not readily stick to them, allowing the particle to penetrate deeper into the aggregate. As a result, there is less void space and the aggregate

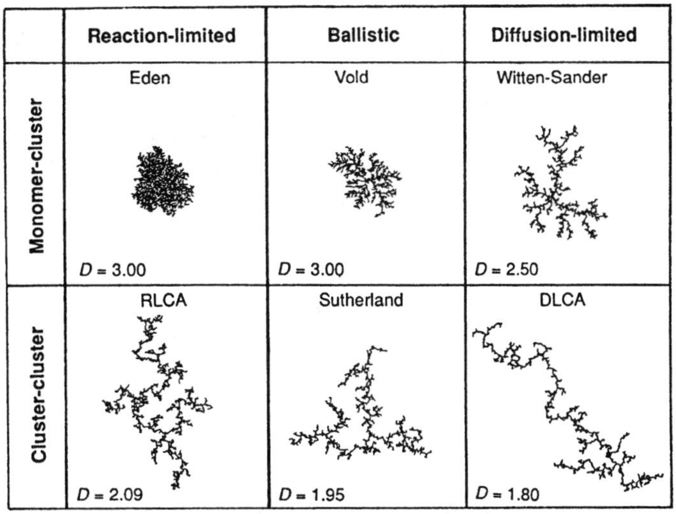

	Reaction-limited	Ballistic	Diffusion-limited
Monomer-cluster	Eden $D = 3.00$	Vold $D = 3.00$	Witten-Sander $D = 2.50$
Cluster-cluster	RLCA $D = 2.09$	Sutherland $D = 1.95$	DLCA $D = 1.80$

Figure 14.13 The general shape of aggregates formed by coagulation is a function of the fractal dimension, D. Different values of D arise from monomer–cluster (particles added one at a time), cluster–cluster (all particles added at the same time), reaction-limited (slow coagulation), diffusion-limited (rapid coagulation), or ballistic (particles added in on a straight-line trajectory) coagulation. (Reprinted with permission from Schaefer 1989. Copyright 1989 American Association for the Advancement of Science.)

becomes more dense for reaction-limited than diffusion-limited conditions. In the reaction-limited case, the fractal dimension is higher, with a value of $D = 2.2$. Experiments with different types of colloidal particles, such as gold, polystyrene, and clay, all produce the same fractal dimensions, indicating the fractal dimension is not a function of particle type (Lin et al., 1989).

Through computer simulations of colloidal aggregation it is possible to dictate the exact mechanism by which particles will be added to an aggregate. These simulations for coagulation by Brownian motion predict a slightly wider range of fractal dimensions than those measured in coagulation tests (Fig. 14.13). The range of D is in part a function of the way the aggregate is grown. When particles are added one at a time to a cluster of particles (monomer–cluster), fractal dimensions range from $D = 3$ for reaction-limited conditions to $D = 2.5$ for diffusion-limited conditions. When the coagulation simulation is conducted with all particles at once, simulating an initially monodisperse suspension, many aggregates are formed, although they all eventually coagulate with each other into a single aggregate (cluster–cluster). Fractal dimensions range from 1.8 to 2.09 for cluster–cluster models depending on whether the aggregation is diffusion- or reaction-limited. Ballistic computer simulations based on particles being "fired" along random paths give values of D intermediate between those for reaction- and diffusion-limited aggregates.

Fractal Collision Kernels

Coagulation of fractal particles proceeds by the same mechanisms described for spheres: by Brownian motion, shear, and differential sedimentation. The rectilinear collision kernels were formulated for aggregates of size d; by analogy, we can rewrite these equations for fractal particles of size l, where l is some characteristic size of these fractal particles that are not necessarily spherical. The collision frequency functions for coagulation of fractal particles by Brownian motion and laminar shear are directly obtained from the rectilinear collision kernels as

$$\beta_{Br} = \frac{2k_B T}{3\mu_w} \frac{(l_i + l_j)^2}{l_i l_j} \tag{14-86}$$

$$\beta_{sh} = \frac{G}{6} (l_i + l_j)^3 \tag{14-87}$$

For the case of differential sedimentation, a fractal-scaling relationship must be used to describe the settling velocity of each particle. The collision function for two spheres (Eq. 14-45) can be written in terms of each particle's diameter and settling velocity as

$$\beta_{ds} = \frac{\pi}{4} (d_i + d_j)^2 |\dot{U}_i - \dot{U}_j| \tag{14-88}$$

Using the settling velocity equation based on a power-law drag coefficient developed in Chapter 13, and replacing diameter d with l in the collision function in Eq. 14-88, we have for differential sedimentation

$$\beta_{ds} = \frac{\pi}{4}\left(\frac{2g\xi_p(\rho_p - \rho_w)b_D}{b_{d1}\rho\xi v^{b_d}}\ l_g^{1+D_2-D}\right)(l_i + l_j)^2\left|l_i^{(D+b_d-D_2)/(2-b_d)} - l_j^{(D+b_d-D_2)/(2-b_d)}\right|$$

(14-89)

While this result for differential sedimentation appears quite complex, the first two terms on the right-hand side of Eq. 14-89 could just be replaced by a constant. The fact that the packing and shape factors would be included within this constant term, and that these aggregate properties are a function of a fractal dimension, indicate that the settling velocities and coagulation rates of fractal particles are a strong function of the fractal dimension.

The contribution of the fractal dimension to the magnitude of the Brownian and shear collision frequency functions does not become apparent until the above equations are written in terms of solid volume. In the previous development of the shear coagulation kernel for spherical particles we did not account for the fact that particles will form loose structures containing large void spaces within the aggregates. Thus we assumed the simple Euclidean relationship that the volume of a particle was related to its size by $v = \dot{\zeta}\xi d^3$, or assuming $\dot{\xi} = \pi/6$, $v = (\dot{\zeta}\pi/6)d^3$, where $\dot{\zeta}$ is the packing factor for a close-packed sphere. If we now assume the fractal scaling relationship developed in Chapter 13, or $v = (\xi l_p^3 b_D l_g^D)l_g^D$, we can rewrite Eqs. 14-84–14.86 in terms of volume and the fractal dimension as

$$\beta_{Br} = \frac{2k_BT}{3\mu_w}(v_i^{-1/D} + v_j^{-1/D})(v_i^{1/D} + v_j^{1/D})$$

(14-90)

$$\beta_{sh} = \frac{G}{6\xi_p b_D^{3/D}}v_p^{1-(3/D)}(v_i^{1/D} + v_j^{1/D})^3$$

(14-91)

$$\beta_{ds} = \frac{\pi}{4}\left(\frac{2g(\rho_p - \rho_w)}{b_{d1}\rho_w\xi_2 v^{b_d}}\right)^{1/(2-b_D)}\xi_p^{-(1/3)}b_D^{-\{2+[(b_d-D_2)/(2-b_d)]\}}v_p^{1/3-1/D\{2+[(b_d-D_2)/(2-b_d)]\}}$$

$$\times \left|v_i^{(1/D)[(D+b_d-D_2/(2-b_d)]} - v_j^{(1/D)[(D+b_d-D_2)/(2-b_d)]}\right|(v_i^{1/D} + v_j^{1/D})^2$$

(14-92)

These equations are rather lengthy, but they demonstrate the important contribution of the fractal geometry of porous aggregates to collision efficiencies.

Comparison of Fractal and Rectilinear Collision Functions The collision functions for fractal particles can be larger than those for Euclidean particles for several reasons. First, if particles are measured in terms of solid volume, using, for example, a Coulter counter, then the sizes of the particles will actually be much larger than the solid equivalent diameter reported by the instrument (see Section 13.4 or the example below). The larger the size of the particle, the larger the collision area between it and other particles. Second, the highly porous nature of a fractal aggregate may allow flow through the interior of the aggregate. Third, the highly branched structure may allow areas of very low repulsion, relative to a larger, more solid sphere, due to the low numbers of particles on a branch. Thus branches of fractal aggregates may allow more frequent collisions compared to spherical particles.

The collision functions between one particle of fixed diameter and other particles up to 1000 times as large, are compared for the three different coagulation mechanisms in Figure 14.14. Collision frequencies are compared as the dimensionless ratio of the collision frequency for a fractal aggregate, having fractal dimensions of 1.5, 2, and 2.5. This comparison assumes drag coefficients for spheres for both cases, rectilinear collision functions, and impermeable particles. The large size of the fractal aggregate produces many more collisions than predicted by the solidsphere model. Collision frequencies for fractal particles by Brownian and differential sedimentation could be up to 8.4 and 774 times as large as those predicted for solid spheres. For shear coagulation, collisions between highly amorphous fractal particles could occur 774 times more frequently than those for solid spherical particles of the same mass. These calculations likely overestimate the true collision frequencies because rectilinear collision functions are used and aggregate permeability is not considered.

Example 14.7

Demonstrate that the volume of fractal particles is much larger than that of a solid sphere by comparing the size of a 20-μm-diameter sphere with that of a fractal aggregate with $D = 2.2$. Assume that all shape factors are equal to $\pi/6$, and the packing factors for the sphere and fractal generator are $\acute{\zeta} = 0.73$ and $\zeta = 0.5$, the fractal generator has a diameter of 10 μm, and the primary particles forming the generator have a diameter of 1 μm.

The volume of a sphere is just $v = \dot{\zeta}\acute{\zeta}d_p^3$, while that of a fractal is $v = (\xi l_p^\beta b_D l_g^D)l_{ag}^D$. Equating, substituting in the definition of b_D, and solving for l_{ag}, we have

$$l_{ag} = \left(\frac{\dot{\zeta}\acute{\zeta}d_p^3 l_g^D}{\xi_p l_p^\beta b_D} \right)^{1/D}$$ (14-93)

Figure 14.14 Collision frequencies of fractal particles, $\beta(D)$, with fractal dimensions of 1.5, 2.0, and 2.5, compared to those of solid spheres, $\beta(3)$, for coagulation mechanisms of Brownian motion, shear, and differential sedimentation: v_i is the volume of the aggregate colliding with a particle of volume v_o. (Reprinted with permission from Jiang and Logan 1991. Copyright 1991, American Chemical Society.)

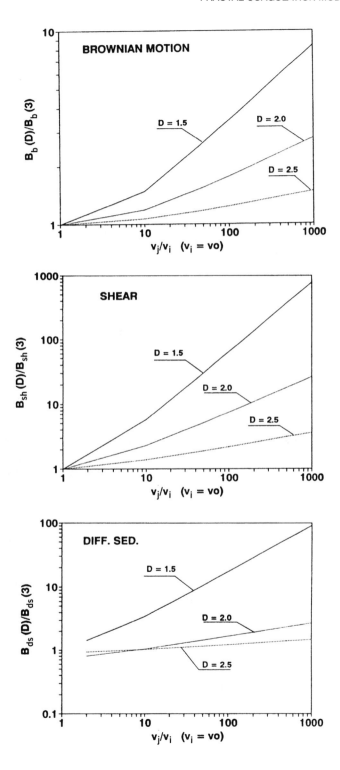

Setting all shape factors to $\pi/6$, we have

$$l_{ag} = \left(\frac{\dot{\zeta} l_g^D}{l_p^3 \zeta_g^{D/3}} \right)^{1/D} d_p^{3/D} = \left(\frac{(0.73)(10)^{2.2}}{(1)^3(0.5)^{2.2/3}} \right)^{1/2.2} (20)^{3/2.2} = 626 \ \mu m \qquad (14\text{-}94)$$

This size of 626 μm is very much larger than that of the 20-μm spherical particle. The main reason for the extraordinarily large difference in particle sizes is that the fractal aggregate probably has a shape factor different from that of a sphere.

Measured Collision Frequencies for Settling and Sheared Fractal Aggregates   Most experiments to test coagulation rate data are conducted with perfect spheres, and yet many applications of interest to environmental engineers involve aggregated particles. Settling fractal particles have an open structure that permits water to flow through the aggregate interior. Particles entrained in this flow can either strike the exterior or the interior of the aggregate. The pores in a fractal aggregate can be quite large, and it is possible that not all particles that enter the aggregate will be removed. This filtration efficiency of porous materials will be further considered in the next chapter.

Collision functions for fractal particles have only recently begun to be investigated. Li and Logan (1997a) grew fractal aggregates in a jar-test apparatus (paddle mixer) using red latex microspheres. They measured the collision rates between these aggregates (200 to 1000 μm in size) and much smaller fluorescent microspheres (1 μm diameter) during settling, and used these data to calculate collision functions. The curvilinear collision frequency function for large settling aggregates and small beads, predicted using Han and Lawler's semiempirical model, was an order of magnitude lower than observed (Fig. 14.15). A rectilinear collision function predicts that all particles below a sinking aggregate (within the projected areas of the two particles) would collide with the aggregate. The observed collision function for the fractal aggregate was two orders of magnitude smaller than predicted using the rectilinear model.

Collision functions calculated in similar experiments using fractal aggregates and fluorescent microspheres were conducted in a jar-test apparatus at shear rates of 2.6, 7.3, and 14.7 s^{-1} (Li and Logan, 1997b). For very dissimilar-sized particles, the calculated curvilinear and rectilinear collision frequency functions are separated by almost seven orders of magnitude (Fig. 14.16). The collision function based on measured collision frequencies was much larger than predicted by curvilinear models (approximately five orders of magnitude), but smaller than that predicted using a rectilinear model (by two orders of magnitude). Thus under conditions of both shear and differential sedimentation, the collision functions for fractal aggregates fall somewhere between those predicted by the rectilinear and curvilinear models.

Steady-State Size Distributions for Fractal Particles

Dimensional analysis can be used to predict the characteristics of particle size spectra for fractal particles using the same approach developed for spherical particles. The

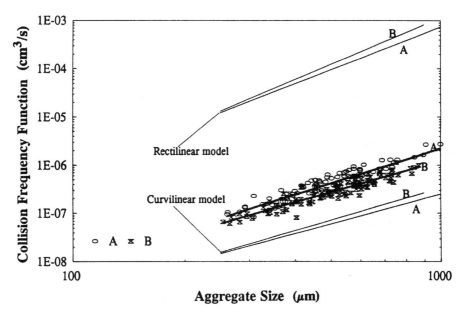

Figure 14.15 Measured collision frequencies between settling aggregates of latex microspheres and smaller (1.48 μm) microspheres versus collision functions predicted using rectilinear and curvilinear models. Group A and B aggregates had fractal dimensions of 1.81 and 2.33. (Reprinted with permission, from Li and Logan 1997a. Copyright 1997 American Chemical Society.)

steady-state particle flux through the size distribution, J_V $[l^3 \ L^{-3} \ t^{-1}]$ must now be considered as the fractal solid volume flux, but otherwise the derivation proceeds as before via a size distribution function of form $n(v, \ J_V, \ \Lambda_{Br}, \ \Lambda_{sh}, \ \Lambda_{ds})$. The same assumptions that were necessary for spherical particles must be made here: The size distribution is steady, and particles enter as very small particles and leave as large particles, sticking coefficients are constant, only one coagulation mechanism is dominant, and a single parameter can characterize each mechanism. The characteristic parameters for Brownian, shear, and differential sedimentation are chosen (Jiang and Logan, 1991) as

$$\Lambda_{Br}[L^3 \ t^{-1}] = \frac{k_B T}{\mu_w} \tag{14-95}$$

$$\Lambda_{sh}[L^3 \ l^{-9/D} \ t^{-1}] = G v_p^{1-(3/D)} \tag{14-96}$$

$$\Lambda_{ds}\{l^{-(3/D)[(4+D-b_d-D_2)/(2-b_d)]}L^3 \ t^{-1}\} = \left(\frac{g(\rho_p - \rho_w)}{v^{b_d}\rho_w}\right)^{1/(2-b_d)} v_p^{(1/3)-(1/D)\{2+[(b_d-D_2)/(2-b_d)]\}} \tag{14-97}$$

where the units of each parameter are shown. Selecting a size interval where only

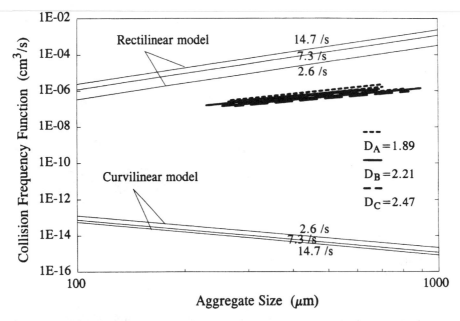

Figure 14.16 Measured collision frequencies between sheared fractal aggregates (fractal dimensions of 1.89, 2.21, and 2.47) of latex microspheres and smaller (1.48 μm) microspheres versus collision functions predicted using rectilinear and curvilinear models at shear rates of 2.6, 7.3, and 14.7 s^{-1}. (Reprinted with permission from Li and Logan 1997b. Copyright 1997, American Chemical Society.)

one coagulation mechanism is assumed to be dominant, $n(v)$ is a function of four variables and three units [l, L, and t]. Using the Buckingham π theorem, the size distribution functions can be shown (Jiang and Logan, 1991) to be

$$n(v)|_{Br} = B_{Br} \left(\frac{J_V}{\Lambda_{Br}} \right)^{1/2} v^{-(3/2)} \tag{14-98}$$

$$n(v)|_{sh} = B_{sh} \left(\frac{J_V}{\Lambda_{sh}} \right)^{1/2} v^{-(3/2)(1+1/D)} \tag{14-99}$$

$$n(v)|_{ds} = B_{ds} \left(\frac{J_V}{\Lambda_{dsr}} \right)^{1/2} v^{-(3/2)-(1/2D)\{1+[(2+D-D_2)/(2-b_D)]\}} \tag{14-100}$$

where B_{br}, B_{sh}, and B_{ds} are constants. Notice for the case of coagulation by Brownian motion that the volume is raised to a power that is not a function of the fractal dimension.

Size distribution functions can also be derived on the basis of particle lengths. The volume and length distributions are related by

$$n(l) = n(v) \frac{dv}{dl} \tag{14-101}$$

Taking the derivative of v, defined in Eq. 3-18 as $v = (\xi_p l_p^\beta b_D l_g^D) l^D$, the above relationship becomes

$$n(l) = \xi_p l_p^\beta b_D D l^{D-1} n(v) \tag{14-102}$$

Using this result, the size distribution functions can be derived based on length as shown in Table 14.6.

Calculating Fractal Dimensions from Steady-State Size Distributions   If size distributions meet the assumed conditions for steady state, then it is possible to calculate fractal dimensions from the slopes of the size distributions. Taking the log of both sides of Eq. 14-97 for coagulation by differential sedimentation, we have the linear equation

$$
\log [n(v)|_{ds}] = \log \left[B_{ds} \left(\frac{J_V}{\Lambda_{ds}} \right)^{1/2} \right]
$$
$$
+ \left[-\frac{3}{2} - \frac{1}{2D} \left(1 + \frac{2 + D - D_2}{2 - b_d} \right) \right] \log v \tag{14-103}
$$

so that on a plot of log $n(v)$ versus log v, the slope s_v would be

$$s_v = -\frac{3}{2} - \frac{1}{2D} \left(1 + \frac{2 + D - D_2}{2 - b_d} \right) \tag{14-104}$$

Slopes from either volume or length distributions can be used to calculate D. The slopes of these different distributions are related by $s_v = (s_l - 2)/3$, where s_l is the slope based on particle–length log–log plots.

Are size distributions at steady state? Using size distributions reported in the literature for different sizes of particles in the ocean, Hunt (1980) compared the slopes from these measured size distributions to those predicted by dimensional analysis. The dominant coagulation mechanism was assigned to each size range based on values of the collision functions using the rectilinear kernels. Hunt found good agreement between those measured and those predicted for spherical particles (Table 14.7). For example, slopes for coagulation by differential sedimentation were predicted for a length distribution to have a slope of -4.5 on log–log plots, while reported slopes were -4.65 to -5.3. Thus it appeared from this analysis that oceanic size distributions were at steady state.

TABLE 14.6 Steady-State Discrete Size Distribution Functions for Fractal Aggregates in Terms of Volume and Length Derived from a Dimensional Analysis

Mechanism	$n(v)$	B'	$n(l)$
Brownian	$B_{Br}\left(\dfrac{J_V}{\Lambda_{Br}}\right)^{1/2} v^{-(3/2)}$	$B_{Br}D\left(\xi_{sp}l_p^{3-D}b_D\right)^{-(1/2)}$	$B'_{Br}\left(\dfrac{J_V}{\Lambda_{Br}}\right)^{1/2} l^{-[1+(D/2)]}$
Shear	$B_{sh}\left(\dfrac{J_V}{\Lambda_{sh}}\right)^{1/2} v^{-(3/2)[1+(1/D)]}$	$B_{sh}D\left(\xi_{sp}l_p^{3-D}b_D\right)^{-(1/2)[1+(3/D)]}$	$B'_{sh}\left(\dfrac{J_V}{\Lambda_{sh}}\right)^{1/2} l^{-(1/2)(D+5)}$
Differential sedimentation	$B_{ds}\left(\dfrac{J_V}{\Lambda_{ds}}\right)^{1/2} v^{-(3/2)-(1/2D)\{1+[(2+D-D_2)(2-b_d)]\}}$	$B_{ds}D\left(\xi_{sp}l_p^{3-D}b_D\right)^{-(1/2)\{1+(1/D)\{1+[(2+D-D_2)(2-b_d)]\}\}}$	$B'_{ds}\left(\dfrac{J_V}{\Lambda_{ds}}\right)^{1/2} l^{-(1/2)\{3+D+[(2+D-D_2)(2-b_d)]\}}$

Source: Jiang and Logan, 1991.

TABLE 14.7 Three-Dimensional Fractal Dimensions Calculated as a Function of the Slope of Different Size Distributions Assuming the System is at Steady State

Size Distribution Type	Brownian	Shear	Differential Sedimentation
Discrete volume, $n(v)$	$—^a$	$\dfrac{-3}{2s_v + 1}$	$\dfrac{D_2 + b_d - 4}{s_v(4 - b_d) - 3b_d + 7}$
Cumulative volume, $N(v)$	$—^a$	$\dfrac{-3}{2S_v + 1}$	$\dfrac{D_2 + b_d - 4}{S_v(4 - b_d) - b_d + 3}$
Discrete length, $n(l)$	$-2(s_l + 1)$	$-(2s_l + 5)$	$\dfrac{s_l(2b_d - 4) + 3b_d + D_2 - 8}{3 - b_d}$
Cumulative length, $N(l)$	$-2S_l$	$-(2S_l + 3)$	$\dfrac{S_l(2b_d - 4) + b_d + D_2 - 4}{3 - b_d}$

^{a}D cannot be calculated since the slope is not a function of D for these two cases for coagulation by Brownian motion where $s_v = -3/2$, and $S_v = -1/2$.

Source: Logan and Kilps, 1995.

Deviations of slopes from those predicted by Hunt (1980) could occur if the particles had fractal dimensions. Rearranging Eq. 14-101, the fractal dimension can be obtained from the slope of a discrete size distribution, s_v, as

$$D = \frac{D_2 + b_d - 4}{s_v(4 - b_d) - 3b_d + 7} \tag{14-105}$$

relationships for D for cumulative size distributions and for Brownian and shear coagulation mechanisms are summarized in Table 14.8. Using Eq. 14-102, we calculate that fractal dimensions of 1.61 to 2.27 could account for deviations of measured slopes (s_v from -2.22 to -2.43) for particles >40 μm in diameter by Hunt for spherical particles (Jiang and Logan, 1991; Table 14.7). This calculation is based on using fractal scaling relationships, such as those in Eq. 14-101, assuming $b_d = 1$, and that $D_2 = D$ for $D < 2$ and $D_2 = 2$ when $D \geq 2$. This range of D calculated for differential sedimentation is reasonable, compared to other values of fractal dimensions reported in the literature. However, fractal dimensions greater than 3 were calculated for particles formed by shear coagulation. This impossible range of D indicates some error either in this calculation or in the assumptions used as a basis for the calculation. Possible reasons for $D > 3$ are errors in particle size measurements, nonsteady conditions, or inaccurate assignment of the coagulation mechanism. Calculations by Han and Lawler (1992) suggest that collision frequencies predicted by shear coagulation are overestimated (at least for spherical particles), and so it

TABLE 14.8 Fractal Dimensions Calculated from Slopes of Discrete Size Distributions Reported in Terms of Either Volume or Length

Distribution	Parameter	Brownian (<2 μm)	Shear (2–40 μm)	Differential Sedimentation (>40 μm)
Length	s_l fractals	$-\left(1 + \dfrac{D}{2}\right)$	$-\left(\dfrac{D + 5}{2}\right)$	$-\dfrac{1}{2}\left(3 + D + \dfrac{2 + D - D_2}{2 - b_d}\right)$
	s_l spheres	-2.5	-4	-4.5
	s_l reported	-2.56	-3.7 to -4.4	$-4.65, -5.2, -5.3$
Volume	s_v fractals	-1.5	$-\dfrac{3}{2}\left(1 + \dfrac{1}{D}\right)$	$-\dfrac{3}{2} - \dfrac{1}{2D}\left(1 + \dfrac{2 + D - D_2}{2 - b_d}\right)$
	s_v spheres	-1.5	-2	-2.17
	s_v measured[a]	—	-1.9 to -2.13	$-2.22, -2.40, -2.43$
	$D(v)$	—	2.4 to 3.75	$2.27, 1.67, 1.61$

[a]Slopes were reported in terms of length, but were actually measured in terms of volume using a resistance-type particle counter. Therefore, the reported slopes in terms of length distributions were converted back to a solid volume distributions assuming the same spherical geometries assumed by particle counter software.

Source: Jiang and Logan, 1991.

may be that shear coagulation was not a dominant coagulation mechanism in this system.

If particle size distributions are measured in terms of both length and solid volume, then the slopes of the size distributions can be used to test whether the system is at steady state by comparing the magnitudes of the fractal dimension obtained from each size distribution. For example, for shear coagulation the slope of the cumulative steady-state size distribution, S_l, can be used to calculate the fractal dimension in terms of length as

$$D = -2S_l - 3 \qquad (14\text{-}106)$$

In terms of solid volume, the analogous relationship is

$$D = \frac{9}{2S_v + 3} \qquad (14\text{-}107)$$

If the system is at steady state, then D calculated using Eqs. 14-106 and 14-107 should be equal. In many instances, however, this has not been found to be the case. A comparison based on steady-state methods and those using the PCT demonstrated that in two different batch coagulation devices steady conditions were not achieved (Table 13.8). For example, fractal dimensions of 1.31 and 2.37 were calculated from size distributions for microsphere aggregates formed in a jar-test apparatus, but the more accurate two-slope method indicated that $D = 1.92$. Similarly, field samples analyzed by Li et al. (1998) using the PCT indicated that $D = 1.77 \pm 0.34$, but an analysis of the same data assuming steady-state conditions produced fractal dimensions of 0.66 and 2.61. Therefore, although the general shape of particle size spectra is consistent with slopes derived from a steady-state analysis, fractal analysis suggests that these systems are not truly at steady state.

14.6 COAGULATION IN THE OCEAN

The ocean contains "particles" that range in size from angstroms (molecules) to meters (whales). It is estimated that nearly 50% of the organic matter in the ocean is in the dissolved form (<0.2 μm). Sheldon et al. (1972) found that over a wide range in particle size the slope of the cumulative size distribution, $N(v)$, remained nearly constant at −3. In order for this to be true, as particle size increases, particle concentration and particle mass must decrease. To what extent does coagulation play a role in the shape of the size distribution of marine "particles"? For larger animals, predator–prey relationships are certainly a factor. Coagulation for these particles is likely unimportant, although the success of a predator–prey interaction (the mass of the two initially combine when the prey is eaten) certainly has an analogy with coagulation in terms of the "efficiency of collision" (particles stick to each other).

Coagulation is likely an important factor in the fate of submicrometer- and micrometer-sized particles. We have already seen in the previous section that the shape of the size distribution of small particles is similar to that expected if coagulation controls the shape of the distribution under steady-state conditions of a constant flux of material through the size distribution. The extent that coagulation is important in determining the shape of the distribution, and even whether marine size spectra can be said to ever be at a steady-state condition, are subjects that are unresolved. In the deep ocean, there are few particles and fluid shear rates are low ($G \ll 1 \text{ s}^{-1}$). McCave (1984) argued that because the rate of particle coagulation is proportional to particle concentration squared, in the deep ocean coagulation would proceed too slowly to affect particle size distributions compared to microbiological and biological rates. Areas in the ocean with high particle concentrations, for example, turbulent mixing near sediments and in coastal areas with high primary production (and therefore high particle concentrations), were noted as possible exceptions to this generalization. Some of the data analyzed by Hunt (1980), demonstrating good agreement between observed and measured slopes of particle size distributions, were measured on sewage sludge for the purposes of understanding the role of coagulation in the fate of particles in ocean outfalls.

The fate of organic matter in the ocean is an important component of global carbon cycles. Two-thirds of the planet is covered with water, and the cycle of carbon fixation and release (through respiration) by phytoplankton is a major factor in this overall balance. Carbon dioxide concentrations in the atmosphere have continued to escalate over the past century. In order for this process to slow down (or be reversed), carbon must be fixed, or converted to particulate organic matter, faster than it is now being liberated from organic forms to the atmosphere by power generation. It was even proposed that by seeding the ocean with iron, a metal that is thought to limit primary production in the open ocean, that algal growth could be stimulated and the rate of carbon fixation accelerated. However, merely stimulating carbon fixation does not assure that the loss of carbon from the atmosphere will be accelerated. Freely suspended algae, or phytoplankton, may not settle fast enough to reach deep ocean sediments before they are consumed by herbivorous animals. Carbon fixed in the form of phytoplankton may merely enhance the growth of small animals that will ultimately convert the carbon back to gaseous CO_2 through respiration.

It is now known that an important factor in the loss of organic matter to deep sediments is due to the net downward flux of large, rapidly sinking particles called marine snow. Marine snow consists of virtually every type of small particle in the water column, including dead and living phytoplankton, bacteria, animal fecal pellets, coagulated debris, and abandoned gelatinous "houses" of small animals called appendicularians (Alldredge and Silver, 1988). The existence of these large (millimeter- to even centimeter-sized) particles was discovered by divers off the coast of California, and later it was shown using underwater cameras that these large, rapidly settling particles were a major contributor to the mass added to sediments (Honjo et al., 1984; Fowler and Knauer, 1986; Alldredge and Silver, 1988). Thus the link between sediment formation and the loss of fixed carbon from the photic zone was shown to be due to large aggregate sedimentation.

Phytoplankton Coagulation   What role does physical coagulation play in the formation of marine snow aggregates? Calculations by Jackson (1990) were instrumental in demonstrating that coagulation could result in the formation of large aggregates from phytoplankton if phytoplankton reached sufficiently high numbers. He modeled the growth of phytoplankton with a 10-μm radius assuming that cells grew exponentially (with a growth rate of 1 d^{-1}), but that the algae were particles that coagulated and therefore that the major losses of cells were through cell aggregation and sinking. The results indicated that large, marine-snow-sized aggregates, would appear to be very rapidly formed (Fig. 14.17). Although aggregation was occurring

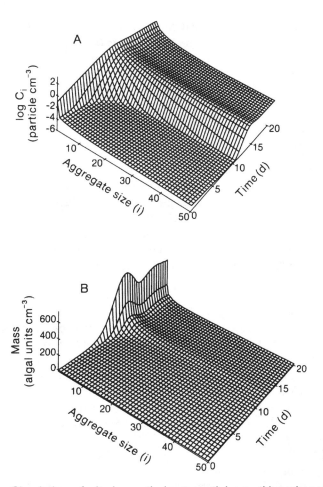

Figure 14.17 Simulation of algal growth (exponential growth) and coagulation. (A) Particle concentrations, and (B) mass concentrations. Parameters are: d_p = 10 μm, μ = 1 d^{-1}, w = 7.7 μm s^{-1}, z = 20 m, α = 0.25, and G = 2 s^{-1}. (Reprinted from Jackson, 1990, with kind permission from Elsevier Science Ltd, The Boulevard, Landford Lane, Kidlington, OX5 1GB.)

constantly, aggregates larger than ~0.2 mm would appear to be formed suddenly when aggregates grew to a size that was visible to the human eye.

As the concentration of algae increases during a phytoplankton bloom in the ocean, the concentration of unattached cells in the water should reach a maximum. Assuming that the dominant loss of cells is through collisions with other cells, then the concentration of unattached cells, N_T, can be modeled as a balance between growth and aggregation (Jackson, 1990) as

$$\frac{dN_T}{dt} = \mu_s N_T - \alpha\beta N_T^2 \qquad (14\text{-}108)$$

where μ_s is the growth rate modified for settling, or $\mu_s = \mu - (w/z)$, w is the cell sinking rate, and z is the depth of the mixed layer (the depth of the layer the cells must settle out to be removed by coagulation). The maximum cell concentration is therefore obtained from Eq. 14-108 when $dN_T/dt = 0$. Assuming coagulation proceeds mainly due to laminar shear coagulation, the maximum cell concentration is

$$N_{max} = \frac{3\mu_s}{4\alpha G d_p^3} \qquad (14\text{-}109)$$

For $d_p = 10$ μm, and typical conditions of $\mu = 1$ d^{-1}, $w = 7.7$ μm s^{-1}, $z = 20$ m, $\alpha = 0.25$, and $G = 2$ s^{-1}, N_{max} is calculated as 5600 cells mL^{-1}. Single-species concentrations of ~10^3 mL^{-1} do occur in the ocean during blooms of some species of phytoplankton, and therefore Eq. 14-109 provides a reasonable estimate of some maximum algal cell concentrations in the ocean (Kiørboe et al., 1994). The observation that marine snow is often dominated by a single species of algae supports calculations that indicate that physical coagulation can produce certain types of marine snow in the ocean.

Marine Snow Formation by TEP   Laboratory experiments by Alldredge et al. (1993) found there could be more to marine snow formation in the ocean than just the coagulation of single species of phytoplankton. Laboratory-grown cultures of a floc-forming diatom, *Chatoceros gracilis*, were coagulated in a laminar shear device in the laboratory. By measuring the rate of formation of dimers, trimers, etc. of cells in the device, it was estimated that $\alpha = 8 \times 10^{-5}$ for this species, and under the coagulation conditions that $t_{1/2} > 24$ h (Eq. 14-62), or that the appearance of large aggregates would take days. However, marine snow aggregates appeared in just minutes! It was discovered that nearly invisible exopolymeric material produced by these cells had accumulated in the culture and that it was this material that was actually coagulating. By measuring the sizes and concentrations of these transparent exopolymer particles (TEP), it was shown that large (>0.5 mm) aggregates could actually form in just a few minutes. It was actually these TEP that were sweeping out the phytoplankton cells into large aggregates. When the aggregate was formed, it appeared to be primarily composed of the cells since they were visible and the polymeric material was not.

Further investigation into phytoplankton coagulation in the ocean has shown that TEP concentrations can be quite high during blooms of some species of phytoplankton. TEP reached concentrations of 590 mL^{-1} in coastal waters off Santa Barbara (Passow et al., 1994), resulting in TEP half-lives of ~1 d, and the appearance of high concentrations of marine snow aggregates only two days later. The appearance of large aggregates was also found to be correlated to TEP concentrations in a freshwater system. In Lake Constance (in Germany) TEP reached a concentration of 860 mL^{-1} and an average size of 20 μm, resulting in a half-life of ~1 day. A few days after this high TEP concentration (and short half-life) was achieved, only 70 to 90 TEP mL^{-1} were subsequently measured in these same surface waters but macroscopic aggregate concentrations had increased eightfold (Logan et al., 1995). Thus these studies demonstrate that different types of particles in the water column (phytoplankton and TEP) can be important in driving aggregate formation.

The finding that different types of particles in the water column can be major contributors to marine snow formation is not unexpected. Less than half of the particulate organic carbon in seawater may be associated with living materials (Sheldon et al., 1972). Although marine scientists tend to describe large increases in individual species of algae as a bloom, there are usually many different types of phytoplankton species in the water at one time and many particles that are not viable. During coagulation, many different particles in the water column may coagulate and form marine snow aggregates.

Submicrometer Particles in the Ocean   The concentration of submicrometer particles (5 to 120 nm) in coastal surface seawaters off southern California has been estimated to be >10^9 mL^{-1} (Wells and Goldberg, 1992). Although high in numbers, the mass concentration of submicrometer particles is estimated to range from 0.03 to 0.09 mg L^{-1}. The high surface area of these colloidal particles (>8 m^2 m^{-3}) suggests they are reactive intermediates in the marine geochemistry of metals and that their fate is important in models of ocean carbon cycling (Moran and Buesseler, 1992). Many of the submicrometer particles <120 nm are in fact aggregates of granules 2–5 nm in size. Larger micrometer-sized particles analyzed from mid-depth and deep waters in the Atlantic and Pacific Oceans have also been found to be the most abundant macroparticles (> 1 μm), with concentrations on the order of 10^5 colloidal aggregates per ml (Wells and Goldberg, 1993). These aggregates have structures that are characteristic of both reaction- and diffusion-limited processes. Models based on ^{234}Th activity (half-life of 24.1 d) suggest that high-molecular-weight components of dissolved organic carbon in the surface open ocean is cycled on the order of ˜7 to 14 d, and that the mean residence time of the colloidal pool is about six times longer (estimates range from 30–40 d and 10–160 d) (Moran and Buesseler, 1992). Not all the dissolved pool is actively cycled to larger particles. Some material in the dissolved pool is apparently quite long-lived (>1000 yr based on ^{14}C techniques). Thus the transfer of material from the dissolved pool to colloidal to micrometer-sized and larger particles by coagulation is apparently variable for different components in the same size fractions.

Example 14.8

The abundance of submicrometer particles in seawater samples preserved with 1 M $HgCl_2$ does not appear to change significantly even after 4 d (Wells and Goldberg, 1992). Compare this observation with a calculation of particle half-lives assuming a sticking coefficient of $\alpha = 0.1$ and 10^9 particles per mL.

Assuming that only collisions generated by Brownian motion are important for the coagulation of small particles, the half-life can be calculated using equations given in Table 14.2 as

$$t_{1/2} = \frac{3\mu \ln 2}{4\alpha k_B T N_{T0}} \tag{14-110}$$

(Notice that this rate is not a function of particle size.) Assuming conditions for seawater of 5°C and $\mu = 0.0164$ g cm^{-1} s^{-1}, and the given conditions, the half-life is

$$t_{1/2} = \frac{3(0.0164 \text{ g cm}^{-1} \text{ s}^{-1})(\ln 2)}{4(0.1)(1.31 \times 10^{-16} \text{ g cm}^2 \text{ s}^{-2} K^{-1})(278\ K)(10^9 \text{ mL}^{-1})} \frac{1 \text{ min}}{60 \text{ s}} \tag{14-111}$$

$$= 39 \text{ min} \tag{14-112}$$

This coagulation rate is much faster than that implied by the observation that particles are stable over long time periods (days). It can take several hours to size-fractionate samples. Thus, it may be that any highly reactive particles (very sticky particles) would have coagulated during sample collection and processing, and that the remaining particles are relatively stable (not sticky). Exactly why particles in highly saline waters (when electrostatic repulsive layer thicknesses are very small) are stable over long time periods is not well understood.

Fractal Dimensions from Marine Size Distributions   If all particles were perfect spheres, and coagulation controlled particle size distributions, then cumulative particle size distributions based on diameter would have slopes of $-3/2$ (Brownian), -3 (shear), and $-7/2$ (differential sedimentation) (Table 14.4). Slopes of particle size distributions have been used in examples and presented in tables at several places in Chapters 13 and 14. Slopes of the cumulative size distributions are typically -3, as shown at three different sites in Pacific coastal waters in Table 14.9. Particle size distributions are not consistent with steady-state conditions since fractal dimensions calculated assuming steady conditions, $D(SS)$, are not the same as those calculated using the TSM or PCT. If the systems were at steady conditions, the fractal nature of the particles would make the slopes more negative. For example, for shear coagulation, the slope of -3.33 implies a fractal dimension of 2.61 if the system is at steady conditions.

Fractal dimensions of marine particles >15 μm are generally in the range of 1.7 to 2.4, a conclusion based on using several different methods to calculate fractal

TABLE 14.9 Fractal Dimensions Measured from the Average Slopes of Particle Size Distributions Using Different Techniques (D_L, Fractal Dimension Based on Length; D_d, Fractal Dimension Based on Solid Equivalent Diameter; SS, Steady-State–One-Slope Method; TS, Two-Slope Method; and PCT, Particle Concentration Technique)

| Site | Average Slopes | | | Fractal Dimensions | | |
	$-S_l$	$-S_d$	$D_L(SS)$	$D(SS)$	$D(TS)$	$D(PCT)$
Monterey Bay, CA Site 1	1.83 ± 0.25	3.33 ± 0.48	0.66 ± 0.49	2.61 ± 0.58	1.67 ± 0.28	1.77 ± 0.34
Monterey Bay, CA Site 2	1.49 ± 0.14	2.59 ± 0.26	-0.01 ± 0.27	4.02 ± 0.54	1.71 ± 0.22	2.12 ± 0.16
East Sound, WA	1.96 ± 0.23	2.49 ± 0.27	0.93 ± 0.46	5.35 ± 2.33	2.42 ± 0.18	2.59 ± 0.17

dimensions. The fractal dimensions of colloidal aggregates is less studied, and therefore less well known. If the shape of colloidal size distributions is controlled by coagulation, then the slope of a cumulative distribution, S_l, should be $-(D + 3)/2$ (Table 14.7). Wells and Goldberg (1992) analyzed particles in seawater isolated by ultracentrifugation and measured using transmission electron microscopy (TEM). They found that $S_l > 5$ (Fig. 14.18), suggesting either nonsteady coagulation conditions for the size fraction or incomplete (or even inaccurate) measurement of particle sizes using the TEM preparation method. However, the high curvature of their size distributions makes any conclusion about the slopes of the distributions uncertain.

Coagulation is certainly an important factor in the size distribution of particles in the ocean, but the extent to which coagulation sets particle size spectra is likely quite variable for different sites. Even at a specific site, the different nature of similar-sized particles (single phytoplankton versus coagulated debris in the water column) makes it difficult to unravel the role of coagulation in setting the size distribution.

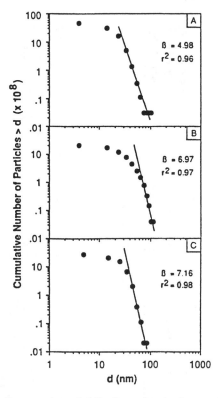

Figure 14.18 Cumulative number distribution of submicrometer particles (<120 nm) taken from surface coastal seawater: (A) 17 May 1988, (B) 25 June 1990, (C) 29 June 1990. (Reprinted from Wells and Goldberg, 1992, with kind permission of Elsevier Science-NL, Sara Burgerhartstraat 25, 1055 KV Amsterdam, The Netherlands.)

As the example below shows, there is much more organic carbon in the submicrometer and molecular-sized pool of organic matter in the ocean than would be expected if coagulation controlled the particle size distribution in the ocean.

Example 14.9

In natural systems, phytoplankton fix CO_2 and bacteria degrade dead matter in order to grow and form new cells. This process can release dissolved organic matter (DOM) into the surrounding water. DOM can be assimilated by bacteria, but some DOM can polymerize into larger molecules (such as polysaccharides, which still may be assimilated by bacteria), while other DOM may be bound up in relatively recalcitrant forms (humic and fulvic acids). How does the production and consumption of this material affect the size distributions? The slopes of particle size distributions in the ocean of particles larger than ~ 1 μm are relatively flat and consistent with particle coagulation. But is the slope of this size distribution constant for particles smaller than 1 μm? And what is the role of coagulation processes in setting the size distribution of submicrometer particles? The answers to these questions are not known, but we can explore these questions using molecular and particle size data obtained from seawater off the coast of California.

Hansell et al. (1993) measured the size distribution of dissolved organic carbon in seawater from Southern California and found a total organic carbon concentration (TOC) of 1.2 mg/L, with 13% of this material retained on a 0.45-μm filter. Samples were size fractionated using ultrafiltration membranes, resulting in 70% of the material being distributed into the <1 K, 1–10 K, and >10 K size fractions, but 17% was not recovered. Let us assume that this unrecovered fraction is in the size range of 100 K to <0.45 μm, and that the >10 K fraction represents only the 10 K–100 K size fraction. Data on the molecular size fractions are given in Figure 14.19. A particle size distribution measured using a resistance-type particle counter on July 29, 1993, in Monterey Bay is shown in the Figure 14.20 (open circles; data from Li et al., 1998).

Assuming that the shape of the particle size distribution in Figure 14.20 is constant, compare the mass distribution measured by Hansell et al. (1993) with that expected from the given size distribution.

In order to calculate the distribution of mass in the seawater, we must use the data from the size distribution to develop an equation relating size and mass. The slope of the size distribution in Figure 14.20 is -4; thus we obtain the size function $n(d_p) = b_d d_p^{-4}$, where $b_d = 2.29 \times 10^5$ [μm^3 mL^{-1}], d_p [μm], and therefore n [# mL^{-1} μm^{-1}]. Given this information, the volume distribution is

$$\frac{dV_p}{d(\log d_p)} = \frac{2.3\pi}{6} d_p^4 n(d_p) = \frac{2.3\pi b_d}{6} \tag{14-113}$$

Assuming a density very close to that of water, or $\rho = 1$ g cm^{-3}, the mass in a

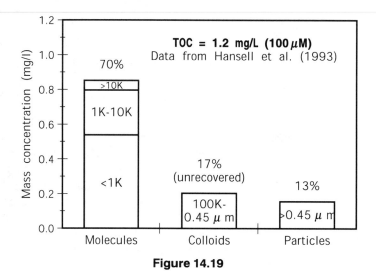

Figure 14.19

log size interval is

$$\Delta m[\text{mg L}^{-1}] = \rho_p \ [\text{g cm}^{-3}] \ \Delta V_p \ [\mu\text{m}^3 \ \text{L}^{-1}] \times 10^{-6} \tag{14-114}$$

$$= \frac{2.3 \times 10^{-6} \pi b}{6} \ \Delta \ \log(d_p)$$

Simplifying

$$\Delta m \ [\text{mg L}^{-1}] = 0.28 \ \Delta \ \log(d_p) \tag{14-115}$$

In order to relate molecule sizes with molecular weights (M), we use data in Table 3.2 to obtain molecule sizes of 0.0013, 0.0029, and 0.0132 μm for molecular weights of 1 K, 10 K, and 100 K. Using this and the above equation, we can plot the results shown in Figure 14.21 assuming an upper limit of particle size of 500 μm.

The predicted size distribution consists of 2.49 mg L^{-1} of all material (inorganic and organic) versus a concentration based on carbon only measured by Hansell et al. of 1.2 mg DOC L^{-1}. If the result of Hansell et al. is converted from a carbon to total organic matter concentration, we would have 3.0 mg L^{-1} assuming a 180:72 = 2.5 ratio (for glucose). Thus the two results on a total concentration basis (if we assume little inorganic matter) are comparable. What we observed from Figure 14.21 is that there is little material predicted to be <1 K, while the measured distribution shows a large concentration of material in this size range. It is likely that a large fraction of small-molecular-weight material is not coagulating, and is therefore stable over long time periods. A second major difference

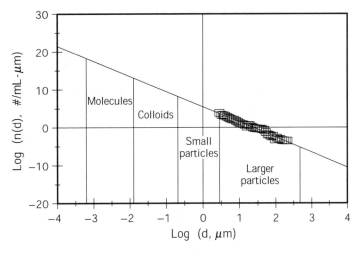

Figure 14.20

is the large fraction of material predicted to be in the 100 K to 0.45-μm size fraction. The measurements made by Hansell et al. indicate very little material in this size range. Thus it would appear that either the existing methods of measuring particle size distributions is not accurate and/or the size distribution is not constant over the molecular size ranges.

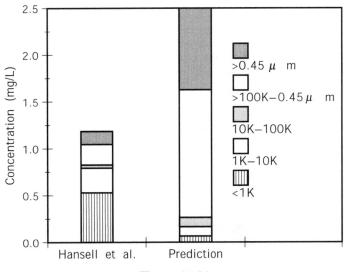

Figure 14.21

PROBLEMS

14.1   The Hamaker constant for particles in aqueous suspensions can range from 0.3×10^{-20} to 10×10^{-20} J, but this constant is often unknown. Compare the interaction forces between two particles 1 μm in diameter having surface potentials of -40 mV, for Hamaker constants of 1×10^{-20} and 8×10^{-20} J, assuming the following constants: ionic strength = 0.1 M; water permitivity, $\varepsilon = 7.1 \times 10^{-12}$ C^2 J^{-1} cm^{-1}; 298 K; Hamaker constant, $\mathcal{A} = 0.8 \times 10^{-21}$ J.

14.2   Repeat the calculation done in Example 14.2, but this time assume a shear rate of $G = 0.1$ s^{-1}. Based on this comparison, what do you conclude about the effect of fluid shear on coagulation?

14.3   Show a detailed derivation of the equation given in Table 14.2 for the half-life of a monodisperse suspension coagulating by Brownian motion.

14.4   A first-order expression was derived for particle coagulation by shear motion for a monodisperse suspension of particles. Derive an analogous first-order expression for particle coagulation by Brownian motion, and compare the particle concentrations predicted for short periods of time using the first- and second-order Brownian motion expressions for conditions of 20°C, $\alpha = 1$, $N_0 = 1 \times 10^7$ mL^{-1}, and $d_p = 1$ μm, over a period of 120 min.

14.5   Measuring particle sizes using a resistance-type instrument, such as a Coulter counter can underestimate the true size of fractal particles because the instrument reports solid diameters. Fractal particles can therefore collide more frequently with other particles than solid particles of the same mass because the fractal particle can sweep out a larger volume of fluid. Assume for this problem that a fractal particle 5 μm in diameter is formed from 13 particles, each 1 μm in diameter. (a) What is the diameter that would be reported by a resistance-type particle counter? (b) Calculate the collision rate of fractal particles with smaller 1-μm particles by calculating the ratio of the collision functions, $\beta(D)/\beta(3)$, for Brownian, shear, and differential sedimentation.

14.6   Dissolved air flotation (DAF) is often used to removed suspended particles in water and wastewater reactors. This problem considers the fate of particles relative to coagulation with bubbles versus coagulation with each other. Let us assume that all collisions of particles with other particles or bubbles are successful. Particles are monodisperse with a diameter of 0.8 μm, while bubbles have a diameter of 50 μm. The cylindrical reactor is 4 m in diameter and is filled with water to a depth of 2 m. The particle concentration is 10^7 mL^{-1}. Assume that bubbles do not coagulate with each other and neglect Brownian coagulation. (a) What is the rise velocity of a 50-μm bubble? (b) How long does it take a single bubble to reach the top of the reactor if the bubble is undisturbed by fluid motion or other bubbles in the reactor? (c) Starting with the general coagulation rate equation, calculate the coagulation rate (#/cm^3 s) of a bubble with other particles assuming there is only 1 bubble in the whole reactor. (d) If the shear rate in the reactor is 100 s^{-1}, what would the overall coagulation rate of the monodisperse particles be with each other? (e) Using your answers in parts (c) and (d), calculate the gas holdup H for the reactor [where H is defined as $H = V_g/(V_g + V_l)$, where V_g is the total volume of gas and V_l is the total liquid volume in the reactor] necessary to produce a coagulation rate by bubbles to be equal to that by particle–particle collisions [calculated in part (d)].

14.7   The coagulation rate of transparent exopolymer particles (TEP) has been speculated to drive coagulation in waters off the coast of California. Let us assume that immediately before a diatom bloom there are 3.5×10^5 cells mL^{-1} and 590 TEP mL^{-1} in the water. The average diameters of TEP and the phytoplankton cells are 26 and 12 μm. Assume cells coagulate only by fluid shear ($G = 1$ s^{-1}). (a) What is the half-life of TEP and the phytoplankton cells assuming only like particles stick to each other, and that for TEP–TEP and cell–cell collisions $\alpha = 1$? (b) If the cells have a much lower sticking coefficient for each other of 10^{-4}, what would be the half-life of the cells? (c) At what α for the phytoplankton would the coagulation rates of the cells be the same as that of TEP with each other, if the α for TEP is unity?

14.8   During a storm in an estuary sediments were resuspended, producing a total solids concentration in the 30-m-deep water column of 40 mg L^{-1}. A particle size distribution revealed $n(d_p) = 10^{5.7} d_p^{-3}$ where n [μm^{-1} mL^{-1}] was measured over the size range of $5 \le d_p \le 50$ μm. Resuspended particles had a density of 1.5 g cm^{-3}, were completely destabilized (the sticking coefficient, α, was unity), and an average shear rate for this section of the estuary is 2 s^{-1}. (a) What fraction of the total mass concentration (40 mg L^{-1}) can be accounted for by the 5 to 50 μm size range? (b) Calculate an average particle size in the 5 to 50 μm size range by (1) particle number concentration, and (2) particle volume concentration. [*Hint*: To see this method, use simple algebra to calculate an average size for 3 balls each of different size by number and volume.] (c) Assuming these particles can be considered to be a monodisperse suspension, calculate the half-life ($t_{1/2}$) of these particles based on the total particle volume if they coagulate only by fluid shear. (d) Assuming an average particle size of 10 μm, compare the rate these monodisperse particles coagulate to the rate that they sink using the dimensionless ratio

$$\sigma = \frac{t_{1/2}}{H/U_p} \qquad (14\text{-}116)$$

where H is the height of the water column and U_p is the particle settling velocity. What does this imply about the importance of coagulation in determining the settling velocities of particles?

14.9   (a) For a monodisperse particle suspension, derive an equation specifying the particle diameter when coagulation by Brownian motion and laminar fluid shear are equal. (b) Calculate this diameter when $G = 20$ s^{-1}, and $T = 20°C$. (c) What shear rate would be necessary to coagulate virus particles ($d_p = 0.1$ μm) equally by shear and Brownian motion? (From Hunt, 1997.)

14.10   Particle coagulation in lakes can be an important process for particle removal, particularly during the spring time, when phytoplankton "bloom" or increase in particle number concentration. Assume the lake can be modeled as a completely mixed reactor of volume V with continuous flow Q through the system. (a) Given an influent particle concentration V_{in}, derive an expression for the effluent particle concentration from the lake assuming particles coagulate according to second-order kinetics, or $dV_p/dt = -bVp^2$. (b) For a lake of surface area A, derive an equation for the rate of sediment accumulation (cm/d) with time. (From Hunt, 1997.)

14.11   Let us examine a monodisperse polysaccharide suspension of 500 μg/L ($d = 0.00132$ μm, $M_P = 10^3$) that coagulate to form particles 1 μm in diameter, assuming monodisperse coagulation kinetics apply throughout the coagulation process. (a) What is N_0? (b) Derive an expression for a first-order rate equation, and solve, obtaining an expression of $t_{1/2} = b/N$. (c) How many doublings are required to obtain 1-μm-sized particles? (d) How long will it take to form these 1-μm-sized particles?

REFERENCES

Adler, P. M. 1981. *J. Coll. Interface Sci.* **83**(1):106.

Alldredge, A. L., U. Passow, and B. E. Logan. 1993. *Deep-Sea Res.* **40**(6):1131–40.

Alldredge, A. L., and W. M. Silver. 1988. *Prog. Oceanogr.* **20**:41–82.

Amirthrajah, A., and C. R. O'Melia. 1990. In: *Water Quality and Treatment*, 4th ed, American Water Works Association, McGraw-Hill, pp. 269–365.

Birkner, F. B., and J. J. Morgan. 1968. *J. Amer. Water Works Assoc.* 174–91.

Camp, T. R., and P. C. Stein. 1943. *J. Boston Soc. Civil Engrs.* **30**:219.

Claesson, P. M., and H. K. Christenson. 1988. *J. Colloid Interface Sci.* **103**:542–53.

Davis, R. H. 1984. *J. Fluid Mech.* **145**:179.

Derjaguin, B. V., and L. D. Landau. 1941. *Acta Physicochim. URSS* **14**:733–62.

Elimelech, M., J. Gregory, X. Jia, and R. Williams. 1995. *Particle Deposition and Aggregation: Measurement, Modeling and Simulation*. Butterworth Heinemann.

Feke, D. L., N. D. Prabhu, and J. A. Mann, Jr. 1984. *J. Phys. Chem.* **88**:5735–39.

Fowler, S. W., and G. A. Knauer. 1986. *Prog. Oceanogr.* **16**:60–68.

Friedlander, S. K. 1977. *Smoke, Dust and Haze*. Wiley, New York.

Gregory, A. 1981. *J. Colloid Interface Sci.* **83**:138–45.

Hamaker, H. C. 1937. *Physica* **4**:1058–72.

Han, M., and D. F. Lawler. 1992. *J. Amer. Water Works Assoc.* **84**(10):79–91.

Hansell, D. A., P. M. Williams, and B. B. Ward. 1993 *Deep Sea Res. 1* **40**(2):219–34.

Hogg, R., T. W. Healy, and D. W. Fuerstenau. 1966. *Trans. Faraday Soc.* **62**:1638–51.

Honjo, S., K. W. Doherty, Y. C. Agrawal, and V. L. Asper. 1984. *Deep-Sea Res. I* **31**:67–76.

Hunt, J. R. 1980. *Adv. Chem. Ser.* **189**:243–57.

Hunt, J. R. 1982. *J. Fluid Mech.* **122**:169–85.

Hunt, J. R. 1997. Lecture notes for Water Treatment Engineering 211, Univ. Calif., Berkeley.

Israelachvili, J. N., and G. E. Adams. 1978. *J. Chem. Soc. Faraday I* **74**:975–1001.

Israelachvili, J. N., and R. M. Pashley. 1978. *J. Colloid Interface Sci.* **98**:500–14.

Jackson, G. A. 1990. *Deep-Sea Res.* **37**(8):1197–211.

Jiang, Q., and B. E. Logan. 1991. *Environ. Sci. Technol.* **25**:2031–38.

Kiørboe, T., C. Lundsgaard, M. Olesen, and J. L. S. Hansen. 1994. *J. Mar. Res.* **52**:297–323.

Li, X., and B. E. Logan. 1997a. *Environ. Sci. Technol.* **31**(4):1229–36.

Li, X., and B. E. Logan. 1997b. *Environ. Sci. Technol.* **31**(4):1237–42.

Li, X., U. Passow, and B. E. Logan. 1998. *Deep Sea Res. I* **45**(1):115–31.

Lin, M. Y., et al. 1989. *Nature* **339**:360–62.

Logan, B. E., and J. R. Kilps. 1995. *Wat. Res.* **29**(2):443–53.

Logan, B. E., U. Passow, A. L. Alldredge, H.-P. Grossart, and M. Simon. 1995. *Deep-Sea Res. II* **42**(1):203–14.

McCave, I. N. 1984. *Deep-Sea Res.* **31**(4):329–52.

Moran, S. B., and K. O. Buesseler. 1992. *Nature* **359**:221–23.

Overbeek, J. Th. G. 1977. *J. Coll. Inter. Sci.* **38**:408–22.

Passow, U., A. L. Alldredge, and B. E. Logan. 1994. *Deep Sea Res.* **41**:335–57.

Ruckenstein, E., and D. C. Prieve. 1976. *AIChE J.* **22**:276–83.

Ryan, J. N., and P. M. Gschwend. 1994. *J. Colloid Inter. Sci.* **164**:21–34.

Schaefer, D. W. 1989. *Science* **243**:1023–27.

Sheldon, R. W., A. Prakash, and W. H. Sutcliffe. 1972. *Limnol. Oceanogr.* **17**:327–40.

Spielman, L. A. 1970. *J. Colloid Interface Sci.* **33**:352.

Swift, D. L., and S. K. Friedlander. 1964. *J. Colloid Sci.* **19**:621–47.

Valioulis, I. A., and E. J. List. 1984. *Adv. Colloid Interface Sci.* **20**:1.

van de Ven, T. G. M., and S. G. Mason. 1977. *Colloid Polymer Sci.* **255**:468.

Verwey, E. J. W., and J. Th. G. Overbeek. 1948. *Theory of the Stability of Lyophobic Colloids.* Elsevier, Amsterdam.

Wells, M. L., and E. D. Goldberg. 1992. *Mar. Chem.* **40**:5–18.

Wells, M. L., and E. D. Goldberg. 1993. *Mar. Chem.* **41**:353–58.

CHAPTER 15

PARTICLE TRANSPORT IN POROUS MEDIA

15.1 INTRODUCTION

Particle removal by filtration during water treatment is critical for improving drinking water quality and protecting human health. Particles contribute to water turbidity and poor water quality. Typical particles in surface waters can be inorganic (clay and minerals), organic and nonviable particles (dead algae and bacteria, detritus and coagulated water column debris), viable cells (bacteria, algae, yeast, amoebae, etc.), spores, and viruses. The large mass of particles in surface waters usually makes direct filtration of this type of water source impractical for economic reasons. Also, bacteria-sized (~ 1 μm) particles are poorly removed in conventional water treatment process filters. Therefore, the water must be treated by a series of processes, starting with a coagulation process to increase particle size and settling velocity. These larger particles can then be efficiently removed by sedimentation; filtration is then used to remove the remainder of the particles in the water. Groundwaters contain few particles and can therefore be filtered without coagulation and sedimentation processes. If direct filtration is used, it is often necessary to add a coagulant aid to make sure that particles are very sticky and more easily removed in the filter.

The transport of microorganisms in soil aquifers can be both beneficial and harmful. It is undesirable, for example, for pathogenic bacteria to migrate long distances from discharge points in the ground such as septic tanks. However, the remediation of soils contaminated with organic pollutants can be enhanced via bioaugmentation, or the injection of certain strains of bacteria able to degrade these pollutants. In order for these bacterial strains to travel over long distance, their adhesion for the soil must be reduced. Bacteria and other colloidal-sized particles should be removed by collisions with soil particles in a groundwater aquifer in the same manner as in a water treatment filter, but a surprising array of particles have been found to be transported in groundwater aquifers. Mineral and organic colloids are usually found suspended in groundwater, indicating that they have become stable, likely through adsorption of dissolved organic matter to their surfaces. Some bacteria and spores can be transported over long distances. Indigenous bacteria appear to be able to travel long distances in aquifers relative to laboratory-derived strains. Bacteria can also move in a stepwise fashion, by attachment and growth, followed by daughter cell production, release, and transport over a short distance. Through repetition of this growth and cell release process, cells can be found at relatively large distances from their origins. Bacteria and viruses have been found to travel hundreds of meters from septic tanks, and viable cells have been found thousands of meters below the surface. Thus it appears that porous media systems, such as groundwater aquifers, are imperfect filters.

In this chapter we will explore the mechanisms of particle transport and removal in porous media. In order to quantify particle transport, a clean-bed filtration model will be developed by reducing the three-dimensional general transport equation to obtain a description of particle transport in one dimension. The assumptions necessary to derive a filtration equation are quite different from those made to describe chemical transport. Therefore, it will be interesting to contrast the important differences between an equation that describes chemical (molecule) transport from nearly identical equations, at least in general appearance, that describe particle transport.

15.2   A MACROSCOPIC PARTICLE TRANSPORT EQUATION

Equations describing chemical transport in porous media have already been specified based on the generalized transport equation (Eq. 12-31) by defining the arbitrary function ξ as c, the chemical concentration. In one dimension, nonsteady transport of chemical C in an infinite medium with reaction,

$$\frac{\partial c}{\partial t} + u\,\frac{\partial c}{\partial x} - E_L\,\frac{\partial^2 c}{\partial x^2} - R = 0 \tag{15-1}$$

where u is the average velocity in the direction of flow, E_L is a longitudinal dispersion coefficient, and R includes any processes related to particle reaction or attachment and detachment. The same governing equation applies to particle transport. For a distribution of particle sizes, we have by direct substitution of n, the size distribution function for the chemical concentration, c

$$\frac{\partial n(d)}{\partial t} + u\,\frac{\partial n(d)}{\partial x} - E_L\,\frac{\partial^2 n(d)}{\partial x^2} - R = 0 \tag{15-2}$$

Restricting our attention for the moment to just one type and size class of particles, we can replace n with N, the number concentration of a single-sized particle to obtain

$$\frac{\partial N}{\partial t} + u\,\frac{\partial N}{\partial x} - E_L\,\frac{\partial^2 N}{\partial x^2} - R = 0 \tag{15-3}$$

The term R must now be defined for particle transport in porous media. Should the loss of particles during transport, R, be considered as a chemical reaction, or does R represent a loss due to mass transport? Assume for a moment that mass transport controls particle removal. Equation 15-3 can then be written in terms of a mass transport coefficient as

$$\frac{\partial N}{\partial t} + u \frac{\partial N}{\partial x} - E_L \frac{\partial^2 N}{\partial x^2} + \mathbf{k}a_v(N - N_s) = 0 \tag{15-4}$$

where $\mathbf{k}$ is a mass transport coefficient, a_V is the surface area per volume, and N_s is a surface concentration. For a "clean bed," or soil particles containing few deposited particles, we can assume $N \gg N_s$

$$\frac{\partial N}{\partial t} + u \frac{\partial N}{\partial x} - E_L \frac{\partial^2 N}{\partial x^2} + \mathbf{k}a_V N = 0 \tag{15-5}$$

If particle transport is mass transport controlled, then $\mathbf{k}$ in Eq. 15-5 should scale with fluid velocity, particle diffusivity, etc. For mass (and heat) transport, we have already seen that we can correlate mass transport data for a packed bed using the dimensionless factor j_D, or Colburn factor, defined as

$$j_D = \frac{Sh}{Re \ Sc^{1/3}} \tag{15-6}$$

For chemical transport, Pfeffer and Happel (1964) predicted for mass transport in a packed bed

$$Sh = B \ Pe^{1/3} \tag{15-7}$$

where B is a constant. Since $Pe = Re \ Sc$, we can rewrite this as

$$j_D = B \ Re^{-2/3} \tag{15-8}$$

Equation 15-8 was shown to describe the transport of a variety of organic solvents in air and chemicals (such as benzoic acid and 2-naphthol) in water (Pfeffer, 1964). Subsequently, Cookson (1970) also demonstrated that Eq. 15-8 was equally valid for very small particles by measuring the removal of virus particles ($E. \ coli$ bacteriophage T_4) in packed beds of activated carbon.

Dispersion coefficients of particles are often neglected in calculating particle filtration rates. Assuming a pore velocity of 0.034 cm s^{-1}, $D_p = 4.9 \times 10^{-9}$ cm^{2}s^{-1} (calculated using the Stokes–Einstein equation for a 1-μm-diameter particle), and soil grain diameter of $d_c = 0.012$ cm, the correlation of Blackwell (1959) produces $E_L = 2 \times 10^{-2}$ cm^2 s^{-1}. Dispersion coefficients for different types of bacteria-sized particles and viruses in porous media range from $\sim 10^{-2}$ to $\sim 10^{-5}$ cm^2 s^{-1} (Table 15.1).

TABLE 15.1 Dispersion Coefficients Measured in Colloid Transport Experiments

Colloid Type and System	E_L (cm^2 s^{-1})	Reference
Bacteriophage (MS-2)—adsorption phase	1.27×10^{-2} 8.9×10^{-2} 2.95×10^{-2}	Bales et al. (1991)
Bacteriophage (MS-2)—desorption phase	5.3×10^{-4} 7.78×10^{-3} 1.6×10^{-4}	Bales et al. (1991)
Bacteriophage (PRD-1)— adsorption phase	3.1×10^{-5} 2.7×10^{-5}	Bales et al. (1991)
Hydrophilic bacterium (S5), quartz medium	$(1.92 \pm 0.48) \times 10^{-4}$	McCaulou et al. (1994)
Hydrophobic bacterium (S139), quartz medium	$(1.65 \pm 0.36) \times 10^{-4}$	McCaulou et al. (1994)
Pseudomonas sp. KL2 on aquifer sand	4.4×10^{-4}	Tan et al. (1994)
Pseudomonas putida PRS2000, motile, sand	3.5×10^{-5}	Barton and Ford (1995)
Deep subsurface isolate AO500, sediments from the Ringold formation, Hanford, WA	2.0×10^{-2}	McCaulou et al. (1995)

Source: Logan et al., 1997.

Assuming steady conditions, and neglecting dispersion, Eq. 15-5 becomes

$$u \frac{dN}{dx} = -\mathbf{k}a_v N \tag{15-9}$$

The concentration of particles declines exponentially during transport through a column so that N can be defined as a log-mean average

$$N_{lm} = \frac{N_{\text{in}} - N_{\text{out}}}{\ln(N_{\text{in}}/N_{\text{out}})} \tag{15-10}$$

and the mass transport coefficient can then be calculated as

$$\mathbf{k} = \frac{u(N_{\text{in}} - N_{\text{out}})}{a_v L N_{lm}} \tag{15-11}$$

Using Eq. 15-10, Cookson (1970) obtained for virus removal

$$j_D = 1.72 \, Re^{-0.65} \tag{15-12}$$

for column heights of 7.50 to 12.40 cm. These j_D values were ~30% higher than those obtained for longer columns (25.4 cm).

It is also possible to define particle removal in porous media as a kinetically controlled process (Bai et al., 1997). Defining R as a first-order rate of reaction, Eq. 15-3 becomes

$$\frac{\partial N}{\partial t} + u\frac{\partial N}{\partial x} - E_L\frac{\partial^2 N}{\partial x^2} + k_{1,a}N = 0 \tag{15-13}$$

where $k_{1,a}$ is a first-order rate constant describing the adsorption of particles onto media surfaces. Thus we can see that if $k_{1,a} = \mathbf{k}a_V$, particle transport can be specified either in terms of a kinetic or mass transport model since the governing equations would be the same. Particle desorption can also be incorporated in either model in terms of a first-order desorption rate constant, $k_{1,d}$. For the kinetic model, this produces

$$\frac{\partial N}{\partial t} + u\frac{\partial N}{\partial x} - E_L\frac{\partial^2 N}{\partial x^2} + k_{1,a}N - k_{1,d}\frac{\rho_b}{\theta}S = 0 \tag{15-14}$$

where S is the number of adsorbed particles per mass of the media surface.

The disadvantage of using a kinetic model such as Eq. 15-13 to describe particle removal in porous media is that $k_{1,a}$ is defined as a constant, and therefore is not a function of column hydraulic conditions. The finding by Cookson (1970) that virus removal obeyed the same scaling relationships used for heat and mass transport is proof that the rate of particle transport is a function of fluid hydrodynamics, and therefore that particle removal is mass-transport controlled. Particle transport to a surface reaches a maximum rate when the surface concentration is zero, or when $\mathbf{k}a_V(N - N_s) \rightarrow \mathbf{k}a_V N$. Because particles can only be transported at a certain rate to a surface, there are physical limits to the magnitude of the reaction term R and, therefore, the mass transport coefficient $\mathbf{k}$, that are included in a mass transport model. However, there are no physical limits to the reaction rate constant. Thus, a mass transport model provides a more physically realistic picture of particle transport to a surface than a kinetic model.

15.3 CLEAN-BED FILTRATION THEORY

Particle removal in packed beds and other porous media is usually described using clean-bed filtration theory. The term *clean bed* refers to the assumption that the media grains, or collectors, are homogeneous and do not contain enough deposited particles to affect the subsequent deposition of additional particles. This clean-bed assumption mathematically translates to $N_s \approx 0$, as assumed in deriving Eqs. 15-5 and 15-9. The filtration equation can be obtained by integration of Eq. 15-9 for particle removal over a distance L as

$$\frac{N}{N_0} = \exp\left(-\frac{\mathbf{k}a_v L}{u}\right) \tag{15-15}$$

Equation 15-15 is the maximum possible rate of particle removal. However, not all particles transported to the surface of the packing media stick, and therefore this rate can be reduced depending on the relative affinity of the particle for the media grains. Defining α as the sticking coefficient, where $0 \leq \alpha \leq 1$, and defining a filtration constant as $\lambda = \mathbf{k}a_v/u$, we obtain the one-dimensional clean-bed filtration equation as

$$\frac{N}{N_0} = e^{-\alpha\lambda L} \tag{15-16}$$

If the kinetic model basis was assumed for calculating Eq. 15-16, we would have used $\lambda = k_{1,a}/u$ as the rate constant assuming first-order particle removal kinetics.

Mass transport and kinetic constants are rarely used in the filtration literature to describe particle removal and transport in porous media. Instead, a filtration equation such as Eq. 15-16 is used, where all the factors that affect particle transport, such as fluid velocity, temperature, media, and particle size, are contained in the filtration constant λ. There are two steps involved in deriving a filtration equation: First, a mass balance is conducted over a control volume based on fluid flow around a collector (a sand grain in our case); second, analytical and empirical correlations are used to predict the frequency of particle–collector collisions. There is no single, correct approach for deriving a filtration equation, and even the form of the equation resulting from a mass balance can vary depending on the assumptions made during the mass balance. As will be shown, the semiempirical filtration equation developed by Rajagopalan and Tien (1976), referred to here as the RT model, has been shown to provide the best agreement with data, and therefore this equation is preferred by aquasol scientists for predicting particle removal in soil columns and groundwater aquifers. In order better to understand the formulation of this equation, and filtration models in general, the RT and other filtration models are derived in the sections that follow. We begin with a development of the best-known filtration model for aquasols, the Yao model.

The Yao Filtration Model

In order to describe particle removal in packed beds, Yao et al. (1971) derived a one-dimensional clean-bed filtration equation from a mass balance around a single collector using the differential slice shown in Figure 15.1. Notice that the control volume is not truly a differential slice, since several particles are contained within the slice. Starting with our mass balance equation

$$\text{accumulation} = \text{in} - \text{out} + \text{reaction} \tag{15-17}$$

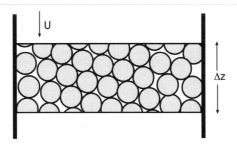

Figure 15.1   Differential slice of thickness Δz used as the control volume in deriving a one-dimensional clean bed filtration equation. The approach (superficial, or Darcy) velocity is U, while the velocity inside the packed bed is the pore velocity, $u = U/\theta$. This increased flow results from the media reducing the available area for flow from A to $A\theta$.

Particle accumulation within the control volume is

$$\text{accumulation} = \Delta N(A\theta) \, \Delta z \tag{15-18}$$

where the control volume, ΔV, has a thickness Δz, cross-sectional area A, and porosity θ. Parentheses are used around the $(A\theta)$ term to emphasize that the porosity reduces the cross-sectional area of flow.

Particles will be assumed to enter into the control volume by advection and dispersion. The number of particles entering into the control volume during a time t by advection are

$$\text{in(advection)} = N_z(A\theta)u \, \Delta t \tag{15-19}$$

where u is the interstitial or pore velocity. The rate of particles entering by dispersion is

$$\text{in(dispersion)} = -E_L(A\theta) \, \Delta t (dN_z/dz) \tag{15-20}$$

The rate particles leave the control volume by both dispersion and advection is therefore

$$\text{out} = -[N_{z+\Delta z}(A\theta)u \, \Delta t] - [-E_L(A\theta) \, \Delta t (dN_{z+\Delta z}/dz)] \tag{15-21}$$

The number of particles removed in the control volume is calculated for a single packing grain, referred to here as a collector, and multiplied by the number of collectors, N_c. Based on the column porosity, N_c is calculated for a spherical collector of diameter d_c, as the total solid volume divided by the volume of an individual collector, or

$$N_c = \frac{A(1 - \theta)\, \Delta z}{\frac{\pi}{6}\, d_c^3} \tag{15-22}$$

Particle accumulation on each collector, a_c, is

$$a_c = r_s \alpha N_c\, \Delta t \tag{15-23}$$

The rate particles that strike a collector, r_s, is calculated in terms of the single collector efficiency, defined as

$$\eta = \frac{r_s}{r_f} \tag{15-24}$$

where r_f is the rate that particles flow toward the collector. In the Yao model, r_f is calculated for an isolated collector, and, as a consequence, the rate particles flow toward the collector is at the superficial, or approach, velocity, $U = u\theta$. Yao et al. (1971) used the superficial velocity rather than the pore velocity because their solution of η relied upon an analytical solution of flow around an isolated collector (see below). This made their mass transport model consistent with their mass balance. The rate that particles strike a single collector is therefore

$$r_s = \eta U N \frac{\pi}{4}\, d_c^2 \tag{15-25}$$

and from Eq. 15-23, the particle accumulation on the collector becomes

$$a_c = \frac{3\alpha\eta(1 - \theta)NAU\, \Delta z\, \Delta t}{2 d_c} \tag{15-26}$$

Combining all terms into Eq. 15-17, and rearranging, we have

$$-u\theta\, \frac{(N_{z+\Delta z} - N_z)}{\Delta z} + E_L \theta \left(\frac{\partial N_{z+\Delta z}}{\partial z} - \frac{\partial N_z}{\partial z} \right) \Big/ \Delta z$$

$$= \theta\, \frac{\Delta N}{\Delta t} + \frac{3}{2 d_c}\, (1 - \theta)\alpha\eta N U \tag{15-27}$$

Taking the limit as Δz and Δt go to zero, dividing by θ, and substituting in $U = u\theta$

$$\frac{\partial N}{\partial t} + u\, \frac{\partial N}{\partial z} - E_L\, \frac{\partial^2 N}{\partial z^2} + \left(\frac{3}{2 d_c}\, (1 - \theta)\alpha\eta u \right) N = 0 \tag{15-28}$$

This result is in the same form as Eq. 15-5, except here the dependence of the loss term is explicitly defined in terms of flow in the porous medium. Neglecting dispersion, under steady conditions the particle removal is

$$\frac{dN}{dz} = -\frac{3}{2d_c} (1 - \theta)\alpha\eta N \qquad (15\text{-}29)$$

Integrating, the Yao filtration equation becomes

$$\frac{N}{N_0} = \exp\left(-\frac{3}{2d_c} (1 - \theta)\alpha\eta L\right) \qquad (15\text{-}30)$$

Defining the filter coefficient in terms of these constants, or

$$\lambda = \frac{3}{2d_c} (1 - \theta)\eta \qquad (15\text{-}31)$$

we obtain Eq. 15-16, the one-dimensional, clean-bed filtration equation.

The Collector Efficiency, η   While a macroscopic balance produces a governing transport equation, mass transport calculations at the microscopic level are all hidden in the formulation of the collector efficiency. In the Yao model, each grain in the porous medium is considered to be an isolated collector with flow lines unaffected by adjacent collectors. Thus the fluid approach velocity is the free stream velocity, U. Particles can collide with an isolated collector by three different mechanisms: diffusion, interception, and gravitational sedimentation (Fig. 15.2). Particles suspended in water will follow the streamlines. If the particle is on a streamline that takes the particle near enough to the packing grain, the particle intercepts the collector and is removed from the water. Particles can move off a streamline by Brownian motion, by diffusion, or by sedimentation.

The complete solution for mass transport to an isolated collector by diffusion was available to Yao et al. (1971) based on an analysis by Levich (1962). Levich demonstrated that this solution could be expressed as a function of the Peclet number, as

$$\eta_D = 4.04 \, Pe^{-2/3} \qquad (15\text{-}32)$$

where η_D is the collector efficiency due to particle diffusion, $Pe = Ud_c/D_p$, and the particle diffusivity is obtained from the Stokes–Einstein equation as $D_p = k_B T/(3\pi\mu d_p)$, where k_B is the Boltzmann constant.

Particle collisions due to interception occur for particles on a streamline closer than the radius of the particle. The collector efficiency due to interception can be

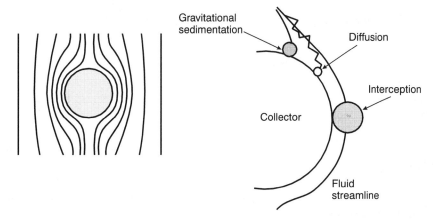

Figure 15.2 (A) Fluid streamlines around an isolated collector in a uniform flow field *U*. (B) Particles moving along streamlines are removed by gravitational sedimentation off a streamline, diffusion off a streamline, or direct interception of the particle with the collector surface.

expressed as a function of the interception number, $R^* = d_p/d_c$, as

$$\eta_I = \tfrac{3}{2} R^{*2} \tag{15-33}$$

Particles with an appreciable settling velocity can move off a streamline and contact the medium grains. The collector efficiency by sedimentation is therefore just the ratio of the particle's settling velocity to the free stream fluid velocity. Defining this ratio as the dimensionless number $S^* = u_{p,s}/U$, we have

$$\eta_S = S^* \tag{15-34}$$

It is usually assumed that particles settle at a velocity given by Stokes' Law, or $u_{p,s}$ = $g\,\Delta\rho\,d_p^2/(18\upsilon\rho)$, where υ, and ρ are calculated for water.

By superposition, or assuming that collisions are additive by each mechanism, the overall collision rate is a sum of the efficiencies by each mechanism, or

$$\eta = \eta_D + \eta_I + \eta_S \tag{15-35}$$

Combining the results from above, we can write the complete solution for η as

$$\eta = 0.9 \left(\frac{k_B T}{\mu d_c d_p U} \right)^{2/3} + \frac{3}{2} \left(\frac{d_p}{d_c} \right)^2 + \frac{g\,\Delta\rho}{18\upsilon\rho U}\, d_p^2 \tag{15-36}$$

or, from the definition of each of the terms, in dimensionless form as

$$\eta = 4.04 \ Pe^{-2/3} + \tfrac{3}{2}R^{*2} + S^* \tag{15-37}$$

The combination of Eqs. 15-30 and 15-37 constitutes what is referred to as the Yao clean-bed filtration model.

Example 15.1

Show the effect of particle size, for particles in the size range of 0.2 to 5 μm, on the collector efficiency for typical groundwater flow conditions of: 120 μm soil particles, Darcy velocity of 1 m d^{-1}, and 10°C.

   The effect of particle size can be shown by plotting η for the individual collision mechanisms as a function of particle size. As an example calculation, consider the case of $d_p = 1$ μm. At 10°C, the dynamic viscosity of water is 0.0103 g cm^{-1} s^{-1}, and η_D is

$$\eta_D = 0.9 \left(\frac{k_B T}{\mu d_c d_p U} \right)^{2/3}$$

$$= 0.9 \left(\frac{(1.38 \times 10^{-16} g \ cm^2 \ s^{-2} \ K^{-1})(283 \ K) \times 86,400 \ s \ d^{-1}}{(0.013g \ cm^{-1} \ s^{-1})(1 \times 10^{-4} \ cm)(120 \times 10^{-4} \ cm)(100 \ cm \ d^{-1})} \right)^{2/3} \tag{15-38}$$

$$= 0.0147 \tag{15-39}$$

For interception

$$\eta_I = \frac{3}{2} \left(\frac{d_p}{d_c} \right)^2 = \frac{3}{2} \left(\frac{1 \ \mu m}{120 \ \mu m} \right)^2 = 1.04 \times 10^{-4} \tag{15-40}$$

and for gravitational settling

$$\eta_S = \frac{g \ \Delta \rho}{18 \upsilon \rho U} d_p^2$$

$$= \frac{(980 \ cm \ s^{-2})(1.07 - 0.9997)g \ cm^{-3}) \times 86,400 \ s \ d^{-1}}{18(0.0131 \ cm^2 \ s^{-1})(0.9997g \ cm^{-3})(100 \ cm \ d^{-1})} (1 \times 10^{-4} \ cm)^2 \tag{15-41}$$

$$= 0.00253 \tag{15-42}$$

Summing up the contributions from all three coagulation mechanisms, we have

$$\eta = \eta_D + \eta_I + \eta_S = (0.0147) + (1.04 \times 10^{-4}) + (0.00253) = 0.0173 \quad (15\text{-}43)$$

Diffusion contributes 84.8% to the overall collector efficiency, while interception and gravitational sedimentation contribute 0.6 and 14.6%. As shown in Figure 15.3, diffusion is the dominant collision diffusion mechanism over only part of the given size range. At 5 μm, gravitational sedimentation is the predominant collision mechanism, contributing 89% of the total collector efficiency.

The Pore Velocity Model

Different forms of the filtration model can result from a mass balance depending on how the fluid environment of the collector is defined in the filtration equation (Logan et al., 1995). If we had just one collector in a very wide tube of water, the approach velocity to the particle in the fluid U would be the same at any radial location as $r \to \infty$. However, medium grains in packed beds are not truly isolated because the adjacent collectors reduce the area available for flow and increase the approach velocity. The average velocity approaching a collector in a packed bed is therefore the pore velocity, where $u = U/\theta$. The fluid velocity around a collector increases from zero at the particle's surface to the maximum flow velocity between adjacent collectors, and then decreases again to zero at another particle's surface. If a mass balance is performed as above for a slice of porous media in a packed bed, except the pore velocity is used to define the approach velocity to the collector, the analysis is identical to that given above except u replaces U in Eq. 15-25. The rate

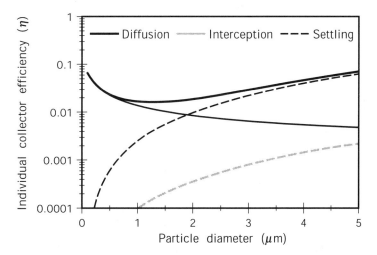

Figure 15.3 Comparison of the different collision mechanisms. The overall collector efficiency (the thick top line) is the sum of the individual efficiencies.

that particles strike the collector becomes

$$r_s = \eta u c \, \frac{\pi}{4} \, d_c^2 \qquad (15\text{-}44)$$

The remainder of the analysis proceeds as before, but this time the final filtration equation includes an additional porosity variable in the exponential term (Flagan and Seinfeld, 1988; Logan et al., 1995), or

$$\frac{N}{N_0} = \exp\left(-\frac{3}{2d_c}\frac{(1-\theta)}{\theta}\alpha\eta L\right) \qquad (15\text{-}45)$$

Equation 15-39, referred to as the pore velocity filtration model, is commonly used to describe aerosol filtration, but Eq. 15-30 is preferred for aquasols. The collector efficiency calculated for the Yao model (Eq. 15-37) can be used in the pore velocity filtration equation, but the velocity used to calculate Pe and S is the pore, and not superficial, velocity.

The Happel Model

Happel (1958) devised a different approach for studying transport in packed beds. He hypothesized that a better approximation of fluid flow around a particle in a packed bed could be obtained by assuming that the flow was completely concentric around the particle (Fig. 15.4). The thickness of the fluid layer around a solid particle was obtained by taking all the fluid in a packed bed and allocating it equally to all

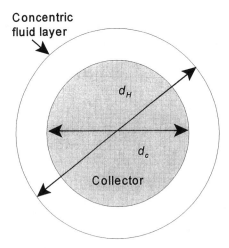

Figure 15.4   Flow around a collector according to the Happel model. Flow is confined to flow within a concentric sphere of diameter d_c around the spherical collector of diameter d_H.

particles. Pfeffer (1964) and Pfeffer and Happel (1964) used this concentric sphere model to obtain a mass transport correlation in the same manner as Levich (1962), for high Pe and low Re, as

$$Sh = 4.04 b_H^{1/3} \, Pe^{1/3} \tag{15-46}$$

where b_H is a constant defined in terms of the column porosity as

$$b_H = \left(\frac{2(1 - \gamma^5)}{2 - 3\gamma + 3\gamma^5 - 2\gamma^6} \right) \tag{15-47}$$

where $\gamma = (1 - \theta)^{1/3}$. As the porosity of the system becomes large, or as $\gamma \to 0$ and $\theta \to 1$, Eq. 15-46 reduces to the Levich solution for an isolated collector (Eq. 15-32), as will be shown below. Comparing this result to our mass transport correlation presented earlier, we see that the constant required in Eq. 15-7 is specified more completely by Eqs. 15-46 and 15-47.

If the mass transport correlation given in Eq. 15-46 is to be used in a mass transport equation (Eq. 15-15), then the interfacial area per volume, a_V, must be calculated in terms of column parameters. Using the Happel cell geometry shown in Figure 15.4, the cross-sectional area for the approaching flow is $(\pi/4)d_c^2$. For a Happel-cell volume of $(\pi/6)d_H^3$, where d_H is the diameter of the sphere containing fluid, the interfacial area per volume is

$$a_V = \frac{(\pi/4)d_c^2}{(\pi/6)d_h^3} = \frac{3}{2} \frac{d_c^2}{d_H^3} \tag{15-48}$$

The Happel cell and collector diameters are related by the media porosity. Because all the void volume is used to calculate the volume of the Happel cell, the void volume is the difference between the Happel cell volume, V_H, and the collector volume, V_C. The porosity, which is defined as the void volume per total volume, can therefore be calculated as

$$\theta = \frac{V_H - V_c}{V_H} = 1 - \frac{(\pi/6)d_c^3}{(\pi/6)d_H^3} = 1 - \frac{d_c^3}{d_H^3} \tag{15-49}$$

Combining the above two equations, the interfacial area can be written in terms of the porosity and collector diameter as

$$a_V = \frac{3(1 - \theta)}{2d_c} \tag{15-50}$$

The interfacial area can now be used in the filtration equation we derived in terms of a mass transport coefficient (Eq. 15-15) to obtain an expression for the filtration rate of particles removed only by diffusive processes as

$$\frac{N}{N_0} = \exp\left(\frac{3(1-\theta)}{2d_c}\frac{k}{u}L\right) \tag{15-51}$$

This can also be written in terms of a Sh correlation. Starting with our mass transport correlation that $Sh = 4.04\, b_H^{1/3}\, Pe^{1/3}$, and the definition of $Sh = \mathbf{k}d_c/D$, we have that

$$\frac{\mathbf{k}}{u} = \frac{D}{d_c u}\, 4.04 b_H^{1/3}\, Pe^{1/3} = 4.04 b_H^{1/3}\, Pe^{-2/3} \tag{15-52}$$

Combining Eqs. 15-51 and 15-52

$$\frac{N}{N_0} = \exp\left(\frac{3(1-\theta)}{2d_c}\, 4.04 b_H^{1/3}\, Pe^{-2/3}\, L\right) \tag{15-53}$$

This result based on the Happel cell is only valid for small particles ($d_p \ll 1\ \mu$m), or when particle removal occurs primarily by diffusion, and therefore interception and sedimentation are negligible.

The Yao–Habibian Filtration Model

In their original paper, Yao et al. (1971) presented data showing that filtration rates measured for completely destabilized particles were larger than those predicted by their isolated collector model (the Yao model). They attributed this greater filtration rate to higher collision frequencies than those predicted by diffusion for an isolated collector. They therefore proposed that the Happel model could be combined with their model by replacing the collector efficiency (η_D) calculated for diffusion by an isolated collector with one based on the Happel cell. Comparing Eq. 15-53 with Eqs. 15-30 and 15-37, the collision frequency for diffusion based on the Happel cells is

$$\eta_D = 4.04 b_H^{1/3}\, Pe^{-2/3} \tag{15-54}$$

The complete Yao–Habibian filtration equation is therefore

$$\eta = 4.04 b_H^{1/3}\, Pe^{-2/3} + \tfrac{3}{2}R^{*2} + S^* \tag{15-55}$$

This result predicts particle removal more accurately than the Yao model, but it has the disadvantage of mixing two different approaches: the Happel cell for diffusive transport, and an isolated collector for removal by interception and gravitational sedimentation.

The RT Model

Recognizing that a model based completely on the Happel cell would be useful for predicting particle filtration rates in porous media, Rajagopalan and Tien (1976) developed a semiempirical model describing particle trajectories. They modeled particle removal by interception and gravitational sedimentation, and assumed diffusion could be included by superposition of removal terms, that is, that the Happel cell solution for diffusion could be added on to their final expression for the collector efficiency. The Rajagopolan and Tien (RT) model include attractive effects due to London–van der Waals forces and reduced collisions resulting from the resistance of an incompressible fluid when it is pushed out from between two colliding particles (the lubrication effect).

All filtration mechanisms in the RT model are based on the Happel cell. A filtration equation based on the Happel cell can be derived from a mass balance in the same manner as previously done for the Yao model. The derivation is identical to that done in Eqs. 15-17–15-30 except that the rate particles flow toward the Happel cell is calculated based on the Happel cell size as $r_f = (\pi/4)d_H^2$ (see Fig. 15.4). The rate particles strike the collector is therefore calculated as

$$r_s = \eta^+ U c \, \frac{\pi}{4} \, d_H^2 \tag{15-56}$$

where the superscript $+$ is added to the collector efficiency based on the projected area for the Happel cell, to avoid confusion with calculations based on η as defined above. Following the derivation through the remaining steps, and neglecting dispersion, the steady-state filtration equation for one-dimensional transport according to the RT model is

$$\frac{N}{N_0} = \exp\left(-\frac{3d_H^2}{2d_c^3}(1 - \theta)\alpha\eta^+ L\right) \tag{15-57}$$

From Eq. 15-49, this can equivalently be written in the form used by Rajagopalan and Tien (1976) as

$$\frac{N}{N_0} = \exp\left(-\frac{3}{2d_c}(1 - \theta)^{1/3}\alpha\eta^+ L\right) \tag{15-58}$$

When the filtration equation is written in this form, the collector efficiency for removal by diffusion must include an additional term, γ^2. The need for this term can be seen by comparing the filtration equation above (Eq. 15-58) with the same equation in mass transfer form (Eq. 15-51). Equating the two terms, we have

$$\frac{3}{2d_c}(1 - \theta)\alpha\,\frac{\mathbf{k}}{u} = \frac{3}{2d_c}(1 - \theta)^{1/3}\alpha\eta^+ \tag{15-59}$$

Substituting in Eq. 15-52 for the mass transport coefficient, and considering the case only of diffusion, or $\eta = \eta_D$, results in

$$\eta_D^+ = 4.04\gamma^2 b_H^{1/3} \, Pe^{-2/3} \tag{15-60}$$

where it was previously explained that $\gamma = (1 - \theta)^{1/3}$. Obviously, it is also true using Eq. 14-49 that $\gamma = d_c/d_H$.

Based on computer simulations of particles transported in porous media, Rajagopalan and Tien (1976) calculated particle removal by interception and gravitational sedimentation as

$$\eta_{I+G}^+ = \gamma^2(b_H \, Lo^{1/8} \, R^{15/8} + 0.00338 b_H S^{*\,1.2} R^{*\,-0.4}) \tag{15-61}$$

where the London number, Lo, is a dimensionless number that accounts for London–van der Waals attractive forces for particle removal, and is defined as

$$Lo = \frac{4\mathscr{A}}{9\pi\mu d_p^2 U} \tag{15-62}$$

and $\mathscr{A}$ is the Hammaker constant. Combining all terms, the collector efficiency for the RT model is

$$\eta^+ = 4.04\gamma^2 b_H^{1/3} \, Pe^{-2/3} + \gamma^2(b_H \, Lo^{1/8} \, R^{*\,15/8} + 0.00338 b_H S^{*\,1.2} R^{*\,-0.4}) \tag{15-63}$$

Because $\gamma^2 = (d_c/d_H)^2 = (1 - \theta)^{2/3}$, the RT filtration equation (Eq. 15-61) can also be used in the Yao model (Eq. 15-30) in the form

$$\eta = 4.04 b_H^{1/3} \, Pe^{-2/3} + b_H \, Lo^{1/8} \, R^{*\,15/8} + 0.00338 b_H S^{*\,1.2} R^{*\,-0.4} \tag{15-64}$$

Comparison of Filtration Models

Each of the different models used to calculate particle filtration rates is summarized in Table 15.2. Of these four models, the RT model provides the most accurate prediction of the number of particles removed in a column. Using filtration data presented in Yao et al. (1971) for completely destabilized latex microspheres in laboratory columns, it can be seen that the RT model most accurately predicts the effect of particle size on fractional removal (Fig. 15.5). The pore velocity model, based on the actual velocity in the column, predicts greater particle attenuation than the Yao model by a factor of $\exp(\theta)$ for removal by interception, and $\exp(\theta^{1/3})$ for removal by diffusion. Because all flow is defined relative to the pore velocity in the PV model, the Peclet number used in the calculation of removal by diffusion is based on a higher velocity in the PV model than in the Yao model.

Sticking coefficients larger than unity have been obtained using these different models. Although the RT model is thought to be the most accurate in predicting the number of particle collisions, sticking coefficients larger than unity may still be

TABLE 15.2 Equations Used for Different Filtration Models

Model	Filtration Equation for N/N_0	Collector Efficiency	Comments
Yao	$\exp\left(-\dfrac{3}{2d_c}(1-\theta)\alpha\eta L\right)$	$\eta = 4.04\,Pe^{-2/3} + \dfrac{3}{2}R^{*2} + S^*$	Isolated collector; characteristic velocity is approach velocity
Pore velocity (PV)	$\exp\left(-\dfrac{3}{2d_c}\dfrac{(1-\theta)}{\theta}\alpha\eta L\right)$	$\eta = 4.04\,Pe^{-2/3} + \dfrac{3}{2}R^{*2} + S^*$	Collector in a packed bed; characteristic velocity is pore velocity
Yao–Habibian (YH)	$\exp\left(-\dfrac{3}{2d_c}(1-\theta)\alpha\eta L\right)$	$\eta = 4.04\gamma^{2/3}\,Pe^{-2/3} + \dfrac{3}{2}R^{*2} + S^*$	Same as Yao model, except diffusion term is based on flow in a concentric sphere (Happel cell)
Rajagopalan and Tien (RT)	$\exp\left(-\dfrac{3}{2d_c}(1-\theta)^{1/3}\alpha\eta^+L\right)$ $\exp\left(-\dfrac{3}{2d_c}(1-\theta)\alpha\eta L\right)$	$\eta^+ = \gamma^2 4.04 b_H^{1/3}\,Pe^{-2/3}$ $\quad + \gamma^2(b_H\,Lo^{1/8}\,R^{*15/8} + 0.00338 b_H S^{*1.2}R^{*-0.4})$ $\eta = 4.04 b_H^{1/3}\,Pe^{-2/3} + b_H\,Lo^{1/8}\,R^{*15/8}$ $\quad + 0.00338 b_H S^{*1.2}R^{*-0.4}$	All terms based on flow in a concentric sphere; includes lubrication theory for all terms except diffusion (lubrication effects are insignificant for collisions by diffusion)

Source: Logan et al., 1995.

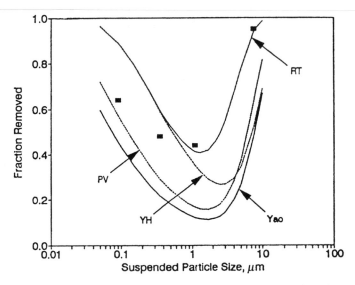

Figure 15.5   Comparison of particle removals predicted by four filtration models versus data reported by Yao et al. (1971). Models are Rajagopalan and Tien, RT; Yao et al., Yao; Yao–Habibian, YH; pore velocity, PV. Constants used in calculations were: $\alpha = 1$, $U = 0.136$ cm s^{-1}, $d_c = 397$ μm, $\rho_p = 1.05$ g cm^{-3}, $\theta = 0.36$, $L = 13$ cm, $\mathcal{A} = 10^{-13}$ erg, and water at 23°C (from Logan et al., 1995).

required when using the RT model. For the data shown in Figure 15.5, α would need to vary from 0.46 to 1.1 to fit these data exactly using the RT model. However, this range is smaller than that of $\alpha = 1.7$–4.6 required by the Yao model, indicating the RT model is a more accurate predictor of particle filtration rates than the Yao model.

Filtration of Particles Having a Distribution of Sizes

Water samples that are filtered contain a range of particle sizes. If the water sample has a particle distribution $n_0(d_p)$, then the effluent particle size according to Eq. 15-16 will be $n(d_p) = n_0(d_p)e^{-\lambda L}$. In terms of the volume distribution function, the influent size distribution is

$$\frac{dV_p}{d(\log d_p)} = \frac{2.3\pi}{6} d_p^4 n_0(d_p) \qquad (15\text{-}65)$$

and the effluent distribution is therefore

$$\frac{dV}{d(\log d_p)} = \frac{2.3\pi}{6} d_p^4 n_0(d_p)e^{-\lambda L} \qquad (15\text{-}66)$$

From the influent particle size spectrum, we can calculate the influent and effluent particle size distributions from a plot of the volume distribution function as a function of particle size, d_p. If particle sizes can be expressed in terms of an analytical expression, such as $n_0(d_p) = b_1 d_p^{-b}$, then the total influent and effluent particle volume concentrations, and mass concentrations, can sometimes be calculated from direct integration as shown in the example below.

Example 15.2

Calculate the mass influent and effluent particle concentrations, and percent removal, from a 1-m-long filter operating at a flow velocity of 0.4 cm s^{-1} and containing 500-μm-diameter media at 20°C for particles of an average density of 1.05 g cm^{-3}, $\theta = 0.35$ and $\alpha = 0.01$. Assume all particles are in the size range of 0.1 to 1 μm and that particles are only removed by Brownian motion. The particle sizes are characterized by the size distribution $n(d_p) = 5.2 \times 10^6 \, d_p^{-3}$.

The influent particle volume distribution is

$$\frac{dV_p}{d(\log d_p)} = \frac{2.3\pi}{6} d_p^4 (5.2 \times 10^6 d_p^{-3}) = 6.3 \times 10^6 d_p \frac{\mu m^3}{cm^3} \qquad (15\text{-}67)$$

The effluent volume distribution must be reduced by $e^{-\lambda L}$. To calculate λ, we begin by calculating η. For removal by Brownian diffusion only, $\eta = \eta_B = 4 \, Pe^{-2/3}$, where $Pe = Ud_c/D_{Pw}$. The diffusivity of the particle in water can be calculated using the Stokes–Einstein equation as

$$D_{Pw} = \frac{k_B T}{3\pi\mu_w d_p} = \frac{(1.38 \times 10^{-16} g \, cm^2/K \, s^2)(293K)}{3\pi(0.01 g/cm \, s)(d_p[\mu m])10^{-4} \, cm/\mu m} \qquad (15\text{-}68)$$

$$D_{Pw}[cm^2] = 4.3 \times 10^{-9} d_p^{-1}[\mu m] \qquad (15\text{-}69)$$

The Peclet number is

$$Pe = \frac{Ud_c}{D_{Pw}} = \frac{(0.4 \text{ cm s}^{-1})(0.05 \text{ cm})d_p[\mu m]}{4.3 \times 10^{-9} \text{ cm}^2 \text{ s}^{-1}} = 4.56 \times 10^6 d_p[\mu m] \quad (15\text{-}70)$$

and the collector efficiency is therefore

$$\eta = 4 \, Pe^{-2/3} = 4(4.56 \times 10^6 d_p)^{-2/3} = 1.36 \times 10^{-4} d_p^{-2/3}[\mu m] \qquad (15\text{-}71)$$

The filter coefficient is

$$\lambda[cm^{-1}] = \frac{3(1 - \theta)\alpha\eta}{2d_c} = \frac{3(1 - 0.35)(0.01)(1.36 \times 10^{-4} d_p^{-2/3})}{2(0.05 \text{ cm})}$$

$$= 2.65 \times 10^{-4} d_p^{-2/3}[\mu m] \qquad (15\text{-}72)$$

The volume distribution in the effluent is calculated as

$$\frac{dV_p}{d(\log d_p)} = \frac{2.3\pi}{6} d_p^4 (5.2 \times 10^6 d_p^{-3}) \exp[-2.65 \times 10^{-4} d_p^{-2/3}(100 \text{ cm})] \quad (15\text{-}73)$$

$$= 6.26 \times 10^6 d_p \exp(-0.0265 d_p^{-2/3}) \quad (15\text{-}74)$$

The volume distributions are shown in Figure 15.6. The total volume concentration of the sample can be calculated by taking the area under the curves. One way to do this is simply to multiply the average value of the volume distribution function, $[V_{dist}(1) + V_{dist}(2)]/2$, by $\Delta \log(d_p)$, although there are other, more exact methods. Using this approach, the total influent volume is $2.47 \times 10^6 \text{ } \mu m^3 \text{ cm}^{-3}$, while the outlet concentration is $1.26 \times 10^6 \text{ } \mu m^3 \text{ cm}^{-3}$. The total mass concentration entering the filter is obtained by multiplying the influent particle concentration by the particle density. After unit conversions, we obtain

$$m_0 = \rho_p V_{p,0} = (1.05 \text{g cm}^{-3})(2.47 \times 10^6 \text{ } \mu m^3 \text{ cm}^{-3})$$

$$\times 10^{-12} \frac{\text{cm}^3}{\mu m^3} \times 10^6 \frac{\mu g}{g} = 2.47 \text{ } \mu g \quad (15\text{-}75)$$

The effluent particle mass concentration is obtained in the same manner, producing an effluent mass concentration of 2.59 µg. Thus total particle removal is 49.1%. What error did we introduce by our manual integration approach? Note that the influent particle concentration is easily analytically integrated as

$$m = \int_{0.1}^{1.0} \rho_p V_{p0} n(d_p) \, dd_p \quad (15\text{-}76)$$

Substituting in the given information, for the influent mass this produces 2.57 µg, or an error of about 0.7% compared to the graphical method (a negligible error).

15.4   DISCRETE PARTICLE SIZE DISTRIBUTIONS PREPARED BY FILTRATION

Filters are commonly used in the laboratory to separate suspended particles based on their size. For example, soluble BOD (sBOD) is calculated for samples after their passage through a 0.2 or 0.45 µm filter. Aquatic scientists have taken this concept further, and have attempted to size fractionate particles in water by passing water samples through a series of filters and screens, assuming that particles are removed only by straining. For example, bacteria are separated from phytoplankton in seawater samples using a 0.8 or 1.0 µm polycarbonate filter. To separate larger phytoplankton based on their size, screens with various pore sizes, and made of a variety

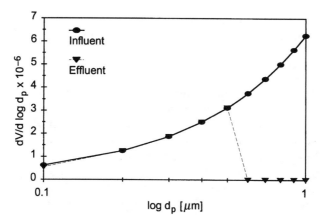

Figure 15.6 Volume distribution of particles in filter influent and effluent.

of different materials such as nylon and polycarbonate, are commonly used. Thick fibers are woven in square patterns in these screens to create pores with known dimensions (Fig. 15.7). The difficulty with preparing size distributions in this manner is that particles smaller than the pore size are removed during filtration. Particles smaller than the pore can collide with screen and fiber material and be removed by filtration processes along with particles larger than the pore. Thus particle separation is not accomplished solely by straining, as assumed. This results in size separations that are not sharp, and particle size distributions are not correct because too much mass is measured in smaller particle size fractions than is actually present in the sample. Filtration theory has been developed to describe particle removals in a variety of screens and filters. As we shall see below, calculations with filtration theory

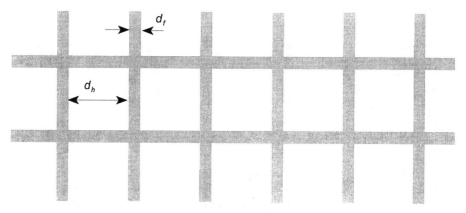

Figure 15.7 Pores are formed in screens of size d_h by cylindrical fibers of width d_f. For monolayers of screens, the filter length is therefore also d_f.

will permit us to examine whether particle size separations using screens and filters can be done to measure discrete particle size distributions.

Types of Laboratory Filters

Laboratory filters can be separated into two classes: surface filters and depth filters. Surface filters are designed to exclude particles larger than the pore size by straining, and commonly are made of either polycarbonate or aluminum. Polycarbonate filters are prepared by using chemicals to broaden holes made by radioactive particles that are directed through a sheet of polycarbonate. The size of the pore is controlled by the etching time. Not all pores are the same size, and not all pores are perpendicular to the sheet surface (Fig. 15.8). Aluminum filters have a honeycomb shape. These pores do not always form separate from each other, resulting in pores larger than the average size.

Depth filters are more commonly used in laboratory separations. These filters can be made of a variety of materials, such as glass fibers, nylon fibers, and cellulose acetate (Fig. 15.9), and they do not have any defined pore size. Instead, particles are removed over the length of the filter via collisions of particles passing through the filter as described above for filters made of spherical collectors. These filters are rated in terms of nominal pore size, set by various techniques such as correlating pressure drops in the filter to pore diameters or by a certain percentage removal of a given particle size. As a result of their variable pore sizes, depth filters can produce wide particle size distributions. Particles much smaller than the pore will be removed by collisions with filter material, and some particles larger than the nominal pore rating will pass through the largest pores in the filter. Depth filters can be used to remove most particles larger than their nominal pore, but they should not be (although they often are!) used to prepare size fractions of particulate material.

Fibrous Filtration Models

A filtration equation can be derived to predict particle removal by long fibers using the same approach employed above to predict removal by spherical collectors. For

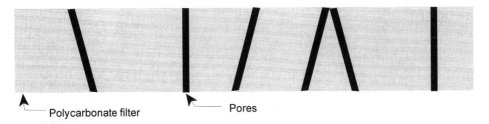

Polycarbonate filter   Pores

Figure 15.8 Paths of pores through a polycarbonate filter. The pores that are formed in these filters are not all perpendicular to the surface of the filter. In addition, some paths are so close that a larger hole is visible on the filter surface. These polycarbonate filters are usually about 10 μm thick, with pore sizes ranging from 0.01 to 10 μm, although 0.2 and 0.45 μm pore sizes are commonly used for filtration.

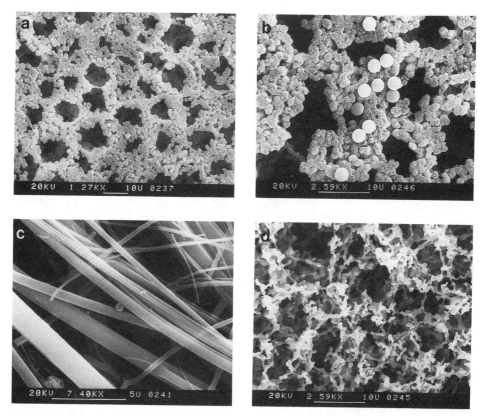

Figure 15.9 Scanning electron photomicrographs of different filters: (a) nylon (5 μm); (b) same, but after capture of 2.2 μm latex microspheres; (c) glass fiber filter (GF/C, 1.2 μm) with captured microsphere; (d) cellulose (8 μm). Given numbers in parentheses are manufacturers' pore diameters. (Reprinted from Logan et al., 1993, with kind permission from Elsevier Science Ltd., The Boulevard, Landford Lane, Kidlington, OX5 1GB.)

fibrous filters, a mass balance approach based on the superficial velocity approach produces the filtration equation

$$\frac{N}{N_0} = \exp\left(-\frac{4}{\pi}\frac{(1-\theta)}{d_f}\alpha\eta L\right) \qquad (15\text{-}77)$$

where the collector diameter d_c is now defined as the fiber diameter, d_f. Another version of this equation can be derived containing θ in the denominator of the exponential term when the mass balance is based on the pore velocity. Thus the only difference between the sphere and fiber models is a geometrical factor of $\frac{3}{2}$ for spheres, and $4/\pi$ for cylinders (see Problem 15.5).

Hinds Model   Several fibrous filtration models have been developed for aerosol filtration, but they have not been extensively tested for aquasol removal. For aquasols the mean path of water molecules relative to fiber diameters is much smaller for liquids than for gases (Fuchs, 1964), and therefore correction factors used for aerosols do not need to be included in aquasol models. Collision efficiencies for aquasol filtration by diffusion, interception, and gravitational sedimentation according to the Hinds (1983) model are therefore

$$\eta_D = 2\,Pe^{-2/3} \tag{15-78}$$

$$\eta_I = \frac{1}{2\,Ku}\left[2(1 + d_p^*)\ln(1 + d_p^*) - (1 + d_p^*) + \left(\frac{1}{1 + d_p^*}\right)\right] \tag{15-79}$$

$$\eta_s = S^*(1 + d_p^*) \tag{15-80}$$

where d_p is the particle diameter, d_f the fiber diameter, and $d_p^* = d_p/d_f$. The Kuwabara number, Ku, is calculated as

$$Ku = -\frac{\ln \phi}{2} - \frac{3}{4} + \phi - \frac{\phi^2}{4} \tag{15-81}$$

where ϕ is the solid volume fraction, or $\phi = 1 - \theta$. Aerosols have a high density relative to air, and their inertia causes them to continue to travel on their path even when air streamlines are deflected by an object. This motion leads to particle removals by a mechanism referred to as impaction. For aquasols, impaction is not assumed to contribute to particle removal. Particle removal for aerosols by impaction can be described as

$$\eta_M = \frac{b_M\,Stk}{2\,Ku^2} \tag{15-82}$$

where the dimensionless Stokes number is $Stk = U_p U/(g d_c)$, and b_M is given by empirical correlations that are a function of d_p^*, according to

$$b_m(d_p^* \geq 0.4) = 2$$

$$b_m(d_p^* < 0.4) = (29.6 - 28\phi^{0.62})d_p^{*2} - 27.5d_p^{*2.8} \tag{15-83}$$

Rubow and Liu Aerosol Model Applied to Aquasols   Based on laboratory experiments of aerosol removal by laboratory fibrous filters, Rubow and Liu (1986) determined only particle capture by diffusion and interception were important. Single fiber collection efficiencies were

$$\eta_I = \frac{b_g \theta d_p^{*2}}{Ku(1 + d_p^*)} \tag{15-84}$$

$$\eta_D = 2.86 b_g \theta^{1/3} Ku^{-1/3} Pe^{-2/3} \tag{15-85}$$

Slip factors developed for aerosols have been assumed here to be negligible for aquasols. The geometrical correction constant $b_g = 1.67$ was introduced to modify the geometrical factor of $4/\pi$ for inhomogeneity of the fibers relative to perfectly aligned fibers.

Fibrous Filter Characteristics   Filter data provided by manufacturers usually include a nominal filter pore diameter and filter thickness, but not the fiber diameter. In order to apply the above models to laboratory filters, all three of these factors must be known. Assuming that all fibers are cylinders stacked in overlapping parallel rows, the fiber diameter, d_c, and pore diameter, d_h, can be related to the filter porosity (Logan et al., 1993) using

$$(1 - \theta) = \frac{\pi}{2\sqrt{3}} \frac{d_c^2}{(d_c + d_h)^2} \tag{15-86}$$

Filter characteristics for nylon, cellulose, and glass fiber filters (GF/C) are given in Table 15.3.

Comparison of Fibrous Filtration Models   Sticking coefficients were measured for nylon and glass fiber filters in laboratory experiments using radiolabeled cells of *Pseudomonas sp.* JS6 by Logan et al. (1993). The percent of cells retained in these filters were 30% (nylon) and 24%(GF/C). Sticking coefficients for the more hydro-

TABLE 15.3   Characteristics of Three Different Filters

Filter Characteristic	Cellulose	Nylon	GF/C
Pore diameter [μm]			
Manufacturer designation	8	5	1.2
d_h, used here	3	5	4.1[a]
Collector diameter [μm]			
Using microscope	1.2	1.5	1.2
d_c, used here	2.17[a]	5.0	2.2
Porosity	0.84	0.77[a]	0.89
Filter length [μm]	120	120	300

[a]Value calculated as unknown.

Source: Logan et al., 1993.

phobic nylon filters ranged from 0.24 to 0.57 for four different filtration models and were an order of magnitude larger than for the glass fiber filters (Table 15.4).

The effect of particle size on particle retention predicted by the models of Yao et al. (1971), Hinds (1983), Rubow and Liu (1986), and Tien and Payatakes (1976) is shown in Figure 15.10. All four models predict a minimum in filter retention efficiency at 0.1 to 0.3 μm. Particle retention increases as particle sizes approach that of the nominal pore size of the filter. The trends shown in particle removals in Figure 15.10 of course require a constant sticking coefficient for all particles. Differences in sticking coefficients for dissimilar-size particles would produce corresponding changes in particle retention.

Filtration Models for Polycarbonate Filters and Screens: Capillary Pore Models

Polycarbonate filters remove particles via two mechanisms: straining at the particle surface, and filtration within the pores, which are analyzed as being tubes, by collisions of particles with the tube walls. We will assume here that all pores for polycarbonate filters and screens are exactly the size indicated by the manufacturer and that these filters remove 100% of particles larger than the pore diameter. Therefore, filtration equations are only necessary to evaluate particle removal for pores smaller than the pore diameter. The models developed to describe particle removal within tubes of polycarbonate filters are known as capillary tube models. The overall efficiency of particle removal in capillary tubes was given by Spurney et al. (1969) as the sum of the individual mechanisms minus the removal that occurs by mechanism overlap, or

$$\frac{N}{N_0} = \alpha(\eta_{Dc} + \eta_{Ic} + \eta_{Mc} - \eta_{Dc}\eta_{Ic} - \eta_{Ic}\eta_{Mc} - \eta_{Dc}\eta_{Mc} + \eta_{Dc}\eta_{Ic}\eta_{Mc}) \quad (15\text{-}87)$$

TABLE 15.4 **Sticking Efficiencies of *Pseudomonas sp JS6* Calculated Using Four Different Filtration Models and Experimentally Determined Removal Efficiencies of 30% and 24% in Nylon (5 μm) and Glass-Fiber (GF/C) Filters**

Filtration Model	Sticking Coefficients	
	Nylon	GF/C
Yao	0.56	0.032
Hinds	0.24	0.033
Rubow and Liu	0.57	0.073
Tien and Payatakes	0.27	0.031

Source: Logan et al., 1993.

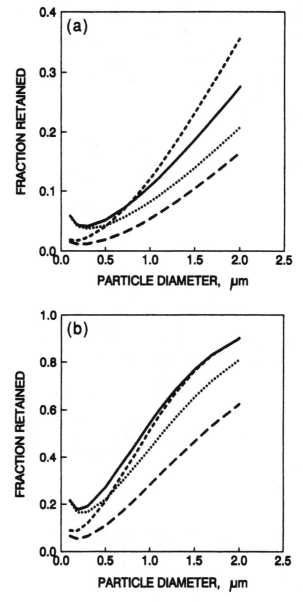

Figure 15.10 Effect of particle size on removal predicted by four different filtration models for (a) nylon (5 μm pore diameter) and (b) glass fiber (GF/C) filters. —— Yao et al. (1971); $\cdots$ Hinds (1982); --- Rubow and Liu (1986); and $---$ Rajagopalan and Tien (1976). See Table 15.3 for filter characteristics. Calculations assume $\alpha = 0.1$, $U = 0.08$ cm s^{-1}, $\rho_p = 1.05$ g cm^{-3}, and water at 20°C. (Reprinted from Logan et al., 1993, with kind permission from Elsevier Science Ltd., The Boulevard, Landford Lane, Kidlington, OX5 1GB.)

where η_{Dc}, η_{Ic}, and η_{Mc} are the individual removal mechanisms for diffusion, interception, and impaction based on the capillary tube model. Particle removal by impaction into the filter surface is important for dense particles in air, but is relatively unimportant for particle removal in aqueous solutions, where particle and fluid densities are similar. For aquasols, terms containing impaction can therefore be dropped, and Eq. 15-75 can be simplified to

$$\frac{N}{N_0} = \alpha(\eta_{Dc} + \eta_{Ic} - \eta_{Dc}\eta_{Ic}) \tag{15-88}$$

Particle removal by diffusion is calculated (Rubow and Liu, 1986) as

$$\eta_{Dc} = 2.56D_c^{*2/3} - 1.2D_c^* - 0.177D_c^{*4/3} \tag{15-89}$$

where D_c^* is a dimensionless number defined as

$$D_c^* = \frac{4LD_p\theta}{d_h^2 U} \tag{15-90}$$

and d_h is the pore diameter and D_p the particle diffusivity. Particles are removed by interception when the fluid streamlines bring a particle sufficiently close to the pore surface. Removal by interception is calculated (John et al., 1978) using

$$\eta_{Ic} = (2d_h^* - d_h^{*2})^{3/2} \tag{15-91}$$

where $d_h^* = d_p/d_h$ is the interception number based on the pore size. These equations have been shown to provide good agreement with experimental work on aerosol removal in Nuclepore filters (Rubow and Liu, 1986). The porosities and fiber sizes of several common screens are given in Table 15.5.

TABLE 15.5   Physical Characteristics of Screens and Filters Used in Calculations

Code	Material	Pore Diameter (μm)	Fiber Diameter (μm)	Porosity
NY10	Nylon	10	45	0.05
NY20	Nylon	20	55	0.14
NY30	Nylon	30	70	0.21
NY60	Nylon	60	55	0.45
PP210	Polypropylene	210	320	0.34
PE230	Polyethylene	230	280	0.42
PP1000	Polypropylene	1000	1020	0.45
GF/C	Glass fibers	4.1	2.2	0.89

Source: Logan, 1993.

Example 15.3

A screen can be analyzed as a sheet of material (with a thickness equal to the fiber thickness) or as a fibrous filter (with only a single layer of fibers). For a 30-μm-pore-diameter nylon screen, compare the fractional removals of 3100 particles 0.1 to 100 μm in diameter equally divided into 31 log-size classes. Assume the following in your calculations: $\alpha = 1$, $U = 0.088$ cm s^{-1}, $\rho_p = 1.05$ g cm^{-3}, $\mu_w = 0.01$ g cm^{-1} s^{-1}, and 20°C.

If particles are separated into 31 size classes over that interval of log 0.1 = -1 to log 100 = log 2, then each size class will be separated by 0.1 log units, or $d(i)$ = $10^{(-1+0.1i)}$ where i = 0 to 30. For the nylon screen with d_h = 30 μm, we have from Table 15.4 that d_c = 70 μm, θ = 0.21, and $L = d_c$ = 70 μm. Let us compare the following two models: the Yao spherical-collector model, using $d_c = d_f$ and using a geometrical factor of $4/\pi$ in the filtration equation (Eqs. 15-30 and 15-36), and the capillary pore model of Rubow and Liu (Eqs. 15-75–15-79). Using these values, we have the result shown in Figure 15.11. The same trend of a broad removal in particle sizes is shown using each model. Notice that the isolated collector model (Yao model) predicts some particles larger than the pore diameter will pass through the pore. For example, 20% of particles 60 μm in diameter, or twice that of the pore size, are calculated to pass through the screen using the Yao model. Based on the ability of the capillary pore models to predict particle

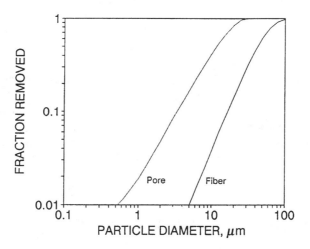

Figure 15.11 Comparison of the capillary pore and Yao models. (Reprinted from Logan (1993) with permission by the American Society of Limnology and Oceanography.)

exclusion for those particles larger than the pore diameter, we conclude that the capillary pore models should be used to predict particle removal by screens.

Example 15.4

Measured size distributions will not always equal the actual size distributions as a result of the removal of particles smaller than the pore diameter by screen fibers. Compare the actual (or given) size distributions in Example 15.3 with those predicted to occur by filtration of the hypothetical sample above (for the given flow conditions) for a size separation accomplished using the following filters and screens sometimes employed in sampling by aquatic scientists: GF/C, NY30, and PE230. Assume two different cases for the sticking coefficient: $\alpha = 1$, where all particles stick to the filter and screen materials, and $\alpha = 0.1$, where particles stick only once for every ten collisions.

Particle removal can be predicted in the fibrous GF/C filter using the Yao model as in Example 15.3. For the screens we use the Rubow and Liu model (Eqs. 15-75–15-79) and the screen characteristics given in Table 15.5.

The results shown in Figure 15.12A for $\alpha = 1$ demonstrate that the numbers of particles actually in the different size fractions are substantially different from those that are predicted to be measured using these filters and screens. In the case of the GF/C filter, we can see that many particles smaller than the nominal pore size of the filter are predicted to be retained by the filter and therefore would not appear in the filtrate. The mass distributions are also predicted to be substantially different from those of the actual sample. As shown in Table 15.6, 50.1% of the total mass of the particles would be retained on the 210-μm-pore-diameter screen, despite the fact that our sample had no particles larger than 100 μm!

As particles become less sticky, or for the case of $\alpha = 0.1$, the size distributions predicted to be obtained using these filters become more similar to those in the actual sample (Figure 15.12B). This makes sense because as particle sticking efficiencies are reduced, particles can collide with the filter and screen fibers but not stick. Thus the measured size distribution becomes more similar to that of the actual sample for particles with low adhesion for the filter material. If highly heterogeneous samples, such as those from natural waters, are size fractionated using different size screens and filters, differences in particle sticking coefficients may therefore be as important as the sizes of the particles in setting the resulting particle size distributions.

15.5   THE DIMENSIONLESS COLLISION NUMBER

Particle removal rates in different porous media systems can be difficult to compare on a common basis when columns vary in length, collector size, and other parameters that affect removal. In order to normalize removal rates to a common length scale,

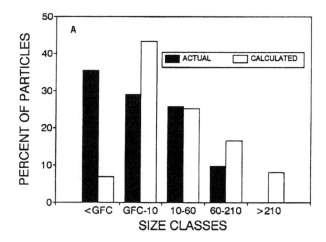

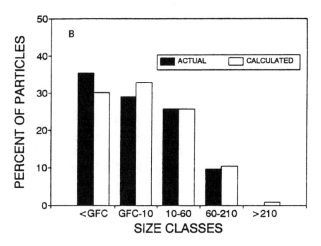

Figure 15.12 (Reprinted from Logan (1993) with permission by the American Society of Limnology and Oceanography.)

a scaling factor called the collision number, ξ, can be defined (Martin et al., 1996) as

$$\xi = \frac{\text{rate particles collide in a column}}{\text{rate particles enter a column}} \tag{15-92}$$

where both rates are evaluated using the same initial particle concentration. The rate particles enter a column is UAc_0, where A is the cross-sectional area of the column and U the superficial velocity. The rate particles collide in the column can either be calculated using a mass balance as above, or it can be calculated as the product of

TABLE 15.6 Particle Size Distributions Based on Number and Mass Concentrations

Mesh Pore Diameter (μm)	Percent of Particles (by Number)			Percent of Particles (by Mass)		
	Calculated	Actual	Difference[a]	Calculated	Actual	Difference[a]
>210	8.0	0.0	—[b]	50.1	0.0	—[b]
60–210	16.6	9.7	171	47.7	87.4	55
10–60	25.2	25.8	98	2.2	12.5	17
GF/C–10	43.4	29.0	149	9.9×10^{-3}	5.0×10^{-2}	20
<GF/C	6.9	35.5	19	3.0×10^{-6}	1.0×10^{-4}	3

[a]Determined using $100 \times (C/A)$, where C and A are the calculated and actual amount of particles (number or mass) in the size classes indicated.
[b]Could not be determined, since actual = 0.

Source: Logan, 1993.

the collision rate and $\alpha = 1$ because the rate particles are removed when $\alpha = 1$ is the rate that they collide. From the chain rule, the rate particles are removed in a column, dc/dt, is

$$\frac{dc}{dt} = \frac{dz}{dt}\frac{dc}{dz} \qquad (15\text{-}93)$$

The first term on the right-hand side is just the fluid velocity, or $dz/dt = U$. The removal rate at the column entrance is calculated by taking the derivative of the filtration equation (Eq. 15-15), evaluated at $c = c_0$, or $dc/dz = -\alpha\lambda c_0$. To obtain the collision rate, we normalize the removal rate by the sticking coefficient. The collision rate is therefore

$$\frac{dc}{dt} = U \left| -\frac{\alpha\lambda c_0}{\alpha} \right| = U\lambda c_0 \qquad (15\text{-}94)$$

Using this result in Eq. 15-65, for a column of volume $V = AL$, the collision number is

$$\xi = \frac{U\lambda c_0 AL}{Uc_0 A} = \lambda L \qquad (15\text{-}95)$$

The dimensionless collision number is therefore just a ratio of the column length to a characteristic travel distance, $1/\lambda$.

The collision number can also be used to compare collision frequencies between different systems. Because it is calculated as the rate of particle collisions at the column entrance, the collision number is equal to the number of collisions a non-attaching particle ($\alpha = 0$) would have to undergo to travel through a column of length L. For a column of length $L = 1/\lambda$, a nonattaching particle would have to undergo one collision on average to reach the end of the column. On the other hand, completely destabilized particles entering a column of length $L = 1/\lambda$ would be reduced in concentration by a factor of 0.37 (e^{-1}). Other scaling factors follow from this analysis. For a 50% reduction in particle concentration, or a particle "half-life" in the column, the length of travel would be

$$L_{50} = \frac{-\ln 0.5}{\lambda} = \frac{\ln 2}{\lambda} \tag{15-96}$$

For a 1-log removal, or an order-of-magnitude reduction in particle concentration

$$L_{1-\log} = \frac{\ln 10}{\lambda} \tag{15-97}$$

Example 15.5

Compare the removals of particles 1 μm in diameter in columns 3 cm long packed with either 150- or 450-μm-diameter spherical beads as a function of column length, and the dimensionless collision number, for conditions of $\alpha = 0.5$, $U = 0.001$ m s^{-1}, and $\theta = 0.37$, at a temperature of 20°C.

The calculated removals within the column can be calculated as a function of column length using the RT model for the given conditions. As shown in Figure 15.13A, fewer particles are deposited in the column packed with the larger (450-μm) particles than in the column packed with the smaller beads (150 μm). To compare these curves on a common basis based on collision frequencies, we can use the collision number as a dimensionless distance as shown in Figure 15.13B.

15.6 PRESSURE DROPS IN FILTERS

There is a decrease in water pressure as water flows through a porous medium in the same way that there is a head loss in flow through pipes. The Darcy–Weisbach equation for flow through a pipe (Cleasby, 1990) is

$$\frac{\Delta h}{\Delta L} = \frac{b_f U^2}{2g d_t} \tag{15-98}$$

where Δh is the head loss over a length of flow ΔL, b_f a friction factor that is a function of the Reynolds number, U the superficial velocity, and d_t the tube or pipe diameter. This equation can be modified to describe flow in porous media if the media is analyzed as a series of capillary tubes with a hydraulic radius r_h, defined as the ratio of the volume of water in the media to the grain surface area (Cleasby, 1990). For a filter of N particles with total bed volume $N v_p/(1 - \theta)$, the hydraulic

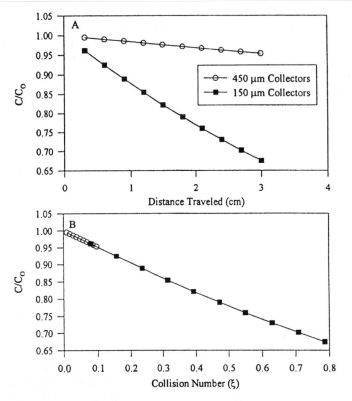

Figure 15.13 Particle removals calculated in Example 15.5. (From Martin et al., 1996, by permission of ASCE.)

radius is $R_h = \theta v_p/[a(1 - \theta)]$, where θ is the bed porosity and a is the surface area per particle. For full pipe flow, $d_t = 4R_h$, $b_f = 64/Re^+$, where $Re^+ = 4Ur_h/\theta v$, Kozeny obtained the result

$$\frac{\Delta h}{\Delta L} = \frac{b_f\mu(1 - \theta)^2 a_f^2 U}{\rho_w g\theta^3} \tag{15-99}$$

where b_f is the Kozeny constant, typically having a value of around 5, U is the superficial velocity, and a_f the specific surface area equal to the surface area of all particles divided by the total bed volume. For spheres, $a_f = 6/d_c$; for nonspherical media, $a_f = 6/b_s d_c$, where b_s is a shape factor and d_c is defined as the equivalent diameter of a sphere of equal volume. The Kozeny equation appears applicable for $Re < 6$, where $Re = Ud_c/v$, and therefore this equation is used for typical filtration velocities of 5 to 12 m h^{-1} (hydraulic loadings of 2 to 5 gpm ft^{-2}), and typical grain sizes of 0.5 to 1 mm, for flow in a clean bed.

At higher velocities the flow may be in the transition flow regime or even turbulent (as in the case of filter backwashing). Under these conditions, additional terms are included in the flow equation and the Ergun equation can be used

$$\frac{\Delta h}{\Delta L} = \frac{4.17\mu(1 - \theta)^2 a_f^2 U}{\rho_w g \theta^3} + \frac{b_{f2}(1 - \theta)^2 a_f^2 U}{g\theta^3} \qquad (15\text{-}100)$$

where b_{f2} =0.48 for crushed media. The head loss in the filter can be interchanged with a pressure drop by recognizing that $\Delta h = \Delta p/\rho_w g$.

15.7 FILTER RIPENING, BLOCKING, AND CLOGGING

As particles accumulate on the surface of filter media, the collector surface geometry and flow around the collector changes, and it is found that equations developed to predict particle removal by clean beds no longer accurately predict particle removal. The transition of a filter from a clean bed to complete clogging can occur in a number of stages as shown in Figure 15.14. Particles introduced into a clean filter will achieve breakthrough at the column exit, reaching a steady state as predicted by the filtration equation. As particles accumulate on the filter media, the amount of clean surface area available for deposition decreases (ripening). If suspended particle collisions with deposited particles are unfavorable, a condition referred to as blocking or unfavorable blocking, the effluent particle concentration will start to rise as a result of less efficient particle retention due to a reduced collector surface area. Eventually, complete particle breakthrough will occur, resulting in no particle removal in the column.

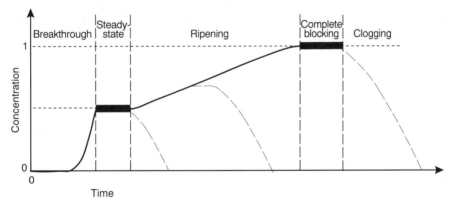

Figure 15.14 Concentration (dimensionless) of particles at the end of a column measured as a function of time. Clogging (dashed lines) of the column can occur at any time depending on the concentration of particles in the influent and the relative affinity of suspended particles for deposited particles. Steady conditions can occur at two different times: after breakthrough, clean bed conditions of few deposited particles, and after long times when all occupied sites on the media are filled with particles that have little affinity for suspended particles.

A more likely situation in a water treatment filter is that all particles are destabilized, and therefore particles striking a collector stick equally well to clean media
or previously deposited particles. When suspended particle and deposited particle
interactions are favorable, a condition referred to here as favorable blocking, the
steady-state filtration rate should change in a predictable manner as a result of growth
in collector size and a decrease in bed porosity. However, it is observed that ripening
leads to very high pressure drops and much more rapid particle removal in filter
beds than that calculated merely from porosity reductions and collector diameter
increases. For example, Hunt et al. (1993) calculated a 15-fold increase in pressure
drop in a section of a filter containing only a 1.25% increase in solid volume of
deposited particles.

Under favorable conditions for continued deposition onto previously deposited
particles, filter ripening can lead to clogging at various times, resulting in a decrease
in particle concentrations in the effluent, as shown in Figure 15.14, and an increase
in the pressure drop in the filter. The rate of ripening and the onset of clogging
depends on the particle deposition rate and the structure of the particle deposits. For
a monodisperse particle suspension, Liu et al. (1995) observed particle concentrations
increased slightly in the effluent after breakthrough at a low concentration of a 1:2
electrolyte (Na_2SO_4), or when particles were incompletely destabilized (Fig. 15.15).
When particles were made more destabilized by the use of higher salt concentrations,
the onset of clogging was immediate, and particle concentrations in the filter effluent
rapidly decreased following breakthrough.

Models of Particle Blocking

Various approaches have been used to model the effect of deposited particles on
filter performance. Rijnaarts et al. (1996) proposed that the sticking coefficient for
the clean bed, α, could be reduced for deposited particles that were blocking the
deposition of additional particles as

$$\alpha = \alpha_0(1 - B_f\theta) \tag{15-101}$$

where α_0 is the clean-bed sticking coefficient, B_f the blocking factor defined as the
ratio of the area blocked by a deposited particles to the area of that particle, and θ
the fraction of surface coverage. The maximum surface coverage occurs when $\theta =
(1/B_f) \equiv \theta_{max}$. It has been observed that deposited particles can block an area larger
than their own surface area. When negatively charged latex microspheres were deposited onto positively charged collectors, unfavorable attractive forces between deposited particles and suspended particles led to a constant number of deposited particles (Fig. 15.16). Repulsion between similarly charged deposited particles results
in open spacing between the particles so that the surface is effectively blocked even
though there is open surface area.

The maximum achievable surface coverage has been measured in packed beds
using monodisperse suspensions. Depending on the suspended particle size, complete
blocking, or jamming, can occur with as little as 3.7% of the surface area occupied

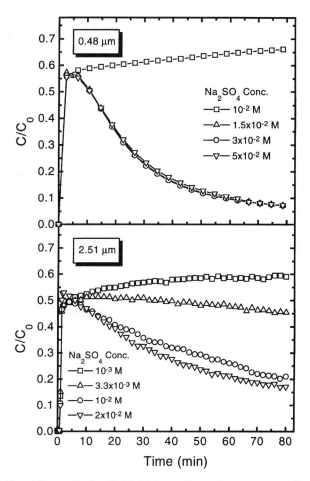

Figure 15.15 Breakthrough of colloidal latex microspheres suspended in a 1:2 electrolyte (Na$_2$SO$_4$) at pH = 5 in glass-bead columns (0.46 mm diameter) when C_0 = 14.9 mg L^{-1} (0.48 μm particles in a 10-cm-long column) or C_0 = 29.8 mg L^{-1} (2.51 μm particles in a 14.2-cm-long column). (Reprinted with permission from Liu et al. (1995). Copyright 1995, American Chemical Society.)

by deposited particles (θ_{max} = 0.037; Fig. 15.17). Experiments conducted using 0.48 and 2.51 μm latex microspheres deposited on uniformly sized (0.46 mm) soda lime glass beads resulted in jamming limits of 4.5 to 59.2% depending on the concentration of the indifferent 1:1 electrolyte (KCl) (Table 15.7). For bacteria deposited on glass beads, Rijnaarts et al. (1996) measured jamming limits of 14 to 34%.

Modification of Filtration Equations for Filter Ripening

The filtration equation for clean beds, $N/N_0 = e^{-\lambda L}$, was originally proposed by Iwasaki (1937) to describe filtration of particles in both clean and dirty beds. This

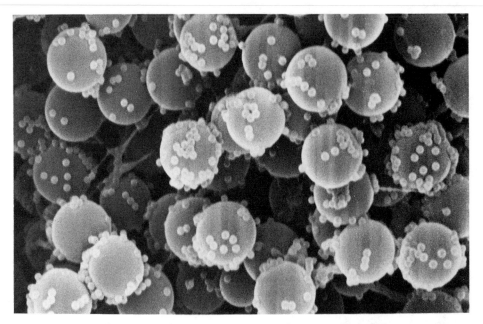

Figure 15.16 Polystyrene latex microspheres (0.32 μm diameter) deposited onto large latex particles (2.17 μm) in a 5 $\times$ 10^{-4} M NaCl solution (photograph courtesy of Brian Vincent).

approach has been modified over the years by including additional terms to account for changes in the collector efficiencies as particles are deposited on the bed, but no single approach has been completely successful. Ives and Shoiji (1965) based their model on the assumption that decreases in the filter coefficient that lead to break-through are a result of deposits that constrict tubular pores. They proposed

$$\frac{N}{N_0} = \exp\left[\left(-\lambda_0 - b_{\lambda 1}\sigma + \frac{b_{\lambda 2}\sigma^2}{\theta - \sigma}\right)z\right] \tag{15-102}$$

where λ_0 is the clean bed filtration coefficient [L^{-1}], $b_{\lambda 1}$ and $b_{\lambda 2}$ filter coefficient constants [L^{-1}], θ the clean-bed porosity, and σ the specific deposit defined as the volume of solids per unit volume of bed. The second term, $b_{\lambda 1}\sigma$, results in a linear increase in the overall filter coefficient during the ripening period, while the third term results in a decrease in the filter coefficient as a result of increasing interstitial velocities caused by deposits. There is evidence that the main difficulty with this model is that the third term fails properly to account for the clogging of the filter (Fox and Cleasby, 1966).

A model proposed by Mintz (1966) includes both attachment and detachment mechanisms for particle transport. According to this model, the decrease in the filter coefficient is a result of deposit scouring, resulting in resuspension of previously

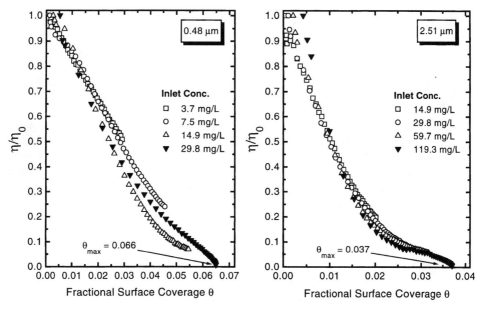

Figure 15.17 Single collector efficiencies (normalized to the clean bed efficiency) in deionized water (6×10^{-6} M). Column conditions given in Figure 15.13. (Reprinted with permission from Liu et al. (1995). Copyright 1995, American Chemical Society.)

deposited particles. In differential form, the Mintz model is

$$\frac{\partial N}{\partial z} = -\left(\lambda_0 N - \frac{b_s \sigma}{U}\right) \tag{15-103}$$

where b_s is a scour coefficient [L T^{-1}] and U the approach velocity. The solution of this equation can only be expressed in an infinite series form.

TABLE 15.7 Jamming Limits ($\theta_{max} = 1/B_f$) and Excluded Areas (B_f) for Latex Particles in Glass Bead Columns as a Function of Ionic Strength of KCl Solutions

Ionic Strength (M)	$d_p = 0.48$ μm		$d_p = 2.51$ μm	
	θ_{max}	B_f	θ_{max}	B_f
6×10^{-6}	0.071	14.08	0.045	22.22
10^{-5}	0.092	10.87	0.078	12.82
10^{-4}	0.173	5.78	0.135	7.41
10^{-3}	0.273	3.66	0.188	5.32
10^{-2}	0.322	3.11	0.576	1.74
10^{-1}	0.592	1.69	—	—

Source: Liu et al., 1995.

O'Melia and Ali (1978) proposed a filter ripening model based on modifying the single collector collision efficiency to include the effect of deposited particles. The product of the single collector efficiency and the sticking coefficient for ripening is

$$\alpha\eta = \alpha_0\eta_0 + \alpha_p\eta_p N_{p,d} R^{*2} \qquad (15\text{-}104)$$

where η_0 and α_0 are the clean-bed collector efficiency and sticking efficiencies, $N_{p,d}$ the number of deposited particles that act as collectors, α_p and η_p the sticking coefficient and collector efficiency of deposited particles, and R^* the interception number defined as the ratio of particle and collector diameters. Increases in filter head loss resulting from ripening can also be calculated by including an additional term to account for deposited particles in the form

$$\frac{h}{L} = \frac{b_f\mu(1-\theta)^2 a_f^2 U}{\rho_w g\theta^3}\left(\frac{1 + b_{f,d}N^* R^{*2}}{1 + N^* R^{*3}}\right) \qquad (15\text{-}105)$$

where $b_{f,d}$ is an empirical factor defined as the fraction of captured particles that contribute to additional head loss, and N_d^* the ratio of the number of deposited particles $(N_{p,d})$ to the number of collectors, N_c. These filtration and head loss equations were examined by Darby et al. (1992). They found that the filtration model worked well when calibrated for a single experiment, but that it did not predict well the effect of different influent particle concentrations. The failures of the model were attributed to changes in deposit morphology possibly due to floc breakoff and re-capture, mechanisms not included in the model formulation. Observed linear changes in head loss during filter ripening were also not well predicted by the model. Although some characterization of filtration efficiency during ripening is possible using several empirical constants, a complete mathematical description of filter ripening is still lacking, likely due to our incomplete understanding of how to model and incorporate deposit morphology, and changes in the morphology, into the governing equations.

15.8 PARTICLE TRANSPORT IN GROUNDWATER AQUIFERS

Particle removal in water treatment has been extensively studied from the perspective of making water treatment filtration efficient and inexpensive. When a water source has few particles (as indicated by low turbidity), the water can be directly filtered avoiding the need for separate flocculation and coagulation processes. Chemicals such as alum, or a cationic polymer, may still need to be added for direct filtration in order to increase particle size and stickiness. As we have seen, particles on the order of 1 μm are the least efficiently removed; so coagulating bacteria-size particles into larger particles increases overall particle removal rates. Without the addition of chemicals, particles might pass through the filter media (a distance of only a meter or so) without being removed.

The transport of particles in groundwater aquifers provides an interesting contrast to colloid filtration in water treatment plants. Bacteria have been isolated that can degrade a variety of common groundwater pollutants. The injection of these bacteria into the ground for pollutant degradation is called bioaugmentation. The same physical factors that make particle removal in water treatment filters practical, however, often preclude the use of the these pollutant-degrading bacteria *in situ:* Bacteria are filtered out before they can travel very far in soils. Under conditions typical of groundwater flow (superficial velocity of 1 m d^{-1}, soil grain diameter of 120 μm, porosity of 0.33, and 10°C), a bacterium (cell diameter of 1 μm) would have to undergo ~500 collisions to be transported a distance of only 1 m. Laboratory-grown bacteria typically have sticking coefficients of ~0.1 to 1.0 for natural soils, indicating that they readily stick to soil grains. These bacterial sticking coefficients for laboratory-grown bacteria need to be reduced by one to three orders of magnitude, or in the range of $10^{-2} < \alpha < 10^{-3}$, to transport successfully them over distances of tens of meters in groundwater aquifers for bioaugmentation (Fig. 15.18). Thus we find that the goals of water treatment and bioaugmentation are exactly the opposite. Instead of needing to make bacteria more sticky, as in water treatment filters, bacteria must be made less sticky for their use in subsurface bioremediation.

Bacterial Transport in Porous Media   Bacterial transport in sandy soils is primarily controlled by electrostatic interactions. At a pH above 2, bacteria carry a net negative charge (Richmond and Fisher, 1973). Soil grains can be considered to be silica surfaces containing metal oxides coated with natural organic matter, consisting of humic and fulvic acids, and soil grains therefore also carry a net negative charge at neutral pHs. Reducing the ionic strength of water injected into a porous medium

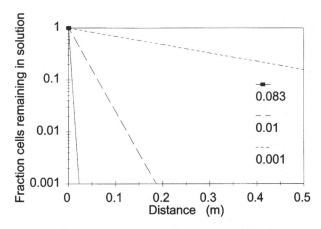

Figure 15.18 Effect of the sticking coefficient (α = 0.083, 0.01, or 0.001) on the transport distance of bacteria in a soil aquifer under typical groundwater conditions of d_c = 127 μm and θ = 0.41 for Arizona soil, groundwater flows of U = 10 m d^{-1} at 10°C, and bacteria with a diameter of 1 μm and density of 1.07 g cm^{-3}. (Reprinted from Li and Logan 1998, with kind permission from Elsevier Science Ltd., The Boulevard, Landford Lane, Kidlington, OX5 1GB.)

can increase the thickness of the electrostatic repulsive layer around both bacteria and soil grains and minimizes bacterial attachment to silica surfaces and sandy soils. Low-ionic-strength solutions have been shown to decrease bacterial attachment to rotating glass disks (Martin et al., 1991), glass beads (Gross and Logan, 1995), glass fiber filters (Logan et al., 1993), and natural soils (Fontes et al., 1991). By using radiolabeled cells, and measuring the deposition of cells with transport distance along a column, it is possible to calculate sticking coefficients both within the column and in the column effluent using liquid scintillation counting. A comparison of different systems indicated that ionic strength has a similar effect of reducing sticking coefficients in all these systems, although the magnitude of the change in the sticking coefficient is not always the same, likely as a result of the different hydrodynamic environments of the flow experiments (Fig. 15.19).

While glass and quartz surfaces are mostly silica dioxide, metals can significantly contribute to increased bacterial retention. When glass surfaces are coated with iron, Johnson and Logan (1996) found that bacterial retention increased from $28 \pm 10\%$ to $73 \pm 14\%$ (Fig. 15.20). The presence of natural organic matter, either sorbed or dissolved, reduces the effect of iron on cell attachment. When iron-coated surfaces

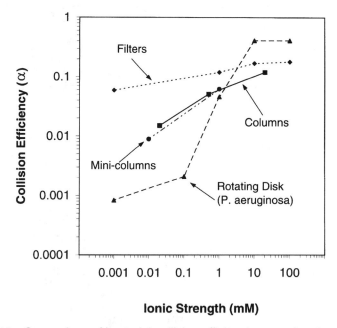

Figure 15.19 Comparison of bacterial collision efficiencies as a function of suspending solution ionic strength for different systems: (glass fiber, GF/C), columns and mini-columns (glass beads), and a rotating disk. *Pseudomonas fluorescens* P17 were used in all experiments except the rotating disk experiment. (Reprinted from Jewett et al. 1995 with kind permission from Elsevier Science Ltd., The Boulevard, Landford Lane, Kidlington, OX5 1GB.)

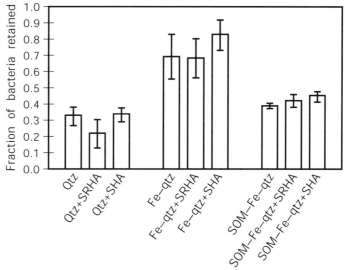

Figure 15.20 The effect of Suwanee River humic acids (SRHA) and soil humic acids (SRA) on bacterial retention in 1-cm-long minicolumns packed with 40-μm quartz (qtz) media. Grouped bars are for cleaned quartz (qtz), iron coated quartz (Fe-qtz), and iron quartz pre-equilibrated with organic matter (SOM-Fe-qtz). (From Johnson and Logan, 1996, with kind permission from Elsevier Science Ltd., The Boulevard, Landford Lane, Kidlington, OX5 1GB.)

were pre-equilibrated with humic acids, cell attachment decreased to $41 \pm 3\%$ in the glass bead columns. In the absence of the iron coating, dissolved humic acids (either soil humic acid or Suwannee River humic acid) decreased bacterial retention an average of 20%. Based on these results, it is likely that either sorbed or dissolved humic acids increase the negative charge of the cells.

Bacterial attachment to surfaces is believed to be controlled both by electrostatic and hydrophobic forces. Van Loosdrecht et al. (1987) were able to correlate adhesion of several different bacteria to different surfaces only by examining both the charge on the bacteria, measured in terms of cell electrophoretic mobility, and cell hydrophobicity, measured from the contact angle of water on a lawn of cells (Fig. 15.21). As cells become more negatively charged, they attach less to hydrophilic surfaces, such as glass, but the magnitude of this increase in attachment is strongly a function of the overall hydrophobicity of the strain. More hydrophobic cells attach more readily to surfaces essentially independent of the cell's electrophoretic mobility.

The adhesion of bacteria to glass surfaces can be reduced by adding surfactants. The addition of Tween 20, a nonionic surfactant, strongly decreased the attachment of pure cultures of *Alcaligenes paradoxus* to glass beads (Table 15.8). The two-order-of-magnitude reduction in the sticking coefficient by Tween 20 was the same as that produced using low-ionic-strength (0.01 mM) solutions. Surfactants can de-

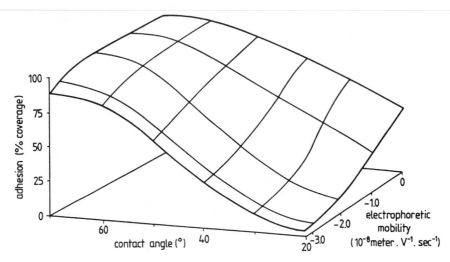

Figure 15.21 Bacterial adhesion is a function of both the bacteria charge (electrophoretic mobility) and relative hydrophobicity (contact angle) based on measurements of 23 strains of bacteria. (Reprinted from van Loosdrecht et al. (1987) with permission of the American Society of Microbiology.)

crease cell attachment either by interfering with cell–surface contact, altering the charge around the cell, or by changing the overall cell hydrophobicity, but exactly which mechanisms dominate are not known. Chemicals that disrupt cell surface structures can decrease attachment, but not to the extent observed for nonionic surfactants or low-ionic-strength solutions. Sticking coefficients were reduced by less than an order of magnitude by EDTA (cell membrane permeabilizer that removes outer membrane lipopolysaccharides), sodium PP_i (surface charge modifier), and proteinase-K (a nonspecific protease that cleaves peptide bonds), and increased when sodium periodate (an oxidizer that cleaves surface polysaccharides) and lysozyme (enzyme that cleaves cell wall components) were added (Table 15.8).

Virus Transport in Porous Media   In contrast to bacteria, viruses are readily transported through porous media. Measurements of human enteric viruses near septic tanks indicate that they can travel over distances of 1 km under certain conditions (Keswick and Gerba, 1980). Sticking coefficients for the bacteriophage MS-2 measured in silica bead columns ranged from 0.0007 to 0.02 and were a function of pH and organic matter (Bales and Li, 1993). Removal at pH = 5 was higher than that at pH = 7, and deposited phage were released by both lowering the ionic strength and using beef extract to increase the ionic strength. Small amounts of organic matter in the porous medium (0.001%) have been shown to retard virus transport.

Injection of Particles in Low-Ionic-Strength Water into the Ground   The suspension of bacteria in low-ionic-strength (IS) water is a method demonstrating

TABLE 15.8   **Effect of Various Chemicals on the Transport of *Alcaligenes paradoxus* in Glass Bead (40-μm-diameter) Minicolumns (1 cm Long). Ionic Strengths of the Solution Varied, and Therefore the Control in Each Case (No Chemical) Are Not the Same. The Effects of the Chemical Can Only Be Evaluated on the Basis of the Change in the Sticking Coefficient, α, in the Presence of a Chemical, and Not Between Categories. The Sticking Coefficient is Calculated Using the RT Model.**

Chemical and Buffer	Chemical Concentration	α
Tween 20 in phosphate buffer	0	0.38
	0.1%	0.0016
NaCl	1 mM	0.014
None (low-ionic-strength water)	0	0.0016
$NaIO_4$ in phosphate buffer	0	0.052
	10	0.10
EDTA in Tris buffer	0	0.61
	1 mM	0.34
EDTA in MOPS buffer	0	0.084
	1 mM	0.17
$Na_4P_2O_7$ in phosphate buffer	0	0.27
	1 mM	0.11
$Na_4P_2O_7$ in MOPS buffer	0	0.20
	1 mM	0.086
Proteinase	0	0.055
	0.1 mg mL^{-1}	0.044
Lysozyme	0	0.0048
	0.1 mg mL^{-1}	0.74

Source: Gross and Logan, 1995.

great promise for increasing bacterial or virus transport in groundwater aquifers, but other effects on the low-IS water on the soil need to be considered. The introduction of solutions containing little organic matter can result in the increased desorption of sorbed organic matter. In addition, loosely held colloidal particles can also desorb and become resuspended, increasing particle concentrations (Ryan and Elimelech, 1996). As water travels through the soil, the ionic strength will increase at a rate dependent on the soil type, and the resuspended particles will again deposit at higher ionic strengths. The use of low-IS solutions can lead to colloid mobilization (Seaman et al., 1995). Although the injection of low-IS water in field studies did not promote colloid migration at one site examined by Wiesner et al. (1996), the presence of high concentrations of colloids in water (due to naturally occurring or mobilized particles due to groundwater pumping) led to aquifer plugging in field studies (Wiesner et al., 1996). Thus the control of colloid transport for increasing particle penetration in porous media may ultimately provide a greater challenge to soil scientists and re-mediation engineers than the relatively easier goal of decreasing particle penetration (making particles more sticky) in water treatment systems.

PROBLEMS

15.1 Cookson (1970) examined virus removal in a column 25.2 cm long with average area of 2.010 cm^2, porosity of 0.3, packing diameter of 0.0987 cm, and interfacial area of 53 cm^2 cm^{-3} for viruses 54 nm in diameter. Using the data in Table P-15.1, calculate a mass transport correlation of the form $j_D = b_1 Re^{b_2}$ assuming Sc = 1.25 × 10^5, and an inlet virus concentration of 1.0 × 10^6 for the first two data sets, and 1.4 × 10^6 for the last data set.

15.2 Particle dispersion should be neglected in column filtration experiments, but solutions to the general transport equation, for a first-order reaction with dispersion in one dimension, can be used to evaluate the magnitude of error inherent in equations containing dispersion. (a) Calculate the value of a first-order rate constant assuming the RT filtration equation and typical laboratory column experimental conditions of U = 0.012 cm s^{-1}, θ = 0.41, d_p = 1 μm, d_c = 120 μm, α = 0.1, L = 20 cm, solution properties of water at 25°C, ρ_p = 1.07 g/cm^3, and $\mathcal{A}$ = 10^{-13} g/cm^2–s^3. (b) Using the RT filtration equation, calculate the effluent particle concentration. (c) Using Eqs. 12-26–12-29, calculate column effluent particle concentrations assuming dispersion coefficients of 10^{-3}, 10^{-2}, and 10^{-1}. (d) Comparing your results in parts (b) and (c), calculate the error made in column effluent concentrations if the dispersion coefficients is neglected. (e) What sticking coefficients would be necessary in the RT equation to produce the sticking coefficients produced by your calculations in part (c)?

15.3 A researcher investigating particle filtration measures particle concentrations at 1-cm intervals in a 20-cm-long column operating at steady-state concentrations. Assume the conditions given in the previous problem, and a dispersion coefficient of 0.01 cm^2 s^{-1}. (a) Using Eqs. 12-26–12-29, plot the steady state effluent concentrations for columns 1 to 20 cm long in the presence of dispersion. (b) What sticking coefficients would be calculated (using the RT model) if dispersion was neglected based on concentrations that would be in the column in the presence of dispersion? Plot the values of α as a function of column length.

TABLE P-15.1   Three Data Sets from Cookson (1970)

Flow Rate (cm^3 s^{-1})	N_{out} × 10^{-5} (mL^{-1})	Flow Rate (cm^3 s^{-1})	N_{out} × 10^{-5} (mL^{-1})	Flow Rate (cm^3 s^{-1})	N_{out} × 10^{-5} (mL^{-1})
4.37	6.7	0.292	1.9	3.77	7.3
3.77	6.4	0.210	1.4	2.89	7.6
3.17	6.5	0.122	0.68	1.72	6.5
2.75	5.7	0.058	0.23	1.13	5.2
2.00	5.7	0.018	0.06		
1.40	5.0				
0.867	4.0				

15.4 Following treatment by coagulation and sedimentation, it is found that the particle size distribution has the following forms prior to entering a filter

$$n(d_p) = 10^7 d_p^{-5/2}, \qquad 0.1 < d_p < 1.0 \ \mu\text{m}$$

$$n(d_p) = 10^7 d_p^{-4}, \qquad 1.0 \leq d_p < 100 \ \mu\text{m}$$

where $n(d_p)$ is in # $\text{mL}^{-1} \ \mu\text{m}^{-1}$ and d_p in μm, and all suspended particles have a density of 1.05 g cm^{-3}. Filter characteristics are: loading rate of 100 L min^{-1} m^{-2} (2.5 gal min^{-1} ft^{-2}), media depth of 100 cm, filter grain sizes of 1.0 mm, bed porosity of 0.4, and a water temperature of 20°C. Assuming a clean-bed efficiency. (a) Calculate the particle volume distribution $dV/d(\log d_p)$ in the effluent from the filter. (b) Calculate the total mass removal efficiency for the filter. (From Hunt, 1997).

15.5 Glass fiber filters are routinely used for filtration of water samples in environmental laboratories. Such filters are given pore diameter ratings, but these filters do not act to strain out particles (like a screen) but actually operate as depth filters. In this problem we analyze the properties of a glass fiber filter (GF/C) rated with a nominal pore diameter of 1.2 μm. (a) Consider the GF/ C filter to be a mat of very long fibers having a diameter of d_c and a porosity of θ. Derive a filter coefficient, λ (assuming you know the form of the single collector efficiency, η, for the cylindrical filter fiber) in the same manner as done for spherical collectors (i.e., by performing a steady particle balance over a small control volume of filter bed). Note that for calculating the number of collectors the volume of a unit length of collector is $\pi d_c^2/4$. (b) Using the Yao et al. model to calculate the single collector efficiency, η, where d_c is the diameter of the fiber (2.2 μm), calculate at room temperature the concentration of bacteria (diameter 1.0 μm, concentration $10^6 \ \text{mL}^{-1}$) passing through the filter assuming a sticking coefficient of 0.1, bed porosity of 0.89, filter bed length of 300 μm, flow velocity 0.1 cm s^{-1}, and assuming collisions due to gravity are negligible. Assume for water, $\rho_w = 1.0$ g cm^{-3}, and $v_w = 0.01 \ \text{cm}^2$ s^{-1}. (c) How many collisions would a nonsticky particle have with the filter fibers during passage through the filter? (d) What does all this imply about sample filtration of bacteria using glass fiber filters?

15.6 Water treatment filters often consist of coarse sand (and also other types of materials in multimedia filters) having a range of particle sizes. In order to make pressure and filtration calculations, the packing must be characterized in terms of a single grain size. Grain sizes can be characterized as d_n, where n is the percent of the media smaller than the diameter. Typical values used are: d_{10}, d_{50} (average), and d_{60}. For this problem, assume that the grains are spheres with density 2.55 g cm^{-3}, and have diameters varying in size from 1 to 4 cm and the grain size distribution is given as

$$n(d_g) = b d_g^{-2}, \qquad 1 \leq d_g \ [\text{cm}] \leq 4$$

where b [cm g^{-1}] = 0.10, and n has units of [$\text{g}^{-1} \ \text{cm}^{-1}$]. (a) Verify that the total mass of media is in the size interval of 1 to 4 cm. (b) Calculate the average sizes of particles based on the three different characteristic diameters based on their mass distribution. (From Hunt, 1997.)

15.7 Let us assume that a water treatment filter 6 ft in depth, and loaded at 4 gpm ft^{-2}, is packed with sand grains of characteristic diameters of $d_{10} = 1.5$ cm and $d_{60} = 3.5$ cm. Martin et al. (1996) recommend for filtration calculations that for mixtures of particle sizes, the media is better characterized in terms of a d_{10} than a d_{60}; this problem explores the effect of these different characteristic sizes on particle removal rates. (a) Calculate the particle removal in this filter of 1-μm-diameter particles of density 1.03 g cm^{-3} for both media diameters, $\alpha = 1$, and $\theta = 0.38$, assuming only collisions by Brownian motion are important ($T = 20°C$). (b) How many collisions would a nonattaching particle undergo in this filter in order to reach the end of the filter assuming media of size d_{10} and size d_{60}? What do you conclude about the removal efficiency of bacteria-sized particles in water treatment filters?

REFERENCES

Bai, G., M. L. Brusseau, and R. M. Miller. 1997. *Appl. Environ. Microbiol.* **63**(5):1866–73.

Bales, R. C., S. R. Hinkle, T. W. Kroeger, K. Stocking, and C. P. Gerba. 1991. *Environ. Sci. Technol.* **25**(12), 2088–95.

Bales, R. C., and S. Li. 1993. *Water Resour. Res.* **29**(4):957–63.

Barton, J. W., and R. M. Ford. 1995. *Appl. Environ. Microbiol.* **61**(9), 3329–35.

Blackwell, J. 1959. Amer. Inst. Chem. Eng. And Soc. Petrol. Eng. 52nd Annual Meeting, San Francisco, Preprint No. 29.

Cleasby, J. L. 1990. In: *Water Quality and Treatment*, 4th Ed., American Water Works Assoc., pp. 455–560.

Cookson, J. T. Jr. 1970. *Environ. Sci. Technol.* **4**(2):128–34.

Darby, J. L., R. E. Attanasio, and D. F. Lawler. 1992. *Wat. Res.* **26**(6):711–26.

Flagan, R. C., and J. J. Seinfeld. 1988. *Fundamentals of Air Pollution Engineering*. Prentice Hall, Englewood Cliffs, N.J.

Fontes, D. E., A. L. Mills, G. M. Hornberger, and J. S. Herman. 1991. *Appl. Environ. Microbiol.* **57**:2473–81.

Fox, D. M., and J. L. Cleasby. 1966. *J. San. Engng. Div. ASCE* **SA5**:4941.

Fuchs, N. A. 1964. *The Mechanics of Aerosols*. Pergamon Press, Oxford.

Gross, M. J., and B. E. Logan. 1995. *Appl. Environ. Microbiol.* **61**(5):1750–56.

Happel, J. 1958. *AIChE J.* **4**(2):197–201.

Harley, S., D. W. Thompson, and B. Vincent. 1992. *Colloids Surf.* **62**:163–76.

Hinds, W. C. 1983. *Aerosol Technology*. Wiley, New York.

Hunt, J. R. 1997. Lecture notes for Water Treatment Engineering 211, Univ. California, Berkeley.

Hunt, J. R., B.-C. Hwang, and L. M. McDowell-Boyer. 1993. *Environ. Sci. Technol.* **27**(6): 1099–107.

Ives, K. J., and I. Sholji. 1965. *J. San. Engrg. Div. ASCE* **SA4**:1–18.

Iwasaki, T. 1937. *J. AWWA* **29**(1):1591–97.

Jewett, D. G., T. A. Hilbert, B. E. Logan, R. G. Arnold, and R. C. Bales. 1995. *Wat. Res.* **29**(7):1673–80.

John, W., G. Reischl, S. Goren, and D. Plotkin. 1978. *Atmos. Environ.* **12**:1555–57.

Johnson, W. P., and B. E. Logan. 1996. *Wat. Res.* **30**(4):923–31.

Keswick, B. H., and C. P. Gerba. 1980. *Environ. Sci. Technol.* **14**:1290–97.

Levich, V. G. 1962. *Physicochemical Hydrodynamics.* Prentice Hall, Englewood Cliffs, N.J.

Li, Q., and B. E. Logan. 1998. *Wat. Res.* In press.

Liu, D., P. R. Johnson, and M. Elimelech. 1995. *Environ. Sci. Technol.* **29**(12):2963–73.

Logan, B. E. 1993. *Limnol. Oceanogr.* **38**(2):372–81.

Logan, B. E., T. A. Hilbert, and R. G. Arnold. 1993. *Wat. Res.* **27**:955–62.

Logan, B. E., D. G. Jewett, R. G. Arnold, E. J. Bouwer, and C. R. O'Melia. 1995. *J. Environ. Eng.* **121**(12):869–73.

Logan, B. E., D. G. Jewett, R. G. Arnold, E. J. Bouwer, and C. R. O'Melia. 1997. *J. Environ. Engng.* **123**(7):730–31.

Martin, R. E., L. M. Hanna, and E. J. Bouwer. 1991. *Environ. Sci. Technol.* **25**(12):2075–82.

Martin, M. J., B. E. Logan, W. P. Johnson, D. G. Jewett, and R. G. Arnold. 1996. *J. Environ. Eng.* **122**(5):407–15.

McCaulou, D. R., R. C. Bales, and R. G. Arnold. 1995. *Water Resour. Res.* **31**(2):271–80.

McCaulou, D. R., R. C. Bales, and J. F. McCarthy. 1994. *J. Contam. Hydrol.* **15**:1–14.

Mintz, D. M. 1966. *Proc. Intl. Water Supply Assn. Seventh Congress* **1**:10.

O'Melia, C. R., and W. Ali. 1978. *Prog. Water Tech.* **10**(5/6):167–82.

Pfeffer, R. 1964. *I&EC Fund.* **3**(4):380–83.

Pfeffer, R., and J. Happel. 1964. *AIChE J.* **10**:605–11.

Rajagopalan, R., and C. Tien. 1976. *AIChE J.* **22**(3):523–33.

Richmond, D. V., and D. J. Fisher. 1973. *Adv. Microbial Physiol.* **9**:1–29.

Rijnaarts, H. H. M., W. Norde, E. J. Bouwer, J. Lyklema, and A. J. B. Zehnder. 1996. *Environ. Sci. Technol.* **30**(10):2877–83.

Rubow, K. L., and B. Y. H. Liu. 1986. In *Fluid Filtration: Gas Vol. 1.* ASTM Publ. 975, American Society for Testing Materials: Philadelphia, pp. 74–94.

Ryan, J. N., and M. Elimelech. 1996. *Colloids Surf. A. Physicochem. Engin. Aspects* **107**:1–56.

Seaman, J. C., P. M. Bertsch, and W. P. Miller. 1995. *Environ. Sci. Technol.* **29**(7):1808–15.

Spurney, K. R., J. P. Lodge, Jr., E. R. Frand, and D. C. Sheesley. 1969. *Environ. Sci. Technol.* **3**:453–64.

Tan, Y., J. T. Gannon, P. Baveye, and M. Alexander. 1994. *Water Resour. Res.* **30**(12):3243–52.

Tien, C., and A. C. Payatakes. 1976. *AIChE J.* **25**:737–59.

Tobiason, J. E., and B. Vigneswaran. 1994. *Wat. Res.* **28**(2):335–42.

van Loosdrecht, M. C. M., J. Lyklema, W. Norde, G. Schraa, and A. J. B. Zehnder. 1987. *Appl. Environ. Microbiol.* **53**(8):1898–901.

Wiesner, M. R., M. C. Grant, and S. R. Hutchins. 1996. *Environ. Sci. Technol.* **30**(11):3184–91.

Yao, K.-M., M. T. Habibian, and C. R. O'Melia. 1971. *Environ. Sci. Technol.* **5**(11):1105–12.

APPENDIX 1

NOTATION

a_n interfacial area in a reactor, where the subscript n varies: where a, air,; b, biofilm; w, water, m, microbe, nw, NAPL–water, v, volumetric coefficient for an overall mass transport coefficient $[L^2\ L^{-3}]$

A area, usually with a subscript such as: b, biofilm; b, bubble; nw, NAPL –water; p, particle; p, packing in an air stripping tower (see Eq. 6-72); p, paddle (see Eq 7-35) $[L^2]$

b constant, with various subscripts and units; special constants are indicated below

$b_{d,n}$ drag coefficient where n is a subscript for a particular system, such as b, bubble; p, paddle; g, aggregate; s, sphere [ ]

b_D fractal constant, defined as $b_D = (\zeta_g \xi_g / \xi_p)^{D/3}$

b_d power of empirical drag correlation, in the form $b_{drag} = b_{dl}\ Re^{b_d}$ [ ]

b_{end} endogenous decay coefficient $[t^{-1}]$

b_f friction constant, as in Eq. 3-75 [ ]

b_l length between atoms in a metal lattice, as in Eq. 3-45 $[L]$

b_H Happel cell constant; see Eq. 15-47 [ ]

B constant, with various subscripts and units

B_{Br} constant used for dimensional analysis of coagulation by Brownian motion; other subscripts are sh for shear and ds for differential sedimentation

c total molar or mass concentration $[M\ L^{-3}]$

c^* dimensionless concentration; for example, $c^* = c_{Cb}/c_{Cb,0}$ [ ]

c_{ij} molar or mass concentration of species i in phase j. When there is only one phase, and one chemical, the subscripts ij may be dropped and other subscripts added as necessary. For example, concentration in or out; eq, equilibrium; ∞, concentration at infinity; etc. $[M\ L^{-3}]$

c_f concentration of all the filtrate collected from an ultrafiltration cell $[M\ L^{-3}]$

c_{min}	minimum substrate concentration necessary to support cells [M L^{-3}]
c_p	concentration in a permeate stream in an ultrafiltration cell [M L^{-3}]
c_r	concentration in the retentate in an ultrafiltration cell [M L^{-3}]
d_n	diameter, where the subscript n varies: ag, aggregate; n, NAPL ganglion; f, fiber; g, fractal generator; h, pore; i, impeller; max, maximum aggregate size [L]
d_p^*	dimensionless diameter of a particle, defined as $d_p^* = d_p/d_f$ [ ]
D	fractal dimension in three dimensions; can also be written as D_3, but the subscript is understood to be 3 when omitted [ ]
D	actual dilution, based on total volume [ ]
D_n	fractal dimension in n dimensions, where n = 1, 2, or 3 [ ]
D_R	relative dilution, based on a ratio of the two volumes [ ]
D_{ib}	overall diffusivity of chemical i in a biofilm [L^2 t^{-1}]
D_{ij}	diffusivity of chemical i in phase j [L^2 t^{-1}]
$D_{ij,h}$	pore diffusivity of chemical i in phase j [L^2 t^{-1}]
$D_{ij,hm}$	overall diffusivity of chemical i in phase j in a porous matrix [L^2 t^{-1}]
$D_{ij,K}$	Knudson diffusivity of chemical i in phase j [L^2 t^{-1}]
$D_{ij,s}$	overall diffusivity of chemical i in phase j (water or air) in soil [L^2 t^{-1}]
$D_{ij,sur}$	diffusivity of chemical i in phase j on the surface of a solid material [L^2 t^{-1}]
$D_{ij,sur+h}$	combined diffusivity of chemical i in phase j due to surface and pore diffusion [L^2 t^{-1}]
$D_{ij,uh}$	micropore diffusivity of chemical i in phase j [L^2 t^{-1}]
e	electrical charge, 1.60219×10^{-19} C
e_x	dispersion coefficient produced by mixing in the x direction (the direction of flow), while subscripts y and z are used for other directions [L^2 t^{-1}]
$e_{\Delta u}$	dispersion coefficient that is produced by a nonconstant velocity profile in the direction of flow [L^2 t^{-1}]
e_{mech}	dispersion coefficient that results from nonconstant velocities in a porous medium in the direction of flow [L^2 t^{-1}]
e_{jump}	jump frequency defined in Eq. 3-45 [t^{-1}]
E	internal energy, usually in cal or joules [M L^2 t^{-2}]
E	ratio of observed nitrification rate to the maximum oxygen transfer rate, assuming a ratio of 4.3 moles of oxygen per mole of ammonia; see Eq. 9-209 [ ]
E	interfacial energy; see Eq. 12-73 [M L^2 t^{-2}]
E_A	energy due to London–van der Waals forces [M L^2 t^{-2}]
E_A	energy due to electrostatic double-layer repulsive forces [M L^2 t^{-2}]
E_L	overall longitudinal dispersion coefficient; the one-dimensional dispersion coefficient in the direction of flow; can also have a subscript indicating transverse (t) or vertical (v) dispersion [L^2 t^{-1}]
E_T	total energy [M L^2 t^{-2}]
E_x	overall dispersion coefficient in the x direction; may also be applied in y and z directions [L^2 t^{-1}]

E_0 tabulated function of dimensionless porosity for flow permeable aggregates [ ]

E dispersion coefficient; potentially a second-order tensor, although in practice just a scalar quantity [$L^2 \, t^{-1}$]

f friction factor [ ]

f_n fraction, where the subscript n varies: a, fraction of cell surface available for transport; c, fractional reduction of concentration at surface of an aggregate relative to bulk concentration; jp, fraction of sBOD remaining in component j after flow over the plate; v, fraction of cells that are viable; T, total, where $f_T = f_a f_v (1 - f_c)$ [ ]

F force [$M \, L \, t^{-2}$]

F Helmholtz free energy, usually in kcal/mol [$L^2 \, t^{-2}$]

F fractional reduction of volume in an ultrafiltration cell, or $F = 1 - (V_F - V_0)$ [ ]

F stability frequency; see Eq. 11-43 [t^{-1}]

F_j fraction of sBOD remaining in each of j components after flow over plastic media modules in a trickling filter [ ]

g gravitational constant [$L \, t^{-2}$]

G shear rate, equal to the velocity gradient of an eddy [L^{-1}]

G Gibbs free energy, usually in kcal/mol [$L^2 \, t^{-2}$]

G^0 Reference state for Gibbs free energy, usually in kcal/mol [$L^2 \, t^{-2}$]

h Debye–Hückel parameter [L^{-1}]

h height of a reactor [L]

H Enthalpy, usually in kcal/mol [$L^2 \, t^{-2}$]

H reactor holdup; see Eq. 7-95 [ ]

H_{ijk} Henry's Law constant defining ratio of chemical concentration in air and water at equilibrium; subscripts j and k refer to units of H for chemical i in air and water: yx, mole fraction basis; px, pressure per mole fraction; cc, concentration units for both phases [various units]

H_v height, with different subscripts: v, rise or vertical height of water in a capillary tube [L]

I Dimensionless integral of velocity in a stream; see Eq. 11-34 [ ]

j_{ij} flux relative to a moving coordinate system; a vector quantity [$M \, L^{-2} \, t^{-1}$]

$j_{ij,x}$ flux relative to a moving coordinate system in the x direction; can also be defined in y or z directions [$M \, L^{-2} \, t^{-1}$]

J_{ij} flux relative to a fixed coordinate system; a vector quantity [$M \, L^{-2} \, t^{-1}$]

$J_{ij,x}$ flux relative to a fixed coordinate system in the x direction; can also be defined in y or z directions [$M \, L^{-2} \, t^{-1}$]

J_V volume flux [$l^3 \, L^{-3} \, t^{-1}$]

k_b rate constant in a biofilm based on first-order kinetics and coagulation theory, defined in Eq. 9-178 [L^2]

k_{ij} reaction rate constant of chemical i in phase j; Units vary depending on reaction order

k_{ij} mass transfer coefficient of chemical i for a single phase j. Usually abbreviated for a specific chemical as just k_j [$L \, t^{-1}$]

k^+	mass transport coefficient that incorporates the specific surface area; used in calculations where interfacial area is not known, such as in NAPL dissolution calculations where $k^+ = k_w a_{nw}$ $[\text{t}^{-1}]$
$k_{H,j}$	thermal heat conductivity, usually in $[\text{cal L}^{-2}\ \text{T}^{-1}]$
K_{eq}	equilibrium constant $[\ ]$
K_{ijk}	overall mass transfer coefficient of chemical i between phases j and k, with the concentration difference written about the first phase subscript, j. Usually abbreviated for a specific chemical as just K_{jk} $[\text{L t}^{-1}]$
K^+	Overall mass transport coefficient that incorporates the specific surface area; used in calculations where interfacial area is not known, such as in NAPL dissolution calculations where $K^+ = K_{nw} a_{nw}$ $[\text{t}^{-1}]$
K_{ijk}	distribution coefficient of chemical i between phases j and k at equilibrium; normally dimensionless, but dependent on the units of concentration in the two phases
K_m	Monod half-saturation growth constant $[\text{M L}^{-3}]$
K_m	Michaelis−Menten half-saturation growth constant $[\text{M L}^{-3}]$
l	length $[\text{L}]$
$L_{nw,h}$	length of a ganglion in the horizontal direction, where n, NAPL; w, water $[\text{L}]$
$L_{nw,v}$	length of a ganglion in the vertical direction, where n, NAPL; w, water $[\text{L}]$
m	mass $[\text{M}]$
m_n	nth moment, where $n = 0$ $[\text{M t L}^{-3}]$, 1 $[\text{t}]$, or 2 $[\text{t}]$
m_p	mass of a particle $[\text{M}]$
M_i	molecular weight of chemical i, usually g mol^{-1} $[\text{M M}^{-1}]$
n	size distribution function describing the particle concentration in a size interval $[\#\ \text{L}^{-3}\ \text{l}^{-1}]$
n_{ij}	moles of chemical i in phase j $[\text{M}]$
N^*	number of particles $[\#]$ or $[\ ]$
N_b^*	number of bubbles in a reactor $[\ ]$
N_n	number concentration, usually with a subscript n, which varies: c, cells; p, particles; S, substrate molecules; usually in units of $\#\ \text{mL}^{-1}$ $[\text{L}^{-3}]$
N_p	number of plates in a trickling filter from top to bottom (for all modules) $[\ ]$
$\dot{N}^*$	number of particles in a Euclidean object such as a sphere $[\#]$
p_c	permeation coefficient in the permeation model for ultrafiltration $[\ ]$
p_{ij}	partial pressure of a gas $[\text{M L}^{-1}\ \text{t}^{-2}]$
p_i^0	vapor pressure of pure species i $[\text{M L}^{-1}\ \text{t}^{-2}]$
P	total air pressure, usually in atm $[\text{M L}^{-1}\ \text{t}^{-2}]$
P	perimeter $[\text{L}]$
P_j	fraction of total sBOD in each of j components $[\ ]$
P	power, usually in watts. May have a subscript such as p, paddle; b, bubbles $[\text{M L}^2\ \text{t}^{-3}]$
q	volumetric flow rate through a permeable aggregate $[\text{L}^3\ \text{t}^{-1}]$
$q_{j,x}$	heat flux in the x direction in phase j $[\text{M t}^{-3}]$

Q	volumetric flow rate, with various subscripts such as in, out, etc. $[L^3 \, t^{-1}]$
Q	reaction coefficient [ ]
r	radial distance $[L]$
r_f	rate particles flow toward a collector $[t^{-1}]$
r_s	rate particles strike a collector $[t^{-1}]$
R	retardation coefficient [ ]
R^*	dimensionless radius [ ]
R^*	interception number used in filtration calculations, defined as the ratio of the particle and collector diameters [ ]
$R_{i,g}$	radius of gyration of chemical i $[L]$
R_i	rate of reaction of species i (a capital letter) $[M \, L^{-3} \, t^{-1}]$
R_n	radius, with subscripts n that vary: ag, aggregate; b, bottom; c, cell; cyl, cylinder; $down$, downstream; f, floc; g, radius of gyration; h, pore radius; h, hydraulic radius (ratio of cross-sectional area to wetted perimeter); h, throat radius (capillary tube radius); i, inner; max, maximum aggregate radius; g, grain (soil grain); o, outer; p, particle radius; pp, primary particle radius; t, top; up, upstream $[L]$
s	separation distance $[L]$
s	sedimentation constant $[t]$
s	paddle velocity $[t^{-1}]$
s^*	dimensionless substrate concentration with various subscripts: ∞, bulk concentration; b, biofilm; b_0, biofilm surface [ ]
s_d	slope of a discrete size distribution in terms of diameter of spherical particles [ ]
s_l	slope of a discrete size distribution in terms of length [ ]
s_v	slope of a discrete size distribution in terms of solid volume [ ]
S^*	gravitational number, defined as the ratio of the settling velocity of a particle to the fluid velocity [ ]
S_d	slope of the cumulative size distribution in terms of diameter of spherical particles [ ]
S_l	slope of the cumulative size distribution in terms of length [ ]
S_v	slope of the cumulative size distribution in terms of solid volume [ ]
T	temperature $[T]$
T_{melt}	melting temperature $[T]$
t	time $[t]$
u_{avg}	average velocity $[L \, t^{-1}]$
u_{ag}	intrafloc velocity $[L \, t^{-1}]$
u_{max}	maximum velocity $[L \, t^{-1}]$
u_{sh}	shear velocity $[L \, t^{-1}]$
$\boldsymbol{u}_j$	velocity of all components in phase j (the phase velocity); a vector quantity $[L \, t^{-1}]$
$u_{i,x}$	velocity of species i in the x direction, u can also be defined in y or z directions $[L \, t^{-1}]$
u_x	velocity of bulk fluid j in x direction, u can also be defined in y or z directions $[L \, t^{-1}]$

$\bar{u}_x$	time-averaged velocity in x direction [L t^{-1}]
u'_x	velocity fluctuation at any instant [L t^{-1}]
$\langle u_x \rangle$	intensity of turbulence, equal to the room-mean-squared velocity of the fluctuations; see Eq. 7-2 [L t^{-1}]
$u_{j,x}$	velocity of species i in the x direction, u can also be defined in y or z direction [L t^{-1}]
U_n	velocity, with different subscripts n as needed [L t^{-1}]
U_b	rise velocity of a bubble [L t^{-1}]
U_s	settling velocity of an aggregate [L t^{-1}]
$U_{s,imp}$	settling velocity of an impermeable aggregate [L t^{-1}]
$U_{s,perm}$	settling velocity of a permeable aggregate [L t^{-1}]
$U_{p,e}$	electrophoretic mobility, or velocity of a particle in an electrical field [L t^{-1}]
v	volume [L^3]
v	for particles and aggregates, solid volume [L^3]
v_{ag}	solid volume of an aggregate [L^3]
$v_{ag,e}$	encased volume of an aggregate [L^3]
v_{max}	maximum growth rate of cells [T^{-1}]
v_p	solid volume of a particle [L^3]
V	volume concentration [l^3 L^{-3}]
V	volume, usually of a reactor [L^3]
V_{dist}	volume distribution function [l^3 L^{-3} l^{-1}]
V_f	filtrate volume in an ultrafiltration cell [L^3]
V_p	solid volume concentration of particles over a certain size range [l^3 L^{-3}]
V_r	retentate volume in an ultrafiltration cell [L^3]
V_T	total solid volume concentration of all particles [l^3 L^{-3}]
$V_{i,b}$	molal volume of pure liquid i at the normal boiling point, usually in cm^3 g mol^{-1} [L^3 M^{-1}]
W	rate of mass transport, sometimes with different subscripts; cell = mass transport to a cell [M T^{-1}]
x	rectangular coordinate [L]
x^*	dimensionless distance, sometimes with a subscript; for example, $x_b^* = x_b/\delta_b$ [ ]
x_i	mole fraction of species i in water [ ]
x_{ij}	mole fraction of species i in liquid phase j [ ]
y	rectangular coordinate [L]
y_i	mole fraction of species i in air [ ]
y_{ij}	mole fraction of species i in gas phase j [ ]
$y_{i,lm}$	log-mean average mole fraction of species i in air [ ]
$Y_{N/c}$	yield coefficient in terms of the number of cells produced per mass of chemical [M^{-1}]
$Y_{X/c}$	Yield coefficient in terms of cell mass produced per mass of chemical [M M^{-1}]
z	rectangular coordinate [L]
z_i	mole fraction of species i in soil [ ]

z_{ij} mole fraction of species i in solid phase j [ ]

Greek

α sticking coefficient [ ]

α dimensionless constant for first-order reaction in a biofilm; see Eq. 9-73 [ ]

α_j thermal diffusivity of phase j [$L^2\ t^{-1}$]

α_{CD} relative volatility of two chemicals, C and D [ ]

α_D dispersivity, defined as the ratio of the dispersion coefficient to fluid velocity, or $\alpha_D = E_L/u$ [L]

β collision function [$L^3\ t^{-1}$]

β_n collision function where the subscript indicates the individual mechanism; Br, brownian; sh, shear; ds, differential sedimentation [ ]

β dimensionless reaction constant for biofilms, as in Eq. 9-45 [ ]

γ solid fraction in a porous medium, where $\gamma = (1 - \theta)$ [ ]

γ_{ij} activity coefficient of chemical i in phase j [ ]

γ_a relative uptake factor based on advective flow [ ]

γ_d relative uptake factor based on diffusion [ ]

Γ variable defined on a case-by-case basis for simplicity of calculations; units vary

δ_n thickness of a layer, with various subscripts for n: b, biofilm; s, stagnant film; S, Stern layer; w, water film [L]

ε permittivity of a solution [$C^2\ cm^{-1}\ J^{-1}$]

ε_{ij} Lennard–Jones parameter, usually in erg [$M\ L^2\ t^{-2}$]

ε_d energy dissipation rate [$L^2\ T^{-3}$]

ζ zeta potential of a particle, commonly [$\mu m\ V\ s^{-1}\ cm^{-1}$]

ζ packing factor of particles in an aggregate [ ]

$\check{\zeta}$ packing factor of particles in an aggregate with Euclidean geometry [ ]

η_a advective effectiveness factor [ ]

η_d effectiveness factor based on diffusion [ ]

η overall collision efficiency [ ]

η_D collision efficiency due to diffusion [ ]

η_I collision efficiency due to interception [ ]

η_S collision efficiency due to gravitational sedimentation [ ]

θ total porosity [ ]

θ_n porosity, where the subscript n varies: a, air; ag, aggregate; n, NAPL; w, water [ ]

κ permeability [L^2]

κ von Karmen coefficient; see Eq. 10-89 [ ]

κ^* dimensionless permeability, $\kappa^* = R/\kappa^{1/2}$ [ ]

λ molal heat of vaporization, usually kcal/mol [$L^2\ t^{-2}$]

λ Kolmogorov microscale, or the size of the smallest eddy [L]

λ_D scale at which molecular diffusion and advection are important transport mechanisms in systems at very low shear rates [L]

Λ dimensionless parameter used for biofilm calculations; see Eq. 9-185 [ ]

Λ^* dimensionless number for solving a concentration profile; see Eq. 5-91 [ ]

Λ_{Br} parameter used to characterize Brownian coagulation [L^3 l^{-3} t^{-1}]; other subscripts are *sh* for shear and *ds* for differential sedimentation [ ]

μ growth rate of bacteria [t^{-1}]

μ_j dynamic viscosity of phase *j*, $\mu_j = \nu_j \rho_j$ [M L^{-1} t^{-1}]

μ_{ij} chemical potential, usually in kcal mole^{-1} [L^2 t^{-2}]

$\hat{\mu}$ maximum growth rate [t^{-1}]

ν_j kinematic viscosity of phase *j*, where $\nu_j = \mu_j/\rho_j$ [$L^2 t^{-1}$]

ν_{ij} stoichiometric coefficient in chemical equation [ ]

ξ shape factor for an aggregate [ ]

ξ dimensionless collision number, defined as the rate particles collide in a column packed with a porous medium to the rate they enter the column [ ]

$\underline{\xi}$ shape factor for an aggregate with Euclidean geometry [ ]

ρ phase density [M L^{-3}]

ρ_c charge density [C L^{-3}]

ρ_p density of a particle [M L^{-3}]

ρ_{ij} mass concentration of chemical *i* in phase *j* [M L^{-3}]

σ_0 surface charge density of a particle [C L^{-2}]

σ_f drainage cross-sectional area [L^2]

σ_{ij} collision diameter, a Lennard–Jones parameter, usually in Å [L]

τ tortuosity in a porous medium [ ]

$\vec{\tau}$ shear stress tensor [M L^{-1} t^{-2}]

τ_0 wall shear stress [M L^{-1} t^{-2}]

τ_f tortuosity factor in a porous medium, where $\tau^2 = \tau_f$ [ ]

τ_B tortuosity defined in the Buckinham equation [ ]

τ_P tortuosity defined in the Penman equation [ ]

τ_{yx} shear stress produced in the *y* direction arising from bulk fluid motion in the *x* direction [M L^{-1} t^{-2}]

ϕ solid fraction [ ]

ϕ_n Thiele moduli with different subscripts: *a*, advective flow; *b*, biofilm; *d*, diffusion [ ]

ϕ_i mass fraction of species *i* in water [ ]

ϕ_{ij} mass fraction of species *i* in liquid phase *j* [ ]

Φ_j association parameter of liquid phase *j* in Wilke–Chang correlation [ ]

ψ_i mass fraction of species *i* in air [ ]

ψ_{ij} mass fraction of species *i* in gas phase *j* [ ]

Ψ_0 surface potential, usually in mV, where $V = W/A$ [M L^2 t^{-3} A^{-1}]

ω angular velocity [T^{-1}]

ω_i mass fraction of species *i* in soil [ ]

ω_{ij} mass fraction of species *i* in solid phase *j* [ ]

APPENDIX 2

TRANSPORT EQUATIONS

TABLE A2.1   The Equation of Continuity

Rectangular coordinates (x, y, z)

$$\frac{\partial \rho}{\partial t} + \frac{\partial}{\partial x}(\rho u_x) + \frac{\partial}{\partial y}(\rho u_y) + \frac{\partial}{\partial z}(\rho u_z) = 0$$

Cylindrical coordinates (r, θ, z)

$$\frac{\partial \rho}{\partial t} + \frac{1}{r}\frac{\partial}{\partial r}(\rho r u_r) + \frac{1}{r}\frac{\partial}{\partial \theta}(\rho u_\theta) + \frac{\partial}{\partial z}(\rho u_z) = 0$$

Spherical coordinates (r, θ, ϕ)

$$\frac{\partial \rho}{\partial t} + \frac{1}{r^2}\frac{\partial}{\partial r}(\rho r^2 u_r) + \frac{1}{r \sin \theta}\frac{\partial}{\partial \theta}(\rho u_\theta \sin \theta) + \frac{1}{r \sin \theta}\frac{\partial}{\partial \phi}(\rho u_\phi) = 0$$

TABLE A2.2   The Convective–Diffusion Equation for a Constant ρ_j and D_{Cj}

Rectangular coordinates (x, y, z)

$$\frac{\partial c_C}{\partial t} + u_x \frac{\partial c_C}{\partial x} + u_y \frac{\partial c_C}{\partial y} + u_z \frac{\partial c_C}{\partial z} = D_C \left(\frac{\partial^2 c_C}{\partial x^2} + \frac{\partial^2 c_C}{\partial y^2} + \frac{\partial^2 c_C}{\partial z^2} \right) + R_C$$

Cylindrical coordinates (r, θ, z)

$$\frac{\partial c_C}{\partial t} + u_r \frac{\partial c_C}{\partial r} + \frac{u_\theta}{r} \frac{\partial c_C}{\partial \theta} + u_z \frac{\partial c_C}{\partial z} = D_C \left[\frac{1}{r} \frac{\partial}{\partial r} \left(r \frac{\partial c_C}{\partial r} \right) + \frac{1}{r^2} \frac{\partial^2 c_C}{\partial \theta^2} + \frac{\partial^2 c_C}{\partial z^2} \right] + R_C$$

Spherical coordinates (r, θ, ϕ)

$$\frac{\partial c_C}{\partial t} + u_r \frac{\partial c_C}{\partial r} + \frac{u_\theta}{r} \frac{\partial c_C}{\partial \theta} + \frac{u_\phi}{r \sin \theta} \frac{\partial c_C}{\partial \phi} = D_C \left[\frac{1}{r^2} \frac{\partial}{\partial r} \left(r^2 \frac{\partial c_C}{\partial r} \right) \right.$$

$$\left. + \frac{1}{r^2 \sin \theta} \frac{\partial}{\partial \theta} \left(\sin \theta \frac{\partial c_C}{\partial \theta} \right) + \frac{1}{r^2 \sin^2 \theta} \frac{\partial^2 c_C}{\partial \phi^2} \right] + R_C$$

TABLE A2.3   Navier–Stokes Equations for a Newtonian Fluid with Constant Density

Rectangular Coordinates (x, y, z)

$$\rho \left(\frac{\partial u_x}{\partial t} + u_x \frac{\partial u_x}{\partial x} + u_y \frac{\partial u_x}{\partial y} + u_z \frac{\partial u_x}{\partial z} \right) = \mu \left(\frac{\partial^2 u_x}{\partial x^2} + \frac{\partial^2 u_x}{\partial y^2} + \frac{\partial^2 u_x}{\partial z^2} \right) - \frac{\partial p}{\partial x} + \rho g_x$$

$$\rho \left(\frac{\partial u_y}{\partial t} + u_x \frac{\partial u_y}{\partial x} + u_y \frac{\partial u_y}{\partial y} + u_z \frac{\partial u_y}{\partial z} \right) = \mu \left(\frac{\partial^2 u_y}{\partial x^2} + \frac{\partial^2 u_y}{\partial y^2} + \frac{\partial^2 u_y}{\partial z^2} \right) - \frac{\partial p}{\partial y} + \rho g_y$$

$$\rho \left(\frac{\partial u_z}{\partial t} + u_x \frac{\partial u_z}{\partial x} + u_y \frac{\partial u_z}{\partial y} + u_z \frac{\partial u_z}{\partial z} \right) = \mu \left(\frac{\partial^2 u_z}{\partial x^2} + \frac{\partial^2 u_z}{\partial y^2} + \frac{\partial^2 u_z}{\partial z^2} \right) - \frac{\partial p}{\partial z} + \rho g_z$$

Cylindrical coordinates (r, θ, z)

$$\rho \left(\frac{\partial u_r}{\partial t} + u_r \frac{\partial u_r}{\partial r} + \frac{u_\theta}{r} \frac{\partial u_r}{\partial \theta} - u_\theta^2 + u_z \frac{\partial u_r}{\partial z} \right)$$

$$= \mu \left[\frac{\partial}{\partial r} \left(\frac{1}{r} \frac{\partial}{\partial r} (r u_r) \right) + \frac{1}{r^2} \frac{\partial^2 u_r}{\partial \theta^2} + \frac{\partial^2 u_r}{\partial z^2} - \frac{2}{r^2} \frac{\partial u_\theta}{\partial \theta} \right] - \frac{\partial p}{\partial r} + \rho g_r$$

$$\partial \left(\frac{\partial u_\theta}{\partial t} + u_r \frac{\partial u_\theta}{\partial r} + \frac{u_\theta}{r} \frac{\partial u_\theta}{\partial \theta} + \frac{u_r u_\theta}{r} + u_z \frac{\partial u_\theta}{\partial z} \right)$$

$$= \mu \left[\frac{\partial}{\partial r} \left(\frac{1}{r} \frac{\partial}{\partial r} (r u_\theta) \right) + \frac{1}{r^2} \frac{\partial^2 u_\theta}{\partial \theta^2} + \frac{\partial^2 u_\theta}{\partial z^2} + \frac{2}{r^2} \frac{\partial u_r}{\partial \theta} \right] - \frac{1}{r} \frac{\partial p}{\partial \theta} + \rho g_\theta$$

$$\rho \left(\frac{\partial u_z}{\partial t} + u_r \frac{\partial u_z}{\partial r} + \frac{u_\theta}{r} \frac{\partial u_z}{\partial \theta} + u_z \frac{\partial u_z}{\partial z} \right)$$

$$= \mu \left[\frac{1}{r} \frac{\partial}{\partial r} \left(r \frac{\partial u_z}{\partial r} \right) + \frac{1}{r^2} \frac{\partial^2 u_z}{\partial \theta^2} + \frac{\partial^2 u_z}{\partial z^2} \right] - \frac{\partial p}{\partial z} + \rho g_z$$

TABLE A2.3 *(Continued)*

Spherical coordinates (r, θ, ϕ)

$$\rho \left(\frac{\partial u_r}{\partial t} + u_r \frac{\partial u_r}{\partial r} + \frac{u_\theta}{r} \frac{\partial u_r}{\partial \theta} + \frac{u_\phi}{r \sin \theta} \frac{\partial u_r}{\partial \phi} - \frac{u_\theta^2 + u_\phi^2}{r} \right)$$

$$= \mu \left[\frac{\partial}{\partial r^2} \left(\frac{1}{r} \frac{\partial}{\partial r} (r^2 u_r) \right) + \frac{1}{r^2 \sin \theta} \frac{\partial}{\partial \theta} \left(\sin \theta \frac{\partial u_r}{\partial \theta} \right) + \frac{1}{r^2 \sin^2 \theta} \frac{\partial^2 u_r}{\partial \phi^2} \right.$$

$$\left. - \frac{2}{r^2 \sin \theta} \frac{\partial}{\partial \theta} (u_0 \sin \theta) - \frac{2}{r^2 \theta} \frac{\partial u_\phi}{\partial \phi} \right] - \frac{\partial p}{\partial r} + \rho g_r$$

$$\rho \left(\frac{\partial u_\theta}{\partial t} + u_r \frac{\partial u_\theta}{\partial r} + \frac{u_\theta}{r} \frac{\partial u_\theta}{\partial \theta} + \frac{u_\phi}{r \sin \theta} \frac{\partial u_\theta}{\partial \phi} + \frac{u_r u_\theta}{r} - \frac{u_\phi^2 \cot \theta}{r} \right)$$

$$= \mu \left[\frac{1}{r^2} \frac{\partial}{\partial r} \left(r^2 \frac{\partial u_\theta}{\partial r} \right) + \frac{1}{r^2} \frac{\partial}{\partial \theta} \left(\frac{1}{\sin \theta} \frac{\partial}{\partial \theta} (u_\theta \sin \theta) \right) \right.$$

$$\left. + \frac{1}{r^2 \sin^2 \theta} \frac{\partial^2 u_\theta}{\partial \phi^2} + \frac{2}{r^2} \frac{\partial u_r}{\partial \theta} - \frac{2 \cot \theta}{r^2 \sin \theta} \frac{\partial u_\phi}{\partial \phi} \right] - \frac{1}{r} \frac{\partial p}{\partial \theta} + \rho g_\theta$$

$$\rho \left(\frac{\partial u_\theta}{\partial t} + u_r \frac{\partial u_\phi}{\partial r} + \frac{u_\theta}{r} \frac{\partial u_\phi}{\partial \theta} + \frac{u_\phi}{r \sin \theta} \frac{\partial u_\phi}{\partial \phi} + \frac{u_\phi u_r}{r} + \frac{u_\theta u_\phi}{r} \cot \theta \right)$$

$$= \mu \left[\frac{1}{r^2} \frac{\partial}{\partial r} \left(r^2 \frac{\partial u_\phi}{\partial r} \right) + \frac{1}{r^2} \frac{\partial}{\partial \theta} \left(\frac{1}{\sin \theta} \frac{\partial}{\partial \theta} (u_\phi \sin \theta) \right) + \frac{1}{r^2 \sin^2 \theta} \frac{\partial^2 u_\phi}{\partial \phi^2} \right.$$

$$\left. + \frac{2}{r^2 \sin \theta} \frac{\partial u_r}{\partial \phi} + \frac{2 \cot \theta}{r^2 \sin \theta} \frac{\partial u_\theta}{\partial \phi} \right] - \frac{1}{r \sin \theta} \frac{\partial p}{\partial \phi} + \rho g_\phi$$

APPENDIX 3

CHEMICAL PROPERTIES

TABLE A3.1 Relative Atomic Weights for the Chemical Elements Based on an Assigned Relative Atomic Weight for $^{12}C = 12$

Element	Symbol	Atomic Number	Atomic Weight	Element	Symbol	Atomic Number	Atomic Weight
Actinium	Ac	89		Mercury	Hg	80	200.6
Aluminum	Al	13	26.98	Molybdenum	Mo	42	95.9
Americium	Am	95		Neodymium	Nd	60	144.2
Antimony	Sb	51	121.7	Neon	Ne	10	20.18
Argon	Ar	18	39.94	Neptunium	Np	93	237.05
Arsenic	As	33	74.92	Nickel	Ni	28	58.7
Astatine	At	85		Niobium	Nb	41	92.91
Barium	Ba	56	137.3	Nitrogen	N	7	14.01
Berkelium	Bk	97		Nobelium	No	102	
Beryllium	Be	4	9.01	Osmium	Os	76	190.2
Bismuth	Bi	83	208.98	Oxygen	O	8	16.00
Boron	B	5	10.81	Palladium	Pd	46	106.4
Bromine	Br	35	79.90	Phosphorus	P	15	30.97
Cadmium	Cd	48	112.40	Platinum	Pt	78	195.1
Calcium	Ca	20	40.80	Plutonium	Pu	94	
Californium	Cf	98		Polonium	Po	84	
Carbon	C	6	12.011	Potassium	K	19	39.1
Cerium	Ce	58	140.12	Praseodymium	Pr	59	140.9
Cesium	Cs	55	132.91	Promethium	Pm	61	
Chlorine	Cl	17	35.45	Protactinium	Pa	91	231.04
Chromium	Cr	24	52.00	Radium	Ra	88	226.03
Cobalt	Co	27	58.93	Radon	Rn	86	
Copper	Cu	29	63.55	Rhenium	Re	75	186.2
Curium	Cm	96		Rhodium	Rh	45	102.91
Dysprosium	Dy	66	162.50	Rubidium	Rb	37	85.468
Einsteinium	Es	99		Ruthenium	Ru	44	101.1
Erbium	Er	68	167.3	Samarium	Sm	62	150.4
Europium	Eu	63	152.0	Scandium	Sc	21	44.96
Fermium	Fm	100		Selenium	Se	34	78.96
Fluorine	F	9	19.00	Silicon	Si	14	28.09
Francium	Fr	87		Silver	Ag	47	107.87
Gadolinium	Gd	64	157.2	Sodium	Na	11	22.99
Gallium	Ga	31	69.72	Strontium	Sr	38	87.62
Germanium	Ge	32	72.5	Sulfur	S	16	32.06
Gold	Au	79	196.97	Tantalum	Ta	73	180.95
Hafnium	Hf	72	178.5	Technetium	Tc	43	98.91
Helium	He	2	4.00	Tellurium	Te	52	127.6
Holmium	Ho	67	164.30	Tebium	Tb	65	158.93
Hydrogen	H	1	1.01	Thelium	Tl	81	204.4
Indium	In	49	114.82	Thorium	Th	90	232.04
Iodine	I	53	126.90	Thulium	Tm	69	168.93
Iridium	Ir	77	192.2	Tin	Sn	50	118.7
Iron	Fe	26	55.85	Titanium	Ti	22	47.9
Krypton	Kr	36	83.80	Tungsten	W	74	183.8
Lanthanum	La	57	138.91	Uranium	U	92	238.03
Lawrencium	Lr	103		Vanadium	V	23	50.941
Lead	Pb	82	207.2	Xenon	Xe	54	131.30
Lithium	Li	3	6.94	Ytterbium	Yb	70	173.04
Lutium	Lu	71	174.97	Yttrium	Y	39	88.91
Magnesium	Mg	12	24.31	Zinc	Zn	30	65.38
Manganese	Mn	25	54.94	Zirconium	Zr	40	91.22
Mendelevium	Md	101					

Source: Thibodeaux, 1996.

TABLE A3.2 Physical Properties of Water

Temperature (°C)	ρ_w, Density (g cm^{-3})	μ_w, Dynamic Viscosity (g cm^{-1} s^{-1})	v_w, Kinematic Viscosity (cm^2 s^{-1})
0	0.99984	0.01793	0.01787
5	0.99999	0.01519	0.01519
10	0.99970	0.01307	0.01307
15	0.99913	0.01139	0.01140
20	0.99821	0.01002	0.01004
25	0.99707	0.008904	0.00893
30	0.99567	0.007977	0.00801
40	0.99222	0.006532	0.00658
50	0.98803	0.005470	0.00553

Source: *CRC Handbook of Physics and Chemistry*, 1990.

TABLE A3.3 Physical Properties of Air

Temperature (°C)	ρ_a, Density (kg m^{-3})	μ_a, Dynamic Viscosity (g cm^{-1} s^{-1})	v_a, Kinematic Viscosity (cm^2 s^{-1})
0	1.292	0.000171	0.132
5	1.269	0.000173	0.136
10	1.247	0.000176	0.141
15	1.225	0.000180	0.147
20	1.204	0.000182	0.151
25	1.184	0.000185	0.156
30	1.165	0.000186	0.160
40	1.127	0.000187	0.166
50	1.109	0.000195	0.176

Source: Thibodeaux, 1979; *CRC Handbook of Physics and Chemistry*, 1990, Lide, 1991.

TABLE A3.4   Dissolved-Oxygen Solubility as a Function of Chloride Concentration (mg L^{-1})

Temperature (°C)	Chloride Concentration (mg L^{-1})				
	0	5,000	10,000	15,000	20,000
0	14.16	13.79	12.97	12.14	11.32
1	13.77	13.41	12.61	11.82	11.03
2	13.40	13.05	12.28	11.52	10.76
3	13.05	12.72	11.98	11.24	10.50
4	12.70	12.41	11.69	10.97	10.25
5	12.37	12.09	11.39	10.70	10.01
6	12.06	11.79	11.12	10.45	9.78
7	11.76	11.51	10.85	10.21	9.57
8	11.47	11.24	10.61	9.98	9.36
9	11.19	10.97	10.36	9.76	9.17
10	10.92	10.73	10.13	9.55	8.98
11	10.67	10.49	9.92	9.35	8.80
12	10.43	10.28	9.72	9.17	8.62
13	10.20	10.05	9.52	8.98	8.46
14	9.98	9.85	9.32	8.80	8.30
15	9.76	9.65	9.14	8.63	8.14
16	9.56	9.46	8.96	8.47	7.99
17	9.37	9.26	8.78	8.30	7.84
18	9.18	9.07	8.62	8.15	7.70
19	9.01	8.89	8.45	8.00	7.56
20	8.84	8.73	8.30	7.86	7.42
21	8.68	8.57	8.14	7.71	7.28
22	8.53	8.42	7.99	7.57	7.14
23	8.38	8.27	7.85	7.43	7.00
24	8.25	8.12	7.71	7.30	6.87
25	8.11	7.96	7.56	7.15	6.74
26	7.99	7.81	7.42	7.02	6.61
27	7.86	7.67	7.28	6.88	6.49
28	7.75	7.53	7.14	6.75	6.37
29	7.64	7.39	7.00	6.62	6.25
30	7.53	7.25	6.86	6.49	6.13

Source: Greenberg et al., 1992.

TABLE A3.5   Henry's Constants of Pesticides and Related Materials (at 20–25°C)

Compound	p_C^o (mm Hg) Vapor Pressure	$\rho_{Cw,eq}$ (μg/L) Water Solubility	$-\log H_{C\rho\rho}$
Toxaphene	3×10^{-1}	10^2–10^3	2.5–3.5
Aroclor 1242-1260	4×10^{-4} to 4×10^{-5}	2.7–240	4.8–5.8
Heptachlor	3×10^{-4}	56	5.3
Aldrin	2.3×10^{-5}	27	6.1
Trifluralin	10^{-4}	580	6.8
p,p'-DDT	1.5×10^{-7}	1.2	6.9
o,p'-DDT	5.5×10^{-6}	85	7.2
p,p'-DDE	6.5×10^{-6}	120	7.3
Dieldrin	2.8×10^{-6}	140	7.7
Helptachlor epoxide	$(10^{-4}$–$10^{-6})$	350	6.5–8.5
Diazinon	2.8×10^{-4}	4×10^4	8.2
Lindane	3.3×10^{-5}	7×10^3	8.3
Endrin	2×10^{-7}	2×10^2	9.0
Parathion	2.3×10^{-5}	2.4×10^4	9.0
Carbofuran	(10^{-5})	2.5×10^5	10.4

Source: Thibodeaux, 1979.

TABLE A3.6   Properties of Selected Hydrocarbons in Water at 25°C and 1 atm

Hydrocarbon	Solubility, $\rho_{Cw,eq}$ (g/m^3)	Vapor Pressure, ρ_C^o (atm)	Activity Coefficient, γ_{Cw}	Henry's Law Constant, H_{cpc} (atm m^3 mol^{-1})
Methane (g)	24.1	269	1.375×10^2	0.665
Ethane (g)	60.4	39.4	7.02×10^2	0.499
Propange (g)	62.4	9.29	4.23×10^3	0.707
n-Butane (g)	61.4	2.40	2.19×10^4	0.947
n-Pentane (l)	38.5	0.675	1.04×10^5	1.26
n-Hexane (l)	9.5	0.205	5.04×10^5	1.85
n-Heptane	2.93	0.0603	1.90×10^6	2.07
n-Octane (l)	0.66	0.0186	9.62×10^6	3.22
n-Nonane (l)	0.22	5.64×10^{-3}	3.24×10^7	3.29
Decane (l)	0.052	1.73×10^{-3}	1.58×10^8	4.93
Dodecane (l)	0.0037	1.55×10^{-4}	2.56×10^9	7.12
Ethene (g)	131	59.91	1.99×10^2	0.214
Propene (g)	200	11.29	9.9×10^3	0.232
1-Butene (g)	222	2.933	5.07×10^{13}	0.268
1-Pentene (l)	148	0.839	2.6×10^4	0.398
1-Hexene (l)	50	0.245	9.35×10^4	0.412
Cyclopentane (l)	156	0.418	250×10^4	0.187
Cyclohexane (l)	55	0.128	8.50×10^4	0.196
Isobutane (g)	48.9	3.52	1.96×10^4	1.24
Isopentane (l)	47.8	0.904	8.39×10^4	1.364
Benzene (l)	1780	0.125	2.4×10^3	5.49×10^{-3}
Toluene (l)	515	0.0374	9.9×10^3	6.66×10^{-3}
Ethyl benzene (l)	152	0.0125	3.88×10^4	8.73×10^{-3}
o-Xylene (l)	175	8.71×10^{-3}	3.37×10^4	5.27×10^{-3}
Naphthalene(s)	34.4	1.14×10^{-4}	7.69×10^{-4}	4.25×10^{-2}
Fluorene(s)	1.90	1.64×10^{-5}	7.95×10^5	2.35×10^{-4}
Anthracene(s)	0.075	5.04×10^{-5}	1.814×10^6	1.65×10^{-3}
Phenanthrene(s)	1.18	4.53×10^{-6}	1.82×10^6	1.48×10^{-4}

Source: Thibodeaux, 1996.

TABLE A3.7   Diffusivities of Selected Chemicals in Air

Substance	Temperature (°C)	D_{Ca} (cm²/s)	Substance	Temperature (°C)	D_{Ca} (cm²/s)
Acetic acid	25	0.133	Ethyl *i*-butyrate	0	0.0591
Acetone	0	0.109	Ethyl ether	25	0.093
Ammonia	25	0.28	Ethyl formate	0	0.0840
Amyl butyrate	0	0.040	Ethyl propionate	0	0.068
Amyl formate	0	0.0543	Ethyl valerate	0	0.0512
i-Amyl formate	0	0.058	Formic acid	25	0.159
Amyl isobutrate	0	0.0419	Hexane	21	0.080
Amyl propinate	0	0.046	Hydrogen	25	0.410
Aniline	25	0.072	Hydrogen cyanide	0	0.173
Anthracene	0	0.0421	Hydrogen peroxide	60	0.188
Benzene	25	0.088	Mercury	0	0.112
n-Butyl acetate	0	0.058	Methyl alcohol	25	0.159
i-Butyl acetate	0	0.0612	Methane	0	0.16
n-Butyl alcohol	30	0.088	Naphthalene	0	0.513
i-Butyl alcohol	0	0.0727	Nitrogen	0	0.13
Butyric acid	0	0.067	*n*-Octane	25	0.060
Caproic acid	0	0.050	Oxygen	0	0.178
Carbon dioxide	25	0.164		25	0.206
Carbon disulfide	25	0.107	*n*-Pentane	21	0.071
Chlorine	0	0.0983	Phosgene	0	0.095
Chlorobenzene	30	0.075	Propane	0	0.088
Chloroform	0	0.091	Propionic acid	25	0.099
o-Chlorotoulene	0	0.059	*n*-Propyl alcohol	25	0.100
p-Chlorotoluene	0	0.051	*i*-Propyl alcohol	30	0.101
Chlorotoluene	25	0.065	*n*-Propyl benzene	25	0.059
Cyclohexane	45	0.086	*i*-Propyl benzene	0	0.0489
Ethane	0	0.108	Sulfur dioxide	0	0.103
Ether (diethyl)	0	0.0778	Toluene	30	0.088
Ethyl acetate	30	0.089	Water	25	0.256
Ethyl alcohol	25	0.119	Xylene	25	0.071
Ethyl benzene	25	0.77			

Source: Thibodeaux, 1996.

TABLE A3.8 Diffusivities of Selected Chemicals in Water

Substance	Temperature (°C)	$D_{Cw} \times 10^5$ (cm^2/s)	Substance	Temperature (°C)	$D_{Cw} \times 10^5$ (cm^2/s)
Acetic acid	20	0.88	Methanol	20	1.28
Acetylene	20	1.56	Nitric acid	20	2.6
Allyl alcohol	20	0.93	Nitrogen	10	1.29
Ammonia	20	1.76		20	1.64
Bromine	20	1.2		25	1.9, 2.01
n-Butanol	25	0.96	Oxygen	10	1.54
Butanol	20	0.77		20	1.80
Caffeine	25	0.63		25	2.5, 2.2
Carbon dioxide	20	1.77		30	3.49
Chlorine	20	1.22	Phenol	20	0.84
Ethanol	20	1.00	n-Propanol	25	1.1
Formic acid	25	1.37		20	0.87
Glucose	20	0.6	Saccharose	25	0.49
	25	0.69	Sodium chloride	20	1.35
Glycerol	20	0.72	Sodium hydroxide	20	1.51
Hydrogen	25	5.85	Succinic acid	25	0.94
	30	5.42	Sucrose	20	0.45
Hydrogen sulfide	20	1.41	Sulfur dioxide	25	1.7
Hydroquinone	20	0.77	Sulfuric acid	20	1.73
Lactose	20	0.43	Urea	20	1.06
Maltose	20	0.43	Urethane	20	0.92

Source: Thibodeaux, 1996.

TABLE A3.9 Diffusivities of Chemicals in the Solid State

System	Temperature (°C)	D_{A3} (cm^2/s)
He in SiO$_2$	20	$(2.4-5.5) \times 10^{-10}$
He in Pyrex	20	4.5×10^{-11}
He in Pyrex	500	2×10^{-8}
H$_2$ in SiO$_2$	500	$(0.6-2.1) \times 10^{-8}$
Elements in glass (borosilicate)	100	$(1.0-100.0) \times 10^{-14}$

Source: Thibodeaux, 1979.

TABLE A3.10 **The Collision Integrals, Ω_μ and Ω_D, Based on the Lennard–Jones Potential**

kT/ε	$\Omega_\mu = \Omega_k$ (Viscosity and Thermal Conductivity)	Ω_D (Mass Diffusivity)	$k_B T/\varepsilon$	$\Omega_\mu = \Omega_k$ (Viscosity and Thermal Conductivity)	Ω_D (Mass Diffusivity)
0.30	2.785	2.662	2.30	1.122	1.026
0.35	2.628	2.476	2.40	1.107	1.012
0.40	2.492	2.318	2.50	1.093	0.9996
0.45	2.368	2.184	2.60	1.081	0.9878
0.50	2.257	2.066	2.70	1.069	0.9770
0.55	2.156	1.966	2.80	1.058	0.9672
0.60	2.065	1.877	2.90	1.048	0.9576
0.65	1.982	1.798	3.00	1.039	0.9490
0.70	1.908	1.729	3.10	1.030	0.9406
0.75	1.841	1.667	3.20	1.022	0.9328
0.80	1.780	1.612	3.30	1.014	0.9256
0.85	1.725	1.562	3.40	1.007	0.9186
0.90	1.675	1.517	3.50	0.9999	0.9120
0.95	1.629	1.476	3.60	0.9932	0.9058
1.00	1.587	1.439	3.70	0.9870	0.8998
1.05	1.549	1.406	3.80	0.9811	0.8942
1.10	1.514	1.375	3.90	0.9755	0.8888
1.15	1.482	1.346	4.00	0.9700	0.8836
1.20	1.452	1.320	4.10	0.9649	0.8788
1.25	1.424	1.296	4.20	0.9600	0.8740
1.30	1.399	1.273	4.30	0.9553	0.8694
1.35	1.375	1.253	4.40	0.9507	0.8652
1.40	1.353	1.233	4.50	0.9464	0.8610
1.45	1.333	1.215	4.60	0.9422	0.8568
1.50	1.314	1.198	4.70	0.9382	0.8530
1.55	1.296	1.182	4.80	0.9343	0.8492
1.60	1.279	1.167	4.90	0.9305	0.8456
1.65	1.264	1.153	5.0	0.9269	0.8422
1.70	1.248	1.140	6.0	0.8963	0.8124
1.75	1.234	1.128	7.0	0.8727	0.7896
1.80	1.221	1.116	8.0	0.8538	0.7712
1.85	1.209	1.105	9.0	0.8379	0.7556
1.90	1.197	1.094	10.0	0.8242	0.7424
1.95	1.186	1.084	20.0	0.7432	0.6640
2.00	1.175	1.075	30.0	0.7005	0.6232
2.10	1.156	1.057	40.0	0.6718	0.5960
2.20	1.138	1.041	50.0	0.6504	0.5756

Source: Welty et al., 1990.

TABLE A3.11 Lennard–Jones Force Constants Calculated from Viscosity Data

Compound	Formula	ε_A/k_B (K)	σ (Å)
Air		97	3.617
Benzene	C_6H_6	440	5.270
Bromine	Br_2	520	4.268
i-Butane	C_4H_{10}	313	5.341
n-Butane	C_4H_{10}	410	4.997
Carbon dioxide	CO_2	190	3.996
Carbon monoxide	CO	110	3.590
Carbon tetrachloride	CCl_4	327	5.881
Chlorine	Cl_2	357	4.115
Chloroform	$CHCl_3$	327	5.430
Cyclohexane	C_6H_{12}	324	6.093
Ethane	C_2H_6	230	4.418
Ethanol	C_2H_5OH	391	4.455
Ethylene	C_2H_6	205	4.232
Fluorine	F_2	112	3.653
Helium	He	10.22	2.576
n-Heptane	C_7H_{16}	282	0.88
n-Hexane	C_6H_{14}	413	5.909
Hydrogen	H_2	33.3	2.968
Hydrogen chloride	HCl	360	3.305
Iodine	I_2	550	4.982
Methane	CH_4	136.5	3.822
Methanol	CH_3OH	507	3.585
Methyl chloride	CH_3Cl	855	3.375
Mercury	Hg	851	2.898
Nitric Oxide	NO	119	3.470
Nitrogen	N_2	91.5	3.681
Nitrous Oxide	N_2O	220	3.879
n-Nonane	C_9H_{20}	240	8.448
n-Octane	C_8H_{18}	320	7.451
Oxygen	O_2	113	3.433
n-Pentane	C_5H_{12}	345	5.769
Propane	C_3H_8	254	5.061
Sulfur dioxide	SO_2	252	4.290
Water	H_2O	356	2.649

Source: Welty et al., 1990.

REFERENCES

Greenberg, A. E., L. S. Clesceri, and A. D. Eaton, eds. 1992. Standard Methods for the Examination of Water and Wastewater, 18th ed. American Public Health Association, Washington, D.C.

Lide, D. R. 1991. Handbook of Chemistry and Physics. CRC Press, Boca Raton.

Thibodeaux, L. J. 1979. *Chemodynamics*. Wiley. New York, NY.

Thibodeaux, L. J. 1996. *Chemodynamics, 2nd Edition*. Wiley. New York, NY.

Welty, J. R., C. E. Wicks, and R. E. Wilson. 1990. *Fundamentals of Momentum, Heat and Mass Transfer*, 3rd ed. John Wiley, New York.

MATHEMATICAL FUNCTIONS AND SOLUTION TECHNIQUES

The sections and tables presented in this section can be referred to when needed to solve the differential equations given in the text. For a more complete review of solutions applicable to mass transport equations, see Mickley et al. (1957).

4.1   SOLUTION OF ORDINARY DIFFERENTIAL EQUATIONS

First-Order Equation   When an equation has the form

$$M(x)\ dx + N(y)\ dy = 0 \tag{A4-1}$$

and M and N are functions of x and y as indicated, then the equation is separable, and the solution is simply

$$\int M(x)\ dx = \int -N(y)\ dy \tag{A4-2}$$

Recall that it is sometimes possible to make the equation separable by a change of variable, such as $u = xy$.

First-Order Nonhomogeneous Linear Equation   For this type of equation, of the form

$$\frac{dy}{dx} + Py = Q \tag{A4-3}$$

where P and Q are either constants or functions only of x, an integrating factor of the form $\exp(\int P\,dx)$ can be used, so that this equation becomes

$$\exp\left(\int p\,dx\right)\frac{dy}{dx} + \exp\left(\int p\,dx\right)Py = \exp\left(\int p\,dx\right)Q \qquad \text{(A4-4)}$$

The left side of the equation is the derivative of $y\exp(\int p\,dx)$, so that the solution is exact, and is

$$y = \frac{\displaystyle\int \exp\left(\int p\,dx\right)Q\,dx + b}{\displaystyle\exp\left(\int p\,dx\right)} \qquad \text{(A4-5)}$$

where b is a constant of integration.

Second-Order Equation with Equidimensional Coefficients When the differential equation is of the form

$$\frac{d^2y}{dx^2} + \frac{a}{x}\frac{dy}{dx} + \frac{b}{x^2}y = 0 \qquad \text{(A4-6)}$$

the coefficients are equidimensional. We guess a substitution of the form $y = cx^s$, where s is calculated as

$$s = -\frac{(a-1) \pm \sqrt{(a-1)^2 - 4b}}{2} \qquad \text{(A4-7)}$$

Depending on the values of the roots of the equation, the solution is

$$y = b_1 x^{s_1} + b_2 x^{s_2}, \qquad s_1 \neq s_2$$
$$\qquad\qquad\qquad\qquad\qquad\qquad\qquad\qquad \text{(A4-8)}$$
$$y = b_1 x^s + b_2 x^s \ln x, \qquad s_1 = s_2$$

Second-Order Equation with Constant Coefficients When the coefficients are constant, the equation has the form

$$\frac{d^2y}{dx^2} + a\frac{dy}{dx} + by = 0 \qquad \text{(A4-9)}$$

We guess a substitution of the form $y = be^{sx}$, where s is calculated as

$$s = -\frac{a \pm \sqrt{a^2 - 4b}}{2} \qquad \text{(A4-10)}$$

which has the solution, depending on the values of the roots

$$y = b_1 e^{s_1 x} + b_2 e^{s_2 x}, \qquad s_1 \neq s_2 \tag{A4-11}$$

$$y = b_1 e^{sx} + b_2 x e^{sx}, \qquad s_1 = s_2$$

Second-Order Equation: General Case   In the general case of a nonhomogeneous function of the form

$$\frac{d^2 y}{dx^2} + P \frac{dy}{dx} + Qy = h \tag{A4-12}$$

where P, Q, and h can all be functions of x. The solution is the sum of the particular and homogeneous solution, where the homogeneous equation has the solution

$$y_{\text{homo}} = b_1 y_1 + b_2 y_2 \tag{A4-13}$$

Thus we must know the homogeneous solution to obtain the full solution for the nonhomogeneous equation. Defining an integrating factor as $G = \exp(\int p \, dx)$, the solution is

$$y(x) = b_1 y_1 + b_2 y_1 \int \frac{dx}{y_1^2 G} + y_1 \int \frac{1}{y_1^2 G} \left(\int h y G \, dx \right) dx \tag{A4-14}$$

Bessel's Equations   Two common forms of the Bessel's equations arise in a variety of transport problems: a regular Bessel equation, and a modified Bessel equation. A *regular* Bessel equation has the form

$$\frac{d^2 y}{dx^2} + \frac{1}{x} \frac{dy}{dx} + \left(1 - \frac{p^2}{x^2} \right) y = 0 \tag{A4-15}$$

The solution to this equation depends on the value of p. When $p \neq 0$ and is not a positive integer

$$y = b_1 J_p(x) + b_2 J_{-p}(x), \qquad p \neq \text{positive}, p \neq 0 \tag{A4-16}$$

where J_p is called the Bessel function of the first kind of order p. When p is 0 or a positive integer n

$$y = b_1 J_n(x) + b_2 Y_n(x), \qquad p = \text{positive integer } n, p = 0 \tag{A4-17}$$

where Y_n is a Bessel function of the second kind of order n.

When the third term is negative, we have a linear second-order equation known as a *modified* Bessel equation

$$\frac{d^2y}{dx^2} + \frac{1}{x}\frac{dy}{dx} - \left(1 - \frac{p^2}{x^2}\right)y = 0 \tag{A4-18}$$

The solution to this equation also depends on the value of p. When $p \neq 0$ and is not a positive integer

$$y = b_1 I_p(x) + b_2 I_{-p}(x), \qquad p \neq \text{positive}, p \neq 0 \tag{A4-19}$$

where I_p is called the modified Bessel function of the first kind of order p. When p is 0 or a positive integer n

$$y = b_1 I_n(x) + b_2 K_n(x), \qquad p = \text{positive integer } n, p = 0 \tag{A4-20}$$

where K_n is a modified Bessel function of the second kind of order n.

The general form of the Bessel equation, useful for solving a variety of problems, is

$$x^2 \frac{d^2y}{dx^2} + x(a + 2bx^r)\frac{dy}{dx} + [e + dx^{2s} - b(1 - a - r)x^r + b^2x^{2r}]y = 0 \tag{A4-21}$$

Through a proper transformation of variables, the solution to this equation is

$$y = x^{(1-a)/2}e^{-(bx^r/r)}\left[b_1 Z_p\left(\frac{\sqrt{|d|}}{s}x^s\right) + b_2 Z_{-p}\left(\frac{\sqrt{|d|}}{s}x^s\right)\right] \tag{A4-22}$$

where

$$p = \frac{1}{s}\left[\left(\frac{1-a}{2}\right)^2 - e\right]^{1/2} \tag{A4-23}$$

and where Z_p denotes a Bessel function that is dependent on the values of the constants: If $\sqrt{d}/s$ is real and p is not zero or an integer, $Z_p \rightarrow J_p$, $Z_{-p} \rightarrow J_{-p}$; if $p = 0$ or an integer n, $Z_p \rightarrow J_n$, $Z_{-p} \rightarrow Y_n$; if $\sqrt{d}/s$ is imaginary and p is not zero or an integer, $Z_p \rightarrow I_p$, $Z_{-p} \rightarrow I_{-p}$; if $p = 0$ or an integer n, $Z_p \rightarrow I_n$, $Z_{-p} \rightarrow K_n$. Special cases of the Bessel equation solutions are given in the section below. A table of zero-order Bessel functions of the first and second kind are given in Section 4.3.

Special Cases and Examples There are some forms of differential equations that occur with great frequency, and serve as good examples of applying the above equations. These equations, and their solutions, are shown in this section.

Spherical Coordinates Transport to or from a spherical particle often involves equations of the form

$$\frac{\partial}{\partial c}\left(r^2\frac{\partial c}{\partial r}\right) - \frac{k}{D}r^2c = 0 \tag{A4-24}$$

This can be rearranged into the form

$$r^2\frac{\partial^2 c}{\partial r^2} + 2r\frac{\partial c}{\partial r} - \left(\frac{k}{D}r^2 + 0\right)c = 0 \tag{A4-25}$$

From comparison with Eq. A4-21, we see that $a = 2$, $b = 0$, $e = 0$, $s = 1$, $\sqrt{d} = i\sqrt{k/D}$. Thus $p = \frac{1}{2}$, and the solution is

$$c = r^{-1/2}[b_1I_{1/2}(\sqrt{k/D}r) + b_2I_{-1/2}(\sqrt{k/D}r)] \tag{A4-26}$$

The properties of the half-order Bessel equation (see Section 4.3) are such that this can be more simply written in terms of hyperbolic sines and cosines (which are readily available on many handheld calculators), as

$$c = r^{-1/2}\left[b_1\sqrt{\frac{2}{\pi r}}\sinh(\sqrt{k/D}r) + b_2\sqrt{\frac{2}{\pi r}}\cosh(\sqrt{k/D}r)\right] \tag{A4-27}$$

Cylindrical Coordinates For transport in a tube or cylinder, a one-dimensional transport equation is

$$\frac{\partial^2 c}{\partial r^2} + \frac{1}{r}\frac{\partial c}{\partial r} + \frac{k}{D}c = 0 \tag{A4-28}$$

This is a zero-order Bessel function, with the solution

$$y = b_1J_0(\sqrt{k/D}r) + b_2Y_0(\sqrt{k/D}r) \tag{A4-29}$$

If the argument $\sqrt{k/D}$ is negative, then the equation would be a zero-order modified Bessel function with a solution containing the Bessel functions I_0 and K_0.

Harmonic Equation The equation

$$\frac{d^2y}{dx^2} + \alpha^2 y = 0 \tag{A4-30}$$

arises in many calculations, and is known as the harmonic equation. The solution of this equation is possible via two approaches. Using the general solution to Bessel's equation, it can be shown that the solution is

$$y = b_1 \sin(\alpha x) + b_2 \cos(\alpha x) \tag{A4-31}$$

which is a solution form used frequently in this text. An alternative approach is to recognize that this equation is second-order equation with constant coefficients, and that its solution is

$$y = b_1 e^{i\alpha x} + b_2 e^{-i\alpha x} \tag{A4-32}$$

When the second term is negative, or Eq. A4-30 becomes

$$\frac{d^2 y}{dx^2} - \alpha^2 y = 0 \tag{A4-33}$$

the solution can be written in its preferable form in terms of hyperbolic sines and cosines as

$$y = b_1 \sinh(\alpha x) + b_2 \cosh(\alpha x) \tag{A4-34}$$

or in terms of exponential functions (less preferable) as

$$y = b_1 e^{\alpha x} + b_2 e^{-\alpha x} \tag{A4-35}$$

4.2 SOLUTION OF PARTIAL DIFFERENTIAL EQUATIONS: NUMERICAL SOLUTIONS

Advances in computer speeds and software capabilities now make it feasible to solve many differential equations through numerical approximations. When a shell balance is set up, the governing equation is initially set up in terms of a change in a variable over a distance (or time). The terms that evolve, for example, $\Delta c/\Delta x$, become, at the limit for small Δx, dc/dx. The basic approach to solving a differential equation is quite simple: The equation is recast in terms of its difference form, with the derivatives approximated using different approaches. Thus an equation of the form

$$\frac{dy}{dx} = F(x, y) \tag{A4-36}$$

can be separated and integrated, and instead written out in difference form with its solution approximated as

$$y = y_0 + \sum_{x_0}^{x} F(x, y)\, \Delta x + O(e) \tag{A4-37}$$

where $O(e)$ indicates there is some error involved in the approximation, of order e = error, which we hope is small.

Numerical solutions are extremely useful to study the properties of a system as a function of time. For example, we can write a mass balance around a CSTR in the absences of reaction as

$$\frac{dc}{dt} = \frac{Q}{V}(c_{\text{in}} - c) \qquad \text{(A4-38)}$$

where Q is the flow into the reactor of volume V. In order to simulate c over time, we approximate the derivative dc/dt over any time interval Δt in terms of a *forward, finite-difference approach*, as

$$\frac{dc}{dt} \cong \frac{c_{t+1} - c_t}{\Delta t} \qquad \text{(A4-39)}$$

We assume that the above two equations are equal, and then calculate the concentration after a change in time as

$$c_{t+1} = c_t + \frac{\Delta t}{\theta}(c_{\text{in}} - c_t) \qquad \text{(A4-40)}$$

where θ is the detention time or $\theta = V/Q$, and where we have chosen to define c over the time interval as c_t. By summing up changes over time, we effectively integrate our initial equation. In addition, if c_{in} is not constant, or is some function of time, we are able to specify its exact form in our numerical integration.

When second-order equations must be solved, approximating the first and second derivatives becomes more difficult, especially if these derivatives are changing rapidly over an interval. In a forward finite-difference approach, dc/dx can be approximated by

$$\frac{d^2c}{dx^2} = \frac{(c_{x+\Delta x} - c_x) - (c_x - c_{x-\Delta x})}{\Delta x^2} = \frac{c_{x+\Delta x} + c_{x-\Delta x} - 2c_x}{\Delta x^2} \qquad \text{(A4-41)}$$

When a partial differential equation is solved, all changes in variables must be approximated in the manner shown above. As an example, consider the equation

$$\frac{\partial c}{\partial t} + u_z \frac{\partial c}{\partial z} = D \frac{\partial^2 c}{\partial x^2} \qquad \text{(A4-42)}$$

This equation can be approximated in finite-difference form as

$$c(x, z, t + \Delta t) = c(x, z, t) - \frac{u_z \Delta t}{\Delta z}[c(z + \Delta z, t) - c(z, t)]$$

$$+ \frac{D \Delta t}{\Delta x^2}[c(x + \Delta x, t) + c(x - \Delta x, t) - 2c(x, t)] \qquad \text{(A4-43)}$$

and can be solved by initializing the equation using the assumed initial and boundary conditions on t, x, and z. At the boundaries of the spatial domain, for example, at points where the physical boundaries of the system exist, it is necessary to modify this equation to reflect these changes. For example, if there is no flux through a wall, then the derivative must be defined to be zero at that point.

The limitation of this finite-difference approach is that the derivative is assumed to be linear over the distance Δx. Thus this approach to approximating the first and second derivatives is just one of many possible approaches that vary based on the techniques used to approximate the derivatives and the methods used to solve the equation. In addition, each time a calculation is made there is a potential for an error in the calculation because each number cannot carry an infinite number of decimal places, and therefore there is roundoff error each time the calculation is completed. Numerical methods will vary in the extent of errors that can be classified together, with errors in derivatives, as numerical integration errors. It is usually prudent to test any new numerical scheme, if possible, against an analytical solution of the equation. It is likely that such an analytical solution is not known (or you would not be using a numerical solution), but a simpler form of the equation may be used as a test of your numerical scheme.

While the finite-difference approach is perhaps the easiest and most direct method of solving differential equations, this method often yields the most errors in solution, and the equations can often lead to unstable solutions. The most common numerical technique now used to solve differential equations is probably the Crank–Nicolson approach. In this method the spatial derivative is calculated at the midpoint of the next and previous time steps, or at $t + \Delta t$ and t. To achieve greater accuracy beyond the linear approximation used above, centered finite-difference equations are used to approximate the first and second spatial derivatives as

$$\frac{dc}{dx} = \frac{1}{2}\left(\frac{c_{x+\Delta x,t+\Delta t} - c_{x-\Delta x,t+\Delta t}}{2\,\Delta x} + \frac{c_{x+\Delta x,t} - c_{x-\Delta x,t}}{2\,\Delta x}\right) \tag{A4-44}$$

$$\frac{d^2c}{dx^2} = \frac{1}{2}\left(\frac{c_{x+\Delta x,t+\Delta t} - 2c_{x,t+\Delta t} + c_{x-\Delta x,t+\Delta t}}{\Delta x^2} + \frac{c_{x+\Delta x,t} - 2c_{x,t} + c_{x-\Delta x,t}}{\Delta x^2}\right) \tag{A4-45}$$

As an example, consider the following equation

$$\frac{\partial c}{\partial t} + u\frac{\partial c}{\partial x} = D\frac{\partial^2 c}{\partial x^2} \tag{A4-46}$$

We can write this equation using the Crank–Nicolson approach as

$$c_{x,t+\Delta t} - c_{x,t} + \frac{u\,\Delta t}{4\,\Delta x}(c_{x+\Delta x,t+\Delta t} - c_{x-\Delta x,t+\Delta t} + c_{x+\Delta x,t} - c_{x-\Delta x,t})$$

$$- \frac{uD}{2\,\Delta x^2}(c_{x+\Delta x,t+\Delta t} - 2c_{x,t+\Delta t} + c_{x-\Delta x,t+\Delta t} + c_{x+\Delta x,t} - 2c_{x,t} + c_{x-\Delta x,t}) = 0 \tag{A4-47}$$

Collecting the unknown concentrations, and combining the known concentrations into a constant e, we have

$$(b - a)c_{x+\Delta x, t+\Delta t} + (1 - 2b)c_{x,t+\Delta t} + (b - a)c_{x-\Delta x, t+\Delta t} = e \qquad \text{(A4-48)}$$

where

$$a = \frac{u\,\Delta t}{4\,\Delta x}, \qquad b = \frac{D\,\Delta t}{2\,\Delta x} \qquad\qquad \text{(A4-49)}$$

$$e = (a - bc_{x+\Delta x, t} + (1 + 2b)c_{x,t} + (a - b)c_{x-\Delta x, t} \qquad \text{(A4-50)}$$

Since $c_{x+\Delta x, t+\Delta t}$, $c_{x,t+\Delta t}$ and $c_{x-\Delta x, t+\Delta t}$ are all unknown, a series of n equations representing n spatial steps must be solved simultaneously. The boundary conditions usually appear as constants in the e term of the first and last model node. The resulting series of equations usually form a tridiagonal matrix, which can be solved efficiently using a method such as the Thomas algorithm.

4.3 SELECTED INTEGRALS

$$\int du = u + c \qquad\qquad \text{(A4-51)}$$

$$\int u\,dv = uv - \int v\,du \qquad\qquad \text{(A4-52)}$$

$$\int u^n\,du = \frac{u^{n+1}}{n+1} + c, \qquad n \neq 1 \qquad \text{(A4-53)}$$

$$\int u^n\,du = \ln u + c, \qquad n = 1 \qquad \text{(A4-54)}$$

$$\int \frac{du}{a + bu} = \frac{1}{b}\ln(a + bu) + c \qquad\qquad \text{(A4-55)}$$

$$\int \frac{u\,du}{a + bu} = \frac{1}{b^2}[a + bu - a\ln(a + bu)] + c \qquad \text{(A4-56)}$$

$$\int \frac{du}{u + bu} = \frac{1}{a}\ln\left(\frac{(a + bu)}{u}\right) + c \qquad\qquad \text{(A4-57)}$$

4.4 TABULATED FUNCTIONS AND FUNCTION PROPERTIES

The Error Function

The error function is a tabulated solution of the equation $(2/\sqrt{\pi}) \int_0^\phi e^{-n^2} dn = \text{erf}(\phi)$ in Table A4.1. Note that $\text{erfc}(\phi) = 1 - \text{erf}(\phi)$.

Bessel Functions

For a complete solution to Bessel functions, consult standard mathematical texts and reference books such as Mickley et al. (1959) and Weast and Selby (1975). However, there are approximate solutions that can be presented here for certain values of x, n, and p. For small values of x, useful Bessel function approximations (Mickley et al., 1959) are

$$J_p(x) \approx \frac{x^p}{2^p p!}, \qquad J_{-p}(x) \approx \frac{2^p x^{-p}}{(-p)!} \tag{A4-58}$$

$$K_n(x) \approx 2^{n-1}(n-1)! x^{-n}, \qquad n \neq 0; \qquad K_0(x) \approx -\ln x \tag{A4-59}$$

TABLE A4.1 The Error Function

ϕ	erf ϕ	ϕ	erf ϕ
0	0.0	0.85	0.7707
0.025	0.0282	0.90	0.7970
0.05	0.0564	0.95	0.8209
0.10	0.1125	1.0	0.8427
0.15	0.1680	1.1	0.8802
0.20	0.2227	1.2	0.9103
0.25	0.2763	1.3	0.9340
0.30	0.3286	1.4	0.9523
0.35	0.3794	1.5	0.9661
0.40	0.4284	1.6	0.9763
0.45	0.4755	1.7	0.9838
0.50	0.5205	1.8	0.9891
0.55	0.5633	1.9	0.9928
0.60	0.6039	2.0	0.9953
0.65	0.6420	2.2	0.9981
0.70	0.6778	2.4	0.9993
0.75	0.7112	2.6	0.9998
0.80	0.7421	2.8	0.9999

$$Y_n(x) \approx -\frac{2^n(n-1)!}{\pi} x^{-n}, \qquad n \neq 0; \qquad Y_0(x) \approx -\frac{2}{\pi} \ln x \qquad \text{(A4-60)}$$

$$I_p(x) \approx \frac{x^p}{2^p p!}, \qquad I_{-p} \approx \frac{2^p x^{-p}}{(-p)!} \qquad \text{(A4-61)}$$

When x is large, or as $x \to \infty$, useful approximations are

$$J_p(x) \approx \sqrt{\frac{2}{\pi x}} \cos\left(x - \frac{\pi}{4} - \frac{p\pi}{2}\right) \qquad \text{(A4-62)}$$

$$Y_n(x) \approx \sqrt{\frac{2}{\pi x}} \sin\left(x - \frac{\pi}{4} - \frac{n\pi}{2}\right) \qquad \text{(A4-63)}$$

$$I_p(x) \approx \frac{e^x}{\sqrt{2\pi x}} \qquad \text{(A4-64)}$$

$$K_n(x) \approx \sqrt{\frac{\pi}{2x}} e^{-x} \qquad \text{(A4-65)}$$

TABLE A4.2   Values of Zero-Order Bessel Functions of the First and Second Kind

x	$J_0(x)$	$Y_0(x)$	$I_0(x)$	$K_0(x)$
0.0	1.0000	$-\infty$	1.0000	∞
0.5	0.9385	-0.4445	1.0635	0.9244
1.0	0.7652	0.0883	1.2661	0.4210
1.5	0.5118	0.3824	1.6467	0.2138
2.0	0.2239	0.5104	2.280	0.1139
2.5	-0.0484	0.4981	3.290	0.06235
3.0	-0.2601	0.3769	4.881	0.03474
3.5	-0.3801	0.1890	7.378	0.01960
4.0	-0.3971	-0.0169	11.302	0.01116
4.5	-0.3205	-0.1947	17.48	0.00640
5.0	-0.1776	-0.3085	27.24	0.00369
5.5	-0.0068	-0.3395	42.69	
6.0	0.1506	-0.2882	67.23	
6.5	0.2601	-0.1732	106.29	
7.0	0.3001	-0.0259	168.6	
7.5	0.2663	0.1173	268.2	
8.0	0.1717	0.2235	427.6	
8.5	0.0419	0.2702	683.2	
9.0	-0.0903	0.2499	1093.6	
9.5	-0.1939	0.1712	1753	
10.0	-0.2459	0.0557		

Source: Mickley et al., 1959.

Bessel function of order $\frac{1}{2}$ can be calculated directly as

$$J_{1/2}(x) = \sqrt{\frac{2}{\pi x}} \sin x, \qquad J_{-1/2}(x) = \sqrt{\frac{2}{\pi x}} \cos x \qquad \text{(A4-66)}$$

$$I_{1/2}(x) = \sqrt{\frac{2}{\pi x}} \sinh x, \qquad I_{-1/2}(x) = \sqrt{\frac{2}{\pi x}} \cosh x \qquad \text{(A4-67)}$$

Zero-order Bessel functions are listed in Table A4.2.

REFERENCES

Weast, R. C., and S. M. Selby, Eds. 1975. *Handbook of Tables for Mathematics*. CRC Press, Boca Raton, FL.

Mickley, H. S., T. K. Sherwood, and C. E. Reed. 1957. *Applied Mathematics in Chemical Engineering*. McGraw-Hill, NY.

INDEX

Activity coefficients, 30, 36–39
Adsorbable organic halides (AOX), 105–106
Aggregates
    advection model, 246
    diffusion model, 241–244
    permeability, 246–250, 483–484
Apparent molecular weight (AMW), 89
Aquifer
    confined, 414
    unconfined, 415
Arnold diffusion cell, 82

Bacterial adhesion to surfaces, 607–608
Bacterial transport in porous media, 567, 604–609
Bessel functions, 638–641, 645–647
Biofilm
    advective flow through, 326–327
    kinetics, 271–286
        deep biofilm, 276–277, 282–285
        fast kinetics, 278
        first–order, 273, 277–280, 282–283
        methanotrophic, 329–330
        Michaelis Menten, 272, 280–285
        shallow biofilm, 274–276, 283
        trickling filter approximation, 303–305
        zero-order, 272, 273–276
    minimum substrate, 285
    minimum thickness, 286
    oxygen profile, 292–293

Biot numbers, 428–429
Blocking, during filtration, 600–601
Boltzman
    constant, 61, 509, 521
    equation, 509
Bond number, 437
Born repulsive forces, 517
Boundary layer theory, 151–153
Brenner, 421–422
Brownian motion, 520
Bubbles, 220–224, 251–252, 260

Camp number, 529
Capillary
    number, defined, 443
    pressure, 433
    rise, 434
Chemical potential, 27–33
Chemical production, 2
Chemical properties
    in groundwater, 440
    many chemicals, Appendix 3
Chilton-Colburn Analogy, 188–189
Clausius-Clapeyron equation, 39
Coagulation
    curvilinear model, 530–533
    fractal, 538–549
    integral form of equation, 507
    monodisperse, 525–530
    phytoplankton model, 550–552
    rectilinear model, 519–530

Coagulation (*Continued*)
  summation form of equation, 507
  TEP model, 552–553
Colburn j-factor, 566
Collector efficiency (*see also* filtration), 572–575
Collision functions
  Brownian, 520–522, 527, 531, 538–539
  differential sedimentation, 523, 532–533, 538–539
  shear, 522–523, 527–528, 531–532, 538–539
  summary for coagulation, 524
Collision integrals, 633
Collision number, 594–597
Constitutive transport equation, defined,
  Cartesian coordinates, 116–117, 123–126
  cylindrical, 125–126, 156–160
  spherical, 125–126, 154–156
Continuity equation, 114
Curvilinear coagulation model, 530–534

Damkohler number, 260
Darcy, units defined, 414
Darcy's Law, 412–413
Davies correlation, 247
Debye-Hückel parameter, 510
Desorption
  TCE from silican gel, 431
DDT in sediments, 22–23
Diffusion coefficient
  bacterial membranes, 76–77
  collision diameter, 61
  correction for temperature, 64–65, 69
  chemicals in
    biofilms, 78–81
    gas, 61–64, 73–75
    sediments, 79–81
    soil, 73–81
    solids, 72
    water, 65–71
  dextran in water, 68
  effect of pores, 73–75
  function of molecular weight, 65, 68, 91, 94, 104
  humic and fulvic acids, 86–89, 93–94
  proteins in water, 68, 91, 105
  surfaces, 75–77
  tortuosity, 78–79
  tortuosity factor, 74–75, 77, 79
Diffusion coefficient measurement,
  gases, 81, 131–138
  liquids,
    capillary tube technique, 89–92
    centrifugal technique, 83–85
    chromatographic, 86–90

field flow fractionation, 92–94
light scattering, 85–86
ultrafiltration, 94
Diffusion through a stagnant gas, (*see* stagnant film)
Diffusion with reaction,
  fast reactions, 143
  wide range of reaction rates, 138–143
Dilute solution, 8–10, 37–38
Dilution definition, 13
Dispersion
  definition, 333–335
  examples, 334
  mechanical, 360–362
  porous media, 358–373, 565–568
  sheared fluids, 342
  size of plume, 345–347, 364
  spike input, 343–345
  Taylor-Aris (tubes), 89–92
  theories in porous media, 364–368
  versus diffusion in porous media, 360–362
  with reaction in a river, 390
Dispersion coefficient
  calculating from velocity fluctuations, 339–340
  channels of different geometries, 362–364
  longitudinal dispersion coefficient, 336, 340–342
    between two flat plates , 352–353
    capillary tube, 89–92
    estuaries, 396
    inclined plane, turbulent flow, 357–358
    ocean, 397
    packed beds, 292, 365–371
    particles in packed beds, 567
    pipe, laminar flow, 353–354
    pipe, turbulent flow, 355–357
    porous media,
      Blackwell, 370
      Function of distance, 373
      Harleman, 371
      Hiby correlation, 370
      Koch and Brady, 367
      Saffman equation, 365–366
    river, 380, 387–390
    typical of different systems, 399
    wetted wall, 375
  transverse dispersion coefficient,
    complete mixing across a river, 385–387
    defined, 380
    estuaries, 396
    lakes (horizontal), 395
    porous media, 335, 365–368
    river, 382–387

Dispersion coefficient (*Continued*)
  vertical dispersion coefficient
    lakes, 394
    porous media, 358–373
    river, 380–381
Dispersivity, 372–373
Dissipation rate
  ocean, 397
Dissolution, pure chemical phase
  rivers, 400–405
  groundwater; *see* ganglia, NAPLs
Dissolved air flotation (DAF), 560
Dissolved oxygen, 628
Distribution coefficients
  gases and liquids, *see* Henry's Law
  multiple liquid phases, 45–46
  soils, 39–40, 47–52
DLVO theory, 514
Double layer thickness around a charged particle, 510
Dynamic equilibrium, 24–25

Effectiveness factor
  advective, 254–255
  diffusive, 252–254
  see also—relative uptake factor
Electrophoretic mobility, 513
Electrostatic forces around a particle, 514
Energy dissipation rate, 194–200, 215
Error function, 645

Fick
  First law, 59, 117, 120
Filters
  fiberous (depth), 586–589
  membrane (surface), 587, 590–594
  pressure drops, 597–599
  screens, 585, 590–594
Filtration
  comparison of models, 580–582
  Happel model, 576–578
  Hinds model, 588
  particle size distributions, 582–584
  pore velocity model, 575–576
  Rubow and Liu model, 588–589
  RT model, 579–580
  theory, 568–583
  Yao-Habibian model, 578
  Yao model, 569–575
Fourier's law, 118
Four-thirds law, 398
Fractal
  dimension defined, 469

  from particle properties 469, 471, 472–473, 477–481
  from size distributions, 493–499, 545–549, 554–559
  characteristic dimensions of different particles
    inorganic, 478, 485
    marine particles, 472, 478, 485
    microbial aggregates, 472–473, 478
    trickling filter biofilm, 485
  defined, 468–470
  generator, 469, 474
  length versus solid equivalent diameter, 540
  packing factor, 474
  permeable aggregates, 482–485
  properties, 470–485
  settling velocity, 476–485
  shape factor, 474
  shapes as a function of D, 537
  summary of properties, 477
Frigon correlation, 68, 70–71
Fugacity, 36–48
  definition, 36
  reference, 36
  simplifying assumptions, 37–38
Fuller correlation, 62–64

Ganglia
  defined, 432
  dissolution rates, 447–460
  lifetimes, 450–455
  maximum length
    saturated media, no flow, 436
    unsaturated media, 441
    with flow, 442
  size distributions of, 437, 439, 444–446
Gel permeation chromatography (GPC), 86–89
General transport equation (GTE), 112–117
Gibbs Free energy, 27–29, 33–35
Gouy-Chapman double layer model, 508
Grashoff number, 221–223

Half–life,
  coagulation, 528
Hamaker constant, 515–516, 560
Harmonic equation, solution of, 640
Hatta number, 143
Hayduk and Laudie correlation, 67–68
Henry's Law, 40–42, 172–177
Hirshfelder equation, 61, 63–64
Holdup in a reactor, 224
Hydraulic
  conductivity, 413
  gradient, 413
  head, 413

Hydrocarbon, 630
Hydrophobic forces, 517–518

Ideal solution, 7–8, 30–31
Isotherms, 50–52

Kinetic theory of gases, 61
Kinetics, microbial
    effect of fluid motion, 233–241
    mass transport model, 233
    see—Biofilm, Michaelis–Menten
Kolmogorov microscale, 196–198
Kozeny-Carmen equation, 247, 251, 261, 414, 443
Knudson diffusion coefficient, 73–74
Kuwabara number, 588

Lennard-Jones force constants, 61, 634
Log mean average, 82
London-van der Waals force, 514–515

Manning equation, 379
Mass balances, 14–19
Mass transport coefficients
    aggregates, 257
    aerated reactors, 224–227
    air stripper, 184–185, 189–190
    analogies used to calculate, 186–189
    biofilm in packed beds, 265–268
    biofilm in RBC, 269
    boundary layer theory, 169–170, 185–186
    bubbles, 221–224
    defined, 165
    droplets, pools and waves in rivers, 400–405
    empirical correlations, 180–186
    function of diffusivity, 169–171
    interphase transport, 171–180
    microbes 236, 239
    modified for area, 448
    NAPLs, 444–460
    oil slick, 405–409
    overall, 173–180
    particle transport, 567
    penetration theory, 168–170
    resistances, 177–179, 220–221
    thin films (*see* wetted walls)
    sheared reactors, 200–204
    sphere in a moving fluid, 180–182, 236–237
    stagnant film, 167–168, 170–171
    stream bottom, 404–405
    volatilization of NAPLs, 460
Michaelis-Menten kinetics, 232, 271–272, 280–282, 292–300

Mixing,
    estuaries, 395–396
    lakes, 390–394
    ocean, 396–399
    rivers—*see* Dispersion coefficient
Molecular size distributions
    see size distributions

NAPL, 432–460
    enhancing removal using
        hot water, 455
        methanol, 456
        steam, 455–456
        surfactants, 456–457, 464
    interfacial area for mass transfer, 447
    volatilization, 459–460
Navier Stokes equation, 114–115
Newton's law, 117–118, 146, 187
Numerical solution techniques, 641–644

Octanol-water partitioning, 425–427
Oxygen transfer
    profile in biofilm, 292–293
    profile in trickling filter liquid film, 323–325
    stirred reactor, 220–227
    trickling filter, 314–326

Packed beds
    biofilm reactor modeling, 291
    longitudinal dispersion, 292 (*see also* dispersion coefficient)
    mass transfer correlation, 265, 268, 449
    NAPLs, 449, 451
    nomographs, 298, 299
    numerical solution, 294–297
Particle
    electrostatics, 508–518
    mass concentration, 488
    measuring concentions, 490–493
    number concentration, 487
    size as a function of shear rate, 215–220
    size spectra, 485
    transport equation, filtration, 565–568
    volume concentration, 487
    volume distribution function, 488
Penetration theory,
    defined, 143–150
    oxygen transfer in trickling filters, 321, 326
    thin films (*see* wetted wall)
Permeability
    aggregates, 246–250, 483–484
    soil, 413
Permeation coefficient model, 17–18, 98–103
Permittivity, of water, 508

Pesticides, 629
Poisson-Boltzman equation,
    particle-particle interactions, 512, 514
    plate-plate interactions, 514
Polson correlation, 68, 70–71
Polymer bridging, 518–519
Pore diffusion modulus, 429
Pore size, 434–435
Porosity, 359
ppm definition, 12
Prandtl number, 119

Radius of gyration, 85–87, 470
Rectilinear collision model, 520
Reddy-Doraiswamy correlation, 67, 69–70
Relative uptake factors, 255–260
Relative volatility, 42–43
Retardation coefficient, 423–425, 428–430, 463
Reynolds Analogy, 186–188
Rotating biological contactors (RBCs), 269, 287,
    289–291
Ripening, during filtration, 599–604

Safe Drinking Water Act, 1, 3
Shear
    function of turbulence, 194–198, 200–201
    wall, 378
Shear rate
    coagulation, 522
    estuaries, 397
    lakes, 393–394
    reactors, 196–198
    steady, 198–200, 201–204
    versus aggregate size, 215–220
Shear rate estimation,
    bubbled reactor, 207–209
    couette device, 209–211
    paddle or mixer, 204–207, 209–215
Shell balance, 154–160
Sherwood number, defined, 166
Size distributions
    analysis of, 485
    filtration of, 582–597
    marine systems, 557–559
    measuring techniques, 84, 490–493
    steady state
        fractal, 542–545
        rectilinear, 534–536
    wastewater samples, 97
    water samples, 88, 96
    versus molecular mass, 65, 84, 303
Size density function, 486
Size exclusion chromatography (SEC), 86–89
Soil distribution coefficient, 423–426

Stability factor, Fuchs, 521
Stagnant film
    biofilms, defined, 264–267
    defined, 130–131, 167–168
    diffusion cell, 26, 131–133, 137–138
    fixed and fluidized bed reactor, 268
    rotating disk reactors, 269
Stanton number, 428–430
Stern layer, 510
Stern-Grahame model, 511
Sticking coefficient
    coagulation, 507, 515, 517, 521, 523
Stokes-Einstein equation, 66, 521
Stokes number, 588
Strange attractor, 24–25
Surface renewal theory, 150–151
Surface tension, 433
Svedburg, defined, 85

Thermal diffusivity, 118–119
Thermodynamic state functions, 28–29
Thiele modulus, 253–254, 258–259, 274, 278,
    309–310, 328–329
Tortuosity
    factor, 360
Transparent exopolymer particles, TEP, 552–553,
    561
Trickling Filters, 300–326
    kinetic limited-removal, 309, 310–314
    Logan models, 300–309, 314–325
    mass transport limited removal, 309–310
    Mehta model, 321–326, 329
    modified-Velz equation, 310
    nitrification efficiencies, 318–321
    oxygen transport, 314–326
    plastic media geometry, 300–302
    removal per plate, 307
    size distributions of wastewater effluent, 95,
        97, 102, 303
    Swilley model 309–310
Trihalomethane(THM), 83, 95

Ultrafiltration
    derivation of mass balance around cell, 17
    permeation coefficient model, 98–103
    size distributions, 94–103

Vapor pressure
    n-alkanes, 408–409
    used in fugacities, 38–39
Velocity
    aggregate settling velocity, 246–249
    average in river, 379

Velocity (*Continued*)
  averaging technique, 338
  average in turbulent flow, 338
  bubble rise, 222, 224
  Darcy, 413
  fractal particle, 476–485
  logarithmic profile, 379
  molar average, 120
  particles, 476–485
  pore, 413
virus transport in porous media, 608
von Karmen constant, 380

Wastewater
  composition, 302–303
wetted wall, 144–150
  average velocity, 147
  entrance length, 183
  fluid thickness, 148
  mass transport correlations, 182–185
  maximum velocity, 147
  transition to turbulent, 183
Wilke-Chang correlation, 66–67, 69–70

Young's equation, 438